Lecture Notes in Computer Science 4958

Commenced Publication in 1973
Founding and Former Series Editors:
Gerhard Goos, Juris Hartmanis, and Jan van Leeuwen

Lecture Notes in Computer Science 4958

Valentin E. Brimkov   Reneta P. Barneva
Herbert A. Hauptman (Eds.)

Combinatorial Image Analysis

12th International Workshop, IWCIA 2008
Buffalo, NY, USA, April 7-9, 2008
Proceedings

 Springer

Volume Editors

Valentin E. Brimkov
SUNY Buffalo State College, Mathematics Department
1300 Elmwood Ave., Buffalo, NY 14222, USA
E-mail: brimkove@buffalostate.edu

Reneta P. Barneva
SUNY Fredonia, Department of Computer Science
Fredonia, NY 14063, USA
E-mail: barneva@cs.fredonia.edu

Herbert A. Hauptman
Hauptman-Woodward Medical Research Institute
700 Ellicott Street, Buffalo, NY 14203-1102, USA
E-mail: hauptman@hwi.buffalo.edu

Library of Congress Control Number: 2008921859

CR Subject Classification (1998): I.4, I.5, I.3.5, F.2.2, G.2.1, G.1.6

LNCS Sublibrary: SL 6 – Image Processing, Computer Vision, Pattern Recognition,
and Graphics

ISSN   0302-9743
ISBN-10   3-540-78274-5 Springer Berlin Heidelberg New York
ISBN-13   978-3-540-78274-2 Springer Berlin Heidelberg New York

Springer is a part of Springer Science+Business Media

springer.com

© Springer-Verlag Berlin Heidelberg 2008
Printed in Germany

Typesetting: Camera-ready by author, data conversion by Scientific Publishing Services, Chennai, India
Printed on acid-free paper   SPIN: 12230853   06/3180   5 4 3 2 1 0

Preface

It is indeed a great pleasure to welcome you to the proceedings of the 12th International Workshop on Combinatorial Image Analysis (IWCIA 2008) held in Buffalo, NY, April 7–9, 2008.

Image analysis is a scientific discipline providing theoretical foundations and methods for solving problems that appear in various areas of human practice, as diverse as medicine, robotics, defense, and security. As a rule, the processed data are discrete; thus, the "discrete," or "combinatorial" approach to image analysis appears to be a natural one and therefore its importance is increasing. In fact, combinatorial image analysis often provides various advantages (in terms of efficiency and accuracy) over the more traditional approaches based on continuous models requiring numeric computation.

The IWCIA workshop series provides a forum for researchers throughout the world to present cutting-edge results in combinatorial image analysis, to discuss recent advances in this research field, and to promote interaction with researchers from other countries. In fact, IWCIA 2008 retained and even enriched the international spirit of these workshops, that had successful prior meetings in Paris (France) 1991, Ube (Japan) 1992, Washington DC (USA) 1994, Lyon (France) 1995, Hiroshima (Japan) 1997, Madras (India) 1999, Caen (France) 2000, Philadelphia (USA) 2001, Palermo (Italy) 2003, Auckland (New Zealand) 2004, and Berlin (Germany) 2006. The IWCIA 2008 Program Committee was highly international as its members are renowned experts coming from 23 different countries, and submissions came from 24 countries from Africa, Asia, Europe, North and South America.

The present volume includes the papers presented at the workshop. Following the call for papers, IWCIA 2008 received 117 submissions. After a preliminary screening of all submissions by the Workshop Chairs, 82 of these were reviewed by Program Committee members or additional reviewers (the others being obviously unacceptable in view of IWCIA standards and technical requirements). Of these 82 papers, 28 were accepted for oral presentation and 10 for poster presentation. The review process was quite rigorous, involving three to four independent double-blind reviews. *OpenConf* provided a convenient platform for smoothly carrying out the review process. The most important selection criterion for acceptance or rejection of a paper was the overall score received. Other criteria included: relevance to the workshop topics, correctness, originality, mathematical depth, clarity, and presentation quality. We believe that as a result, only high-quality papers were accepted for presentation at IWCIA 2008 and for publication in the present volume. We also hope that many of these papers are of interest to a broader audience, including researchers working in areas such as computer vision, image processing, and computer graphics.

The program of the workshop was arranged into ten sessions. These included presentations of contributed papers, as well as invited talks by five distinguished scientists.

An opening talk was given by Herbert Hauptman, Nobel Laureate. He shared with the audience his vision about science in general, and specifically of mathematics and its amazing applicability to other sciences and human practice. Hauptman supported his theses by a number of interesting examples, in particular from his own scientific contributions. "The history of science shows that advances in science, particularly progress in its more basic aspects, including mathematics, have had the most profound influence in serving to improve the quality of life and making possible development of a modern advanced technological society," Hauptman said.

"Can a computer recognize the activities of a person, e.g., one fighting another person or one climbing a fence? Can a computer recognize an illegally parked car?" Jake Aggarwal asked, and continued: "Computer vision has matured to a discipline that addresses societal problems: monitoring public places—what is a person doing? Leaving an unattended bag, climbing a fence or breaking into a car are examples of action recognition." In his invited talk, the speaker presented his recent research on human motion understanding, modeling and recognition of human faces, actions and interactions, as well as human–object interactions.

Polina Golland considered the problem of identifying large co-activating networks in brain based on dynamical imaging, and proposed an approach leading to hierarchical, anatomically meaningful representations of brain activity across experiments and across subjects. "This way we are able to effectively decompose the four-dimensional collection of the activation values into spatial maps that align with our notion of anatomical structure of the brain and the dynamics associated with these maps," Golland said.

Arie Kaufman presented a new research area known as virtual colonoscopy. It is a combination of computed tomography scanning and volume visualization technology, and incorporates a novel pipeline of computer-aided detection of colonic polyps employing segmentation, electronic cleansing, conformal colon flattening, volume rendering with texture, and shape analysis. "Virtual colonoscopy is poised to become the procedure of choice in lieu of the conventional optical colonoscopy for mass screening for colon polyps—the precursor of colorectal cancer," Kaufman said.

The workshop scientific program was completed by the invited closing talk of Gabor Herman. He presented a methodology for translating the problem of classification of heterogeneous microscopic projections into homogeneous subsets into an optimization problem on a graph. He also provided a combinatorial algorithm that achieves a useful solution at a low computational cost. "The proposed methodology makes it possible to visualize the functioning of a biological molecular machine," Herman said.

In addition to the main theoretical track of IWCIA 2008, for the first time a Special Track on Applications was organized. It provided researchers and software developers with the opportunity to present their work and demonstrate working computer systems for image analysis.

Many individuals and organizations contributed to the success of IWCIA 2008. First of all, the Chairs are indebted to IWCIA's Steering Committee for endorsing the candidacy of Buffalo for the 12th edition of the Workshop. Our most sincere thanks go to the IWCIA 2008 Program Committee whose cooperation in carrying out high-quality reviews was essential in establishing a very strong workshop program. We also appreciate the assistance of the additional reviewers who helped us maintain the timeline for the review process and author notification. We express our sincere gratitude to the invited speakers Jake Aggarwal, Polina Golland, Herbert Hauptman, Gabor Herman, and Arie Kaufman for their remarkable talks and overall contribution to the workshop program. We wish to thank everybody who submitted their work to IWCIA 2008. Thanks to their contributions, we succeeded in having a technical program of high scientific quality. We are indebted to all participants and especially to the contributors of this volume.

The success of the workshop would not be possible without the hard work of the local Organizing Committee. We are grateful to Joaquin Carbonara, Dan Cunningham, François de Vieilleville, Peter Mercer, Mike Szocki, Khalid Siddiqui, and João Tavares for their valuable work. We are obliged to SUNY Buffalo State College and SUNY Fredonia for the continuous support through their designated offices. Special thanks go to Muriel Howard, President of SUNY Buffalo State, and Dennis Hefner, President of SUNY Fredonia, for endorsing IWCIA 2008, to Dennis Ponton, Provost of Buffalo State, for his strong support, and to Larry Flood, Dean of the School of Natural and Social Sciences at Buffalo State, for continuously promoting the workshop since its very early stages. We also remember with gratitude the assistance provided by several students from Buffalo State and all who made this conference an enjoyable and fruitful scientific event. Finally, we wish to thank Springer for the pleasant cooperation in the timely production of this volume.

April 2008

Valentin E. Brimkov
Reneta P. Barneva
Herbert A. Hauptman

Organization

IWCIA 2008 was held in Buffalo, NY, USA, April 7–9, 2008.

General Chair

Valentin E. Brimkov   SUNY Buffalo State College, USA

Vice Chair

Reneta P. Barneva   SUNY Fredonia, USA

Steering Committee

Valentin E. Brimkov   SUNY Buffalo State College, USA
Gabor T. Herman   CUNY Graduate Center, USA
Ralf Reulke   Humboldt University, Germany

Invited Speakers

Jake K. Aggarwal   University of Texas at Austin, USA
Polina Golland   Massachusetts Institute of Technology, USA
Herbert A. Hauptman   Hauptman-Woodward Institute, USA
Gabor T. Herman   CUNY Graduate Center, USA
Arie E. Kaufman   SUNY Stony Brook, USA

Program Committee

Til Aach   RWTH Aachen University, Germany
Eric Andres   University of Poitiers, France
Tetsuo Asano   JAIST, Japan
Jacky Baltes   University of Manitoba, Canada
Reneta P. Barneva   SUNY Fredonia, USA
George Bebis   University of Nevada at Reno, USA
Bedrich Benes   Purdue University, USA
Bhargab B. Bhattacharya   Indian Statistical Institute, India
Peter Brass   City College, City University of New York, USA
Srecko Brlek   Université du Québec à Montréal, Canada
Alfred M. Bruckstein   Technion, I.I.T, Israel

Peter Veelaert   Hogeschool Gent, Belgium
Petra Wiederhold   CINVESTAV-IPN, Mexico
Jinhui Xu   SUNY University at Buffalo, USA
Jason You   Cubic Imaging LLC, USA
Richard Zanibbi   Rochester Institute of Technology, USA

Organizing Committee

Reneta P. Barneva, Co-chair   SUNY Fredonia, USA
Joaquin Carbonara, Co-chair   SUNY Buffalo State College, USA
Daniel Cunningham   SUNY Buffalo State College, USA
François de Vieilleville   SUNY Buffalo State College, USA
John Favata   SUNY Buffalo State College, USA
Peter Mercer   SUNY Buffalo State College, USA
Khalid Siddiqui   SUNY Fredonia, USA
Michael Szocki   SUNY Fredonia, USA
João Manuel R. S. Tavares   University of Porto, Portugal

Additional Reviewers

Marco Antonio Alvarez   Mesmoudi Mohammed Mostefa
Gholamreza Amayeh   Yoshihiro Nishimura
Péter Balázs   Benedek Nagy
Sascha Bauer   João José Neto
Amaury Antonio Castro Jr.   Peter Noel
C. Chandrasekar   Alessandro Perina
David Coeurjolly   Yongwu Rong
Marco Cristani   Khalid Siddiqui
Riccardo Gherardi   Isabelle Sivignon
Daniel Hein   Robin Strand
Kamen Kanev   Mohamed Tajine
Kostadin Koroutchev   Alireza Tavakkoli
Elka Korutcheva   Guang Xu
K. Krithivasan   Lei Xu
V. Masilmani   Yongding Zhu
Frederik Meysel

Table of Contents

Digital Geometry and Topology: Curves and Surfaces

Connectivity Preserving Voxel Transformation 1
  Anvesh Komuravelli, Arnab Sinha, and Arijit Bishnu

Thinning on Quadratic, Triangular, and Hexagonal Cell Complexes 13
  Petra Wiederhold and Sandino Morales

Experimental Comparison of Continuous and Discrete Tangent
Estimators Along Digital Curves 26
  François de Vieilleville and Jacques-Olivier Lachaud

Polyhedral Surface Approximation of Non-convex Voxel Sets through
the Modification of Convex Hulls 38
  Henrik Schulz

Weighted Neighborhood Sequences in Non-standard Three-Dimensional
Grids – Parameter Optimization 51
  Robin Strand and Benedek Nagy

Computing Homology Generators for Volumes Using Minimal
Generalized Maps 63
  Guillaume Damiand, Samuel Peltier, and Laurent Fuchs

Digital Segments and Hausdorff Discretization 75
  Mohamed Tajine

Combinatorics in Digital Spaces: Lattice Polygons, Polytopes, Tilings, and Patterns

Scaling of Plane Figures That Assures Faithful Digitization 87
  Valentin E. Brimkov

Computing Admissible Rotation Angles from Rotated Digital Images ... 99
  Yohan Thibault, Yukiko Kenmochi, and Akihiro Sugimoto

On the Number of hv-Convex Discrete Sets 112
  Péter Balázs

Finding the Orthogonal Hull of a Digital Object: A Combinatorial
Approach 124
  *Arindam Biswas, Partha Bhowmick, Moumita Sarkar, and
  Bhargab B. Bhattacharya*

A Discrete Approach for Supervised Pattern Recognition............... 136
    *João P. Papa, Alexandre X. Falcão, Celso. T.N. Suzuki, and
    Nelson D.A. Mascarenhas*

Image Representation, Segmentation, Grouping, and Reconstruction

Robust Decomposition of Thick Digital Shapes 148
    Alexandre Faure and Fabien Feschet

Segmentation of Noisy Discrete Surfaces 160
    Laurent Provot and Isabelle Debled-Rennesson

MRF Labeling with a Graph-Shifts Algorithm 172
    Jason J. Corso, Zhuowen Tu, and Alan Yuille

Label Space: A Multi-object Shape Representation 185
    James Malcolm, Yogesh Rathi, and Allen Tannenbaum

A New Image Segmentation Technique Using Maximum Spanning
Tree.. 197
    Qiang He and Chee-Hung Henry Chu

Applications of Computational Geometry, Integer and Linear Programming to Image Analysis

Reducing the Coefficients of a Two-Dimensional Integer Linear
Constraint 205
    Emilie Charrier and Lilian Buzer

A Branch & Bound Algorithm for Medical Image Registration 217
    Michael Stiglmayr, Frank Pfeuffer, and Kathrin Klamroth

Global Optimization for First Order Markov Random Fields with
Submodular Priors 229
    Jérôme Darbon

Transformation Polytopes for Line Correspondences in Digital
Images 238
    Kristof Teelen and Peter Veelaert

Linear Boundary and Corner Detection Using Limited Number of
Sensor Rows 250
    Bishal Prasad, Arijit Bishnu, and Tetsuo Asano

Fuzzy and Stochastic Image Analysis, Parallel Architectures and Algorithms

A Convergence Proof for the Horn-Schunck Optical-Flow Computation
Scheme Using Neighborhood Decomposition   262
    Yusuke Kameda, Atsushi Imiya, and Naoya Ohnishi

Topologically Correct 3D Surface Reconstruction and Segmentation
from Noisy Samples   274
    Peer Stelldinger

Detecting the Most Unusual Part of a Digital Image   286
    Kostadin Koroutchev and Elka Korutcheva

Labeling Irregular Graphs with Belief Propagation   295
    Ifeoma Nwogu and Jason J. Corso

Grammars and Models for Image or Scene Analysis

Image Registration Using Markov Random Coefficient Fields   306
    Edgar Román Arce-Santana and Alfonso Alba

A Secret Sharing Scheme for Digital Images Based on Two-Dimensional
Linear Cellular Automata   318
    Angel Martín del Rey

Pure 2D Picture Grammars (P2DPG) and P2DPG with Regular
Control   330
    K.G. Subramanian, Atulya K. Nagar, and M. Geethalakshmi

A Deterministic Turing Machine for Context Sensitive Translation of
Braille Codes to Urdu Text   342
    Muhammad Abuzar Fahiem

Rewriting P Systems Generating Iso-picture Languages   352
    S. Annadurai, D.G. Thomas, V.R. Dare, and T. Kalyani

Discrete Tomography, Medical Imaging, and Biometrics

Reconstructing a Matrix with a Given List of Coefficients and
Prescribed Row and Column Sums Is NP-Hard   363
    Yan Gerard

A Reasoning Framework for Solving Nonograms   372
    K. Joost Batenburg and Walter A. Kosters

A Memetic Algorithm for Binary Image Reconstruction   384
    Vito Di Gesù, Giosuè Lo Bosco, Filippo Millonzi, and Cesare Valenti

Personal Identification Based on Weighting Key Point Scheme for Hand
Image 396
    Dongbing Pu, Shuang Qi, Chunguang Zhou, and Yinghua Lu

A Min-Cost-Max-Flow Based Algorithm for Reconstructing Binary
Image from Two Projections Using Similar Images 408
    Vedhanayagam Masilamani and Kamala Krithivasan

Comparison of Local and Global Region Merging in the Topological
Map... 420
    Alexandre Dupas and Guillaume Damiand

Novel Edge Detector 432
    Imran Touqir and Muhammad Saleem

Author Index.. 445

Connectivity Preserving Voxel Transformation

Anvesh Komuravelli[1], Arnab Sinha[2], and Arijit Bishnu[1]

[1] Computer Science and Engineering Department, Indian Institute of Technology,
Kharagpur, Kharagpur-721302, India
{anvesh,bishnu}@cse.iitkgp.ernet.in
[2] Dept. of Electrical Engineering, Princeton University, Princeton,
New Jersey, USA-08544
sinha@princeton.edu

Abstract. A three dimensional digital binary image is $\mathbf{B}_{26}$ connected
if its set of black voxels is 26-connected, i.e. for all black voxels there
exists at least one black voxel among its 26 neighbors. We show that any
two such images I and J of c_1 and c_2 number of connected components
respectively and n voxels each, can be transformed into one another
maintaining the $\mathbf{B}_{26}$ connectivity of the black voxels by $O((c_1 + c_2)n^2)$
interchanges.

1 Introduction

A three (two) dimensional digital binary image (I) is a function $I : \mathbb{Z}^3 \rightarrow \{0,1\}$
$(I : \mathbb{Z}^2 \rightarrow \{0,1\})$. Any element in $\mathbb{Z}^3$ $(\mathbb{Z}^2)$ is called a *voxel (pixel)*. We consider
finitely many lattice points (voxels/pixels) from $\mathbb{Z}^3$ $(\mathbb{Z}^2)$ in I. A voxel (pixel) p
is black (white) if $I(p) = 1$ $(I(p) = 0)$. We call two pixels (x_1, y_1) and (x_2, y_2)
to be 8-neighbors if and only if $(x_1 - x_2)^2 + (y_1 - y_2)^2 \leq 2$. For 4-neighbors,
it is $(x_1 - x_2)^2 + (y_1 - y_2)^2 \leq 1$. We call two voxels (x_1, y_1, z_1) and (x_2, y_2, z_2)
to be 26-neighbors if and only if $(x_1 - x_2)^2 + (y_1 - y_2)^2 + (z_1 - z_2)^2 \leq 3$. This
induces a graph G_{26} whose vertex set is $\mathbb{Z}^3$ and there exist edges between two
lattice points satisfying the above inequality. In a 3-D binary image I, $\mathbf{B}_{26}$ is a
sub-graph of G_{26} induced by the black voxels in I. Similar graphs can be defined
for 2-D binary images also. A pair of neighboring (4 or 8) [4] opposite-valued
pixels in a 2-D binary image I is called *interchangeable* if reversing their values
preserves the topology of the image [5,6]. The interchange does not affect the
number of 0s and 1s in I. We will define interchangeable voxel pair later on.
Two 2-D binary images I and J are called *IP-equivalent* [5,6] if there exists
a sequence of binary images $I = I_0, I_1, \ldots, I_i, \ldots, I_k = J$ such that any I_i
$(1 \leq i \leq k)$ can be obtained from I_{i-1} by reversing an *interchangeable* pixel pair.
Rosenfeld and Nakamura [6] proved the conjecture made in [5] that if two binary
images I and J have two simply connected sets S and T respectively of the same
number of 1s, then I and J are *IP-equivalent*. In a recent comprehensive work
that also deals with the combinatorial bounds on the number of interchanges,
Bose et al. [1] generalized the results in [6]. They showed that for any $(a, b) \in$
$\{(4, 8), (8, 4), (8, 8)\}$, any two $\mathbf{B}_a, \mathbf{W}_b$-connected images I and J each with n

V.E. Brimkov, R.P. Barneva, H.A. Hauptman (Eds.): IWCIA 2008, LNCS 4958, pp. 1–12, 2008.
© Springer-Verlag Berlin Heidelberg 2008

black pixels differ by a sequence of $O(n^2)$ interchanges. The corresponding result for two $\mathbf{B}_4, \mathbf{W}_4$ connected images is $O(n^4)$. A binary image I is called $\mathbf{B}_a, \mathbf{W}_b$ $(a, b \in \{4, 8\})$ connected if its foreground (1) is a-connected and its background (0) is b-connected. The interchanges considered by Bose et al. [1] are also of *interchangeable* pixel pairs such that the connectivity of both foreground and background are maintained as in [5,6]. This sort of transformation problems has motivation in robotics [2] where researchers are interested in the number of moves needed in going from a configuration to another under some restrictions in the movement patterns. Under a more restricted and complex interchange rule, Dumitrescu and Pach [3] show that any two $\mathbf{B}_4$ connected images are apart by $O(n^2)$ interchanges where an interchange takes place between two 8-neighbor pixels such that the image obtained after the interchange is still $\mathbf{B}_4$ connected. Though Dumitrescu and Pach talk of modular metamorphic systems in terms of motion planning in [3], a connection to pixels is straightforward.

In this work, we consider connectivity preserving voxel transformation of 3-D binary images under a very relaxed and simple model of connectivity. To the best of our knowledge, connectivity preserving voxel transformation has not been considered earlier. Section 2 discusses preliminaries needed for our work. Section 3 discusses the main body of our work. In Section 4, we discuss connectivity preserving voxel transformation under the model proposed in [2] for 2-D.

2   Preliminaries

2.1   Definition and Notations

Our model is simple as we do not consider the connectivity of the white voxels. The earlier works in 2-D [5,6,1] consider connectivity of both black and white pixels. We think considering this simple model is worthwhile for an initial study on connectivity preserving voxel transformation.

We call a pair of neighboring (26) opposite-valued voxels in I *interchangeable* if reversing their values preserves the $\mathbf{B}_{26}$ connectivity of the 3-D image. The interchange obviously does not affect the number of 0s and 1s in I. Two $\mathbf{B}_{26}$ connected 3-D binary images I and J of the same number of voxels are called *transformable* if there exists a sequence of 3-D binary images $I = I_0, I_1, \ldots, I_i, \ldots, I_k = J$ such that any I_i $(1 \leq i \leq k)$ can be obtained from I_{i-1} by reversing an *interchangeable* voxel pair. We show in this work that any two 3-D binary images I and J which are $\mathbf{B}_{26}$ connected and have the same number of voxels are transformable. We do this by transforming I to a linear chain of voxels. So, it follows that J can also be transformed to a linear chain of voxels; and the transformation of I to J can be obtained by transforming I to a linear chain of voxels and then retracing the transformation (of J) from the linear chain of voxels back to J.

Following are the definitions of the terms we will be using throughout. See Fig. 1. When a voxel moves because of the interchanges such that its z-coordinate remains unaffected, we use the term *pixel* also. In the body of the text, we interchangeably use the term *voxel* and *pixel*.

Layer: A 3-D object spans over some layers, where each layer contains a 2D structure.

Connectivity-sensitive pixel: Consider the topmost layer (let it be *Layer 1*) of Fig. 1(a). As the object is connected, there must exist at least one pixel P_{layer1} which has a 26-neighbor in the layer just below it. We denote such pixels as *connectivity-sensitive* pixels. For the preservation of connectivity, one of these connectivity-sensitive pixels are not interchanged during the first part of the transformation, as the layers present below are hanging from that particular pixel of the top-layer.

Merge Axis: Merge axis $(\mathcal{M}(P))$ is a coordinate axis passing through pixel P and the pixels lying on it are defined to be *non-interchangeable* throughout the first part of the transformation. A merge-axis contains at least one *connectivity-sensitive pixel*. Figure 1(a) shows $\mathcal{M}(P_{layer1})$, $\mathcal{M}(P_{layer2})$ and $\mathcal{M}(P_{layer3})$ in *Layer 1*, *Layer 2* and *Layer 3* respectively. All the pixels of the given 2D component are finally brought onto or *merged* on this axis using connectivity preserving interchanges. Where P is obvious, we use just $\mathcal{M}$.

Level: In a given layer and in a given connected component in that layer, a *level* is the shortest distance of a pixel of that component, from the *Merge Axis*.

Cut and Non-cut pixels: A pixel whose removal disconnects the originally connected component is a cut-pixel, otherwise it is non-cut.

Coordinate Axes: For any black pixel on a 2D layer, its four coordinate axes determine the direction in which the adjacent black pixels are located. The coordinate axes through pixel P in Fig. 1(b) are the following (i) $A(P)_v$ (vertical axis), (ii) $A(P)_h$ (horizontal axis), (iii) $A(P)_{45}$ (making 45° with $A(P)_h$) and (iv) $A(P)_{-45}$ (making -45° with $A(P)_h$).

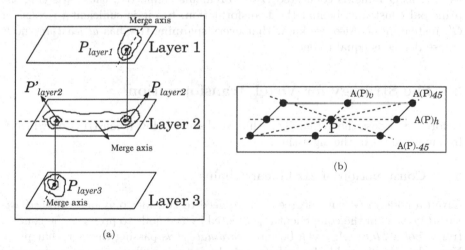

Fig. 1. (a) The 3D object in different layers. The adjacency between P_{layer1} and P_{layer2} (P'_{layer2} and P_{layer3}) maintains the connectivity across *Layer 1* and *Layer 2* (*Layer 2* and *Layer 3*). (b) The coordinate axes through a given P.

Merge Path: Extending the concept of *Merge Axis*, a Merge Path is a path (not necessarily a straight line) on which we finally merge all the pixels.

We use $G = (V, E)$ to denote a graph, where V denotes the set of vertices and E, the set of edges between the vertices. *Complexity Analysis*, wherever used, denotes the number of *interchanges* between black and white voxels required for the particular algorithm.

2.2   Solution Strategy

In this problem, our fundamental strategy is to attack the 2D layers of a $\mathbf{B_8}$-connected finite binary image found in a $\mathbf{B_{26}}$-connected 3D object. In a given 2D layer, a pixel can have at most 8 neighbors. Hence, we borrow from Bose et al. [1] the idea of transforming any 2D binary image into a *vertical* image, ensuring that the object preserves connectivity during the transformation. However, we cannot directly adopt the strategy in [1] since the vertical image produced in the 2D plane is *unique* and in our case might snap the connectivity between two layers. A general case of the problem may have the images I and J such that, each layer has more than one connected component of black pixels.

Define a graph $G = (V, E)$, such that (i) each connected component in any layer corresponds to a node in V, and (ii) for any two connected components, C_1 and C_2, if there is at least one pair of voxels (u, v) which are $\mathbf{B_{26}}$ adjacent, with $u \in C_1$ and $v \in C_2$, then we have an edge. It is easy to see that, as the black pixels in the original image I are connected, G is connected. Let $G' = (V, E')$ be any spanning tree of G. We know from the definition of a spanning tree that, as long as G remains connected G' also remains connected, thus satisfying the principal constraint behind the transformation. So, it is sufficient to consider G', instead of G. Also, we know that every spanning tree has *at least one* node whose degree is equal to one.

3   The Strategy for Voxel Transformation

Before we discuss the actual algorithm we discuss below a construction which is frequently used in the algorithm.

3.1   Construction of 2D Linear Chains

Given a node in G' with degree one, we need to consider only one *connectivity-sensitive pixel* in the component represented by the node to preserve the connectivity. So, a *Merge Axis* can be any *coordinate axis* passing through that pixel. Now, the rest of the black pixels (which do not originally lie on the merge axis) are interchanged preserving the connectivity such that they finally appear as a linearly connected chain along the merge-axis.

The strategy can be outlined as follows. We compress the 2D region, step by step, from the boundary, simultaneously expanding on the merge axis, $\mathcal{M}$.

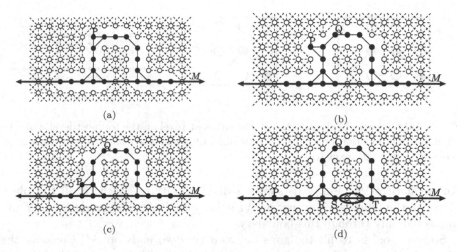

Fig. 2. (a) This shows a connected component in a layer. P is a *non-cut* pixel and M, the *Merge Axis*. (b) This shows P in its new position after the interchange with its adjacent white pixel. (c) P is interchanged with white pixels twice more. (d) P is finally placed on M. This leaves all other pixels on the boundary to be *cut* pixels. Q is one such pixel. The oval region shows the disconnectivity on the *Merge Axis* before collapsing the *cut* pixels.

Ultimately, we have the linear connected chain on M of all the pixels originally in the 2D plane. Now, we describe our algorithm.

Take a *non-cut* pixel (if any) on the boundary, other than those on M. Clearly, its removal doesn't disconnect the rest of the black region. Hence, we move it along the boundary until we first reach M, interchanging with the white pixels that come in the way. This clearly maintains connectivity of the black pixels. Place it on M, by interchanging with the white pixel already present.

We repeat the above process till all the *non-cut* pixels are exhausted. Now, we are left with only *cut* pixels on the boundary (if any).

Figure 2(a) shows the part of the original image, which is of concern (one layer). The movement of the *non-cut* pixel P along the boundary to the merge axis M is shown in Fig. 2(b) to Fig. 2(d).

Lemma 1. *Consider the situation when all the black pixels on the boundary, not on M, are cut pixels. Also consider a part of the boundary which starts and ends on M and let m_a and m_b be a pair of black pixels on M through which the cut pixels on this part of the boundary are connected to M. Now, m_a and m_b are connected only through this part of the boundary. Moreover, this is true for every such part of the boundary.*

Proof. Let us suppose that we have another path connecting m_a and m_b. This clearly implies that there is a *non-cut* pixel on the part of the boundary contradicting the hypothesis. The same argument follows for all such parts of the boundary.   □

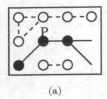 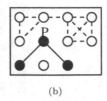 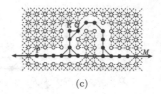

(a)　　　　　　　　　　　　(b)　　　　　　　　　　　　(c)

Fig. 3. (a) One possible location of P, the leftmost pixel on the topmost level. (b) The other possible location of P. (c) The *cut* pixel Q is collapsed onto the previous *level*, exposing a *non-cut* pixel R.

For an illustration, Fig. 2(d) shows a discontinuity on $\mathcal{M}$ with all the black pixels on the boundary and not on $\mathcal{M}$ being *cut* pixels. The pixels S and T are connected only through this boundary.

So, our goal is to fill the gaps between the two ends on $\mathcal{M}$. Consider the leftmost pixel in the topmost level, say P. As this is the topmost level, this pixel has no $\mathbf{B}_8$ neighbors in the level above or to the left of it. Now, considering the remaining possibilities the only two situations where there are no *non-cut* pixels on the boundary are illustrated in Fig. 3. As it is clear from the figure, filling up of the gaps on $\mathcal{M}$ can be clearly done by collapsing P to the *level* below it.

Figure 3(c) shows the collapsing of Q for example. Q is interchanged with the pixel right below it. This is formed from Fig. 2(d). It is easy to see that collapsing preserves connectivity.

We continue collapsing. If this results in a new *non-cut* pixel, we go for the next iteration.

Complexity Analysis: Assume that the total number of pixels in the 2D region is n. Any pixel can be a *cut* or a *non-cut* pixel at any point of time during the transformation. If it is a *non-cut* pixel, and if it is chosen to be moved along the boundary to $\mathcal{M}$, it takes $O(n)$ interchanges to reach $\mathcal{M}$, as the boundary contains at most n pixels. If it is a *cut* pixel, all pixels other than those on $\mathcal{M}$ are *cut* pixels and this particular pixel has been chosen to be collapsed to a *level* below it, then it takes one interchange to do so. There can be at most n such interchanges for any particular pixel. Hence, for any pixel, it takes at most $O(n)$ interchanges and therefore, the complexity is $O(n^2)$.

3.2　Algorithm-Part I

Let u be a node with degree one in G' and also let (u, v) be the edge emerging from u. In other words, the components represented by u and v, say U and V respectively, have at least one $\mathbf{B}_{26}$ adjacent voxel pair (v_u, v_v), with $v_u \in U$ and $v_v \in V$. As the degree of u in G' is one, we develop a strategy to *merge* U with V.

To make the merging easier, we first form a single straight chain of all the pixels on U. The merge axis $\mathcal{M}$ for U can be in any direction. So, let us fix it to be horizontal. We merge all the black pixels in U on $\mathcal{M}$.

Let us suppose that U and V are in a layers i and $i + 1$ $(i - 1)$, respectively. Again, to make merging easier we take $\mathcal{M}$ to such a location on layer i that the top view of these two components U and V looks like, $\mathcal{M}$ protruding out from the boundary of V. So, we translate $\mathcal{M}$ horizontally, in the layer in which U is present, say i, till any further move removes the connectivity between U and V, using the procedure described below.

Translation: The idea behind translation is simple. We move pixel by pixel. Figure 4 shows an example of how we do it. The extreme pixel is moved first followed by the next farthest pixel. A given pixel gets displaced $O(n)$ times. There are $O(n)$ pixels to be moved. Hence, the complexity for translation is $O(n^2)$.

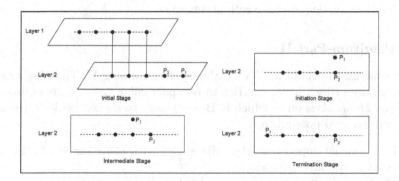

Fig. 4. Here P_1 and P_2 needed to be displaced. In this illustration, the transformation of P_1 is shown. P_1 is moved along the chain (for preserving the connectivity) and brought back to the chain whenever the first white pixel is found. The displacement of P_2 can be similarly done.

If U intersects with any other connected component in the *layer i* either during the process of merging on $\mathcal{M}$ or during the process of translation, we do the following.

1. We stop the process.
2. We consider the compound component formed by U and the component with which it intersects instead of the original components.
3. We build a new G and form the new spanning tree, G'.
4. We go for the next iteration.

Note that, U might have established new links with other components in layers $i - 1$ and $i + 1$. Figure 5 shows an illustration of this part.

Complexity (Part I): Let n_u and n_v be the number of black pixels in U and V respectively. From our earlier discussions, merging all n_u pixels on $\mathcal{M}$ takes $O(n_u^2)$ interchanges. As translating $\mathcal{M}$ horizontally by one pixel takes $O(n_u)$ interchanges, the entire translation phase takes $O(n_u n_v)$ interchanges. If any process has to be stopped in the middle, then a new iteration has to be started after making some changes, mentioned above.

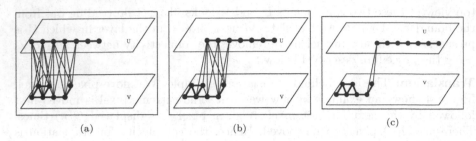

Fig. 5. (a) The pixels in U have been merged on to its *Merge Axis*. This shows all the adjacencies. (b) Two of the pixels have been translated by the procedure described above. (c) The situation after the entire translation.

3.3   Algorithm-Part II

Now, we merge the chain in U with the layer containing V. The pixels can be merged in any order but we restrict to one particular order, namely, from the end of the *Merge Axis* on U which is $\mathbf{B}_{26}$ adjacent to a pixel on V to the other end. There are two possibilities.

Case I: U has developed new $\mathbf{B}_{26}$ adjacencies with some voxels of other components in the layer containing V.

Case II: No such adjacency has been developed.

It is very well possible that some other edge between U and any other component W in G got snapped. Case II is the easiest of the two. All we need to do is, keep interchanging the voxels on $\mathcal{M}$ in U, with the white voxels in the layer containing V starting from either end of $\mathcal{M}$. Figure 6 shows an example. It is easy to see that the complexity in this case is $O(n_u)$. Now, let us consider Case I. Then, there is a possibility that, if we follow the same steps as suggested above for Case II, after certain number of steps, we encounter another connected component. Figure 7 shows an example. If we encounter such a component, we adopt a sequence of steps, similar to those considered in *Part I*.

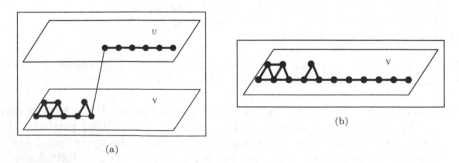

Fig. 6. (a) Starting from the situation in Fig. 5(c) a pixel has been merged with V. (b) All pixels on U have been merged onto V.

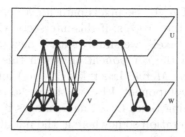

Fig. 7. U develops new adjacencies with W, during translation

1. We stop the process. This may be the end of the process.
2. We consider the compound component formed by V and the other component in the same layer which we encountered, along with the pixels interchanged between U and this layer by now, instead of V and that other component.
3. We update U. U now contains fewer pixels on the chain $\mathcal{M}$.
4. We rebuild G and form the new spanning tree, G'.
5. We start a new iteration.

3.4   Proof of Correctness and Overall Complexity

Lemma 2. *The algorithm suggested above, eventually leads us to the intermediate structure, a single chain containing all the black voxels in the original image I.*

Proof. In one pass through *Part I* of the algorithm, we either merge U with another connected component in the same layer, i, or move the chain, $\mathcal{M}$ to a new location, again in the same layer, i. So, decrease in $|V|$ in *Part I* is less than or equal to one. In one pass through *Part II* of the algorithm, we merge either U with V or V with some other connected component in its layer or both. So, decrease in $|V|$ is either one or two.

Now, if any pass through *Part I* merges two connected components, we don't touch *Part II* until again we pass through *Part I*, as a new iteration is started. If the pass through *Part I* doesn't merge but, simply translates U to a new location, we definitely pass through *Part II* and this guarantees that at least two components will be merged. Hence, each iteration through the algorithm reduces $|V|$ by at least one and therefore, after at most $|V| - 1$ iterations, we are left with a single component. Now, we can form $\mathcal{M}$ for this single component in any direction starting from anywhere and form the single chain. $\square$

Let us find the complexity of an iteration. Assume that *component i* has n_i number of black pixels and that there are c components in total. Note that, components in a particular layer may be disconnected but they can be connected using voxels of layers above and below. Let $\sum_{i=1}^{c} n_i = n$. We divide the complexity calculation into two parts as follows.

1. Merging of all the pixels in a single component.
2. Merging of different components.

Merging a component of n_i black pixels onto its *Merge Axis* takes $O(n_i^2)$ interchanges. Now, during the process, if this intersects with another component with n_j number of black pixels, we simply start a new iteration. Let n_k be the total number of pixels of the component on which this *Merge Axis* has to be merged. Translation of the *Merge Axis* takes $O(n_i n_k)$ interchanges if it doesn't intersect with any other component. Else, we simply start a new iteration. Once translation is done, merging takes $O(n_i)$ interchanges if it's Case II. In Case I, we have to start a new iteration somewhere in the middle.

So, the worst case complexity of an iteration is

$$O(n_i^2) + O(n_i n_k) + O(n_i) = O(n_i^2 + n_i n_k)$$

And the overall worst case complexity is simply a summation of the above complexity over all the iterations. From Lemma 2 it is clear that the number of iterations is at most $c-1$. Note that n_i, n_k may change after every iteration due to merging of components. In any case, n_i and n_k are $O(n)$. So, an upper bound of the complexity is

$$\Sigma_{i,k=1}^{c-1} O(n_i^2 + n_i n_k) = \Sigma_1^{c-1} O(n^2) = O(cn^2)$$

Theorem 1. *Given any two binary images I and J with c_1 and c_2 number of connected components respectively and n voxels each, both can be transformed into one another maintaining the original connectivity of the black voxels by $O((c_1 + c_2)n^2)$ interchanges.*

Proof. The theorem follows from Lemma 2 and the above discussion.   □

4   Voxel Transformation under a Different Connectivity Model

In the model presented till now, a *valid interchange* is taken as such an interchange between any two $\mathbf{B}_{26}$ adjacent black and white voxels, which preserves the connectivity of the image *before* and *after* the interchange. A slightly different model can be obtained if we impose a *single backbone* condition [2] along with our original connectivity model. Dumitrescu and Pach [3] also consider this as an alternative model. In our case, a *backbone* is defined as the set of all black voxels except the one which we currently interchange. The condition is that the *backbone* must be $\mathbf{B}_{26}$ connected at any given point of time. In order to adopt this model, we only need to make small changes in our algorithm.

First, note that, in the algorithm we described, there are only two situations where the *single backbone* condition fails.

1. While collapsing the pixels on the boundary when all the *non-cut* pixels not on *Merge Axis* are exhausted to form the 2D linearly connected chains.
2. While merging the translated *Merge Axis* with a component in an adjacent layer.

First Situation: While forming the 2D linearly connected chains, instead of collapsing the pixels to the *level* below when all the pixels on the boundary are *cut*, we can move the black pixels between m_a and m_b on the *Merge Axis* (refer Lemma 1) to one of the extreme ends of the axis (through interchanges). This is similar to the *Translation*, mentioned in *Part I* in Section 3.2. So, ultimately what we have is a *Merge Path* which is the union of two parts of the original *Merge Axis* and the *cut* black pixels on the boundary. For example, consider Fig. 2(d). The pixels R, S and T have to be translated to the ends of M.

This can be easily adopted to the algorithm discussed. We need to consider only one *connectivity-sensitive pixel* for each 2D connected component. So, we can easily decide which part of the *Merge Axis* is to be extended and which part should be left untouched (depending on which part the *connectivity-sensitive pixel* lies on). For example, suppose that in Fig. 2(d), the pixel R is the *connectivity-sensitive pixel*. We should not move R during the translation mentioned above. So, a possible and easy solution is to translate all the pixels starting from the rightmost end of M till T to the left end of M. Then, translate S. And we end up with the *Merge Path*.

Now, we are left with *bending* the *Merge Path* to a straight line. The only curvy portion is that of the chain of black pixels. Again, in a similar manner, considering the above example, translate each pixel on the chain, starting from the right end to the left end of M.

Second Situation: We only need to change the order in which the pixels on the *Merge Axis* are merged with the component in the other layer. Note that the end of the *Merge Axis* other than the one whose removal disconnects the components, is a *non-cut* pixel. So, we can merge starting from that end, just the opposite way we mentioned in Section 3.3. Now, clearly, interchanging with a *non-cut* pixel maintains the backbone's connectivity. The only problem is with Case I of the Section 3.3. The new adjacencies are at the very end where we have *non-cut* pixels. One possible solution is starting from this end, find the first pixel which is not $\mathbf{B}_{26}$ adjacent with any of the pixels in the new component which the Case I refers to. So, starting from this pixel (which is clearly *non-cut*) keep merging till the other end. The rest of the algorithm follows.

The changes mentioned in both the above mentioned situations do not change the complexity.

5   Conclusion

To conclude, we have shown that two 3D binary images I and J of c_1 and c_2 components differ by a sequence of $O((c_1 + c_2)n^2)$ 26-local interchanges preserving the original black connectivity. We also discussed an alternative approach to fit into a slightly different model in Section 4. One possible extension to the algorithm would be to consider a more general model where we try to preserve the connectivity of the white pixels (background) along with that of the black voxels (foreground).

References

1. Bose, P., Dujmovic, V., Hurtado, F., Morin, P.: Connectivity-Preserving Transformations of Binary Images. In: Computer Vision and Image Understanding, Elsevier, Amsterdam (accepted, 2007)
2. Dumitrescu, A., Suzuki, I., Yamashita, M.: Motion planning for metamorphic systems: feasibility, decidability and distributed reconfiguration. IEEE Transactions on Robotics and Automation 20(3), 409–418 (2004)
3. Dumitrescu, A., Pach, J.: Pushing squares around. Graphs and Combinatorics 22(1), 37–50 (2006)
4. Klette, R. and Rosenfeld, A.: Digital Geometry: Geometric Methods for Digital Picture Analysis. Morgan Kaufman, Elsevier, New Delhi, India, (2005)
5. Rosenfeld, A., Saha, P.K., Nakamura, A.: Interchangeable pairs of pixels in digital images. Pattern Recognition 35(9), 1853–1865 (2001)
6. Rosenfeld, A., Nakamura, A.: Two simply connected sets that have the same area are IP-equivalent. Pattern Recognition 34(2), 537–541 (2002)

Thinning on Quadratic, Triangular, and Hexagonal Cell Complexes

Petra Wiederhold and Sandino Morales

Department of Automatic Control,
Centro de Investigacion y de Estudios Avanzados (CINVESTAV) - IPN
Av. I.P.N. 2508, Col. San Pedro Zacatenco, Mexico 07000 D.F., Mexico
{biene,smorales}@ctrl.mx

Abstract. This paper deals with a thinning algorithm proposed in 2001 by Kovalevsky, for 2D binary images modelled by cell complexes, or, equivalently, by Alexandroff T_0 spaces. We apply the general proposal of Kovalevsky to cell complexes corresponding to the three possible normal tilings of congruent convex polygons in the plane: the quadratic, the triangular, and the hexagonal tilings. For this case, we give a theoretical foundation of Kovalevsky's thinning algorithm: We prove that for any cell, local simplicity is sufficient to satisfy simplicity, and that both are equivalent for certain cells. Moreover, we show that the parallel realization of the algorithm preserves topology, in the sense that the numbers of connected components both of the object and of the background, remain the same. The paper presents examples of skeletons obtained from the implementation of the algorithm for each of the three cell complexes under consideration.

Keywords: parallel thinning, 2D binary images, simple cell, locally simple cell, cellular complex, cell complex, Alexandroff space, Kovalevsky skeleton.

1 Introduction

This paper deals with theoretical aspects of thinning on 2D binary images. Thinning is an important and widely used preprocessing method in digital image processing, in order to facilitate the classification or recognition of objects of interest. In the case of binary digital images, where the set of objects has already been determined, thinning is an iterative procedure which produces a particular subset, named skeleton, from the set of all object elements. The skeleton should represent topological properties like connectedness, as well as geometrical properties related to size and form, of the object. At the same time, the skeleton should have as less as possible elements. During thinning, in each iteration, simple and non-final object elements are deleted from the "frontier" of the remaining object. Due to [17], final elements are situated at the end of arcs, which should be part of the skeleton, and simple elements are those whose deletion preserves the connectedness both of the object and of the background. In the spirit of this idea, we define the following:

V.E. Brimkov, R.P. Barneva, H.A. Hauptman (Eds.): IWCIA 2008, LNCS 4958, pp. 13–25, 2008.
© Springer-Verlag Berlin Heidelberg 2008

Definition 1. *A thinning method, to be applied to an object within a 2D binary image, is said to preserve topology, whenever it preserves both the number of connected components of the object and the number of connected components of the background.*

Important theoretical questions about thinning are,
1) the characterization of simplicity by local properties considered only within certain neighborhood of the element, for example, by a local simplicity.
2) the question if a proposed method can be parallelized, that means, if the parallel implementation of the method preserves topology.

What "connected component", "frontier", "simple", and "final" mean, depends on the mathematical model used for the (domain of the) digital image. Digital images are usually modelled by adjacency graphs. In particular, for 2D digital binary images, the (4,8)- and (8,4)- adjacency graphs are applied, but there are also proposals for adjacency graphs related to the triangular and hexagonal tilings of the plane [2]. Theoretical treatments in a more general setting have been published, for example by Saha et al ([18], [6]), and by Kong [4].

The domain of a digital image can be alternatively modelled by a cell complex, or, equivalently, by an Alexandroff T_0 space. This model has been introduced, justified, defended and applied by Kovalevsky, see for example [7], [8], [9], [10], [11], but also by other authors, see for example [5], [20] and [21]. In this regard, in a short note within [8], a thinning method was proposed, which seems to be the first proposal of a thinning algorithm on cell complexes. The same method was shortly described within [9] and [10], each time with augmented detail, but not enough to cover the theoretical background of this method. Due to our opinion, thinning on cell complexes is a very interesting topic, and the publications [8], [9] and [10] opened many theoretical questions, which provided the motivation of our investigations. The scope of our paper is to give some theoretical foundation of Kovalevsky's thinning method. In this context, we will answer both theoretical questions cited above: First, we prove that a local simplicity is sufficient to satisfy simplicity, and that both are equivalent for certain elements. Second, we show that Kovalevsky's thinning method can be parallelized, by showing that the parallel realization of the method preserves topology.

The paper is organized as follows: In sections 2 and 3, preliminaries about cell complexes and their corresponding Alexandroff topological spaces, and important suppositions for this paper, are presented. In section 4, the relation between simplicity and local simplicity, as well as a characterization of simplicity by a connectivity number, are studied. Section 5 presents and analyzes Kovalevsky's thinning algorithm. Section 6 presents an idea to prove that the parallel implementation of Kovalevsky's algorithm preserves topology. Section 7 shows some examples of the application of Kovalevsky's algorithm, and section 8 contains concluding remarks.

In this conference paper, only the main ideas of the proofs are presented. All properties presented in this paper are proved in detail, under the same suppositions, in [13]. All proofs will be presented in detail, under more general suppositions, within a forthcoming journal paper.

Throughout the paper, $\mathbb{Z}$ denotes the set of integer numbers, and $\mathbb{R}$ the set of real numbers. In a topological space X, for $M \subseteq X$, $cl_X(M)$ denotes the closure of M, $int_X(M)$ is the interior of M, $fr_X(M)$ is the frontier of M, but we omit the index X if we work only in one space X. We denote by $\mathbb{R}^2$ the Euclidean plane, equipped with the standard topology. For a finite set M, $|M|$ denotes the number of its elements. For a subset M of a set X, M^c denotes its complement $X \setminus M$.

2   Cell Complexes

Recall the definition of a cell complex, from [16], as it has been used in many papers of Kovalevsky:

Definition 2. *An **abstract cell complex** is a structure $(X, \leq, \dim)$, where $(X, \leq)$ is a **poset** (partially ordered set, that is, $\leq$ is a binary reflexive transitive and antisymmetric relation on the set X), and $\dim : X \to \mathbb{N} \cup \{0\}$ is a function such that $x \leq y$ implies that $\dim(x) \leq \dim(y)$, for any $x \in X$. The elements of X are called cells, and, for $x, y \in X$, if $\dim(x) = k$, x is named k-**cell**. The **dimension** of $(X, \leq, \dim)$ is defined by $\sup\{\dim(x) : x \in X\}$.*

If $X = (X, \leq, \dim)$ is an abstract cell complex, then a **subcomplex** $M = (M, \leq_M, \dim_M)$ of X is entirely determined by the subset $M \subseteq X$, by defining $\leq_M$ as the restriction of $\leq$ onto $M \times M$, and $\dim_M$ as the restriction of the function $\dim$ onto M.

In this paper, we consider particular abstract cell complexes, related to the three normal tilings of congruent regular convex polygons in the Euclidean plane $\mathbb{R}^2$. Recall that a normal tiling of the plane is a family of (closed) polygons whose union covers the plane, and where any intersection of two polygons, if non-empty, is a common side of both polygons, or a unique common vertex. It is well-known that there exist only three normal tilings of the plane whose elements are congruent convex regular polygons: the quadratic one (where all polygons are congruent to the same square), the triangular one (where the polygons are congruent to an equilateral triangle), and the hexagonal one (where the polygons are congruent to a regular hexagon), see section 8.3 of [15].

There is a natural bounding relation on the set C of all polygons of a tiling $\mathcal{T}$, and all their sides and vertices: let M and N be subsets of $\mathbb{R}^2$ each of which is an element of C, then define $M \leq N$ if and only if $M \subseteq cl_{\mathbb{R}^2}(N)$. Whereas the elements of C are considered as closed subsets of $\mathbb{R}^2$ in [3], we consider C to be a decomposition of $\mathbb{R}^2$, so, the polygons are supposed to be open sets in $\mathbb{R}^2$, and each side is considered without its end points.

There is a natural dimension on the set C of all polygons, sides of polygons, and vertices of polygons of $\mathcal{T}$, given for any $x \in C$, by $\dim(x) = 2$ if x is a polygon, $\dim(x) = 1$ if x is a side, and $\dim(x) = 0$ if x is a vertex. It is easy to see that $(C, \leq, \dim)$ is a two-dimensional abstract cell complex, for each of the three tilings under consideration.

Because we study only topological (and not, geometrical) properties of the cell complex C, we permit the polygons of the triangular and hexagonal tilings to be no regular polygons (but they have to be congruent and convex). In the rest of this paper, the term **cell complex** $C = (C, \leq, \dim)$ **always means an abstract cell complex, constructed as just explained, from the quadratic or triangular or hexagonal normal tiling of the plane.**

We suppose that a two-dimensional binary digital image is modelled by a cell complex C, where an **object** of interest is modelled by a finite subcomplex T, such that the image function assigns the value 1 to any cell of T, whereas each cell of T^c has the value 0; T^c is named the background. To model the digital image by a cell complex C, supposing that the 2D image is defined on a discrete set $D \subseteq \mathbb{R}^2$, there are two principal ideas, as follows:

a) D is identified with the set of 2-cells (or, with the set of 0-cells [20]) of C, and the other cells of C are generated as an additional structure, which serves to describe (topological) properties of the set of 2-cells, or equivalently, of D. Starting with $D = \mathbb{Z}^2$, Kovalevsky constructed the quadratic cell complex by identifying each (pixel) $p \in D$ with the 2-cell given as the unit square centered in p, and then, assigning 1- and 0-cells to the object by a maximum rule [7]; under this construction, the object is modelled by a closed subcomplex. This philosophy is also defended in [3], [5], and [21]. It is possible to identify $D = \mathbb{Z}^2$ with the set of 2-cells of the triangular and the hexagonal cell complex, too.

b) D is identified with C. Taking into account our supposition that the elements of C form a decomposition of $\mathbb{R}^2$, the natural quotient map $\pi : \mathbb{R}^2 \to C$, which assigns to each point x of $\mathbb{R}^2$ the (unique) cell of C which contains x, is an example of a digitization map. For $M \subseteq \mathbb{R}^2$, $\pi(M)$ is the set of all cells which, as subsets of $\mathbb{R}^2$, intersect M, so, this is an analog to the Gauss digitization defined for a set of pixels, see page 56 of [3]. We applied this digitization scheme in order to generate the objects for our experiments.

We are conscious that in practice, a digital image is modelled by a finite portion M of the cell complex C. For our implementations of the thinning algorithm to be correctly working, we suppose that the object of interest T does not touch the boundary of the image domain M, or, equivalently, T is completely surrounded by cells of $M \setminus T$. Our proofs do not need such a supposition, because they work in the complete cell complex C, and any object T is supposed to be finite.

3   Cell Complexes and Alexandroff Spaces

It is well-known that any poset $(X, \leq)$ generates a topological T_0 space, in which any element is contained in a minimal open neighborhood, given as the intersection of all open sets which contain this element. Such spaces were named discrete spaces by Alexandroff [1], but nowadays, they are usually called Alexandroff spaces ([5], [21]). Alexandroff T_0 spaces and posets are equivalent structures: For a given poset $(X, \leq)$, the set $st(x) = \{y \in X : x \leq y\}$, named the *open star* of x, is the minimal open neighborhood of x, and the family $\{st(x) : x \in X\} \cup \{\varnothing\}$ is a base of an Alexandroff T_0 topology τ on X. Conversely, for a given Alexandroff T_0 space (X,τ), the corresponding partial order, called the *specialization order* of (X,τ), is defined by $x \leq y \Leftrightarrow x \in cl(y) \Leftrightarrow y \in st(x)$. For this reason, a cell complex is a topological model for digital images. The specialization order $\leq$ of (X,τ) can be used to describe topological properties in (X,τ), for example as follows (see [1], [21]):

$cl(M) = \{y \in X : y \leq m \text{ for some } m \in M\}$, $cl(\{x\}) = \{y \in X : y \leq x\}$;

$int(M) = \{m \in M : st(m) \subseteq M\} = \{m \in M : m \leq y \text{ implies } y \in M\}$;

$fr(M) = \{y \in X : st(y) \cap M \neq \varnothing \text{ and } st(y) \cap M^c \neq \varnothing\}$

$= \{y \in X : y \leq m \text{ for some } m \in M \text{ and } y \leq m \text{ for some } m \in M^c\}$.

In this work, we will also use a concept dual to the frontier, called open frontier, which was introduced in [8] and is important in order to formulate the thinning algorithm.

Definition 3. *Let (X,τ) be an Alexandroff T_0 space and $M \subseteq X$. The open frontier of M is defined to be the set $of(M) = \{y \in X : cl(\{y\}) \cap M \neq \varnothing$ and $cl(\{y\}) \cap M^c \neq \varnothing\}$.*

Using the properties quoted above, it is clear that $of(M) = \{y \in X : m \leq y$ for some $m \in M$ and $m \leq y$ for some $m \in M^c\}$. In our two-dimensional cell complex C, for any $M \subseteq X$, $fr(M)$ does not contain any 2-cell, and $of(M)$ does not contain any 0-cell. The open frontier of M is the frontier of M in the dual topological space which is determined by the reversed specialization order $\geq$. Another important property is connectedness, which can be described by means of a graph theoretical one, derived from the specialization order, as we will see in what follows.

Definition 4. *Let $(X,\leq)$ be a poset. Two elements $x, y \in X$ are called incident if $x \leq y$ or $y \leq x$. The set $in(x) = \{y \in X : x \text{ is incident with } y\}$, for $x \in X$, is named incidence set of x.*

Note that the incidence relation is reflexive and symmetric. Hence, X with the incidence relation is an undirected graph, called *incidence graph*, which provides a well-known graph theoretical connectedness concept:

Definition 5. *Two elements p, q of X are named connected in X if there exist a finite sequence $\{p_0, p_1, ..., p_{n-1}, p_n\}$ of elements of X such that $p_0 = p$ and $p_n = q$, and p_i is incident with p_{i+1}, $0 \leq i \leq n - 1$; this sequence is called a pq-path. A subset M of X is named connected in the incidence graph if any two $p, q \in M$ are connected in M.*

Now, considering also the topological space (X,τ) corresponding to the poset $(X,\leq)$, recall that X is *(topologically) connected* if there are no two open disjoint non-empty proper subsets A and B of X such that $A \cup B = X$. Then, a subset $M \subseteq X$ is connected if the topological subspace M is connected. The following property is known (see [5]).

Lemma 1. *Let (X,τ) be an Alexandroff T_0 space, $\leq$ its specialization order, and $M \subseteq X$. Then, M is connected if and only if M is connected in the incidence graph.*

This lemma is true for the Alexandroff T_0 space corresponding to any cell complex C, where we have for $x \in C$ that $in(x) = st(x) \cup cl(\{x\})$; in particular, $in(p) = st(p)$ for any 0-cell p, and $in(q) = cl(\{q\})$ for any 2-cell q.

4   Simplicity, Local Simplicity, and Connectivity Number

In analogy to the usual definitions of simple points and end points in adjacency graphs, we define the following:

Definition 6. *Let T be an object in the cell complex C, and $p \in T$.*
(i) The cell p is named simple if T has the same number of connected components as $T \setminus \{p\}$, and, T^c has the same number of connected components as $T^c \cup \{p\}$.
(ii) The cell p is named final if it is incident with exactly one cell $q \in T$, $q \neq p$.

Kovalevsky used in [9] another definition of simplicity, which we will name local simplicity and is defined in the following. We comment that in [10], another (local) simplicity (named IS-simplicity) was defined, which will be taken into account in our forthcoming paper, too.

Definition 7. *Let T be an object in the cell complex C, and $p \in (fr(T) \cup of(T)) \cap T$. The cell p is named locally simple,*
in the case that $p \in fr(T)$, if $(st(p) \setminus \{p\}) \cap T$ and $(st(p) \setminus \{p\}) \cap T^c$ both are non-empty and connected, and,
in the case that $p \in of(T)$, if $(cl(\{p\}) \setminus \{p\}) \cap T$ and $(cl(\{p\}) \setminus \{p\}) \cap T^c$ both are non-empty and connected.

It is easy to see that any simple or final cell of T belongs to $(fr(T) \cup of(T)) \cap T$. In practical thinning algorithms, the simplicity of an element x is checked using local properties, for example, by template matching, or, by calculating some connectivity number, usually based on adjacency graphs, whose value indicates simplicity, see [14]. The first idea was realized by implementations of Kovalevsky's algorithm for the quadratic cell complex in [22], and for the hexagonal cell complex in [19]. The second idea was applied in [13] to implement the same algorithm on a triangular cell complex. In the following, a connectivity number for cell complexes is introduced, which can be used for deducing a characterization of simplicity by the local simplicity. First, observe the following important properties.

Lemma 2. *In any cell complex C, for any cell $p \in C$, the set of cells of $(in(p) \setminus \{p\})$ can be ordered in a cyclic sequence $\{c_0, c_1, ..., c_k\}$ such that, in this sequence,*
(i) any two consecutive cells are incident (including that c_0 is incident with c_k),
(ii) any cell is incident with exactly two other cells,
(iii) the cells are alternating a-cells and b-cells, where $(a, b) = (1, 2)$ for any 0-cell p, $(a, b) = (0, 2)$ for any 1-cell p, $(a, b) = (0, 1)$ for any 2-cell p.
(iv) For any 1-cell p, the sequence has exactly four cells.

Definition 8. *Let T be an object in the cell complex C, and $p \in T$. If $\{c_0, ..., c_k\}$ is a cyclic sequence of the cells of $(in(p) \setminus \{p\})$, which satisfies all properties of lemma 2, and v_i denotes the value of the cell c_i in the binary image represented by C (for $i = 0, 1, ..., k =\mid in(p) \setminus \{p\} \mid$), then define the* **connectivity number** *of p to be the number*

$$cn(p) = \sum_{i=0}^{|in(p)|-2} \mid v_i - v_{(i+1)} \mid,$$

where the sum in the index set is calculated modulo $(\mid in(p) \mid -1)$.

The connectivity number is the number of changes from value 1 to value 0 or vice versa, in the set $(in(p) \setminus \{p\})$, and its definition depends on lemma 2. It is easy to see that $cn(p)$ is independent of the selection of the cyclic sequence, and the following properties are proved in [13]:

Lemma 3. *Let T be an object in the cell complex C, and $p \in T$. Then,*
(i) $cn(p)$ is strictly positive if and only if $p \in fr(T) \cup of(T)$.
(ii) For $p \in fr(T) \cup of(T)$, the number of connected components of $(in(p) \setminus \{p\}) \cap T$ equals the number of connected components of $(in(p) \setminus \{p\}) \cap T^c$, and both are equal to $\frac{1}{2}cn(p)$.

The connectivity number will be used to characterize simplicity, based on the following property:

Theorem 1. *Let T be an object in the cell complex C, and $p \in (fr(T) \cup of(T)) \cap T$. If p is locally simple then p is simple.*

Proof. Consider here only the case that $p \in fr(T)$. Assuming that p is not simple, we prove that $(st(p) \setminus \{p\}) \cap T$ or $(st(p) \setminus \{p\}) \cap T^c$ is not connected: Based on definition 7, we have to study the following two suppositions:
(1) $T \setminus \{p\}$ has strictly more connected components than T.
(2) $T^c \cup \{p\}$ has strictly less connected components than T^c.
Consider (1) (the proof under (2) is similar). Since $p \in fr(T)$, $dim(p) \in \{0, 1\}$.
(1a) If p is a 0-cell, the supposition (1) and lemma 1 imply that there exist $q_1, q_2 \in T \setminus \{p\}$ such that there is no q_1q_2-path in $T \setminus \{p\}$, but there exists a q_1q_2-path w in T. Provided that w is not a path in $T \setminus \{p\}$, it follows that $w = \{q_1 = \gamma_1, ... \gamma_{i-1}, p, \gamma_{i+1}, ..., \gamma_n = q_2\}$, where $2 \leq i \leq n-1$ and $\gamma_{i-1}, \gamma_{i+1}$ are distinct. Supposing that $(st(p) \setminus \{p\}) \cap T$ is connected, there is a $\gamma_{i-1}, \gamma_{i+1}$-path z in $(st(p) \setminus \{p\}) \cap T$, implying that $\{q_1 = \gamma_1, ... \gamma_{i-1}\} \cup z \cup \{\gamma_{i+1}, ..., \gamma_n = q_2\}$ is a q_1q_2-path in $T \setminus \{p\}$, which contradicts our supposition. In consequence, $(st(p) \setminus \{p\}) \cap T$ is not connected.
(1b) If p is a 1-cell, let $\{p_1, p_2\} = cl(\{p\}) \setminus \{p\}$, and $\{c_1, c_2\} = st(p) \setminus \{p\}$. Because $p \in fr(T) \cap T$, $(st(p) \setminus \{p\}) \cap T^c \neq \varnothing$. Furthermore, by studying the incidence set of p, it is not difficult to prove that $(st(p) \setminus \{p\}) \cap T = \varnothing$, which implies by lemma 2 that $(st(p) \setminus \{p\}) \cap T^c = \{c_1, c_2\}$ which is not connected. $\square$

Now we apply theorem 1 in order to prove the following characterization of simplicity by means of the connectivity number.

Proposition 1. *Let T be an object in the cell complex C, and $p \in (fr(T) \cup of(T)) \cap T$. Then p is simple if and only if $cn(p) = 2$.*

Proof. Consider here only the case $p \in fr(T)$. Suppose first that $cn(p) = 2$. By theorem 1, for showing that p is simple, it is sufficient to prove that $(st(p) \setminus \{p\}) \cap T$ and $(st(p) \setminus \{p\}) \cap T^c$ both are non-empty and connected. It is easily seen that both sets are non-empty. Now, if p is a 0-cell then $in(p) \setminus \{p\} = st(p) \setminus \{p\}$, and lemma 3 implies that each of the sets $(st(p) \setminus \{p\}) \cap T$ and $(st(p) \setminus \{p\}) \cap T^c$ consists of exactly one connected component. If p is a 1-cell then $(st(p) \setminus \{p\}) \cap T = \varnothing$, or $(st(p) \setminus \{p\}) \cap T$ has exactly one element. In both situations, p is simple.

Suppose now that $cn(p) \neq 2$, and let us prove that then p is not simple. By lemma 3 and using $p \in fr(T)$, we can assume that $cn(p) \geq 4$. Hence each of the two sets $(in(p) \setminus \{p\}) \cap T$ and $(in(p) \setminus \{p\}) \cap T^c$ has at least two connected components. Choose a cyclic sequence $\{c_0, c_1, ..., c_k\}$ of $(in(p) \setminus \{p\})$ such that the cells c_α, c_β belong to distinct components of $(in(p) \setminus \{p\}) \cap T^c$, and c_γ, c_δ belong to distinct components of $(in(p) \setminus \{p\}) \cap T$, and $\alpha < \gamma < \beta < \delta$. If c_α, c_β belong to distinct components of T^c then p is not simple because c_α, c_β belong to the same component of $T^c \cup \{p\}$. Suppose now that c_α, c_β belong to the same component of T^c. If c_γ, c_δ belong to distinct components of $T \setminus \{p\}$ then again, p is not simple. But if c_γ, c_δ are cells of the same component of $T \setminus \{p\}$ then, by the alternating position of $c_\alpha, c_\beta, c_\gamma, c_\delta$, the cells c_α, c_β belong to distinct components of T^c, which contradicts our supposition. In consequence, $cn(p) \neq 2$ implies that p is not simple.   □

Proposition 1 and lemma 3 imply the following equivalence.

Corollary 1. *Let T be an object in the cell complex C, and let $p \in (fr(T) \cup of(T)) \cap T$ be a 0-cell or a 2-cell or a non-final 1-cell. Then p is simple if and only if p is locally simple .*

5   Kovalevsky's Thinning Algorithm

Recall that any object T is a finite subcomplex of a cell complex C, each cell of T has value 1, and each cell of T^c has value 0. To delete a cell of T means that its value is changed from 1 to 0, so, after its deletion, the cell belongs to T^c. We quote from [9] and [10] the following algorithm which in our paper will be named **Kovalevsky's algoritm:**

Definition 9. *Kovalevsky's algorithm: Let T be an object in a cell complex C. Each iteration consists in the following two steps:*
1) Detect and delete all cells from $fr(T) \cap T$, which are simple and non-final. Count the number of cells which are deleted in this step, and denote it by a. Let T be the remaining object.
2) Detect and delete all cells from $of(T) \cap T$, which are simple and non-final. Count the number of cells which are eliminated in this step, and denote it by b. Let T be the remaining object.

*In the case that $a + b \neq 0$, perform the next iteration, starting with step 1; in the case $a + b = 0$, the algorithm is finished, and the actual remaining object T is considered the result of the algorithm, and will be called the **Kovalevsky skeleton**.*

Kovalevsky considered any subcomplex of an arbitrary cell complex as an input object. Nevertheless, the unique example of [9] and [10] involved a closed object. By our theorem 1, the global property of simplicity is guaranteed by the local simplicity. In [9], Kovalevsky defines a cell to be simple if it is locally simple due to our definition. Hence, his proposal of algorithm previews the detection of locally simple cells. However, our characterization of simplicity by the connectivity number is another useful base for the implementation of the algorithm.

The definition of the algorithm does not specify whether a sequential or a parallel implementation is described. In a sequential implementation, step 1 of the algorithm (and, similarly, step 2) works in the following manner: First, determine $fr(T) \cap T = \{m_1, m_2, ..., m_k\}$. Then, for $i = 1, 2, ..., k$, if m_i is detected to be simple and non-final in T, then delete m_i immediately from T, that is, $T := T \setminus \{m_i\}$, before proceeding to check the next element m_{i+1}. In consequence, the deletion of some m_i from T can have influence on whether m_{i+1} is simple and non-final in T or not. It is evident from the definition of simple elements, that the sequential deletion of simple elements always preserves topology. Hence, in particular, the application of a sequential implementation of Kovalevsky's algorithm, applied to a non-empty connected object, produces a non-empty connected Kovalevsky skeleton.

The situation is distinct for the **parallel implementation of Kovalevsky's algorithm**. In this case, step 1 (and, similarly, step 2) works as follows: First, determine $fr(T) \cap T = \{m_1, m_2, ..., m_k\}$. Then, for $i = 1, 2, ..., k$, if m_i is detected to be simple and non-final in T, then it is marked but not (yet) deleted. When all cells of $\{m_1, m_2, ..., m_k\}$ have been checked, then all marked cells are deleted, that is, $T := T \setminus \{m \in fr(T) \cap T : m \text{ simple and non-final in } T\}$. In consequence, the fact that some cell m_i is marked as to be deleted later, does not have any influence on whether m_{i+1} is simple in T or not. In other words, the simplicity of a cell is checked always with respect to the object which equals the actual remaining whole object at the beginning of the step.

It is well-known that the resulting skeletons obtained from sequential and from parallel implementations of the same thinning algorithm can be quite distinct, and, that it is far from trivial whether a parallel implementation of a thinning algorithm, preserves topology. We will show in the next section that the parallel implementation of Kovalevsky's algorithm preserves topology.

6   Parallel Thinning Due to Kovalevsky's Algorithm

Recall from definition 9 that each iteration of Kovalevsky's algorithm consists in two steps. The following theorem will imply that Kovalevsky's algorithm can be parallelized.

Theorem 2. *Let T be an non-empty object in a cell complex C. Denote by T_k the remaining object after having applied k steps of the parallel implementation of Kovalevsky's algorithm to the input object T, for $k \geq 0$, where $T_0 = T$. Then, for all $k \geq 1$, the number of connected components of T_k is equal to the number of components of T, and also, the number of connected components of T_k^c equals the number of components of T^c.*

Proof. We give here only the main idea of the proof by induction on k. Let us call two objects equivalent if they have the same number of connected components, and if the numbers of connected components of their complements also coincide. In the induction base it is proved that T_1 is equivalent to T, and then, under the induction hypothesis that T_k is equivalent to T, it is proved that T_{k+1} is equivalent to T. Both the induction base and the induction step proof, are based on the following reasoning:

Let R be an object which is is equivalent to T, and whose simple non-final cells $r_1, ..., r_n$ are cells of its frontier [or, analogously, of its open frontier], which are arbitrarily ordered and have been detected in a parallel manner. The latter means that each of these cells was detected to be simple and non-final, as a cell of the whole object R. It is proved using $(n-1)$ steps that $R_1 = R \setminus \{r_1, ..., r_n\}$ is equivalent to T. The l-th step consists in proving that $r_{l+1}, ..., r_n$ are simple in $R \setminus \{r_1, ..., r_l\}$. That $R \setminus \{r_1, ..., r_l\}$ is equivalent to T, is obtained in the $(l-1)$-th step, for $l > 1$. In the case $l = 1$, we apply that R is equivalent to T. In the $(n-1)$-th step, it is proved that the cell r_n is simple in the object $R \setminus \{r_1, ..., r_{n-1}\}$, so, it is proved that R_1 is equivalent to T. Observe that R_1 is the result of having applied to the object R one step of the parallel implementation of Kovalevsky's algorithm, which corresponds to the treating of cells of the frontier [or, analogously, of the open frontier]. □

Corollary 2. *The parallel implementation of Kovalevsky's algorithm preserves topology. In particular, the Kovalevsky skeleton of any non-empty object, is non-empty; and, the Kovalevsky skeleton of any connected object, is connected.*

7   Examples of Kovalevsky Skeletons

In this section we show some examples of skeletons, obtained by the parallel implementation of Kovalevsky's algorithm.

Figure 1 shows the skeletons of a capital "T", in three distinct orientations. Observe the robustness of the skeleton with respect to rotations, on the hexagonal cell complex, which, due to our experiments, has "better" geometrical properties than the other two cell complexes; this topic will be treated in another paper.

Figures 2 and 3 present objects and their skeletons on the quadratic cell complex.

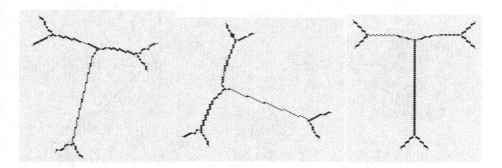

Fig. 1. On the hexagonal complex, the skeletons of a letter T, in three orientations

Fig. 2. On the quadratic cell complex, left: a closed object to be thinned, middle: a remaining object during the thinning process, right: the resulting skeleton

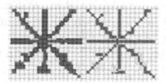

Fig. 3. On the quadratic cell complex, an object (left), which is neither closed nor open, and its corresponding skeleton (right)

The object in figure 3 present the interesting property that the skeleton can contain cells p, q, r, p a 0-cell, q a 1-cell, and r a 2-cell, such that $p \leq q \leq r$. Hence the skeleton is a subcomplex which has a topological dimension, as defined in [21] for Alexandroff spaces, equal to two. This contradicts the intuitive idea, that a skeleton should be "thin" or "curve-like", but similar properties were observed for skeletons on adjacency graphs [12]. Figure 4 presents another example for this fact.

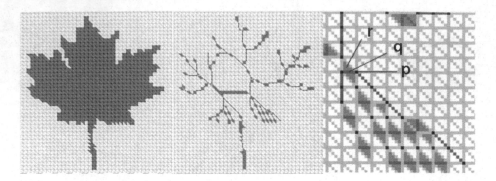

Fig. 4. On the triangular cell complex, left: the original object, middle: its skeleton, right: a detail of the skeleton, where a chain of the form $p < q < r$ in the poset can be observed

8   Concluding Remarks

In this paper, we give a theoretical foundation of Kovalevsky's thinning algorithm, if it is applied to 2D binary images modelled by a quadratic, triangular, or hexagonal cell complex. We proved that local simplicity is sufficient for simplicity, and we characterized simplicity by a connectivity number which is locally computed. Moreover, we show that the parallel realization of the algorithm preserves topology.

It is clear from Kovalevsky's algorithm, that the resulting Kovalevsky skeleton is irreducible in the sense, that it does not contain any simple non-final cell, in analogy to the definition of irreducibility of subgraphs of certain adjacency graphs used in [12]. The Kovalevsky skeleton is unique if obtained from the parallel implementation; but for sequential implementations, there are various ways of tracing the (open) frontier, which can result in different skeletons. The example presented in figure 4 shows that the Kovalevsky skeleton can contain cells p, q, r, pairwise distinct, such that $p \leq q \leq r$. In this case, the topological dimension, as defined in [21] for Alexandroff spaces, of the skeleton is equal to two.

Acknowledgements. The authors thank the referees for their useful comments. Figure 1 is from [19]; the authors thank Alfredo Trejo for having implemented Kovalevsky's algorithm on the hexagonal cell complex.

References

1. Alexandroff, P.: Diskrete Räume. Matematicheskij Sbornik 2(44), 501–519 (1937)
2. Deutsch, E.S.: Thinning Algorithms on Rectangular, Hexagonal, and Triangular Arrays. Communications of the ACM 15(9), 827–837 (1972)
3. Klette, R., Rosenfeld, A.: Digital Geometry - Geometric Methods for Digital Picture Analysis. Morgan Kaufman Publ., Elsevier, CA, U.S.A. (2004)

4. Kong, T.Y.: On the Problem of Determining Whether a Parallel Reduction Operator for N-dimensional Images Always Preserves Topology. In: Vision Geometry II. SPIE Proc. Series, vol. 2060, pp. 69–77 (1993)
5. Kong, T.Y., Kopperman, R., Meyer, P.R.: A topological approach to digital topology. Amer. Math. Monthly 98, 901–917 (1992)
6. Kong, T.Y., Saha, P.K., Rosenfeld, A.: Strongly Normal Sets of Contractible Tiles in N Dimensions. Pattern Recognition 40, 530–543 (2007)
7. Kovalevsky, V.A.: Finite Topology as Applied to Image Analysis. Computer Vision, Graphics and Image Processing 46, 141–161 (1989)
8. Kovalevsky, V.A.: Finite Topology and Image Analysis. In: Hawkes, P.W. (ed.) Advances in Electronics and Electron Physics. Image Mathematics and Image Processing, vol. 84, pp. 197–259. Academic Press, San Diego (1992)
9. Kovalevsky, V.: Algorithms and Data Structures for Computer Topology. In: Bertrand, G., et al. (eds.) Digital and Image Geometry. LNCS, vol. 2243, pp. 38–58. Springer, Heidelberg (2001)
10. Kovalevsky, V.: Algorithms in Digital Geometry Based on Cellular Topology. In: Klette, R., Žunić, J. (eds.) IWCIA 2004. LNCS, vol. 3322, pp. 366–393. Springer, Heidelberg (2004)
11. Kovalevsky, V.: Axiomatic Digital Topology. Journal of Mathematical Imaging and Vision 26, 41–58 (2006)
12. Latecki, L., Eckhardt, U., Rosenfeld, A.: Well-Composed Sets. Computer Vision and Image Understanding 61(1), 70–83 (1995)
13. Morales Chávez, P.S.: Topología del Adelgazamiento sobre el complejo celular cuadrático y sobre el complejo celular triangular, Master thesis, Dept. of Automatic Control, CINVESTAV-IPN, Mexico City (2007), http://www.ctrl.cinvestav.mx/~biene/Thesis.htm
14. Parker, J.R.: Algorithms for Image Processing and Computer Vision. John Wiley and Sons, U.S.A./Canada (1997)
15. Quaisser, E.: Diskrete Geometrie, Spektrum Akademischer Verlag, Heidelberg Berlin Oxford (1994)
16. Rinow, W.: Lehrbuch der Topologie, VEB Deutscher Verlag der Wissenschaften, Berlin (1975)
17. Rosenfeld, A.: A Characterization of Parallel Thinning Algorithms. Information and Control 29, 286–291 (1975)
18. Saha, P.K., Rosenfeld, A.: Local and Global Topology Preservation on Locally Finite Sets of Tiles. Information Sciences 137, 303–311 (2001)
19. Trejo Martínez, A.: Geometría del Adelgazamiento sobre el complejo celular cuadrático y sobre el complejo celular hexagonal, Master thesis, Dept. of Automatic Control, CINVESTAV-IPN, Mexico City (2007), http://www.ctrl.cinvestav.mx/~biene/Thesis.htm
20. Webster, J.: Cell Complexes and Digital Convexity. In: Bertrand, G., et al. (eds.) Digital and Image Geometry. LNCS, vol. 2243, pp. 272–282. Springer, Heidelberg (2001)
21. Wiederhold, P., Wilson, R.G.: The Alexandroff dimension of digital quotients of Euclidean spaces. Discrete and Computational Geometry 27, 273–286 (2002)
22. Zempoalteca Ramírez, M.d.R.: Adelgazamiento de subcomplejos celulares 2D, Master thesis, Dept. of Automatic Control, CINVESTAV-IPN, Mexico City (2004), http://www.ctrl.cinvestav.mx/~biene/Thesis.htm

Experimental Comparison of Continuous and Discrete Tangent Estimators Along Digital Curves

François de Vieilleville[1] and Jacques-Olivier Lachaud[2,*]

[1] SUNY State College at Buffalo
1300 Elmwood Avenue, Buffalo, NY 14222-1095, U.S.A.
devieill@math.buffalostate.edu
[2] Laboratoire de Mathématiques, UMR CNRS 5127
Université de Savoie, 73776 Le-Bourget-du-Lac, France
jacques-olivier.lachaud@univ-savoie.fr

Abstract. Estimating the geometry of a digital shape or contour is an important task in many image analysis applications. This paper proposes an in-depth experimental comparison between various continuous tangent estimators and a representative digital tangent estimator. The continuous estimators belong to two standard approximation methods: least square fitting and gaussian smoothing. The digital estimator is based on the extraction of maximal digital straight segments [9,10]. The comprehensive comparison takes into account objective criteria such as isotropy and multigrid convergence. Experiments underline that the proposed digital estimator addresses many of the proposed objective criteria and that it is in general as good - if not better - than continuous methods.

1 Introduction

The proper detection of significant features along digital curves often relies on an accurate estimation of the geometry of the underlying curve that has been digitized. Local geometric quantities such as the curvature at given points can lead to corner detection [17], more generally curvature and tangent estimation lead to the detection of dominant points on digital curves [14]. Correct tangent estimation allows length computation by simple integration.

Estimating local geometric quantities on digitized shapes is a difficult task in itself for at least four major reasons:

(1) Given a digitized shape there exists infinitely many continuous Euclidean shapes that have the same digitization.
(2) Given a digital point and a point on the continuous curve, determining the required size of the computation window to achieve a good estimation is tricky.

* Work supported by the GeoDIB ANR project and LAVOISIER fellowships. Part of this work has been done in France at LaBRI, Université Bordeaux 1.

V.E. Brimkov, R.P. Barneva, H.A. Hauptman (Eds.): IWCIA 2008, LNCS 4958, pp. 26–37, 2008.
© Springer-Verlag Berlin Heidelberg 2008

(3) The digitized curve can be noisy or damaged, worsening the preceding problems.

(4) The time spent on computations may be limited.

The first problem implies that, given a digital shape, additional hypotheses are required to define its reference shape, such as smoothness, compactness, convexity, minimal perimeter or maximal area. For instance, given a digital disk, a reasonnable hypothesis is that the underlying shape is an Euclidean disk, and not some kind of gears with small cogs. The second problem involves the adaptability of computation windows to the local geometry of the shape, e.g. curves with huge curvature variations require different sizes for the computation windows. Sizes of computation windows have a huge impact on the multi-grid convergence (see [4]). The third problem is a common problem which is efficiently addressed in the continuous world, but lacks proper definitions in the digital world. This entails that continuous methods are generally preferred for the extraction of geometric quantities. The fourth problem arises when the computation windows are too large, while narrowing their sizes has a direct impact on the precision of the method. These issues are related to many interesting topics on digital curves such as multi-grid convergence [3,8], digitization problems and topology issues [12], combinatorial properties of digitized shapes [1] and new models for digital straight segments taking into account some distortions [6].

As mentioned earlier usual geometric estimators are based on approximation techniques in the continuous Euclidean space. They forget the specificities of subsets of the digital plane. By this way, they address problem (3) considering that it is the main issue. The noise is then handled by tuning some external parameters. In fact the external parameters often reduce to the choice of the size of the computation window, handling problems (2), (3) and (4) at the same time with a trade-off. The continuous methods can be of various type with different aims with respect to the digital curve: interpolation, reconstruction or fit. The choice of the underlying curve in problem (1) is then often made explicitly with the method itself, e.g. using cubic-splines to interpolate points along a digital curve lead to degree three polynomials as the underlying curve. The numerical methods required to extract the chosen solution can be costly and may even require parameters themselves, this is particularly true when the chosen underlying curve is the solution of a non trivial optimisation problem. As a result (1) and (2) have a direct impact on (4).

On the contrary, standard digital estimators based on digital straight segment recognition estimate local geometric quantities like tangent or curvature with an adaptive computation window and, at the same time, they do not require any external parameters [7,9,19]. Recently, an evaluation of digital tangent estimators was performed in [9] and the λ-MST was shown to outperform the others on many criteria like precision, maximal error, isotropy, convergence, convexity. The tangent orientation is determined using digital straight segment recognition, which entails a computation window adapted to the local curve geometry (addressing problem (2)) and without assumptions on the underlying curve (addressing problem (1)). The average size of the computation window is known and is roughly in $\Theta\left(h^{-1/3}\right)$ where h is the grid step (see [4] for technical proofs). As a result, the asymptotic convergence — or multigrid convergence — of the λ-MST estimator is proved for smooth and convex curves [10]. This estimator is also the best among digital ones at rough scale [9,10]. Its computation on the whole digital curve, i.e. the computation of the tangent orientation field, may be done in time linear with the number of digital points (optimal time, addressing (4)). This estimator is yet to be shown as good as standard continuous methods.

This is precisely the goal of this paper which is achieved by experimental comparison between the λ-MST estimator and two representative classes of continuous estimators. We naturally examine classical criteria like the average absolute error. Furthermore, we propose to use the product precision by computational cost to compare them as objectively as possible. Besides, our aim is not only to compare these estimators but to see if they can benefit from one another. This is the case here where we show how an optimal computation window (problem (2)) can be chosen for the Gaussian derivative technique. The obtained improvements are illustrated experimentally. These experiments indicate that even with the best possible window, the continuous estimators are outperformed by the λ-MST according to the product precision by cost. We stress that we treat only the ideal case where digital contours are perfect digitizations of continuous shapes, without any perturbation or noise. Indeed, a first evaluation must be carried out before in the ideal case, for instance to identify the best precision an estimator may achieve. Secondly, the λ-MST estimator is easily extensible to maximal *blurred* digital straight segments [6], which can accommodate local perturbations in the digital contour. An experimental evaluation of continuous versus discrete estimators in the presence of noise could then be carried out similarly, and would be the object of a future work.

The paper is organized as follow. First we describe continuous tangent estimator methods, more specifically the ones based on least square fitting with polynomials and the ones using convolution with a gaussian derivative. Their main drawbacks are also recalled. In a second time we briefly recall the definition of the λ-MST estimator and its main properties (Section 3). In Section 4, an experimental evaluation between the different estimators is presented, following some of the objective criteria proposed in [9,10]. We also propose several improvements of the two continuous methods, which are then underlined experimentally. The criterion precision times computational cost shows that, even with these improvements, the digital estimator competes with the best possible continuous methods. Our conclusion is thus that digital straight segments are a powerful tool to analyse the geometry of digital curves.

2   Continuous Tangent Estimators

This section presents two continuous classes of methods that are used to extract geometric information from curves. Both methods need external parameters to achieve the best possible accuracy. In the remaining of the paper the considered digital curves are digital 4-curves, that is a 4-connected closed sequence of points in $\mathbb{Z}^2$ such that each of them has exactly two 4-neighbors: a predecessor and a successor (given an orientation). Such curves arise naturally from the cellular decomposition of the Gauss digitization of simple Euclidean objects, provided they are well-composed [11]. The obtained digital curve is denoted C and its points are ordered increasingly with a counterclockwise order, C_i denotes the i-th point of the digital curve and $C_{i,j}$ is the digital path from the i-th point to the j-th point.

2.1 Least Square Methods Using Polynomials

The aim of these methods is to find a polynomial of finite degree which minimizes a positional squared error from a set of (possibly noisy) samples. More precisely, let us denote by $(s_i = (x_i, y_i))_{1 \leq i \leq M}$ a set of M samples obtained from a planar curve parameterized as $y = f(x)$. We thus seek to minimize the functional
$E(a_0, \ldots, a_N) = \sum_{i=1}^{M} \left(y_i - \sum_{j=0}^{j=N} a_j x_i^j \right)^2$.

In the general case, the problem can be reduced to a matrix inversion problem. At least one solution exists and can be efficiently computed using QR factorisation [16]. For small degree polynomials, direct computation is possible as it involves square matrices of order two and three. It is not compulsory that the polynomial be the supposed underlying curve itself. It can also be its local Taylor expansion as explained in [13] for implicit parabola fitting, an approach which is generalized by the *n-jets* of [2].

Once the optimal polynomial for E is determined, the coefficient associated to its X monomial may be used to estimate the tangent orientation. We naturally focus on low order polynomials. That is the linear regression (LR, Eq. (1)), implicit parabola fitting (IPF, Eq. (2)), and explicit parabola fitting (EPF, Eq. (3)). When used for approaching the tangent orientation at the point of interest C_0, considered as the origin, with a computation window ranging from C_{-q} to C_q, those three methods give very similar results (see Figure 1).

$$E_{LR}(a, b) = E(a, b, 0, \ldots) , \tag{1}$$
$$E_{IPF}(a, b) = E(0, a, b, 0, \ldots) , \tag{2}$$
$$E_{EPF}(a, b, c) = E(a, b, c, 0, \ldots) . \tag{3}$$

A refinement of this method is the *weighted least square fitting*, where each sample has a variable importance in the fitting process: the heavier the weight, the more important the fit. However, it is not easy to find meaningful weights within our context. Another refinement is to use independent coordinates, that is a fit on each coordinates with respect to a given parameterization of the curve. Usually centered windows are considered: C_0 is the point of interest, $M = 2q + 1$ is the size of the computation window going from C_{-q} to C_q. When using independent coordinates, the arc-length from C_0 to C_i, is computed as $\sum_{k=0}^{i-1} d_1(C_k, C_{k+1})$ if $i > 0$ and $-\sum_{k=0}^{i-1} d_1(C_k, C_{k+1})$ otherwise.[1]

---

[1] d_1 denotes the distance obtained from the $|| \cdot ||_1$ norm.

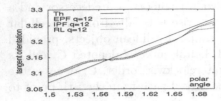 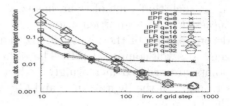

Fig. 1. We represent the tangent orientation estimated with IPF,EPF and LR methods. The test shape is a circle of radius 1. Computation window equals $2q + 1$. (Left) Grid step equals 0.01, x-axis represents the polar angle, the y-axis represents the orientation of the tangent. (Right) The plot is in log-space and represent the average absolute error between true tangent and estimated tangent as a function of the grid step. For each grid step 50 experiences are made with a random shift on the center of the shape.

2.2 Reconstruction Using Gaussian Smoothing

The use of gaussian filters is a common technique for improving the quality of noisy images. This filter can also be used when trying to analyze a digital curve, and has been used in the pattern recognition community for almost 30 years. It is essentially a weighted averaging over a finite window. The obtained smoothed continuous curve is considered to be a good approximation of the underlying curve. Its derivatives are easily computed yielding geometric quantities of the first and second order. This reconstruction has one major drawback, which is the choice of the parameter σ. This tuning parameter is often chosen for the whole curve, but it is not satisfying if the curve has huge curvature variations, entailing then over-smoothing for some region and under-smoothing for others. As a result techniques using scale-space were proposed [15,20] to achieve a better localization of the dominant points across the different σ values. From a discrete point of view we will consider that the estimated derivative at the digital point C_0, say $\hat{\mathbf{C}}'_0$, is obtained as : $\hat{\mathbf{C}}'_0 = \sum_{i=-q}^{q} G'_{\sigma_q}(-i)\mathbf{C}_i$, with $\sigma_q = \frac{2q+1}{3}$ and where $G'_\sigma(t)$ is the first derivative of the Gaussian function $G_\sigma(t) = \frac{1}{\sigma\sqrt{2\pi}} \exp\left(\frac{-t^2}{2\sigma^2}\right)$.

2.3 Common Drawbacks

In the context of digital geometry, the methods presented above share similar drawbacks, which we try to analyze here. First of all, if we consider the digitization of convex shapes, we see that the analysis of its border with the preceding techniques may lead to false concavity/convexity detection, even in the simplest case of the circle as shown on Figure 2. This is particularly true when the size of the computation window is not large enough.

The false convexity/concavity detection can be alleged to a wrong size of the computation window. Experimentally on digitized circles it seems that if the size of the computation window exceeds some value being a functional of the radius and the grid step, there is no false convexity/concavity points. More precisely, this phenomenon is related to the maximal curvature of the shape under study.

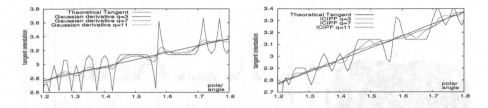

Fig. 2. Test shape is a circle of radius 1, digitized with a grid step equal to 0.01. Tangent orientation is plotted as a function of the polar angle. The x-axis represents the polar angle, the y-axis represents the orientation of the tangent. The size of the computation window equals $2q + 1$. (Left) Tangent orientation obtained using convolutions by the gaussian derivative σ_q. (Right) Tangent orientation obtained using implicit parabola fitting with independent coordinates (ICIPF).

Another fundamental problem related to fixed size computation windows is that one parameter, even if suited for some regions, cannot adapt to the geometry of a digitized shape with huge curvature variations. This statement is particularly underlined on Figure 3. Moreover, a fixed parameter prevents the multigrid convergence of continuous estimators, since it limits the number of data taken into account in the fitting or smoothing process, thus limiting the number of possible local geometries. This is illustrated on Figure 4, where the size on the computation window has a direct impact on the average error.

Last but not least, the computed curvilinear abscissa obtained from the summation of the elementary steps on digital curve is a poor estimation (see [18] for a proof of non convergence for length estimators using fixed-size windows on euclidean segments). Thus the problems of parametrization induce displacements and errors in the continuous proposed methods. A way to solve this problem would be to use an estimation of the elementary steps ds along the curve using the tangent orientation computed with a convergent estimator.

3   Discrete Tangent Estimators

This section recalls the definitions of elementary objects regarding digital straight segments. We then briefly present the λ-MST estimator and its properties.

3.1   Properties and Definitions

Digital straight lines can be simply seen as the digitization of euclidean straight lines. More formally, a *standard line* of characteristics $(a, b, \mu) \in \mathbb{Z}^3$ is the subset of $\mathbb{Z}^2$ $\{(x, y) \in \mathbb{Z}^2 \mid \mu \leq ax - by < \mu + |a| + |b|\}$. They form 4-connected sequences of digital points. We say that a set of successive points $C_{i,j}$ of the digital curve C is a *digital straight segment (DSS)* iff there exists a standard line (a, b, μ) containing them. The predicate "$C_{i,j}$ is a DSS" is denoted by $S(i, j)$. When $S(i, j)$, the characteristics associated with the digital straight segment

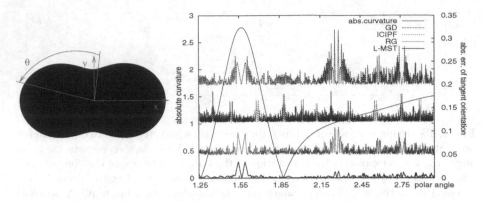

Fig. 3. (Left) The digitized shape is the Gauss digitization at grid step 0.001 of a flower with two extremities, maximum radius 1.4 and minimum radius 0.6. (Right) We plot the absolute error between theoretical tangent orientation and estimations obtained using window size of 21 points, except for the λ-MST. To identify difficult points, we also superposed the theoretical curvature with a dash-dotted plot, but in an other scale. Plots are shifted for presentation. The different estimators are: gaussian derivative (GD - dashed plot), parabola fitting with independent coordinates (ICIPF - dotted plot), linear regression (RG - small dotted plot), λ-MST (L-MST - solid thick plot). We see that the three methods are less precise on the part of the border of the shape which has the fastest curvature variation because of the size of the computation window which is not adapted to the local geometry of the shape. On the contrary, the λ-MST which has an adaptive window size behave much better.

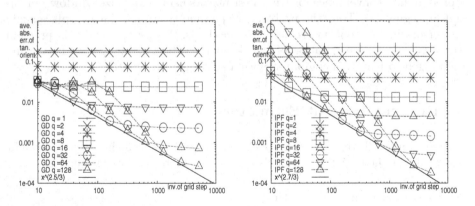

Fig. 4. Experimental multigrid convergence analysis drawn in log-space: x-axis is the inverse of the grid step, y-axis is the average of the absolute error between theoretical tangent and estimated tangent, the shape of reference is a circle of radius one. At each grid step 50 experiences are made and the center is shifted randomly. (Left) Gaussian derivative (GD) with various window size. (Right) Implicit parabola fitting (IPF). In both cases, fixed parameters cannot achieve convergence.

(extracted with the **DR95** algorithm [5]) are the characteristics (a, b, μ), which minimize $|a| + |b|$.

The slope a/b of a DSS provides a coarse estimation of the slope of the underlying tangent. Upon the many existing classes of DSS, we choose to focus on a particular class, the one that contains all the other DSS:

Definition 1. *We say that a portion $C_{i,j}$ of C is a maximal digital straight segment (MS) iff $S(i,j) \wedge \neg S(i-1,j) \wedge \neg S(i,j+1)$.*

Maximal segments can be numbered with increasing indices on the digital curve, $M^i = C_{b_i, f_i}$ denoting the i-th maximal segment. With an incremental version of the **DR95** algorithm (see [7,9,10]), the set of all the maximal segments on a finite digital curve can be extracted in linear time with respect to the number of points of the curve. As maximal segments generally overlap, we introduce the set of all the maximal segments traversing a point.

Definition 2. *The pencil of maximal segments of C_k, denoted $\mathcal{P}(k)$ is the set of MS containing C_k.*

Since every DSS can be extended to form a MS, the pencil of any point is never empty. We also define the *eccentricity* of a point C_k with respect to a maximal segment M^i in $\mathcal{P}(k)$ as: $e_i(k) = \frac{\|C_k - C_{b_i}\|_1}{L_i} = \frac{k - b_i}{L_i}$ with $L_i = \|C_{f_i} - C_{b_i}\|_1$. This value indicates if a digital point is centered within a maximal segment: it is perfectly centered if the value equals $1/2$, limit values are 0 and 1 for extremal points of a maximal segment.

3.2   The λ-MST Tangent Estimator

The λ-MST tangent estimator at one point is designed to take into account the various orientations of the MS in the pencil weighted by a functional of their respective eccentricity with respect to the point of interest:

Definition 3. *The λ-maximal segment tangent direction at point C_k (λ-MST) is defined as $\hat{\theta}(k) = \frac{\sum_{i \in \mathcal{P}(k)} \lambda(e_i(k)) \theta_i}{\sum_{i \in \mathcal{P}(k)} \lambda(e_i(k))}$, where θ_i is the angle of the slope of the i-th MS with the x-abscissa.*

Considering the properties of the eccentricity and the non-emptyness of pencils, this value is always defined and may be computed locally. For particular λ functions the λ-MST estimator satisfies the convexity/concavity property[2] (see Theorem 8 of [10]).

This implies that the border of digitally convex shapes analysed with the λ-MST estimator under the conditions of the preceding theorem does not contain any false concavity. In practice the triangle function is used as the λ function: it matches the preceding conditions and brings good results.[3] Other nice properties are a good isotropic behaviour, multigrid convergence and computation of the tangent field in time linear with respect to the number of curve points (see [9,10] and Figure 5).

---

[2] Estimated tangent directions are monotone for digitization of convex shapes.
[3] The triangle function is defined as $x \to x$ if $x \in [0, \frac{1}{2}]$ and $x \to 1 - x$ if $x \in [\frac{1}{2}, 1]$.

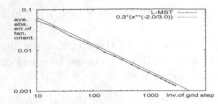 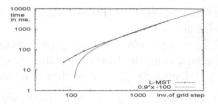

Fig. 5. The test shape is a circle of radius 1, the x-axis represents the inverse of the grid step. For each grid step, fifty experiences were launched with uniform random shift of the center of the shape. Plots are drawn in log-space. (Left) Average absolute error between true tangent and estimated tangent with the λ-MST, the law seems to be in $\mathcal{O}(h^{-2/3})$. (Right) Time spent on computing the tangent orientation field with the λ-MST, the law follows $\mathcal{O}(1/h)$, the same magnitude as the number of points constituting the border of the digitized shape.

4   Experimental Evaluation

The multigrid convergence of the λ-MST estimator is shown on Figure 5 and its good behaviour with respect to huge curvature variations is exemplified on Figure 3. On the contrary the non multigrid convergence of the proposed estimators using fixed size computation window is shown on Figure 4 with the measure of the average absolute error as a function of the grid step. Though the precision of an estimator is important the time spent on the computation has also to be taken into account, a criterion measuring these two parameters at the same time is proposed in the next subsection, yielding the same conclusion.

4.1   A New Criterion Balancing Precision and Computation Time

This subsection introduces a new criterion to compare local tangent estimators, called AAEBT: we measure the product of the average absolute error of tangent direction estimation by the computation time for the whole curve. The lower the quantity as the grid step decreases, the better. As problem (2) penalizes estimators using fixed size windows on curves with huge curvature variations we ran the experiments on digitizations of a disk. The experiments on Figure 6 clearly show that criterion AAEBT for the GD estimator of fixed size window becomes linear with the inverse of the grid step after some rank. For each window size, there is thus a bound to the maximum reachable precision (Figure 4 also illustrates this matter). However, judging from the experiments, the λ-MST estimator has a much better AAEBT which seems to be in $\mathcal{O}((1/h)^{1/3})$. This behaviour is consistent with the average absolute error of the tangent orientation in $\mathcal{O}(h^{2/3})$ and the computational cost in $\mathcal{O}(1/h)$ (see Figure 5).

4.2   Improving Continuous Estimators Using Fixed-Size Windows

Figure 4 clearly suggests that there is a somewhat best window size to pick for each grid step. Judging from experiments on the circle for the GD estimator,

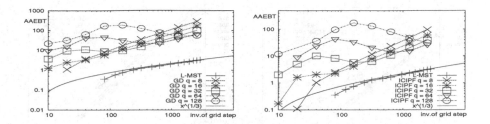

Fig. 6. Test shape is a circle of radius 1, the x-axis is the inverse of the grid step. We represent the time spent on computing the tangent field multiplied by the average absolute error between true tangent orientation and estimation with particular estimators. Plots are drawn in the log-space. (Left) Comparison between gaussian derivative with various window sizes and the λ-MST estimator. For the GD estimator, curves tend to be linear once the maximum precision is reached. (Right) Comparison between implicit parabola fitting with independent coordinates with various window sizes and the λ-MST estimator. For the λ-MST estimator, the law seem to be in $\mathcal{O}((1/h)^{1/3})$.

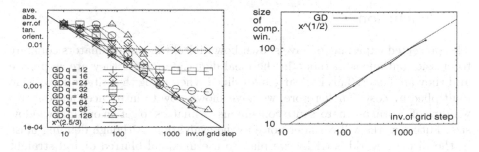

Fig. 7. (Left) Suggested best possible average absolute error with GD as being some $\mathcal{O}((1/h)^{5/6})$, with parameter $\sigma = (2q + 1)/3$. (Right) The suggested size of the computation window to achieve best possible accuracy is in $\mathcal{O}((1/h)^{1/2})$.

the best possible accuracy is in $\mathcal{O}(h^{5/6})$ provided the size of the computation window follow $\mathcal{O}((1/h)^{1/2})$ as shown on Figure 7. The parameter σ of GD is set to one third of the computation window size.

Let us use an adaptive window defined as the maximum distance between the point of interest and the ends of its pencil. The defined size of the computation window increase as a functional of the inverse of the grid step, and even though on average it only grows in $\mathcal{O}((1/h)^{1/3})$ this size brings multigrid convergence for both fits and gaussian derivative, as exemplified on Figure 8 (H-GD and H-ICIPF). The size can also be set globally using the average size of the maximal segments, again multigrid convergence is observed, see Figure 8 (HG-GD and HG-ICIPF).

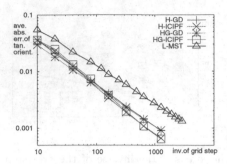 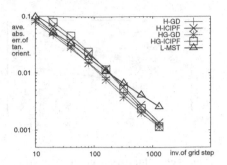

Fig. 8. Average absolute error between true tangent and estimated tangent. The plotted estimators use computation window whose size is determined with maximal segments. Hybrid estimators (H-GD and H-ICIPF) use the maximal distance between the point of interest and the ends of its pencil as q parameter. Hybrid global estimators (HG-GD and HG-ICIPF) use the average size in terms of number of points of the maximal segments as q parameter. (Left) The test shape is a circle of radius one. (Right) Test shape is a flower with two extremities, maximal radius 1.4 and minimal radius 0.6.

5   Conclusion

The presented experiments have shown how digital tangent estimators compare to classic continuous methods in the ideal digitization case: they are as precise and they are faster. This is clearly underlined when using the criterion precision multiplied by cost. Furthermore, we have shown how to introduce the adaptive window of digital estimators into continuous estimators to get an optimal window size. Future works will consider noise in the evaluation. Although defining noise in the discrete world is tricky, we plan to use maximal blurred digital straight segments to take into account distortion in the digital curve.

References

1. Balog, A., Bárány, I.: On the convex hull of the integer points in a disc. In: Symposium on Computational Geometry (SCG 1991), pp. 162–165. ACM Press, New York (1991)
2. Cazals, F., Pouget, M.: Estimating differential quantities using polynomial fitting of osculating jets. Computer Aided Geometric Design 22, 121–146 (2005)
3. Coeurjolly, D., Klette, R.: A comparative evaluation of length estimators of digital curves. IEEE Trans. on Pattern Analysis and Machine Intelligence 26(2), 252–258 (2004)
4. de Vieilleville, F., Lachaud, J.-O., Feschet, F.: Convex digital polygons, maximal digital straight segments and convergence of discrete geometric estimators. Journal of Mathematical Imaging and Vision 27(2), 139–156 (2007)
5. Debled-Renesson, I., Reveillès, J.-P.: A linear algorithm for segmentation of discrete curves. International Journal of Pattern Recognition and Artificial Intelligence 9, 635–662 (1995)

6. Debled-Rennesson, I., Feschet, F., Rouyer-Degli, J.: Optimal blurred segments decomposition of noisy shapes in linear times. Computers and Graphics 30(1), 30–36 (2006)
7. Feschet, F., Tougne, L.: Optimal time computation of the tangent of a discrete curve: Application to the curvature. In: Bertrand, G., Couprie, M., Perroton, L. (eds.) DGCI 1999. LNCS, vol. 1568, pp. 31–40. Springer, Heidelberg (1999)
8. Klette, R., Zunic, J.D.: On discrete moments of unbounded order. In: Kuba, A., Nyúl, L.G., Palágyi, K. (eds.) DGCI 2006. LNCS, vol. 4245, pp. 367–378. Springer, Heidelberg (2006)
9. Lachaud, J.-O., Vialard, A., de Vieilleville, F.: Analysis and comparative evaluation of discrete tangent estimators. In: Andrès, É., Damiand, G., Lienhardt, P. (eds.) DGCI 2005. LNCS, vol. 3429, pp. 240–251. Springer, Heidelberg (2005)
10. Lachaud, J.-O., Vialard, A., de Vieilleville, F.: Fast, accurate and convergent tangent estimation on digital contours. Image and Vision Computing 25(10), 1572–1587 (2007)
11. Latecki, L.J., Eckhardt, U., Rosenfeld, A.: Well-composed sets. Computer Vision and Image Understanding 8, 61–70 (1995)
12. Latecki, L.J., Conrad, C., Gross, A.: Preserving topology by a digitization process. Journal of Mathematical Imaging and Vision 8(2), 131–159 (1998)
13. Lewiner, T., Gomes Jr., J.D., Lopes, H., Craizer, M.: Curvature and torsion estimators based on parametric curve fitting. Computers and Graphics 29, 641–655 (2005)
14. Marji, M.: On the detection of dominant points on digital planar curves. PhD thesis, Wayne State University, Detroit, Michigan (2003)
15. Mokhtarian, F., Mackworth, A.K.: Scale-based description and recognition of planar curves and two-dimensional shapes. IEEE Transactions on Pattern Analysis and Machine Intelligence 8(1), 34–43 (1986)
16. Press, W.H., Flannery, B.P., Teukolsky, S.A., Vetterling, W.T.: Numerical Recipes in C: The Art of Scientific Computing, 2nd edn. Cambridge University Press, Cambridge (1992)
17. Ray, B.K., Pandyan, R.: Acord — an adaptive corner detector for planar curves. Pattern recognition 36, 703–708 (2003)
18. Tajine, M., Daurat, A.: On local definitions of length of digital curves. In: Nyström, I., Sanniti di Baja, G., Svensson, S. (eds.) DGCI 2003. LNCS, vol. 2886, pp. 114–123. Springer, Heidelberg (2003)
19. Vialard, A.: Geometrical parameters extraction from discrete paths. In: Miguet, S., Ubéda, S., Montanvert, A. (eds.) DGCI 1996. LNCS, vol. 1176, Springer, Heidelberg (1996)
20. Witkin, A.P.: Scale-space filtering. In: 8th Int. Joint Conf. Artificial Intelligence, Karlsruhe, vol. 2, pp. 1019–1022 (August 1983)

Polyhedral Surface Approximation of Non-convex Voxel Sets through the Modification of Convex Hulls

Henrik Schulz

Forschungszentrum Dresden - Rossendorf
Department of Information Technology
Dresden, Germany
h.schulz@fzd.de

Abstract. In this paper we want to introduce an algorithm for the creation of polyhedral approximations for objects represented as strongly connected sets of voxels in three-dimensional binary images. The algorithm generates the convex hull of a given object and modifies the hull afterwards by recursive repetitions of generating convex hulls of subsets of the given voxel set or subsets of the background voxels. The result of this method is a polyhedron which separates object voxels from background voxels. The objects processed by this algorithm and also the background voxel components inside the convex hull of the objects are restricted to have genus 0.

1 Introduction

An often arising problem in the field of three-dimensional image analysis is the efficient encoding of the surface of a digital object which is given as a set of voxels. The most popular approach is that of the triangulation, not only because of the simplicity of triangles but also because of the existing hardware support for tasks in the field of computer graphics.

A widely used approach to triangulate voxel objects is the Marching Cubes Algorithm by Lorensen and Cline [9]. It has a very low time complexity, i.e. it is linear in the number of voxels, which makes this algorithm applicable in practical tasks. But it has also two important drawbacks. First, the number of generated triangles is in most cases greater than the number of surface elements (faces) of the original voxel image and second, the orientation of the triangles is limited to a few directions. This is not desirable when an approximation of the original object (before digitization) is needed, which has a smooth surface with a constant curvature, for example.

Other triangulation methods use a divide-and-conquer approach. The algorithm described in [3] is applicable not only in 2-dimensional spaces to produce Delaunay triangulations, but also in higher dimensions, i.e. also in the 3-dimensional case. The algorithm separates the input data into two subsets and constructs the triangulation of the subsets recursively.

V.E. Brimkov, R.P. Barneva, H.A. Hauptman (Eds.): IWCIA 2008, LNCS 4958, pp. 38–50, 2008.

Another possibility consists in creating the Voronoi diagram [11] of a set of points using the duality between Delaunay triangulations and Voronoi diagrams. Efficient algorithms for constructing Voronoi diagrams in 2D are well known [10]. The concept can be easily adopted to 3D.

Sometimes there exists the necessity to generate a more economical surface than a triangulation. Especially when triangulations would have lots of coplanar triangles, a polyhedral surface (a surface containing faces with more than three edges) would be much more efficient. The problem of approximating polyhedral surfaces is also well studied in the field of computational geometry [1,2,4].

In [8] we have already shown that for a convex voxel set (see Definition DCS below) the convex hull is such a polyhedral surface. In this paper we present an improvement of this algorithm to approximate non-convex objects, too.

The formal task to be solved is the following: Given a set V of voxels we want to create a closed polyhedral surface H containing V with the minimum surface area. Since there can be more than one polyhedron with the minimum surface area, we search the one with the minimum number of faces.[1] The polyhedral surface H shall separate object voxels (interior) from background voxels (exterior) in such a way that no object voxel lies outside H and no background voxel lies inside H. Voxels lying on H, especially the vertices of the polyhedron, have to be marked as being object voxels or background voxels. For the first criterion (minimum surface area) we will only present an approximation here. The second criterion (separation) is stated to realize the possibility to exactly restore the original voxel set V from the polyhedron H. An efficient data structure to store the polyhedral surface is the cell list [5].

The vertices of the polyhedron H are the voxels. There also exists the possibility to create a polyhedral surface separating the 0-cells and with the 0-cells as its vertices. These two approaches are dual.

In the next Section we present the basic definitions used in this paper. In Section 3 we discuss the algorithm for the construction of the polyhedron. In Section 4 we show some example images and experimental results. The paper is closed with a conclusion in Section 5 and a bibliography.

2 Basic Definitions

The algorithm presented here is based on the theory of abstract cell complexes (AC complexes) [5]. Most of the basic notions of this theory relevant to the topic of polyhedral surfaces are gathered in the Appendix to [8].

Let V be a given set of voxels in a Cartesian three-dimensional space. The voxels of V are specified by their coordinates. Our aim is to construct the convex hull K of V and a modification H of K which represents the polyhedral surface of V. We consider the convex hull and the modified hull as abstract polyhedra according to the following definition [8]:

---

[1] Example: The faces of a cube (squares) can be subdivided into coplanar triangles. The surface area does not change, but the number of faces increases.

Definition AP: An *abstract polyhedron* is a three-dimensional AC complex containing a single three-dimensional cell whose boundary is a two-dimensional combinatorial manifold without boundary. The two-dimensional cells (2-cells) of the polyhedron are its *faces*, the one-dimensional cells (1-cells) are its *edges* and the zero-dimensional cells (0-cells) are its *vertices* or points.

An abstract polyhedron is called a *geometric* one if coordinates are assigned to each of its vertices. We shall call an abstract geometric polyhedron an AG-polyhedron. Each face of an AG-polyhedron PG must be planar. This means that the coordinates of all 0-cells belonging to the boundary of a face F_i of PG must satisfy a linear equation $H_i(x, y, z) = 0$. If these coordinates are coordinates of some cells of a Cartesian AC complex A then we say that the polyhedron PG is embedded into A or that A contains the polyhedron PG.

Definition CP: An AG-polyhedron PG is called *convex* if the coordinates of each vertex of PG satisfy all the linear inequalities $H_i(x, y, z) \leq 0$ corresponding to all faces F_i of PG. The coefficients of the linear form $H_i(x, y, z)$ are the components of the outer normal of F_i.

A cell c of the complex A containing the convex AG-polyhedron PG is said to *lie in* PG if the coordinates of c satisfy all the linear inequalities $H_i(x, y, z) \leq 0$ of all faces F_i of PG.

Definition CH: The *convex hull* of a finite set V of voxels is the smallest convex AG-polyhedron PG containing all voxels of the set V. "Smallest" means that there exists no convex AG-polyhedron different from PG which contains all voxels of V and whose all vertices are in PG.

For the differentiation between voxel sets being convex or not, we need to define what a convex voxel set actually is.

Definition DCS: A *digital half-space* is the set of all voxels whose coordinates satisfy a linear inequality. A *digital convex subset* of the space is a non-empty intersection of digital half-spaces.

3   The Algorithm

It is well known that every non-convex set can be considered as the sum or the difference of convex sets. We use this property of non-convex sets to extend the method of [8] to construct an abstract polyhedron from a non-convex set of voxels by subtracting small convex hulls from an initial convex hull. This is motivated by the imagination of modelling the surface through pressing faces of the convex hull onto the voxel object.

3.1   Constructing the Convex Hull

The first step to build a non-convex abstract polyhedron consists in creating the convex hull of the given voxel object V as described in [8]. The algorithm for constructing the convex hull consists of two parts: in the first part a subset of vectors v pointing to voxels must be found which are candidates for the vertices of the convex hull. The coordinates of the candidates are saved in an array L. The second part constructs the convex hull of the set L.

From the point of view of AC complexes the given set V is the set of three-dimensional cells (3-cells) of a subcomplex M of a three-dimensional Cartesian AC complex A. The complex A represents the topological space in which our procedure is acting. It is reasonable to accept that M is homogeneously three-dimensional. This means that each cell of M whose dimension is less than 3 is incident to at least one 3-cell of M. With other words, M has no "loose" cells of dimensions less than 3.

The problem of finding the vectors v can be defined as follows: A 0-cell is called a *convex* 0-cell iff it is incident to exactly one 3-cell of M (Figure 1). All 3-cells incident to at least one convex 0-cell are the candidate vectors v. The vectors v are stored in L.

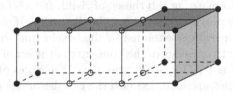

Fig. 1. Four voxels and their convex 0-cells (depicted as black disks). Non-convex 0-cells are depicted as circles. The voxel in the center of the front row is not incident to any convex 0-cell and thus it is not a candidate vector.

As already mentioned, the second part of our algorithm is that of constructing the convex hull of the set L of the candidate vectors v found by the first part.

To build the convex hull of L we first create a simple convex polyhedron spanning four arbitrary non-coplanar voxels v of L. It is a tetrahedron. It will be extended step by step until it becomes the convex hull of L. We call it the *current polyhedron* CP.

The surface of the current polyhedron is represented with the data structure called the *two-dimensional cell list* [5]. The cell list of a two-dimensional complex consists in the general case of three sublists. The kth sublist contains all k-dimensional cells (k-cells), $k = 0, 1, 2$. The 0-cells are the vertices, the 1-cells are the edges, the 2-cells are the faces of the polyhedron. Each entry in the kth sublist corresponds to a k-cell c^k. The entry contains indices of all cells incident to c^k. The entry of a 0-cell contains also its coordinates.

The cell list according to this definition contains much redundancy, because it contains for a pair of two incident cells c^k and c^m both the reference from c^k to c^m and from c^m to c^k. The redundancy makes the computation faster, because cells incident to each other may be found immediately, without a search. However, for the exact reconstruction of the voxel set from the cell list the redundancy can be eliminated to make the encoding more economical.

The next step in constructing the convex hull is to extend the current polyhedron while adding more and more voxels, some of which become vertices of the convex hull. When the list L of the candidate vectors is exhausted, the current polyhedron becomes the convex hull of M. The extension procedure is based on the notion of visibility of faces which is defined as follows.

Definition VI: The face F of a convex polyhedron is *visible* from a cell c, if c lies in the outer open half-space bounded by the face F, i.e. if the scalar product (N, w) of the outer normal N of the face F and the vector w pointing from a point Q in F to c, is positive. If the scalar product is negative then F is said to be *invisible* from c. If the scalar product is equal to zero then F is said to be *coplanar* with c.

To extend the current polyhedron the algorithm processes one voxel after another. For any voxel v it computes the visibility of the faces of the polyhedron from v. Consider first the simpler case when there are no faces of the current polyhedron, which are coplanar with v. The algorithm labels each face of the current polyhedron as being visible from v or not. If the set of visible faces is empty, then the voxel v is located inside the polyhedron and may be discarded. If one or more faces are visible, then the polyhedron is extended by the voxel v and some new faces. Each new face connects v with one edge of the boundary of the set of visible faces. A new face is a triangle having v as its vertex and one of the edges of the said boundary as its base. All triangles are included into the cell list of the current polyhedron while all visible faces are removed. Also each edge incident to two visible faces and each vertex incident only to visible faces is removed.

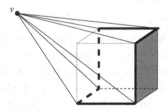

Fig. 2. The current polyhedron (a cube) being extended by the voxel v as a new vertex

In Figure 2 the boundary of the visible subset is shown by bold lines (solid or dashed). The edges shown by dotted lines must be removed together with the three faces visible from v. The algorithm repeats this procedure for all voxels in L.

Consider now the problem of coplanar faces. There are three variants to treat a face coplanar with a new voxel v. Two of them are quite easy, i.e. coplanar faces can be treated as visible or as invisible ones. The third variant is a little bit more sophisticated and treats coplanar faces neither as visible nor as invisible.

After having tested all three variants we came to the decision that the best solution consists in treating coplanar faces as visible [8]. In this case the program creates sometimes many coplanar triangles which must be merged together. But the procedure of merging triangles is rather simple and fast.

The procedure of adding new faces to the current polyhedron ends after processing all candidate vectors in L. With this step the convex hull is completely constructed and the first part of creating a polyhedral surface of the given voxel object V ends.

3.2 Finding Concavities

As already mentioned, the convex hull is a good means to encode the surface of a convex object. The convex hull of a convex set of voxels (according to Definition DCS) never contains voxels of the background. If the given voxel object is not convex, we have to find components of the set of background voxels included into the convex hull. These components can be cavities, concavities and tunnels [12].

Definition CO: A *concavity* is a component of background voxels inside the convex hull incident to exactly one connected set of faces of the convex hull.

Due to this definition a concavity does not have to be a convex protrusion of the background, i.e. it does not have to be convex.

Definition CA: A *cavity* is a component of background voxels inside the convex hull, which is not incident to any face of the convex hull.

Definition TU: A *tunnel* is a component of background voxels inside the convex hull incident to more than one connected set of faces of the convex hull. If the tunnel is incident to exactly two connected sets it is called a *non-branched* tunnel. If the number of connected sets of faces of the convex hull incident to the component is greater than two, the tunnel is called a *branched* one.

In this paper we only deal with concavities. Cavities are a trivial problem and will be mentioned later on. Tunnels are part of future work.

We use the algorithm described in [7] to find the components of the set of background voxels inside the convex hull and classify them by the number of the connected sets of faces of the convex hull which are incident to the component.

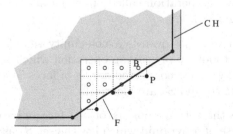

Fig. 3. Convex hull CH of an object (gray shaded). Some of the 0-cells of the background voxel component are located outside CH and thus there exists a visible face.

To check whether a component of background voxels is incident to a set of faces of the convex hull, we just have to compute the visibility of faces from the 0-cells of the background component. If every 0-cell of a component has no visible faces then the component is entirely located inside the hull and thus it is a cavity. If one or more 0-cells have visible faces or one or more 0-cells are coplanar with a set of faces then the component is incident to the hull, i.e. it is a concavity or a tunnel (see Figure 3).

The components are labeled and thus we can modify the convex hull by treating one component of background voxels after another.

3.3   Modification of the Convex Hull

After all components of background voxels inside the convex hull are found, we can modify the convex hull. As already mentioned, the convex hull is a convex polyhedron. After the first modification we can no longer speak of the convex hull, because it is no longer a convex polyhedron. Hence we call the modified hull *current polyhedron* again until it becomes a polyhedron with the desired properties.

To modify the current polyhedron we first need to know which faces are incident to the current background voxel component. This can be determined by using again the notion of visibility. In the previous step we have labeled a component if its 0-cells have visible faces and now we label the faces which are visible from the 0-cells of the current background voxel component. This means that a face F becomes labeled if the following criteria are all satisfied:

1. The face F is visible from some of the 0-cells of the background voxel component or some 0-cells are coplanar with the face F.
2. If there is a 0-cell P outside H and a background voxel B inside or on H with $P \in Cl(B)$, then the projection of B onto the face F is inside the boundary of F.
3. The 0-cell P does not lie on the boundary of the face F.

We want to mention that the steps of labeling the background voxel components and labeling the corresponding faces can be merged together.

A topology preserving operation called "pressing-in" is applied to the set of labeled faces (see Figure 4).

Definition PR: *Pressing-in* towards a non-empty set of cells located inside a polyhedron H is a topology preserving operation which replaces a connected set S_1 of faces of H by a new connected set S_2 of faces in such a way that the boundaries of the sets S_1 and S_2 are identical.

In the simplest case the set S_1 consists of one face only and thus it can be interpreted as the base of a pyramid which has the set S_2 as its sides. The apex of the pyramid (P in Figure 4) is located inside the polyhedron. In the general case the set S_1 consists of several faces and the destination of the pressing-in is not necessarily a single cell.

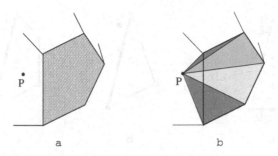

Fig. 4. A cell P inside the polyhedron (a) and the resulting polyhedron after pressing-in (b)

We perform the pressing-in by constructing a polyhedron around the background voxel component. As mentioned in the Introduction, we want to apply our convex hull algorithm recursively to modify the convex hull of the voxel object. This means that we now create the convex hull of the background voxel component and modify it again and again until there are no object voxels outside the polyhedron and no background voxels inside it. To do so we have to combine the cell lists of the current polyhedron and that of the current polyhedron of the background voxel component. But this is not trivial. Definition PR implies that we can identify faces of both polyhedra, but this is not possible in the general case (see Figure 5a).

To avoid identifying faces of these two polyhedra, which do not have identical boundaries, we do not construct the convex hull of the background voxel component independently. A more precise approach consists in spanning the convex hull by starting with the labeled faces of the current polyhedron. This means that we span an initial polyhedron with this set of faces and a voxel of the background component, which lies on the inside of the set of labeled faces (Figure 5b). This initial polyhedron can be extended in the same way as the tetrahedron being the initial convex hull. The result is the convex hull of the set of labeled faces and the set of voxels of the background voxel component lying inside the polyhedron before applying the pressing-in operation (Figure 5c).

Another problem is that of the recognition of 0-cells outside a non-convex polyhedron, because faces of such a polyhedron can be visible from 0-cells inside a non-convex polyhedron. Therefore we need another method than the visibility approach to decide for each 0-cell whether it is located in the inside or not. An easy approach is to compute a ray from the current 0-cell to a point which is certainly outside the polyhedron (i.e. a point on the boundary of the space) and just to count the number of intersections of the ray with the polyhedron. If this number is odd then the current 0-cell lies inside. This method is of course only applicable for polyhedra with no self-intersections, but since we only deal with binary voxel images, self-intersections do not appear.

At this stage of the modification we have the current polyhedron and a second smaller polyhedron which has a connected set of faces in common with the first

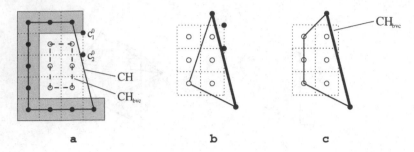

Fig. 5. (a) Convex hull CH of a given set of voxels and the convex hull CH_{bvc} of the background voxel component. Two 0-cells c_1^0 and c_2^0 of the background voxel component are located outside CH. (b) Initial convex hull of the set of background voxels created from a fixed face (bold line). (c) Resulting convex hull CH_{bvc} of the background voxels.

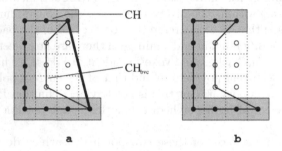

Fig. 6. CH and CH_{bvc} with a common face (a) and the resulting polyhedron after the modification step (b).

one (see Figure 6a). Now we perform our idea of recursivity by interpreting the second polyhedron as the current polyhedron and we change the roles of object and background voxels. Now we can apply the same algorithm to the new current polyhedron and thus we search for object voxels inside this polyhedron for which we have to do a pressing-in. This leads to a protuberance to the outside of the first polyhedron.

It is always assured that the algorithm ends after a finite number of steps since every step deals with a current polyhedron which includes a finite set of voxels and a smaller polyhedron which includes a subset of the set of voxels inside the first polyhedron. Hence the second polyhedron is smaller than the first one in the sense that it includes a fewer number of voxels. This means that by reducing the number of voxels in the current polyhedron the number of recursive steps cannot be greater than the number of voxels inside the concavity.

As already mentioned in Section 3.2, the algorithm can also deal with cavities, because it is a trivial problem. According to Definition CA cavities have no connection to the outer surface of the polyhedron and thus we can independently compute the convex hull of this background voxel component and modify it, if the cavity is a non-convex one. After applying the algorithm to the cavity we have to change the orientation of the normal vectors of the faces to ensure that they point to the outside of the surface of the voxel object, which means that they point to the inside of the cavity.

We want to mention that our algorithm has an important drawback. At this level of development it is not able to deal with a class of objects whose surfaces have a genus greater than 0 or whose background voxel components have a surface with a genus greater than 0, such as tori or mushrooms. This is justified by the fact that the pressing-in does not work for a set of faces composing a cycle. We are currently working on an improvement of our algorithm to solve this problem.

4   Results of Computer Experiments

We have implemented the described algorithm and we have tested it with several objects of different size and complexity.

One simple example is presented below in Figure 7-10 to show how intermediate and final results look like.

Figure 7 (left) shows the voxel object. The voxels are interpreted as cells of an Euclidean complex and thus they are represented as small cubes. The object consists of 4949 voxels. On the right hand side of Figure 7 there is the convex hull CH surrounding the voxels, i.e. the centers of the cubes. The convex hull consists of 288 faces, 530 edges and 244 vertices.

The convex hull CH of the given voxel object includes one component of background voxels. The convex hull CH_{bvc} of these background voxels is shown on the left in Figure 8. It is created by starting with a set of 9 connected faces of CH. Finally it consists of 32 faces and it is surrounding 835 background voxels, but also 424 object voxels.

In this situation we apply our algorithm to the polyhedron shown in Figure 8 (left). We interpret the set of background voxels now as the set of object voxels with the 424 object voxels as the set of background voxels inside the polyhedron. The background voxels are composing two components. The convex hulls of these components are shown on the right of Figure 8.

The left image in Figure 9 shows the polyhedral surface after applying only the first modification of the convex hull, which corresponds to the unification of CH and CH_{bvc}, where the set of labeled faces is removed. This polyhedron includes no voxels of the background, but two sets of object voxels are located on the outside of the polyhedron.

On the right hand side of Figure 9 the resulting polyhedron after all modification steps is presented. The resulting polyhedron consists of 300 faces, 566 edges and 268 vertices.

Figure 10 shows the triangulation of this object with the Marching Cubes method. It consists of 5184 triangles.

One of our primary goals concerning this algorithm is the efficiency of the encoding. Therefore we have compared the results of our algorithm with that of

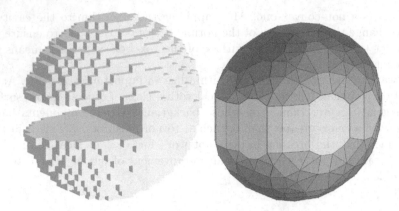

Fig. 7. Example "pac-man". Left: The voxel object. Voxels are depicted as small cubes. Right: Convex hull of of the centers of the small cubes.

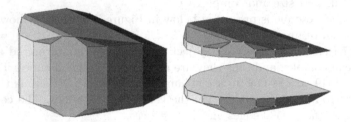

Fig. 8. Left: Convex hull of the background voxels lying inside the convex hull of the given object. Right: Convex hulls of the object voxels lying inside the convex hull of the background voxels.

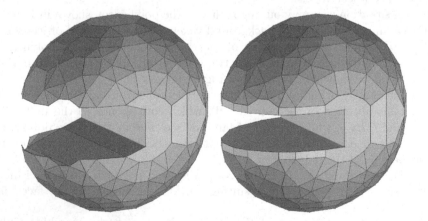

Fig. 9. Left: Polyhedron after subtracting the convex hull of the background voxels only. Right: Resulting polyhedron after all modification steps.

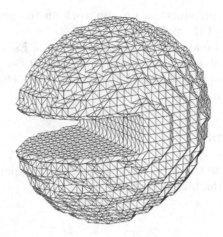

Fig. 10. Marching Cubes triangulation of the object of Figure 7

Table 1. Comparison of the memory requirements of the cell list and those of the triangulation for the example of Figure 7

	faces	vertices	integers to be saved
MC–triangulation	5184	~2592	23328
modified convex hull	300	268	1936

the Marching Cubes triangulation method, which is very often used in practical applications. As already mentioned in [8] we assume that 4.5 integers have to be saved for each triangle within this method. Our algorithm produces a non-redundant cell list containing 3 integers per vertex (its coordinates) and N_{F_i} integers per face, where N_{F_i} is the number of vertices of face i. Table 1 presents the obtained values for the memory requirements.

5 Conclusion

In this paper we present a new algorithm for computing a polyhedral surface approximating a 3-dimensional digital object represented as a set of voxels. The resulting polyhedral surface is an abstract polyhedron which is a particular case of an abstract cell complex. The polyhedron is encoded by the non-redundant version of the well-known 2-dimensional cell list which is a good tool to save topological and geometric information efficiently and without redundancy. The cell list also provides the possibility to exactly reconstruct the voxel object.

The algorithm presented in this paper is still under progress. There are some drawbacks, especially the one mentioned in section 3.3, concerning surfaces with a genus greater than 0. Also the labeling criteria are still under investigation.

As mentioned in the Introduction, we also work on the proof of the minimum surface area property of the polyhedral surface.

The algorithm can be applied in a variety of tasks. Especially in the field of 3-dimensional image analysis and computer graphics it can be used to visualize voxel sets by polyhedra and to store large sets of voxels efficiently and without any loss of information.

Acknowledgement

The author is pleased to thank Vladimir Kovalevsky for many useful discussions on earlier versions of the manuscript.

References

1. Agarwal, P.K., Suri, S.: Surface Approximation and Geometric Partitions. SIAM Journal on Computing 27(4), 1016–1035 (1998)
2. Brönnimann, H., Goodrich, M.T.: Almost Optimal Set Covers in Finite VC-Dimension. Discrete and Computational Geometry 14, 263–279 (1995)
3. Cignoni, P., Montani, C., Scopigno, R.: DeWall: A Fast Divide & Conquer Delaunay Triangulation Algorithm in E^d. Computer Aided Design 30(5), 333–341 (1998)
4. Das, G., Goodrich, M.T.: On the Complexity of Optimization Problems for 3-dimensional Convex Polyhedra and Decision Trees. Computational Geometry: Theory and Applications 8, 123–137 (1997)
5. Kovalevsky, V.A.: Finite Topology as Applied to Image Analysis. Computer Vision, Graphics and Image Processing 45(2), 141–161 (1989)
6. Kovalevsky, V.A.: Algorithms and Data Structures for Computer Topology. In: Bertrand, G., Imiya, A., Klette, R. (eds.) Digital and Image Geometry. LNCS, vol. 2243, pp. 37–58. Springer, Heidelberg (2001)
7. Kovalevsky, V.A.: Algorithms in Digital Geometry Based on Cellular Topology. In: Klette, R., Žunić, J. (eds.) IWCIA 2004. LNCS, vol. 3322, pp. 366–393. Springer, Heidelberg (2004)
8. Kovalevsky, V.A., Schulz, H.: Convex Hulls in a 3-dimensional Space. In: Klette, R., Žunić, J. (eds.) IWCIA 2004. LNCS, vol. 3322, pp. 176–196. Springer, Heidelberg (2004)
9. Lorensen, W.E., Cline, H.E.: Marching Cubes: A High-Resolution 3D Surface Construction Algorithm. Computer Graphics 21(4), 163–169 (1987)
10. O'Rourke, J.: Computational Geometry in C. Cambridge University Press, Cambridge (1994)
11. Preparata, F.P., Shamos, M.I.: Computational Geometry - An Introduction. Springer, Heidelberg (1985)
12. Svensson, S., Arcelli, C., Sanniti di Baja, G.: Characterising 3D Objects by Shape and Topology. In: Nyström, I., Sanniti di Baja, G., Svensson, S. (eds.) DGCI 2003. LNCS, vol. 2886, pp. 124–133. Springer, Heidelberg (2003)

Weighted Neighborhood Sequences in Non-standard Three-Dimensional Grids – Parameter Optimization

Robin Strand[1] and Benedek Nagy[2]

[1] Centre for Image Analysis, Uppsala University,
Box 337, SE-75105 Uppsala, Sweden
[2] Department of Computer Science, Faculty of Informatics, University of Debrecen,
PO Box 12, 4010, Debrecen, Hungary
robin@cb.uu.se, nbenedek@inf.unideb.hu

Abstract. Recently, a distance function was defined on the face-centered cubic and body-centered cubic grids by combining weights and neighborhood sequences. These distances share many properties with traditional path-based distance functions, such as the city-block distance, but are less rotational dependent. We introduce four different error functions which are used to find the optimal weights and neighborhood sequences that can be used to define the distance functions with low rotational dependency.

1 Introduction

When using non-standard grids such as the face-centered cubic (fcc) grid and the body-centered cubic (bcc) grid for 3D images, less samples are needed to obtain the same representation/reconstruction quality compared to the cubic grid [10]. This is one reason for the increasing interest in using these grids in, e.g., image acquisition [10], image processing [14,6,15], and visualization [2,17].

Measuring distances on digital grids is of great importance both in theory and in many applications. Because of its low rotational dependency, the Euclidean distance is often used as distance function. In digital grids, however, the Euclidean distance may not be the best option [9]. Both from a theoretical point of view and for several applications path-based digital distances are better options. For example, when minimal cost-paths are computed, a distance function defined as the minimal cost path between any two points is better suited, see, e.g., [3], where the constrained distance transform is computed using the Euclidean distance resulting in a complex algorithm. The corresponding algorithm using a path-based approach is simple, fast, and easy to generalize to higher dimensions [21,18]. Examples of path-based distances are weighted distances, where weights define the cost (distance) between neighboring grid points [1,6,14], and distances based on neighborhood sequences, where the cost is fixed but the adjacency relation is allowed to vary along the path [13,15]. These path-based distance functions are generalizations of the well-known city-block and chessboard distance function defined for the square grid in [12]. We will abbreviate neighborhood sequence with *ns*, distance based on neighborhood sequences with *ns-distances*, and weighted distances based on neighborhood sequences with *weighted ns-distances* or just *wns-distances*.

Many approaches where the deviation from the Euclidean distance is minimized in order to find the optimal ns (ns-distances) or weights (weighted distances) have been proposed for $\mathbb{Z}^2$. In most papers, error functions minimizing the asymptotic maximum difference of a Euclidean ball and a ball obtained by using ns-distances [23,5,4] or weighted distances [1,22,6] are minimized. Other approaches have also been considered for ns-distances. In [7], optimal ns for the 2D hexagonal and triangular grids are found using a compactness ratio – the ratio between the squared perimeter and the area of the convex hull of the disks obtained by using ns. In [8], the symmetric difference is used for ns in $\mathbb{Z}^2$ and in [11], the following error functions are considered for ns on the fcc and the bcc grids: absolute error, relative error, compactness ratio, maximal inscribed ball, and minimal covering ball.

In [23], a general definition allowing both weights and ns was presented. The full potential of using both weights and ns was discovered in [19], where ns and weights were together used in the sense of [23], but with the well-known natural neighborhood structure of $\mathbb{Z}^2$. In [19], the basic theory for weighted ns-distances on the square grid is presented including a formula for the distance between two points, conditions for metricity, optimal weight calculation, and an algorithm to compute the distance transform. In [16], some results for weighted ns-distances on the fcc and bcc grids were presented. The theory for weighted ns-distances on the fcc and bcc grids was further developed in [20] by presenting sufficient conditions for metricity and algorithms that can be used to compute the distance transform and a minimal cost-path between two points.

The asymptotic error using the compactness ratio was used to find the optimal weights and ns for weighted ns-distances on the fcc and bcc grids in [16]. The analysis presented in [16] is extended in this paper by considering the relative error, the compactness ratio, the maximal inscribed ball, and the minimal covering ball. Also, we analyze the behavior when ns of finite length is used.

Note that the results presented here also applies to weighted distances and ns-distances, since they are both special cases of the proposed distance function. The distance function proposed here is used to find optimal weights for the weighted distance and optimal ns for ns-distances.

2   Basic Notions and Previous Results

The following definitions of the face-centered cubic (fcc: $\mathbb{F}$) and body-centered cubic (bcc: $\mathbb{B}$) grids are used:

$$\mathbb{F} = \{(x, y, z) : x, y, z \in \mathbb{Z} \text{ and } x + y + z \equiv 0 \pmod{2}\}. \tag{1}$$
$$\mathbb{B} = \{(x, y, z) : x, y, z \in \mathbb{Z} \text{ and } x \equiv y \equiv z \pmod{2}\}. \tag{2}$$

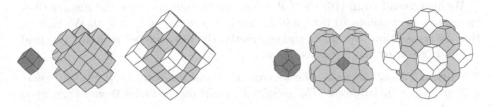

Fig. 1. The grid points corresponding to the dark and the light grey voxels are 1-neighbors. The grid points corresponding to the dark grey and white voxels are (strict) 2-neighbors. Left: fcc, right: bcc.

For results that are valid for both the fcc grid $\mathbb{F}$ and the bcc grid $\mathbb{B}$, the notation $\mathbb{G}$ is used. Two distinct grid points $\mathbf{p}_1 = (x_1, y_1, z_1), \mathbf{p}_2 = (x_2, y_2, z_2) \in \mathbb{G}$ are ρ-neighbors, $1 \leq \rho \leq 2$, if

1. $|x_1 - x_2| + |y_1 - y_2| + |z_1 - z_2| \leq 3$ and
2. $\max\{|x_1 - x_2|, |y_1 - y_2|, |z_1 - z_2|\} \leq \rho$

The points $\mathbf{p}_1, \mathbf{p}_2$ are *adjacent* if $\mathbf{p}_1$ and $\mathbf{p}_2$ are ρ-neighbors for some ρ. The 2-neighbors which are not 1-neighbors are called *strict* 2-neighbors. The neighborhood relations are visualized in Figure 1 by showing the Voronoi regions, i.e. the voxels, corresponding to some adjacent grid points.

A ns B is a sequence $B = (b(i))_{i=1}^{\infty}$, where each $b(i)$ denotes a neighborhood relation in $\mathbb{G}$. If B is periodic, i.e., if for some fixed strictly positive $l \in \mathbb{Z}_+$, $b(i) = b(i + l)$ is valid for all $i \in \mathbb{Z}_+$, then we write $B = (b(1), b(2), \ldots, b(l))$. A *path*, denoted $\mathcal{P}$, in a grid is a sequence $\mathbf{p}_0, \mathbf{p}_1, \ldots, \mathbf{p}_n$ of adjacent grid points. A path is a B-*path* of length n if, for all $i \in \{1, 2, \ldots, n\}$, $\mathbf{p}_{i-1}$ and $\mathbf{p}_i$ are $b(i)$-neighbors. The notation 1- and (strict) 2-steps will be used for a step to a 1-neighbor and step to a (strict) 2-neighbor, respectively.

Definition 1. *Given the ns B, the ns-distance $d(\mathbf{p}_0, \mathbf{p}_n; B)$ between the points $\mathbf{p}_0$ and $\mathbf{p}_n$ is the length of (one of) the shortest B-path(s) between the points.*

Let the real numbers α and β (the *weights*) and a path $\mathcal{P}$ of length n, where exactly l ($l \leq n$) adjacent grid points in the path are strict 2-neighbors, be given. The *length of the (α, β)-weighted B-path $\mathcal{P}$* is $(n - l)\alpha + l\beta$. The B-path $\mathcal{P}$ between the points $\mathbf{p}_0$ and $\mathbf{p}_n$ is a *minimal cost (α, β)-weighted B-path between the points $\mathbf{p}_0$ and $\mathbf{p}_n$* if no other (α, β)-weighted B-path between the points is shorter than the length of the (α, β)-weighted B-path $\mathcal{P}$.

Definition 2. *Given the ns B and the weights α, β, the weighted ns-distance $d_{\alpha,\beta}(\mathbf{p}_0, \mathbf{p}_n; B)$ is the length of (one of) the minimal cost (α, β)-weighted B-path(s) between the points.*

The following notation is used:

$$1_B^k = |\{i : b(i) = 1, 1 \leq i \leq k\}| \text{ and}$$
$$2_B^k = |\{i : b(i) = 2, 1 \leq i \leq k\}|.$$

We now recall from [16] the following two theorems giving the distance between two grid points $(0,0,0)$ and (x,y,z), where $x \geq y \geq z \geq 0$. We remark that by translation-invariance and symmetry, the distance between any two grid points is given by the formulas below.

Theorem 1. *Let the ns B, the weights α, β and the point $(x,y,z) \in \mathbb{F}$, where $x \geq y \geq z \geq 0$, be given. The weighted ns-distance between $\mathbf{0}$ and (x,y,z) is given by*

$$d_{\alpha,\beta}\left(\mathbf{0},(x,y,z);B\right) = \begin{cases} \frac{x+y+z}{2} \cdot \alpha & \text{if } x \leq y+z \\ (2k-x) \cdot \alpha + (x-k) \cdot \beta & \text{otherwise,} \end{cases}$$

where $k = \min_{k} : k \geq \max\left(\dfrac{x+y+z}{2}, x - 2_B^k\right).$

The value of k is the least integer that is not less than $\frac{x+y+z}{2}$ such that $2_B^k + k \geq x$.

Theorem 2. *Let the ns B, the weights α, β, and the point $(x,y,z) \in \mathbb{B}$, where $x \geq y \geq z \geq 0$, be given. The weighted ns-distance between $\mathbf{0}$ and (x,y,z) is given by*

$$d_{\alpha,\beta}\left(\mathbf{0},(x,y,z);B\right) = (2k-x) \cdot \alpha + (x-k) \cdot \beta, \text{ where}$$

$$k = \min_{k} : k \geq \max\left(\frac{x+y}{2}, x - 2_B^k\right).$$

Here k is the least integer which is not less than $\frac{x+y}{2}$ such that $2_B^k + k \geq x$.

Not all weights and ns give metric distance functions. The following sufficient conditions for metricity were derived in [20].

Theorem 3. *If*

$$\sum_{i=1}^{N} b(i) \leq \sum_{i=j}^{j+N-1} b(i) \quad \forall j, N \geq 1 \quad and$$

$$0 < \alpha \leq \beta \leq 2\alpha$$

then $d_{\alpha,\beta}(\cdot,\cdot;B)$ is a metric on the fcc and bcc grids.

3   Optimization of Weights and Neighborhood Sequences

The optimization is carried out in $\mathbb{R}^3$ by finding the best shape of polyhedra corresponding to balls of constant radii using the proposed distance functions. To do this, the distance functions presented for the fcc and bcc grids in the previous section are stated in a form that is valid for all points $(x,y,z) \in \mathbb{R}^3$, where $x \geq y \geq z \geq 0$. Note that this gives the asymptotic shape of the balls. The following distance functions are considered:

$$d_{\alpha,\beta}^{fcc}\left(\mathbf{0},(x,y,z);\gamma\right) = \begin{cases} \frac{x+y+z}{2} \cdot \alpha & \text{if } x \leq y+z \\ (2k-x) \cdot \alpha + (x-k) \cdot \beta & \text{otherwise,} \end{cases}$$

$$\text{where } k = \min_{k} : k \geq \max\left(\frac{x+y+z}{2}, \frac{x}{2-\gamma}\right)$$

and

$$d^{bcc}_{\alpha,\beta}\left(\mathbf{0},(x,y,z);\gamma\right)=(2k-x)\cdot\alpha+(x-k)\cdot\beta,\text{ where}$$

$$k=\min_{k}:k\geq\max\left(\frac{x+y}{2},\frac{x}{2-\gamma}\right),$$

where $k\in\mathbb{R}$ and $\gamma\in\mathbb{R}$, $0\leq\gamma\leq 1$ is the fraction of the steps where 2-steps are *not* allowed (so $\mathbf{1}^k_B$ and $\mathbf{2}^k_B$ corresponds to γk and $(1-\gamma)k$, respectively). Note that $k\geq x/(2-\gamma)$ if and only if $(1-\gamma)k+k\geq x$, which is analogous to the condition $\mathbf{2}^k_B+k\geq x$ of Theorems 1 and 2. In this way we obtain a generalization

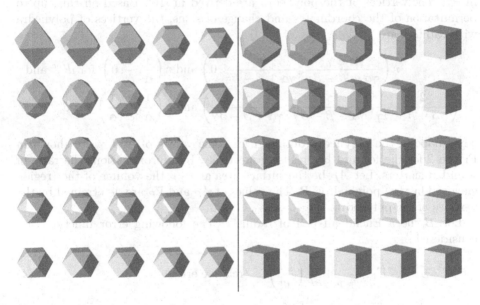

Fig. 2. Shapes of balls for $d^{fcc}_{\alpha,\beta}(\cdot,\cdot;\gamma)$ (left 5×5 block) and $d^{bcc}_{\alpha,\beta}(\cdot,\cdot;\gamma)$ (right 5×5 block) for a fixed radius r, $\alpha=1$, and (left to right) $\gamma=0,0.25,0.5,0.75,1$ and (top to bottom) $\beta=1,1.25,1.5,1.75,2$

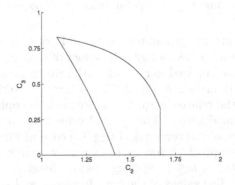

Fig. 3. The domain for C_2 and C_3 in Table 1

of the distance functions in discrete space $\mathbb{G}$ valid for all points (x, y, z) where $x \geq y \geq z \geq 0$ in continuous space $\mathbb{R}^3$. By considering

$$d_{\alpha,\beta}^{fcc}\left(\mathbf{0}, (x, y, z); \gamma\right) = r \text{ and } d_{\alpha,\beta}^{bcc}\left(\mathbf{0}, (x, y, z); \gamma\right) = r, \tag{3}$$

for some radius r, the points on a sphere of constant radius r are found. When $\gamma \in \mathbb{R}$ ($0 < \gamma < 1$) the functions d^{fcc} and d^{bcc} can be understood as asymptotic approximations to the distance functions of Theorems 1 and 2 for large values of x.

For any triplet α, β, γ ($\alpha, \beta > 0$ and $0 \leq \gamma \leq 1$), (3) defines polyhedra P in $\mathbb{R}^3$. The vertices of the polyhedra are derived in [16]. Based on this, up to permutation of the coordinates and change of signs, the vertices of polyhedra with radius r are

$$r\left(\frac{2 - \gamma}{\gamma\alpha + \beta - \beta\gamma}, \frac{\gamma}{\gamma\alpha + \beta - \beta\gamma}, 0\right) \text{ and } r\left(\frac{1}{\alpha}, \frac{1}{\alpha}, 0\right) \text{ for } d^{fcc} \text{ and}$$

$$r\left(\frac{2 - \gamma}{\gamma\alpha + \beta - \beta\gamma}, \frac{\gamma}{\gamma\alpha + \beta - \beta\gamma}, \frac{\gamma}{\gamma\alpha + \beta - \beta\gamma}\right) \text{ and } r\left(\frac{1}{\alpha}, \frac{1}{\alpha}, \frac{1}{\alpha}\right) \text{ for } d^{bcc}$$

The shape of the polyhedra obtained for some values of α, β, γ are shown in Figure 2 for the fcc and bcc grids, respectively. In approximations the ratio of α and β matters. Let A_P be the surface area and V_P the volume of the (region enclosed by the) polyhedron P. The values of A_P and V_P are determined by the vertices of the polyhedra.

Let B_r be a Euclidean ball of radius r. The following error functions are considered

$$E_1 = \max_{p,q \in \partial P} \left(\frac{|p|}{|q|}\right) \quad \text{(relative error)} \tag{4}$$

$$E_2 = \frac{\frac{A_P^3}{V_P^2}}{36\pi} \quad \text{(compactness ratio)} \tag{5}$$

$$E_3 = \min_{r:B_r \subset P} (V_P/V_{B_r}) \quad \text{(maximal inscribed ball)} \tag{6}$$

$$E_4 = \min_{r:P \subset B_r} (V_{B_r}/V_P) \quad \text{(minimal covering ball)} \tag{7}$$

These error functions attain their minimum value 1 when A is the surface area and V is the volume of a Euclidean ball. The values of α, β, and γ that minimize the error functions are computed numerically. All error functions attain a minimum value within the domain $0 < \alpha \leq \beta \leq 2\alpha$, $0 \leq \gamma \leq 1$, so the computation is straight-forward. For the relative error on the fcc grid, the optimum is obtained on a region as shown in Figure 3. The optimal values are found in Table 1 and visualized by the shape of the corresponding polyhedra in Figure 4.

In Table 2, the asymptotic behavior is shown by letting $\alpha = 1$ and β be constant and for each k, $1 \leq k \leq 1000$ using a ns B of length k that approximates the optimal fraction γ. The values of the error functions and k are plotted in the figures.

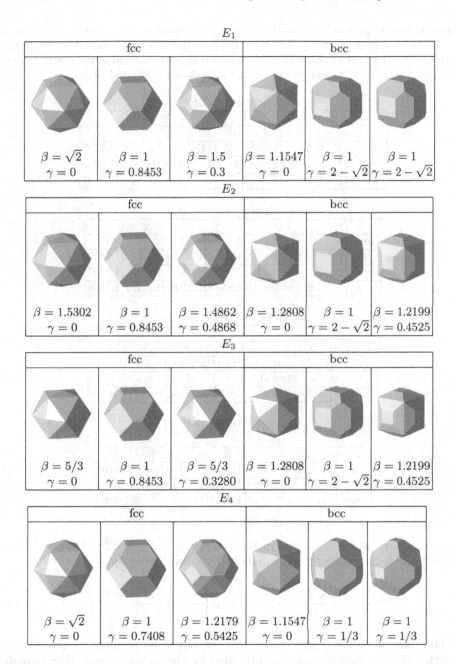

Fig. 4. Shapes of balls using $\alpha = 1$ and values of β and γ that minimize E_1–E_4, see Table 1

In Table 1, C_1 can be any value in the range

$$\sqrt{2} \leq C_1 \leq 5/3. \tag{8}$$

Table 1. Performance of wns-, weighted- (w), and ns-distances using the error functions $E_1 - E_4$ defined in the text. The optima of the error functions are attained whenever t is a strictly positive real number. The values shown in bold are fixed in the optimization. The parameters C_1, C_2, and C_3 are related as is shown in (8) and (9).

Relative error, E_1

	fcc				bcc			
Name	α	β	γ	E_1	α	β	γ	E_1
w	t	$C_1 t$	0	1.2247	t	$1.1547t$	0	1.2393
ns	1	1	0.8453	1.2393	1	1	$[1/3, 2-\sqrt{2}]$	1.2247
wns	t	$C_2 t$	C_3	1.2247	t	t	$[1/3, 2-\sqrt{2}]$	1.2247

Compactness ratio, E_2

	fcc				bcc			
Name	α	β	γ	E_2	α	β	γ	E_2
w	t	$1.5302t$	0	1.1367	t	$1.2808t$	0	1.1815
ns	1	1	0.8453	1.2794	1	1	$2-\sqrt{2}$	1.2147
wns	t	$1.4862t$	0.4868	1.1267	t	$1.2199t$	0.4525	1.1578

Maximal inscribed ball, E_3

	fcc				bcc			
Name	α	β	γ	E_3	α	β	γ	E_3
w	t	$(5/3)t$	0	1.1578	t	$1.2808t$	0	1.1815
ns	1	1	0.8453	1.2794	1	1	$2-\sqrt{2}$	1.2147
wns	t	$(5/3)t$	0.3280	1.1563	t	$1.2199t$	0.4525	1.1578

Minimal covering ball, E_4

	fcc				bcc			
Name	α	β	γ	E_4	α	β	γ	E_4
w	t	$\sqrt{2}t$	0	1.4234	t	$1.1547t$	0	1.5708
ns	1	1	0.7408	1.4448	1	1	$1/3$	1.3860
wns	t	$1.2179t$	0.5425	1.3272	t	t	$1/3$	1.3860

Moreover, C_3 can be any value in the range $0 \leq C_3 \leq 2(\sqrt{2} - 1)$ and C_2 any value satisfying the following inequalities (see also Figure 3):

$$\frac{\sqrt{C_3^2 + 2 - 2C_3} - C_3}{1 - C_3} \leq C_2 \leq \min\left(\frac{\sqrt{3} - C_3\left(\frac{1}{2}\sqrt{3} + 1\right)}{1 - C_3}, \frac{5}{3}\right). \tag{9}$$

The plots in Table 2 show how the error functions perform for ns of finite lengths. Neighborhood sequences obtained by the following recursive formula are used

$$b(k+1) = \begin{cases} 1 & \text{if } 1_B^k < \gamma k, \\ 2 & \text{otherwise.} \end{cases}$$

The value of γ is shown in Table 1. For the cases when γ is not uniquely defined, we use constant values within the allowed interval. The same thing applies to β.

Table 2. Optimal values of E_1–E_4 (vertical axis) on the fcc and bcc grid for neighbourhood sequences of length k ($0 < k \leq 1000$, horizontal axis showing $\log k$) with $\alpha = 1$. See Table 1 for asymptotic optima.

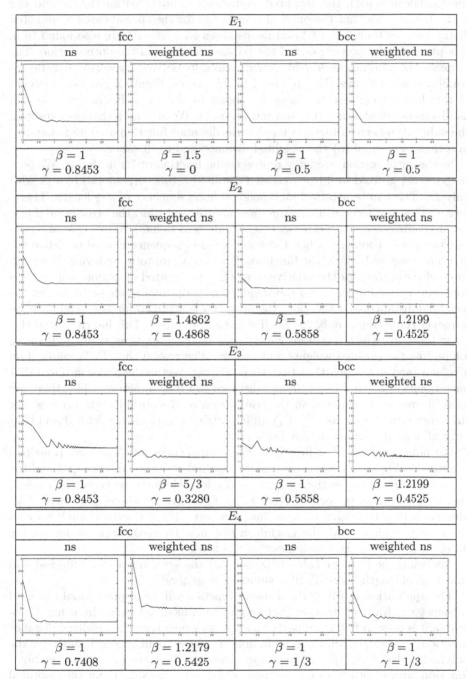

E_1			
fcc		bcc	
ns	weighted ns	ns	weighted ns
$\beta = 1$ $\gamma = 0.8453$	$\beta = 1.5$ $\gamma = 0$	$\beta = 1$ $\gamma = 0.5$	$\beta = 1$ $\gamma = 0.5$
E_2			
fcc		bcc	
ns	weighted ns	ns	weighted ns
$\beta = 1$ $\gamma = 0.8453$	$\beta = 1.4862$ $\gamma = 0.4868$	$\beta = 1$ $\gamma = 0.5858$	$\beta = 1.2199$ $\gamma = 0.4525$
E_3			
fcc		bcc	
ns	weighted ns	ns	weighted ns
$\beta = 1$ $\gamma = 0.8453$	$\beta = 5/3$ $\gamma = 0.3280$	$\beta = 1$ $\gamma = 0.5858$	$\beta = 1.2199$ $\gamma = 0.4525$
E_4			
fcc		bcc	
ns	weighted ns	ns	weighted ns
$\beta = 1$ $\gamma = 0.7408$	$\beta = 1.2179$ $\gamma = 0.5425$	$\beta = 1$ $\gamma = 1/3$	$\beta = 1$ $\gamma = 1/3$

4   Conclusions

By introducing a number of error functions that all favor "round" balls (in the Euclidean sense), the weighted ns-distance is analyzed for the fcc and bcc grids. It turns out that the optimal parameters for the special cases of weighted distances ($\gamma = 0$ or $B = (2)$) and ns-distances ($\alpha = \beta = 1$) are also found from this procedure by keeping one of the parameters fixed in the optimization. The same weights and neighborhood sequences as were derived for weighted distances [14,6] and ns-distances [15] are found in this paper. Figure 2 gives an overview of this fact – weighted distances are shown in the left columns ($\gamma = 0$) and ns-distances are shown in the top rows ($\beta = 1$). We also note that, as expected, the value of the error functions for the wns distance function are lower than (or, in some cases, equal to) the weighted distance and ns-distance.

In the optimization, we let γ represent the fraction of 1:s in the ns. We note that when γ is fixed to 0 (for weighted distances), this corresponds to a ns with only 2:s. This can be attained for a neighborhood sequence of *any* length. Therefore, in this case the optimum is not asymptotic and thus, the error is valid also for short distances (between e.g. neighboring grid points). When γ is also subject to optimization (i.e. when the neighborhood sequence is used to define the distance function), the error functions have an asymptotic behavior. However, some of the optima for the relative error (E_1) are located on regions where E_1 is constant. For example, E_1 for the weighted ns is optimal when $\gamma = 0$, i.e. for the weighted distance, and therefore the optimum is attained for any neighborhood sequence consisting of only 2:s. See Table 2 and Figure 3. This indicates that this error function, which has been widely used in the literature, is not well-suited for finding the optimal weights and ns here. The reason that E_1 is minimal on a region (and not a point) is that there are two vertices and two surfaces that have points that can be at minimal distance (up to symmetry). Thus, there are more degrees of freedom than the restrictions in the optimization process. For the other error functions (E_2–E_4), differentiable functions are defined and they have all a single minimum, see Table 1.

We note that the error functions E_2 (compactness ratio) and E_3 (maximal inscribed ball) give the same asymptotic optimal result for the bcc grid and the same for ns-distances on the fcc grid, see Table 1. However, as is seen in Table 2, the error functions perform differently for finite neighborhood sequences. This illustrates that the different error functions *are* different, even though they all are used to approximate the Euclidean distance. Different applications require different aspects of the "roundness" of the balls.

Analysing the plots in Table 1, we see that the error converges quite fast and that a ns of length (period) 10 is sufficient in general.

The application in which the distance function will be applied should be used to select which error function that should be considered. Also, by using Theorem 3, it is easy to find neighborhood sequences such that the resulting distance function is a metric, which is preferable in many applications. Intuitively, the polyhedra that best approximate the Euclidean ball are given by a distance function where both γ and β are non-trivial, see Figure 2. From the "optimal shapes" in Figure 4, we see that this is what, e.g., the compactness ratio E_2 favors. Thus, without any specific application in mind, we suggest the parameters $B = (1, 2)$, $\beta = 1.4862\alpha$ for the fcc grid and $\beta = 1.2199\alpha$ for the bcc grid. This gives $E_2 = 1.1276$ for the fcc grid (the optimum is 1.12757 and the approximated value is 1.12760) and $E_2 = 1.1591$ (the optimum is 1.1578) for the bcc grid. We conclude that we get a good approximation of the optimal values also with a short neighborhood sequence.

Acknowledgements

The authors thanks for the reviewers for their useful comments. The research is supported by *Digital geometry with applications in image processing in three dimensions (DIGAVIP)* at Center for Image Analysis, Uppsala University.

References

1. Borgefors, G.: Distance transformations in digital images. Computer Vision, Graphics, and Image Processing 34, 344–371 (1986)
2. Carr, H., Theussl, T., Möller, T.: Isosurfaces on optimal regular samples. In: Bonneau, G.-P., Hahmann, S., Hansen, C. (eds.) Proceedings of the symposium on Data visualisation 2003, Eurographics Association, pp. 39–48 (2003)
3. Coeurjolly, D., Miguet, S., Tougne, L.: 2D and 3D visibility in discrete geometry: an application to discrete geodesic paths. Pattern Recognition Letters 25(5), 561–570 (2004)
4. Das, P.P., Chatterji, B.N.: Octagonal distances for digital pictures. Information Sciences 50, 123–150 (1990)
5. Das, P.P.: Best simple octagonal distances in digital geometry. Journal of Approximation Theory 68, 155–174 (1992)
6. Fouard, C., Strand, R., Borgefors, G.: Weighted distance transforms generalized to modules and their computation on point lattices. Pattern Recognition 40(9), 2453–2474 (2007)
7. Hajdu, A., Nagy, B.: Approximating the Euclidean circle using neighbourhood sequences. In: Proceedings of 3rd Hungarian Conference on Image Processing, Domaszék, pp. 260–271 (2002)
8. Hajdu, A., Hajdu, L.: Approximating the Euclidean distance using non-periodic neighbourhood sequences. Discrete Mathematics 283, 101–111 (2004)
9. Klette, R., Rosenfeld, A.: Digital Geometry: Geometric Methods for Digital Image Analysis (The Morgan Kaufmann Series in Computer Graphics). Morgan Kaufmann, San Francisco (2004)
10. Matej, S., Lewitt, R.M.: Efficient 3D grids for image reconstruction using spherically-symmetric volume elements. IEEE Transactions on Nuclear Science 42(4), 1361–1370 (1995)
11. Nagy, B., Strand, R.: Approximating Euclidean distance using distances based on neighbourhood sequences in non-standard three-dimensional grids. In: Reulke, R., Eckardt, U., Flach, B., Knauer, U., Polthier, K. (eds.) IWCIA 2006. LNCS, vol. 4040, pp. 89–100. Springer, Heidelberg (2006)
12. Rosenfeld, A., Pfaltz, J.L.: Sequential operations in digital picture processing. Journal of the ACM 13(4), 471–494 (1966)

13. Rosenfeld, A., Pfaltz, J.L.: Distance functions on digital pictures. Pattern Recognition 1, 33–61 (1968)
14. Strand, R., Borgefors, G.: Distance transforms for three-dimensional grids with non-cubic voxels. Computer Vision and Image Understanding 100(3), 294–311 (2005)
15. Strand, R., Nagy, B.: Distances based on neighbourhood sequences in non-standard three-dimensional grids. Discrete Applied Mathematics 155(4), 548–557 (2007)
16. Strand, R.: Weighted distances based on neighbourhood sequences in non-standard three-dimensional grids. In: Ersbøll, B.K., Pedersen, K.S. (eds.) SCIA 2007. LNCS, vol. 4522, pp. 452–461. Springer, Heidelberg (2007)
17. Strand, R., Stelldinger, P.: Topology preserving marching cubes-like algorithms on the face-centered cubic grid. In: Proceedings of 14th International Conference on Image Analysis and Processing (ICIAP 2007), Modena, Italy, pp. 781–788 (2007)
18. Strand, R., Malmberg, F., Svensson, S.: Minimal cost-path for path-based distances. In: Petrou, M., Saramäki, T., Erçil, A., Loncaric, S. (eds.), Proceedings of 5th International Symposium on Image and Signal Processing and Analysis (ISPA 2007), pp. 379–384. Istanbul, Turkey (2007)
19. Strand, R.: Weighted distances based on neighbourhood sequences. Pattern Recognition Letters 28(15), 2029–2036 (2007)
20. Nagy, B., Strand, R.: Weighted neighbourhood sequences in non-standard three-dimensional grids – metricity and algorithms, DGCI 2008 (accepted, 2008)
21. Verwer, B.J.H., Verbeek, P.W., Dekker, S.T.: An efficient uniform cost algorithm applied to distance transforms. IEEE Transactions on Pattern Analysis and Machine Intelligence 11(4), 425–429 (1989)
22. Verwer, B.J.H.: Local distances for distance transformations in two and three dimensions. Pattern Recognition Letters 12(11), 671–682 (1991)
23. Yamashita, M., Ibaraki, T.: Distances defined by neighborhood sequences. Pattern Recognition 19(3), 237–246 (1986)

Computing Homology Generators for Volumes Using Minimal Generalized Maps*

Guillaume Damiand[1], Samuel Peltier[2], and Laurent Fuchs[3]

[1] LaBRI, Université Bordeaux 1, UMR CNRS 5800, 33405 Talence cedex, France
damiand@labri.fr
[2] IFP, 92852 Rueil-Malmaison Cedex, France
samuel.peltier@ifp.fr
[3] SIC, Université de Poitiers, 86962 Futuroscope Chasseneuil Cedex, France
fuchs@sic.univ-poitiers.fr

Abstract. In this paper, we present an algorithm for computing efficiently homology generators of 3D subdivided orientable objects which can contain tunnels and cavities. Starting with an initial subdivision, represented with a generalized map where every cell is a topological ball, the number of cells is reduced using simplification operations (removal of cells), while preserving homology. We obtain a minimal representation which is homologous to the initial object. A set of homology generators is then directly deduced on the simplified 3D object.

Keywords: topological features, homology generators, generalized maps.

1 Introduction

In this paper, we present an algorithm for computing efficiently the three dimensional *minimal generalized map* homologous to a given 3D object. Then we show how cells that belong to homology group generators can be directly characterized onto this minimal object.

Homology is a topological invariant, classically studied in algebraic topology [3], which characterizes an object by its "holes" in each dimension. This corresponds to connected components in dimension 0, tunnels in dimension 1, cavities in dimension 2; this notion of hole can be generalized in any dimension. For each dimension d, the number of $d-$dimensional holes of a given object is called its dth Betti number. Homology group generators are d-dimensional paths (edges, faces) that surround the d-dimensional holes.

Generalized maps [8] are a combinatorial cellular structure which can be used to represent both topological and geometrical information of a three dimensional subdivision, with particular properties that makes it a good model for features extraction. In this work, generalized maps are used to compute a minimal cell decomposition (called *minimal map*) of a 3-manifold in $\mathbb{R}^3$ with the same homology as the initial 3D object. For that, we extend in 3D the work of [2]. Starting

---
* Partially supported by the ANR program ANR-06-MDCA-008-05/FOGRIMMI.

V.E. Brimkov, R.P. Barneva, H.A. Hauptman (Eds.): IWCIA 2008, LNCS 4958, pp. 63–74, 2008.
© Springer-Verlag Berlin Heidelberg 2008

from the initial subdivision, where every cell is equivalent to a topological ball, the number of cells is progressively reduced using removal operations [1]. At the end of the simplification, we show that the minimal obtained object is homologous to the initial subdivision. Moreover, this minimal map allows us to directly characterize cells of the subdivision that belong to homology generators.

This paper is organized as follows. In Section 2, basic notions related to generalized maps and homology groups are recalled. The removal operations, which are used to compute a minimal map are introduced. In Section 3, the algorithm for computing a minimal map is detailed, and its complexity is discussed. We give the arguments to show that our algorithm provides a minimal object with the same homology as the initial one. Finally, Section 4 concludes and gives some perspectives.

2   Preliminaries

In this section some basic notions are presented. Our algorithm deals with subdivisions of 3D topological spaces. A subdivision is a partition into 4 subsets whose elements are $\{0, 1, 2, 3\}$-*cells* of dimension 0, 1, 2 and 3 (respectively called vertices, edges, faces and volumes). The border of an i-cell is a set of $(j<i)$-cells. Two cells are *incident* if one belongs to the border of the other, and two i-cells are *adjacent* if they are both incident to a common $(j<i)$-cell. The *cell degree* of an i-cell c is the number of distinct $(i+1)$-cells incident to c. We only consider quasi-manifolds[1].

Generalized Maps. For 3D quasi-manifolds, incidence and adjacency relations can be represented using 3-dimensional generalized maps (3-*G-maps*) [7]. Intuitively, a 3D generalized map can be obtained by successive (from volumes to vertices) decompositions of a 3D object into elementary elements called *darts*. Then, adjacency relations between i-cells are reported onto darts (denoted α_i). *Involution*[2] α_i connects the two darts incident to the two adjacent i-cells incident to the darts (see [7] for a formal definition).

Within the generalized map framework, all cells are implicitly represented through the notion of *orbit*. Given, $\{p_1, \ldots, p_j\}$, a set of involutions, and a dart d, an orbit $< p_1, \ldots, p_j > (d)$ is the set of darts that can be reached with a breadth-first search algorithm, starting with d, and using all combinations of p_i $\forall k, 1 \leq k \leq j$.

Removal Operations. Removal operations are the basic operations used during our algorithm. The removal of an i-cell c (called i-removal of c) leads to the merging of the two $(i+1)$-cells incident to c. For 3D subdivisions, i-removal

---

[1] A n-dimensional quasi-manifold is an nD space subdivision which can be obtained by gluing together n-dimensional cells along $(n$-$1)$-dimensional cells. In such subdivision, an $(n$-$1)$-cell cannot belong to the boundary of more than two n-cells. This notion is weaker that the manifold property, see [7].

[2] An involution f on S is a one to one mapping from S onto S such that $f = f^{-1}$.

operations are defined for $i = 0, 1, 2$ (see [1] for the definition of removal operations). The i-removal operation consists mainly in locally modify the α_i relation for each dart that belongs to the neighborhood of the removed cell.

Homology. In this part, basic homology notions are recalled; interested readers can find more details in [3] for algebraic approach and [4] for more computational approach. The notion of homology is defined in an algebraic way using the sets of i−cells used to describe the 3D manifold. Within this context, a p-chain (i.e. a chain of dimension p) is a formal sum of p-cells. From this, the group C_p of p-chains is defined. The boundary of a p-chain is defined as the sum of the boundaries of its p-cells. Note that the boundary of a p-chain must be a $(p-1)$-chain. The set of p-chains which have a null boundary (i.e. p-cycles) is a subgroup of C_p (denoted Z_p). The set of p-chains which are boundaries of a $(p+1)$-chain (i.e. p-boundaries) form a subgroup of C_p (denoted B_p).

An essential property is that the boundary of any boundary is null. Hence, every boundary is a cycle and B_p is contained in Z_p. Two p-cycles z_1 and z_2 are homologous if their difference is a boundary, i.e. there is a $p + 1$-chain f such that $z_1 = z_2 + \partial f$. From this, an equivalence relation can be defined and the homology class of z is the set $\{z + b \mid b \in B_p\}$. The homology group of dimension p, denoted H_p, is defined as the quotient group Z_p/B_p, and its elements are the homology classes. For a group G, a set of generators is a maximal subset S of elements of G, such that every element of G can uniquely be defined as a linear combination of elements of S.

3   Computation of Homology Generators

In this section, we present an algorithm for computing the generators of homology groups of 3D orientable objects with cavities, i.e. an object bounded by one or more orientable surfaces.

3.1   Related Works

In [2], the authors propose an algorithm for computing a minimal representation of a 2D surface. It is shown that the homology generators H_1 can be directly deduced from this minimal representation.

In [9], the authors study the homology of 3D manifolds object X bounded by several surfaces. Indeed the considered objects are 3D balls with tunnels and cavities. For example, Fig. 1(a) illustrates a 3D object which contains three tunnels and two cavities. The authors then show how to compute homology generators H_1 and H_2 of such objects. Moreover, it is shown that if X is bounded by $j + 1$ surfaces $s_0, ..., s_j$, then the set $\{s_1, ..., s_j\}$ is a basis of $H_2(X)$. They also introduce the notion of longitudinal and latitudinal generators of a surface and show that if s_0 denotes the external boundary of X, then the set of latitudinal generators of $\{s_1, ..., s_j\}$ together with the longitudinal generators of s_0 forms a basis of $H_1(X)$ (s_0 is called external surface, others s_i are called internal surfaces). For example, on Fig. 1(b), the homology generators of the object

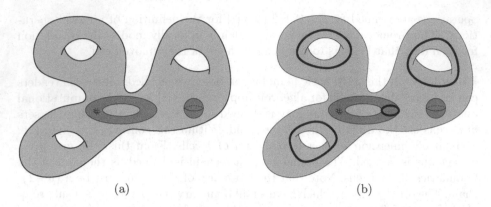

Fig. 1. (*a*): a 3D object with 3 tunnels and 2 cavities, (*b*): the latitudinal generator of the torus cavity together with the longitudinal generators of the external boundary forms a basis of the first homology group of the object

represented in (*a*) is made of the 3 longitudinal generators of the external surface together with the latitudinal generator of the torus cavity. Note that the first homology groups of a sphere is trivial.

3.2   Simplification Algorithm

In this section, we extend the algorithm presented in [2] to 3D regions with cavities and tunnels. Our algorithm, given in Algorithm 1, provides the same result as in [9] but by working only with the initial subdivision of the object, and use basic simplification operations and combinatorial characteristics of cells.

Starting from a subdivision of a 3D object, where each cell is homeomorphic to a topological balls, we simplify progressively the subdivision, by decreasing cell dimension. First we remove faces (i.e. 2-cells) while keeping the volume homeomorphic to a topological ball. For that, we keep *fictive faces*, i.e. faces that are "inside" the volume, and whose removal involves map disconnection. We obtain a representation made of only one volume. To compute the minimal representation for other cells, we use the algorithm of [2] on each surface of the map. But for that, it is necessary to remove all the fictive faces in order to obtain 2D objects. After having computed the minimal representation of each surface, we need to reconstruct the minimal representation of the 3D object. This is achieved by adding the minimal number of fictive faces in order to obtain a connected volume homeomorphic to a topological ball.

This minimal subdivision is homologous to the initial object, and allows to directly compute the homology generators of the initial object by simple cell characterization. Moreover, this principle gives some perspectives to generalize our approach to *n*-dimensional objects.

Now we detail more precisely each step of our algorithm. The first simplification step (line 1 of Algorithm 1) is similar to the algorithm described in [2], which provides a minimal representation (in term of cells) in the case of 3D objects

---

Algorithm 1. Simplification of a 3D subdivision in its minimal homologous form.

Input: A generalized map M representing an orientable subdivision of a 3D object such that each cell is homeomorphic to a topological balls

Output: The minimal subdivision homologous to M

---

1 **foreach** *face f of the map* **do**

    if *the degree of f is 2* **then**

        ⌊ Remove f;

    else if *f is a dangling face* **then**

        push(P, f);

        repeat

            $f \leftarrow$ pop(P);

            push in P all the dangling faces adjacent to f;

            Remove f;

        until *empty(P)* ;

    else Mark f as fictive face;

2 Mark external surface, without considering fictive faces;

3 Remove all fictive faces;

4 Compute the H_1 generators of each surface;

5 **if** *the external surface is a sphere* **then**

    ⌊ $ext \leftarrow$ the only edge of the external surface;

    else for *one edge out of two e of the external surface* **do**

        Add a fictive face along e;

        ⌊ $ext \leftarrow e$;

    foreach *internal surface s* **do**

        if *s is a sphere* **then**

            ⌊ $int \leftarrow$ the only edge of s;

        else for *one edge out of two e of s* **do**

            Add a fictive face along e;

            ⌊ $int \leftarrow e$;

        Add a fictive face between int and ext;

---

bounded by only one surface. The difference concerns the dimension of processed cells, since we need here to consider in first volumes (3-cells) by removing faces (2-cells), while in [2] faces (2-cells) are processed by removing edges (1-cells).

During this step, we remove either degree two faces (i.e. faces between two different volumes), or dangling faces (i.e. faces inside a volume). Indeed, both faces can be removed without modifying the homology of the subdivision. When we remove a dangling face f, we need to reconsider dangling faces adjacent to f. Indeed, some faces adjacent to f can be non-dangling before f was removed and become dangling after its removal, hence non minimal subdivision can be obtained. To reconsider these faces, we use a stack of dart which contains one dart of each reconsidered face, and when we remove a dangling face, we put in this stack one dart for each dangling adjacent faces.

The subdivision obtained after applying the first step is composed by only one volume homeomorphic to a 3D ball, some real faces (faces with each dart

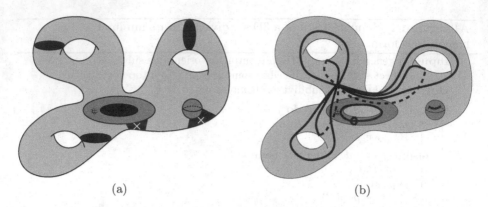

(a)   (b)

Fig. 2. (a) After the first simplification step, the 3D object contains some fictive faces. Black faces are fictive faces which stop up the tunnels, and black faces marked with crosses are fictive faces that link the different surfaces. (b) Result obtained after having disconnected surfaces and compute H_1 generators for each surface. External surface is represented by 6 edges and 1 vertex, the torus cavity is represented by 2 edges and 1 vertex and the sphere by 1 edge and 2 vertices. The obtained map is disconnected in three connected components.

3-free) and some fictive faces (faces with each dart not 3-free). Fig. 2(a) illustrates a possible subdivision that can be obtained after the first simplification step. In this subdivision, there is exactly one volume, and thus the non 3-free darts belong necessarily to fictive faces (i.e. degree one faces incident twice to the volume). Note that there are two types of fictive faces: the ones which link different surfaces in order to keep only one connected component, and the ones which stop up the tunnels, and that allow to keep the volume homeomorphic to a topological ball.

In the second step (line 2 of Algorithm 1), we work on each surface. Firstly, we mark the external surface. This step is necessary since H_1 generators are longitudinal for external surface whereas they are latitudinal for internal surfaces. Finding a dart of the external surface can be achieved directly by searching among all the darts the one associated with the smaller 3D coordinates. Starting from this dart, we can run through all the darts of the external surface by using a breadth first search algorithm which uses involutions α_0, α_1, and α_2, and which jumps over darts of fictive faces. Darts not 3-free belong to fictive faces, and darts 3-free and non-marked belong to internal surfaces.

Then, all the fictive faces are removed in order to continue the simplifications in smaller dimension (line 3 of Algorithm 1). The next step (line 4 of Algorithm 1) consists in computing independently the H_1 generators for each surface. This step is not detailed here since it is achieved by using the work of [2].

After this step, we obtain the minimal representation of each surface, composed with one face, one edge and two vertex if the corresponding surface is a sphere, and composed with one face, $2k$ edges and one vertex if the corresponding surface is a torus with k holes. Each edge of the minimal representation

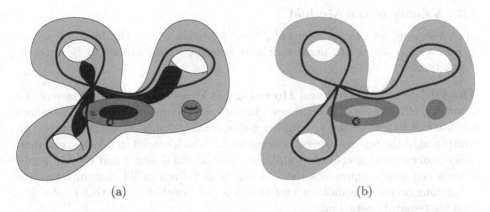

Fig. 3. (a) The minimal map obtained at the end of our algorithm (partial representation, fictive faces which link internal surfaces and the external surface are not represented.). This map is homologous to the initial subdivision. (b) H_1 generators are composed with all the edges non-incident to fictive faces, and non-incident to spheres.

belong to an H_1 generator of the 2D object, except if the corresponding surface is a sphere. Indeed, in such a case, there is no H_1 generator (see Fig. 2(b)).

Now, we have all necessary information to reconstruct the minimal representation of the initial 3D subdivision. This is the goal of the last step of our algorithm (line 5 of Algorithm 1). We firstly add fictive faces along one edge out of two of each surface, except for surfaces which represent spheres. This step is necessary to stop up the tunnels, and then obtain a volume homeomorphic to a topological ball. There are two cases to consider: if the surface is a sphere, there is no fictive face to add, otherwise the surface is a torus with k holes, and in this case there are k fictive faces to add in order to cut all the tunnels. Since the surface is composed with $2k$ edges in its minimal representation, to add k fictive faces, we just need to add one fictive face along one edge out of two. Then, fictive faces are added in order to connect each internal face with the external face (Fig. 3(a)).

After these two steps, we obtain a map where each cell is homeomorphic to a topological ball. This is necessary in order to ensure that this map is homologous to the initial subdivision. Moreover, this map is composed with $j + 1$ real faces, one for each surface (external and internal surfaces) of the initial subdivision.

This minimal map M gives directly the homology generators:

1. H_2 generator is composed with all the real faces that belong to internal surfaces of M;
2. H_1 generator is composed with all the edges of M not incident to fictive faces, and which are not incident to a sphere (see Fig. 2);
3. H_0 generator is always isomorphic to $\mathbb{Z}$ since we only consider one connected object.

3.3 Validity of the Method

In this section, we use the works of [5] and [9] to show that the resulting object is *minimal*; homology is preserved; and homology generators can be directly characterized.

The Object is Minimal and Homology is Preserved. The first step of the algorithm (the simplification process) preserves the homology. In [5], the authors use *interior face reduction* to simplify a simplicial complex. These reductions are equivalent to the removal operations used in [2] which consist in taking a common i-face c of exactly two $(i + 1)$-simplices a and b, and delete c and replace b and c by a cell which represents their union. It is proven in [5] that interior face reductions preserve homology and thus we can conclude that this is also the case for removal operations.

After the removal of fictive faces, each surface is simplified in its minimal form while preserving homology (step 3 and 4 which use [2]). The last step build a minimal representation of the initial 3D object. This is done into 3 steps:

1. each internal generator of the external surface is filled;
2. each longitudinal generator of internal surfaces is filled;
3. each internal surface is connected to the external surface.

It is shown in [6] that adding faces into each latitudinal generator *cut* the volume corresponding to the external surface into a topological ball. For each internal surface homeomorphic to a torus with g holes, each latitudinal generator (tunnels of internal surfaces) is filled by a face as these generators are no longer generators when the internal surface is considered as a cavity [9]. Lastly, each cavity is connected to the external surface and the obtained volume is minimal as we have added the minimal number of fictive faces. Moreover, each cavity is homeomorphic to a topological ball as each tunnel has been filled.

Direct Characterization of Homology Generators. As mentioned before, if a 3-manifold X is bounded by $j + 1$ surfaces $s_0, ..., s_j$, then the set $\{s_1, ..., s_j\}$ is a basis of $H_2(X)$ (see [9]). This set corresponds to all non fictive internal faces.

Moreover, the set of longitudinal generators of s_0 together with all the latitudinal generators of all internal surfaces forms a basis of H_1. Once the minimal form of each surface has been computed (step 4 of the algorithm), all the edges are either a longitudinal or a latitudinal generator. As seen before, all the latitudinal generators of s_0 and all the longitudinal generators of the internal surfaces are incident to a fictive face. Thus homology generators of the 3D object are all the edges that are not incident to a fictive face. Note that detecting an edge incident to a fictive face is done in a combinatorial way, thus we do not need the linking numbers or perturb the generators as it is done in [9].

3.4 Complexity

The complexity of Algorithm 1 is equal to $O((3 - \chi) \times n)$ with $\chi = \sum_{i=1}^{k} \chi_i$, χ_i being the Euler characteristic of surface s_i, and n is the number of darts of the subdivision.

The first step is linear in number of faces of the map. Firstly, each face is consider at most twice, a first time during the loop around all the faces of the map, and a second time during the second loop which remove dangling faces. When the face is reconsidered, it is removed and thus it will be never reconsidered later.

To test the face degree, we use union-find trees [10] allowing to represent efficiently disjointed sets. This structure is handled by two operations: *find* which returns, given an element, the representative of the set, and *union* which allows to merge two sets. The amortized cost of a series of m union-find operations on n elements can be done in time $O(n.\alpha(m, n))$ with $\alpha(m, n)$ being the inverse Ackermann function which grows extremely slowly, and which is less than 5 in practical cases (see [10] for the demonstration about the complexities).

We link each dart of a volume of the initial subdivision with an union-find tree representing the volume. When we remove a face, we merge both corresponding trees by using the *union* operation. The test if d and $\alpha_3(d)$ belong to the same volume is simply achieved by testing if *find(d)* is equal to *find($\alpha_3(d)$)*. Since we only consider one 3D object, subdivided in several volumes in the initial subdivision, we are sure that if d and $\alpha_3(d)$ belong to the same volume, the corresponding face is a degree one face (i.e. incident twice to the volume) and otherwise the face is a degree two face.

The face removal is achieved locally, by running through each edge incident to the face to remove and by modifying locally α_2 involutions. Moreover, to test if a face f is dangling or not, we have to run through each edge incident to f, and test if the edge is only incident to f, i.e. if d a dart of the edge is such that $\alpha_{23}(d) = d$.

To summarize the first step, the cost of the test on the face degree can be bounded by 5, the cost of the dangling face test is linear in number of edges of the face, and the face removal is also achieved linearly in number of edges of the face. This shows that the first step of Algorithm 1 is linear in number of darts of the map (indeed, the number of darts is always greater than the number of cells).

The second step (mark external surfaces) is also achieved linearly in number of darts of the map. Indeed, find the smaller dart of the map need to run through all the darts. Then, mark the surface is achieved by using for example a breadth first search algorithm by using involutions α_0, α_1, and α_2, and jumping over darts of fictive faces.

The fictive faces removal is achieved linearly in number of darts of the map, since we need to consider each dart, and modify locally involution α_2 for those that belong to fictive faces. Testing if a dart belongs to a fictive face is achieved in constant time (if d is 3-free or not).

Compute H_1 generators of a surface s_i is achieved in $O((3-\chi_i) \times n_i)$ where n_i is the number of darts of surface s_i, and χ_i is the Euler characteristic of the surface (see [2]). Since we compute H_1 generators for each surface, we obtain the final complexity by adding the complexity of each surface, which gives $O((3 - \chi) \times n)$ with $\chi = \sum_{i=1}^{k} \chi_i$ and n is the number of darts of the map.

The last step is achieved linearly in number of darts of the map since we just run through all the edges of the map, using the mark on darts to distinguish longitudinal and latitudinal generators, and distinguish external and internal surfaces. Moreover, adding a face along a loop is achieved in constant time, and adding a face between two edges that belong to two distinct surfaces is also achieved in constant time.

This gives the global complexity of our method: $O((3 - \chi) \times n)$. Indeed, all steps are linear in number of darts, except the step which allows to compute H_1 generators of each surface which depends on the number of darts multiply by the sum of the Euler characteristics of all surfaces.

3.5   Geometry of Generators

Algorithm 1 gives a method to compute the minimal representation homologous to a given 3D subdivision. This minimal representation allows to characterize directly the homology generators of the object. Depending on the need of applications, it is sometime necessary to embed the generators onto the original subdivision, for example to draw the H_1 generators onto the surfaces.

This is directly possible for H_2 generators since they are composed with all the non-fictive faces of the minimal subdivision, and each non-fictive face corresponds exactly to one surface into the original subdivision. This surface can be retrieved easily by running through the original subdivision and following the boundary (i.e. darts 3-free) and jumping over darts not 3-free.

But the problem is more complex for H_1 generators. Indeed, in the last step of Algorithm 1, fictive faces are added along one edge out of two, without particular properties on chosen edges. This is possible because all the configurations obtained by adding fictive faces along one edge out of two are homologous. However, configurations are not equivalent if we take into account the links with the initial subdivision. Indeed, in such a case, we need to distinguish latitudinal and longitudinal generators since they do not have the same role for the homology of the 3D object.

Algorithm 1 can easily be modified in order to compute a minimal subdivision which is homologous to the original subdivision, and which take into account these two kinds of generators. It is only necessary to make two modifications:

1. after having computed H_1 generators, we determine the class (latitudinal or longitudinal) of these generators by using the algorithm given in [9];
2. when we add fictive faces along edges, we need to chose edges which are not H_1 generators. These edges are either latitudinal generators that belong to the external surface, or longitudinal generators that belong to an internal surface. Since edges are distinguish by the previous modification, this modification can be directly added in the last step of Algorithm 1, by replacing the loop "For one edge out of two e of the external surface" with "ForEach latitudinal edge e of the external surface", and replacing the second loop "For one edge out of two e of s" with "ForEach longitudinal edge e of s".

With these two basic modifications, the computed minimal subdivision is not only homologous to the initial subdivision but H_1 generators can also be embedded onto the initial surfaces. Note that compute this embedding is not so straightforward than for H_2 generators since we need to keep links between edges of the minimal representation and edges of the subdivision during the whole simplification process.

However, these modifications involve complexity modifications. Indeed, to distinguish longitudinal and latitudinal generators, we use the method given in [9] which is in $O(n^2\bar{g})$ with $\bar{g} = max_{1\le i\le k}g_i$, and g_i is the genus of the surface s_i. This step is thus the more expensive part of the modified method, and so gives the global complexity of the method. One perspective of this work is to either improve this part, or remove it by computing generators by using only the combinatorial structure.

4 Conclusion

In this paper, we have presented an algorithm that computes the minimal generalized map homologous to a given 3D object, orientable and with or without cavities. Thanks to this minimal form, we can characterize easily and directly the cells that belong to homology generators. This gives a new method to compute efficiently the generators of a 3D object.

The main interest of our approach is to use a simple method which simplify the given subdivision. Moreover, the method is efficient since the complexity of our algorithm is in $O((3-\chi)\times n)$ with $\chi = \sum_{i=1}^{k}\chi_i$, where each χ_i is the Euler characteristic of surface s_i, and where n is the number of darts of the subdivision.

We have proposed a modified version of our algorithm which allows to embed the H_1 generators onto the initial subdivision. To do that, the method given in [9] to distinguish longitudinal and latitudinal generators is used. However, with this additional step, the complexity of the method become in $O(n^2\bar{g})$ with $\bar{g} = max_{1\le i\le k}g_i$, and g_i is the genus of the surface s_i.

Another main interest of our approach is that the simplification is made by decreasing cell dimension. This allows to reuse previous work in 2D [2]. We disconnect the 3D object into several 2D surfaces by removing fictive faces. After having 2D minimal representations, we insert back fictive faces in order to obtain the 3D minimal representation. This principle can be generalized in nD, where we simplify an nD object into a representation with only one n-cell, then $(n-1)$ fictive cells are removed to process independently $(n-1)$D objects. Lastly, $(n-1)$ fictive cells are inserted back to obtain the minimal representation of the initial object. However, there is a lot of work to do to validate the method. We need to show that the homology is preserved, and show the link between the minimal representation and the homology generators. This is one perspective of this work: extend our method to deal with nD objects. Moreover, we can also study how to consider orientable or non-orientable objects, with or without boundaries.

Another perspective is to improve the step which allows to distinguish longitudinal and latitudinal generators. Indeed, this step is necessary if we need to keep a link between the minimal representation and the original object, but using the method of [9] leads to increase the complexity of our algorithm. We want to study the possibility to compute this information directly onto the subdivision, by using the fictive faces to characterize the different edges of the original object, and propagate these information during the simplification steps.

References

1. Damiand, G., Lienhardt, P.: Removal and contraction for n-dimensional generalized maps. In: Nyström, I., Sanniti di Baja, G., Svensson, S. (eds.) DGCI 2003. LNCS, vol. 2886, pp. 408–419. Springer, Heidelberg (2003)
2. Damiand, G., Peltier, S., Fuchs, L.: Computing homology for surfaces with generalized maps: Application to 3d images. In: Bebis, G., Boyle, R., Parvin, B., Koracin, D., Remagnino, P., Nefian, A., Meenakshisundaram, G., Pascucci, V., Zara, J., Molineros, J., Theisel, H., Malzbender, T. (eds.) ISVC 2006. LNCS, vol. 4292, pp. 235–244. Springer, Heidelberg (2006)
3. Hatcher, A.: Algebraic Topology. Cambridge University Press, Cambridge (2002), http://www.math.cornell.edu/~hatcher/AT/ATpage.html
4. Kaczynski, T., Mischaikow, K., Mrozek, M.: Computational Homology. In: Applied Mathematical Sciences, vol. 157, Springer, Heidelberg (2004)
5. Kaczynski, T., Mrozek, M., Slusarek, M.: Homology computation by reduction of chain complexes. Computers & Math. Appl. 34(4), 59–70 (1998)
6. Kartasheva, E., Gasilov, V.: Computer aided analysis of the polyhedrons topology. In: GraphiCon 1997, pp. 60–66 (1997)
7. Lienhardt, P.: N-dimensional generalized combinatorial maps and cellular quasi-manifolds. International Journal of Computational Geometry and Applications 4(3), 275–324 (1994)
8. Lienhardt, P.: Topological models for boundary representation: A comparison with n-dimensional generalized maps. Commputer Aided Design 23(1) (1991)
9. Tamal, K.D., Sumanta, G.: Computing homology groups of simplicial complexes in R^3. Journal of the ACM 45(2), 266–287 (1998)
10. Tarjan, R.: Efficiency of a good but not linear set union algorithm. Journal of the ACM 22(2), 215–225 (1975)

Digital Segments and Hausdorff Discretization

Mohamed Tajine

LSIIT CNRS UMR 7005, Université Louis Pasteur (Strasbourg 1),
Pôle API, Boulevard Sébastien Brant, 67400 Illkirch-Graffenstaden, France
tajine@dpt-info.u-strasbg.fr

Abstract. In this paper we investigate some properties of digital segments in floor and Hausdorff discretizations. We characterize the Hausdorff discretization of straight lines and we prove that the frequency of digital segment in a digital straight line is continuous and piecewise affine function relatively to the slope. It allows to prove some combinatorial properties of digital segments. In particular we give a new proof of the results in [3,2,8] corresponding to the frequencies and the numbers of digital segments of size m.

Keywords: Hausdorff discretization, cellular metric, continuous piecewise affine function, digital segment, frequencies of digital segment.

1 Introduction

Digital straight lines are classical objects of discrete geometries. Their properties have been studied in a lot of papers, for recent review on the subject, see [4]. In this paper we investigate some properties of digital segments in two frameworks of discretization: floor discretization and Hausdorff discretization. we prove that the frequency of digital segment in a digital straight line is continuous and piecewise affine function relatively to the slope. It allows to prove some combinatorial properties of digital segments. In particular we give a new proof of the results in [3,2,8] corresponding to the frequencies and the numbers of digital segments of size m. This paper is organized as the following. In the second section we give some metrical notions. In the third section we introduce the notion of Hausdorff discretization and recall some properties of this framework of discretization. The Hausdorff discretization is introduced and studied in series of papers [9,11,12] and the results of the third section are proved in these papers. The last section contain new results about Hausdorff discretization of straight lines and digital segments.

2 Some Metrical Notions

In this section, we give the metrical notions used in this paper.

Definition 1. *Let $(\mathcal{E}, d)$ be a metric space and let $E \subseteq \mathcal{E}$.*

- *Let $p \in \mathcal{E}$ and $r \in \mathbb{R}^+$, $\mathcal{B}_r^d(p) = \{x \in \mathcal{E} \mid d(x,p) \leq r\}$. $\mathcal{B}_r^d(p)$ is called the* ball *of center p and of radius r relatively to the metric d.*

V.E. Brimkov, R.P. Barneva, H.A. Hauptman (Eds.): IWCIA 2008, LNCS 4958, pp. 75–86, 2008.

- $int(E) = \{p \in E \mid \exists r > 0, \mathcal{B}_r^d(p) \subset E\}$, $int(E)$ is called the interior of E.
- $cl(E)$ is the intersection of all closed sets containing E, $cl(E)$ is called the closure of E.
- A metric d on $\mathbb{R}^n$ is said to be invariant under translation if

$$\forall(x, y, z) \in (\mathbb{R}^n)^3, \ d(x + z, y + z) = d(x, y).$$

Examples: Consider $\mathcal{E} = \mathbb{R}^n$ and let $x = (x_1, x_2, ..., x_n) \in \mathbb{R}^n$. $\forall p \geq 1$, $|x|_p = \sqrt[p]{|x_1|^p + ... + |x_n|^p}$ and $|x|_\infty = max\{|x_i| \mid 1 \leq i \leq n\} = \lim_{p \to \infty} |x|_p$ are norms over $\mathbb{R}^n$. The metrics d_p and d_∞ induced by these norms are invariant under translation.

2.1 Hausdorff Metric

The definitions and results presented in this subsection can be found for example in [1].

Definition 2. Let $(\mathcal{E}, d)$ be a metric space, $\mathcal{H}(\mathcal{E})$ is the set of the non-empty compact subsets of $\mathcal{E}$.

On $\mathcal{H}(\mathcal{E})$, we will define a metric H_d, such that if $(\mathcal{E}, d)$ is a complete metric space then $(\mathcal{H}(\mathcal{E}), H_d)$ is a complete metric space.

Definition 3. Let $(\mathcal{E}, d)$ be a metric space.

- Let $A \subset \mathcal{E}$ and $x_0 \in \mathcal{E}$; $d(x_0, A) = inf\{d(x_0, y) \mid y \in A\}$.
- We define the oriented Hausdorff metric from a set $A \in \mathcal{H}(\mathcal{E})$ to a set $B \in \mathcal{H}(\mathcal{E})$ by $h_d(A, B) = sup\{d(a, B) \mid a \in A\}$.
- The Hausdorff distance between two non-empty compact sets $A, B \in \mathcal{H}(\mathcal{E})$ is defined by $H_d(A, B) = max(h_d(A, B), h_d(B, A))$.

Remark: Let $\mathcal{F}'(\mathcal{E})$ be the set of non-empty closed sets of $\mathcal{E}$. Then, the functions h_d and H_d can be extended in a natural way as a function from $\mathcal{F}'(\mathcal{E}) \times \mathcal{F}'(\mathcal{E})$ to $\mathbb{R}^+ \cup \{+\infty\}$.

H_d is a 'generalized metric' on $\mathcal{F}'(\mathcal{E})$ in the sense that it satisfies the axioms of a metric, but can take infinite values.

3   Hausdorff Discretization

In this section, we study a framework of discretization of closed sets based on Hausdorff metric. First, we present the theory of Hausdorff discretization in the general case. In the second subsection, we characterize the Hausdorff discretization for a subclass of metrics.

Definition 4. Let d be a metric on $\mathbb{R}^n$. The covering radius of the metric d is

$$r_c(d) = sup\{d(x, \mathbb{Z}^n) \mid x \in \mathbb{R}^n\}.$$

3.1 Characterization of Hausdorff Discretization

Let F be a non-empty closed subset of $\mathbb{R}^n$, $S \subseteq \mathbb{Z}^n$ is a Hausdorff discretization of F if it minimizes the Hausdorff distance to F. In this subsection, we study the properties of Hausdorff discretizations.

Definition 5. *Let $F \in \mathcal{F}'(\mathbb{R}^n)$.*

- *A set $S \subseteq \mathbb{Z}^n$ is a* Hausdorff discretization *of F if $H_d(F,S) = inf\{H_d(F,S') \mid S' \subseteq \mathbb{Z}^n\}$.*
- *$\mathcal{M}_H(F) = \{S \subseteq \mathbb{Z}^n \mid H_d(F,S) = inf\{H_d(F,S') \mid S' \subseteq \mathbb{Z}^n\}\}$ is the set of Hausdorff discretizations of F in $\mathbb{Z}^n$.*
- *$\Delta_H(F) = (\bigcup_{S \in \mathcal{M}_H(F)} S)$ is called the* maximal Hausdorff discretization *of F.*
- *The value $r_H(F) = sup\{d(x,\mathbb{Z}^n) \mid x \in F\}$ is called the* Hausdorff radius *of the closed set F for the metric d in the discrete space $\mathbb{Z}^n$.*

We will now characterize the Hausdorff discretization.

Theorem 1 ([12]). *Let $F \in \mathcal{F}'(\mathbb{R}^n)$; then*

- *$\mathcal{M}_H(F)$ is non-void and if $S \in \mathcal{M}_H(F)$ then $H_d(F,S) = r_H(F)$,*
- *$\Delta_H(F) = \{p \in \mathbb{Z}^n \mid d(p,F) \le r_H(F)\} \in \mathcal{M}_H(F)$,*
- *if $(S_i)_{i \in I}$ is a family of members of $\mathcal{M}_H(F)$, then $\bigcup_{i \in I} S_i \in \mathcal{M}_H(F)$,*
- *if $(S_n)_{n \in \mathbb{N}}$ is a decreasing sequence in $\mathcal{M}_H(F)$ (relatively to the set inclusion) then $\bigcap_{n \in \mathbb{N}} S_n \in \mathcal{M}_H(F)$ and*
- *$r_H(F) \le r_c(d)$.*

Property 1 ([12]). Let $F \in \mathcal{F}'(\mathbb{R}^n)$, $r \in \mathbb{R}^+$ and let $S \subseteq \mathbb{Z}^n$ such that $F \subseteq \bigcup_{p \in S} \mathcal{B}_r^d(p)$ and $\forall p \in S$, $\mathcal{B}_r^d(p) \cap F \ne \emptyset$. Then $H_d(F,S) \le r$. So if $r = r_H(F)$ then $S \in \mathcal{M}_H(F)$.

3.2 Homogeneous Metric and Hausdorff Discretization

In this subsection we introduce the notion of a homogeneous metric. We present some properties of a homogeneous metric, we refine the characterization of Hausdorff discretizations for homogeneous metric and we compare Hausdorff discretizations to other discretization schemes.

Notation
Let $p \in \mathbb{Z}^n, \mathcal{W}(p) = \mathcal{B}_{\frac{1}{2}}^{d_\infty}(p)$ *(i.e. $\mathcal{W}(p)$ is the square of size 1 centered on p).*

Definition 6. *A metric d over $\mathbb{R}^n$ is called* cellular *if*

$$\forall x \in \mathbb{R}^n, \; \forall p,q \in \mathbb{Z}^n, \quad x \in \mathcal{W}(p) \implies d(p,x) \le d(q,x).$$

In particular, if $x \in \mathcal{W}(p) \cap \mathcal{W}(q)$, then $d(p,x) = d(q,x)$.

All the usual metrics are cellular: d_p is cellular for all $p \ge 1$ and for $p = \infty$.

Let $F \in \mathcal{F}'(\mathbb{R}^n)$ and $M \subseteq \mathbb{Z}^n$. If $\forall p \in M$, $F \cap \mathcal{W}(p) \ne \emptyset$ and $F \subseteq \bigcup_{p \in M} \mathcal{W}(p)$ then M is called a covering discretization of E. So the popular supercover discretization is the maximal covering discretization.

Property 2. Let d be a cellular metric. If $F \in \mathcal{F}'(\mathbb{R}^n)$ and S is a covering discretization of F then $S \in \mathcal{M}_H(F, \rho)$, in particular the supercover discretization of F is a Hausdorff discretization of F.

Proof. If $x \in F$, then there exists $s \in S$ such that $x \in \mathcal{W}(s)$ and thus, $d(x, s) = d(x, \mathcal{D}_\rho) \leq r_H(F, \rho)$, so $h_d(F, S) \leq r_H(F, \rho)$.

If $s \in S$, then there exists $x \in F$ such that $x \in \mathcal{W}(s)$ and thus, $d(x, s) = d(x, \mathcal{D}_\rho) \leq r_H(F, \rho)$, so $h_d(S, F) \leq r_H(F, \rho)$. Therefore $S \in \mathcal{M}_H(F, \rho)$.   □

Definition 7.   • *A norm N on $\mathbb{R}^n$ is* homogeneous *if* $\forall (x_1, ..., x_n) \in \mathbb{R}^n$, $\forall (\varepsilon_1, ..., \varepsilon_n) \in \{-1, 1\}^n$, *for every permutation σ of* $\{1, ..., n\}$, $N(\varepsilon_1 x_{\sigma(1)}, ..., \varepsilon_n x_{\sigma(n)}) = N(x_1, ..., x_n)$. *So, if $n = 2$, then N is* homogeneous *iff* $\forall (x_1, x_2) \in \mathbb{R}^2$, $N(x_1, x_2) = N(-x_1, x_2) = N(x_2, x_1)$.
   • *A metric induced by a homogeneous norm is called a* homogeneous *metric.*
   • *A metric d on $\mathbb{R}^2$ is called* strictly homogeneous *if d is homogeneous and* $\mathcal{B}^d_{r_c(d)}(0, 0) \cap \mathcal{B}^d_{r_c(d)}(1, 1) = \{(\frac{1}{2}, \frac{1}{2})\}$.

For the strictly homogeneous metrics, the balls of covering radius centered about diagonally adjacent discrete points intersect only at their corners. For example d_p is strictly homogeneous for all $p > 1$ and for $p = \infty$.

Definition 8. *Let d be a metric on $\mathbb{R}^n$ and $F \in \mathcal{F}'(\mathbb{R}^n)$, the skeleton of $\Delta_H(F)$ is the set*

$$\mathcal{S}k(F) = \bigcap_{S \in \mathcal{M}_H(F)} S$$

Definition 9. *Let F be a subset of $\mathbb{R}^2$ and $\mathcal{S}$ be a square in $\mathbb{R}^2$, we say that F* crosses $\mathcal{S}$ *if $\exists p, q \in F$ such that p, q belong to two distinct faces of $\mathcal{S}$, the segment $[p, q]$ is not in a face of $\mathcal{S}$ and p, q belong to a same connected component of $F \cap \mathcal{S}$ (i.e. in particular $p \neq q$, $p, q \in (F \cap (\mathcal{S} \setminus int(\mathcal{S}))$ and $[p, q] \cap int(\mathcal{S}) \neq \emptyset$).*

Property 3. Let d be a homogeneous metric on $\mathbb{R}^2$, and let $F \in \mathcal{F}'(\mathbb{R}^2)$. If F is connected and $r_H(F) < r_c(d)$, then

$$F \text{ crosses } \mathcal{W}(p) \Longrightarrow p \in \mathcal{S}k(F)$$

4   Hausdorff Discretizations of Straight Lines

Let $a, b \in \mathbb{N}$ and $a \leq b$. The discrete interval $\{a, a + 1, \ldots, b - 1, b\}$ is denoted $[\![a, b]\!]$. For $x \in \mathbb{R}$, $\lfloor x \rfloor$ (resp. $\langle x \rangle$) denotes the integral part (resp. the fractional part) of x. So, $x = \lfloor x \rfloor + \langle x \rangle$ with $\lfloor x \rfloor \in \mathbb{Z}$, $\lfloor x \rfloor \leq x < \lfloor x \rfloor + 1$ and $0 \leq \langle x \rangle < 1$.

   Let $\alpha, \beta \in \mathbb{R}$, $\mathcal{L}(\alpha, \beta) = \{(x, \alpha x + \beta) \mid x \in \mathbb{R}\}$ is the straight line of slope α and 0-value β and $\mathfrak{D}(\mathcal{L}(\alpha, \beta)) = \{(x, \lfloor \alpha x + \beta \rfloor) \mid x \in \mathbb{Z}\}$ is the floor discretization of the straight line $\mathcal{L}(\alpha, \beta)$.

   So, $\mathfrak{D}(\mathcal{L}(\alpha, \beta)) = \{(x, y) \in \mathbb{Z}^2 \mid 0 \leq \alpha x + \beta - y < 1\}$.

   In all the following, we consider only the straight lines with slopes $\alpha \in [0, 1]$, the other cases can be obtained by symmetries.

Property 4. Let $x, r \in \mathbb{R}$. Then,

$$\lfloor x + r \rfloor - \lfloor x \rfloor = \begin{cases} \lfloor r \rfloor & \text{if } \langle x \rangle < 1 - \langle r \rangle, \\ \lfloor r \rfloor + 1 & \text{otherwise} \end{cases}$$

Property 5. [6] Let α be an irrational number and $\beta \in \mathbb{R}$, then $\{\langle \alpha n + \beta \rangle \mid n \in \mathbb{Z}\}$ is dense in $[0, 1]$.

Corollary 1. *Let d be a cellular distance invariant by translation and $\alpha, \beta \in \mathbb{R}$:*

1. *If α is an irrational number, then $\{(x - \lfloor x + \frac{1}{2} \rfloor, \alpha x + \beta - \lfloor \alpha x + \beta + \frac{1}{2} \rfloor) \mid x \subset \mathbb{R}\}$ is dense in $\mathcal{W}(0, 0)$ and thus $r_H(\mathcal{L}(\alpha, \beta)) = r_c(d)$.*
2. *If α is a rational number of the form $\frac{p}{q}$ where p and q are coprime then $r_h(\mathcal{L}(\alpha, \beta)) = max\{max(d(x_0(i, j), (i, j)), d(x_1(i, j), (i, j)) \mid 0 < i < q, \text{ int} (\mathcal{W}(i, j)) \cap \mathcal{L}(\alpha, \beta) \neq \emptyset\}$ where $x_0(i, j)$ and $x_1(i, j)$ are the two intersection points of $\mathcal{L}(\alpha, \beta)$ with the boundary of $\mathcal{W}(i, j)$.*

In all the following the distance d is strictly homogeneous.

Property 6. Let $\alpha, \beta \in \mathbb{R}$. Then:

1. If $\alpha = 0$ and $\beta = \frac{1}{2} + n$ for $n \in \mathbb{Z}$, then $\mathcal{S}k(\mathcal{L}(\alpha, \beta)) = \emptyset$ and thus $S \in \mathcal{M}_H(\mathcal{L}(\alpha, \beta))$ if and only if $S \subseteq (\{(m, n) \mid m \in \mathbb{Z}\} \cup \{(m, n + 1) \mid m \in \mathbb{Z}\})$ and for all $m \in \mathbb{Z}$, there exists n_m such that $(m, n_m) \in S$.
2. $\mathcal{S}k(\mathcal{L}(\alpha, \beta)) = \{p \in \mathbb{Z}^2 \mid int(\mathcal{W}(p)) \cap \mathcal{L}(\alpha, \beta) \neq \emptyset\}$ and if $\alpha \neq 0$ or $\beta \neq n + \frac{1}{2}$ for $n \in \mathbb{Z}$, then $\mathcal{S}k(\mathcal{L}(\alpha, \beta)) \in \mathcal{M}_H(\mathcal{L}(\alpha, \beta))$,
3. $\mathcal{S}k(\mathcal{L}(\alpha, \beta)) = \{(x, y) \in \mathbb{Z}^2 \mid -\frac{\alpha+1}{2} < \alpha x + \beta - y < \frac{\alpha+1}{2}\}$

Definition 10. Let $\alpha, \beta \in \mathbb{R}$ and $r = r_H(\mathcal{L}(\alpha, \beta))$. $\mathcal{O}p_H(\mathcal{L}(\alpha, \beta)) = \{p \in \mathbb{Z}^2 \mid \mathcal{B}_r^d(p) \cap \mathcal{L}(\alpha, \beta) \neq \emptyset$ and $int(\mathcal{W}(p)) \cap \mathcal{L}(\alpha, \beta) = \emptyset\}$.

Corollary 2. *Let $\alpha, \beta \in \mathbb{R}$ then $S \in \mathcal{M}_H(\mathcal{L}(\alpha, \beta))$ if and only if there exists $S' \subseteq \mathcal{O}p_H(\mathcal{L}(\alpha, \beta))$ such that $S = \mathcal{S}k(\mathcal{L}(\alpha, \beta)) \cup S'$.*

So, if $\alpha \neq 0$ or $\beta \neq n + \frac{1}{2}$ for $n \in \mathbb{Z}$, then any Hausdorff discretization of $\mathcal{L}(\alpha, \beta)$ contains inevitably all the points of $\mathcal{S}k(\mathcal{L}(\alpha, \beta))$ and 'optional' points in $\mathcal{O}p_H(\mathcal{L}(\alpha, \beta))$. Then in all the following, if $\alpha \neq 0$ or $\beta \neq n + \frac{1}{2}$ for $n \in \mathbb{Z}$, then we chose $\mathcal{S}k(\mathcal{L}(\alpha, \beta))$ as the Hausdorff discretization of $\mathcal{L}(\alpha, \beta)$ and we chose $\{(m, n) \mid m \in \mathbb{Z}\}$ as Hausdorff discretization of $\mathcal{L}(0, n + \frac{1}{2})$ for $n \in \mathbb{Z}$ and, in the two cases, we denoted this choice by $\mathfrak{C}_H(\mathcal{L}(\alpha, \beta))$.

Definition 11. Let $\alpha, \beta \in \mathbb{R}$, $m \in \mathbb{N}$ and $n \in \mathbb{N}^*$.

1. The segment *of size m at abscissa n is the set $S(m, n, \alpha, \beta) = \{(x, \lfloor \alpha x + \beta \rfloor) \mid x \in [\![n, n + m]\!]\}$. The point $p_0 = (n, \lfloor \alpha n + \beta \rfloor)$ is called the starting point of the digital segment $S(m, n, \alpha, \beta)$.*
   The set $S'(m, n, \alpha, \beta) = \{p - p_0 \mid p \in S'(m, n, \alpha, \beta)\}$ is called a digital segment of size m. So $S(m, n, \alpha, \beta)$ (respectively $S'(m, n, \alpha, \beta)$) can be viewed as function from $[\![0, m]\!]$ to $\mathbb{Z}$.

2. The segment for Hausdorff discretization *of size m at abscissa n is the set* $S_H(m, n, \alpha, \beta) = \{p \in S_H(\mathcal{L}(\alpha, \beta)) \mid p_x \in [\![n, n+m]\!]\}$ *where* $p = (p_x, p_y)$. *The point* $p_0 = (n, s) \in S_H(m, n, \alpha, \beta)$ *such that s is minimal is called* the starting point *of the digital segment for Hausdorff discretization* $S_H(m, n, \alpha, \beta)$. *The set* $S'_H(m, n, \alpha, \beta) = \{p - p_0 \mid p \in S'(m, n, \alpha, \beta)\}$ *is called* a H-digital segment *of size m.*

Notation: Consider the sequence $(B_k^\alpha)_{0 \le k \le m}$ as the sequence $(1 - \langle \alpha k \rangle)_{0 \le k \le m}$ reordered increasingly *(notice that* $B_m^\alpha = 1 - 0\alpha = 1$*)* and put by convention $B_{-1}^\alpha = 0$.

Remarks: Let $\alpha, \beta \in [0, 1]$, $m \in \mathbb{Z}$ and $n \in \mathbb{N}^*$.

1. Let $k \in [\![0, m]\!]$. Then $S'(m, n, \alpha, \beta)(k) = \lfloor \alpha(n + k) + \beta \rfloor - \lfloor \alpha n + \beta \rfloor = \lfloor \alpha k \rfloor$ if $\langle \alpha n + \beta \rangle < 1 - \langle \alpha k \rangle$, $\lfloor \alpha k \rfloor + 1$ otherwise.

   $S'(m, n, \alpha, \beta)$ only depend on the position of the number $\langle \alpha n + \beta \rangle$ relatively to the elements of the increasing sequence $(B_k^\alpha)_{-1 \le k \le m}$. Then, for $n_1, n_2 \in \mathbb{Z}$, $S'(m, n_1, \alpha, \beta) = S'(m, n_2, \alpha, \beta) \iff \exists i \in [\![-1, m-1]\!]$ such that $\langle \alpha n_1 + \beta \rangle, \langle \alpha n_2 + \beta \rangle \in [B_i^\alpha, B_{i+1}^\alpha)$.

2. Let $p = (p_x, p_y)$ be the starting point of $S_H(m, n, \alpha, \beta)$ (respectively $S(m, n, \alpha, \beta)$). Then by considering the translation of the straight line $\mathcal{L}(\alpha, \beta)$ by the vector $-p$ to obtain the straight line $\mathcal{L}(\alpha, \beta')$ where $\beta' = \alpha p_x + \beta - p_y$ we have $\beta' \in (-\frac{1}{2}, 1)$ and $S'_H(m, n, \alpha, \beta) = S_H(m, 0, \alpha, \beta')$ (respectively $\beta' \in (-\frac{1}{2}, 1)$ and $S'(m, n, \alpha, \beta) = S_H(m, 0, \alpha, \beta')$) which has $(0, 0)$ as starting point.

Definition 12. *Let* $m \in \mathbb{N}^*$. *The set* $\mathcal{F}_m = \{\frac{p}{q} \mid 0 \le p \le q, \ 0 < q \le m \text{ and } gcd(p, q) = 1\}$ *is called the* set of Farey numbers of order m. *The elements of* $\mathcal{F}_m$ *are called* m-Farey numbers.

For example, $\mathcal{F}_4 = \{\frac{0}{1}, \frac{1}{4}, \frac{1}{3}, \frac{1}{2}, \frac{2}{3}, \frac{3}{4}, \frac{1}{1}\}$.

The m-Farey numbers have several properties, for example, if $f < f'$ and $f = \frac{p}{q}, f' = \frac{p'}{q'}$ are two consecutive m-Farey numbers then $p'q - pq' = 1$, $gcd(p + p', q + q') = 1$, $q + q' > m$ and $\frac{p}{q} < \frac{p+p'}{q+q'} < \frac{p'}{q'}$ [7].

Property 7. [10] Let $\alpha, \beta \in [0, 1]$ and $m \in \mathbb{N}^*$.

1. for any $i \in [\![-1, m-1]\!]$, if $B_i^\alpha < B_{i+1}^\alpha$ then there exists $n \in \mathbb{Z}$ such that $\langle \alpha n + \beta \rangle \in [B_i^\alpha, B_{i+1}^\alpha)$.
2. card($\{S'(m, n, \alpha, \beta) \mid m \in \mathbb{Z}\}$) = card($\{i \mid -1 \le i < m \text{ and } B_i^\alpha < B_{i+1}^\alpha\}$) $\le m + 1$.
3. Let $\frac{p}{q} \in \mathcal{F}_m$ with $0 \le p \le q$ and p, q are coprime, then card($\{S'(m, n, \alpha, \beta) \mid m \in \mathbb{Z}\}$) = q.
4. If $\alpha \in ([0, 1] \setminus \mathcal{F}_m)$, then all the elements of the sequence $(B_k^\alpha)_{-1 \le k \le m}$ are distinct and thus card($\{S'(m, n, \alpha, \beta) \mid m \in \mathbb{Z}\}$) = $m + 1$.

Definition 13. *Let* $m \in \mathbb{N}^*$

- $\mathcal{C}(m, \alpha, \beta) = \{S'(m, n, \alpha, \beta) \mid n \in \mathbb{Z}\}$ which correspond to the set of digital segment of size m in $\mathfrak{D}(\mathcal{L}(\alpha, \beta))$.

- $\mathfrak{S}_m = \{S(m, 0, \alpha, \beta) \mid (\alpha, \beta) \in [0,1]^2\}$ *is the set of all* digital segments, *of size* m. *So,* $\mathfrak{S}_m = \bigcup_{\alpha, \beta \in [0,1]} \mathcal{C}(m, \alpha, \beta)$.
- *Let* $\alpha \in [0,1], \beta \in (-\frac{1}{2}, 1)$, $\mathcal{C}_H(m, \alpha, \beta) = \{S'_H(m, n, \alpha, \beta) \mid n \in \mathbb{Z}\}$ *which correspond to the set of H-digital segment of size* m *in* $\mathfrak{C}_H(\mathcal{L}(\alpha, \beta))$.
- $\mathfrak{S}_{H,m} = \{S_H(m, 0, \alpha, \beta) \mid (\alpha, \beta) \in [0,1] \times (-\frac{1}{2}, 1)\}$ *is the set of all* digital segments, *of size* m, *for Hausdorff discretization. So,* $\mathfrak{S}_{H,m} = \bigcup_{\alpha \in [0,1], \beta \in (-\frac{1}{2}, 1)} \mathcal{C}_H(m, \alpha, \beta)$.

Remark: As the sequence $(B_k^\alpha)_{-1 \leq k \leq m}$ is independent of β then Property 7 imply that $\mathcal{C}(m, \alpha, \beta)$ is independent on β, so in the following, we use the notation $\mathcal{C}(m, \alpha)$ instead $\mathcal{C}(m, \alpha, \beta)$.

Definition 14. *Let* $m \in \mathbb{N}^*$

- *Let* $S \in \mathfrak{S}_m$. $\mathcal{I}m(S) = \{(\alpha, \beta) \in [0,1]^2 \mid S = S(m, 0, \alpha, \beta)\}$ *is called* the dual region corresponding to S.
- *Let* $S \in \mathfrak{S}_{H,m}$. $\mathcal{I}m_H(S) = \{(\alpha, \beta) \in [0,1] \times (-\frac{1}{2}, 1) \mid S = S_H(m, 0, \alpha, \beta)\}$ *is called* the dual region corresponding to S.

Property 8. Let $S \in \mathfrak{S}_m$ (respectively $S \in \mathfrak{S}_{H,m}$). Then $\mathcal{I}m(S)$ (respectively $\mathcal{I}m_H(S)$) is a convex polygon of $\mathbb{R}^2$.

Proof.   • Let $S \in \mathfrak{S}_m$. Then $\mathcal{I}m(S) = \{(\alpha, \beta) \in [0,1]^2 \mid 0 \leq \alpha x + \beta - y < 1$ for $(x, y) \in S\}$. So $\mathcal{I}m_H(S)$ is a convex polygon.
- Let $S \in \mathfrak{S}_{H,m}$. Then $\mathcal{I}m_H(S) = \{(\alpha, \beta) \in [0,1] \times (-\frac{1}{2}, 1) \mid -\frac{\alpha+1}{2} < \alpha x + \beta - y < \frac{\alpha+1}{2}$ for $(x, y) \in S\}$. So $\mathcal{I}m_H(S)$ is a convex polygon.   □

4.1   Frequencies of Digital Segments

Let $\alpha, \beta \in [0,1], m \in \mathbb{Z}$ and $n \in \mathbb{N}^*$.

Definition 15. *The* β-frequency *of a digital segment* S *for the slopes* α *(denoted* $\text{freq}_\alpha(S)$*) is the length of the interval* $I^\alpha(S) = \{\beta \in [0,1] \mid (\alpha, \beta) \in \mathcal{I}m(S)\}$. *(so the function* $TP : \mathcal{I}m(S) \to \mathbb{R}$ *such that* $TP(\alpha) = \text{freq}_\alpha(S)$ *is the tomographic projection of* $\mathcal{I}m(S)$ *w.r.t. the second coordinate direction).*

Remark: $I^\alpha = \{\beta \in [0,1] \mid (\alpha, \beta) \in \mathcal{I}m(S)\} = \{\beta \in [0,1] \mid \langle 0\alpha + \beta \rangle \in [B_i^\alpha, B_{i+1}^\alpha)\} = [B_i^\alpha, B_{i+1}^\alpha)$ where $i \in [\![0, m-1]\!]$ and $[B_i^\alpha, B_{i+1}^\alpha)$ is the interval corresponding to the digital segment S.

Definition 16. The overlapping frequency *of a digital segment* S *in the digital line* $\mathfrak{D}\mathcal{L}(\alpha, \beta)$ *is*

$$\lim_{N \to +\infty} \frac{\text{card}(\{n \in [\![-N, N]\!] \mid S'(m, n, \alpha, \beta) = S\})}{(2N + 1)}$$

if the limit exists. It is denoted $\text{overfreq}_{\alpha, \beta}(S)$.

So, $\text{overfreq}_{\alpha, \beta}(S) = \lim_{N \to +\infty} \frac{\text{card}(\{n \in [\![-N, N]\!] \mid \langle \alpha n + \beta \rangle \in I^\alpha(S)\})}{(2N+1)}$

We have the following properties:

Proposition 1. *For any $\alpha \in [0,1]$ and $\beta \in \mathbb{R}$ we have:*

1. $S \in \mathcal{C}(m,\alpha)$ *if and only if* $\text{freq}_\alpha(S) > 0$.
2. $\text{overfreq}_{\alpha,\beta}(S) = \text{freq}_\alpha(S)$

Proof. 1. If $S \in \mathcal{C}(m,\alpha)$, then there exists $n \in \mathbb{Z}$ and $\beta \in \mathbb{R}$ such that $S = S'(m,n,\alpha,\beta)$. So, $\langle \alpha n + \beta \rangle \in I^\alpha(S)$. Then $\text{freq}_\alpha(S) = \mu(I^\alpha(S)) > 0$ because $I^\alpha(S)$ is a non-empty interval of the form $[A, A')$.

Conversely if $\text{freq}_\alpha(S) > 0$ then $I^\alpha(S) \neq \emptyset$. So, by Property 7, for all $\beta \in \mathbb{R}$ there exists $n \in \mathbb{Z}$ such that $\langle \alpha n + \beta \rangle \in I^\alpha(S)$, which implies that $S \in \mathcal{C}(m,\alpha)$.

2. We prove now that $\text{overfreq}_{\alpha,\beta}(S) = \text{freq}_\alpha(S)$ for any $\alpha \in [0,1]$ and $\beta \in \mathbb{R}$.
   (a) Suppose first that α is rational and let $\beta \in \mathbb{R}$, then $\alpha = \frac{p}{q}$ where p,q are co-prime.

   Put $k_0 = \lfloor \gamma q \rfloor$ and $\beta' = \beta - \frac{\lfloor \gamma q \rfloor}{q}$. Then $0 \leq \beta' < \frac{1}{q}$.

   As p,q are co-prime, then $\{\langle \alpha x + \beta \rangle \mid x \in \mathbb{Z}\} = \{\beta' + \frac{i}{q} \mid i \in [\![0, q-1]\!]\}$.
   Let $z \in \mathbb{Z}$ and consider the set $E(z) = [\![z, z+q-1]\!]$. Then $\{\langle \alpha x + \beta \rangle \mid x \in E(z)\} = \{\beta' + \frac{i}{q} \mid i \in [\![0, q-1]\!]\}$ because p,q are co-prime. So, if $i,j \in E(z)$ and $i \neq j$ then $\langle \alpha i + \beta \rangle \neq \langle \alpha j + \beta \rangle$. As for all $k \in [\![-1, m-1]\!]$, $B_k^\alpha = B_{k+1}^\alpha$ or $B_{k+1}^\alpha - B_k^\alpha = \frac{1}{q}$, then for all $k \in [\![-1, m-1]\!]$ such that $[B_k^\alpha, B_{k+1}^\alpha) \neq \emptyset$ there exists only one $i \in E(z)$ such that $\langle \alpha i + \beta \rangle \in [B_k^\alpha, B_{k+1}^\alpha)$. Let $N \in \mathbb{N}$ and put $E_N = [\![-N, N]\!]$.
   Then $H_N = \left(\bigcup_{i \in [\![0, \lfloor \frac{2N+1}{q} \rfloor - 1]\!]} E(-N + qi) \right) \bigcup H_N'$ where $H_N' = [\![-N + q(\lfloor \frac{2N+1}{q} \rfloor - 1), N]\!]$. So H_N is partitioned on $\lfloor \frac{2N+1}{q} \rfloor$ segments of the form $E(s)$ and H_N'.
   Then

   $$\frac{\text{card}(\{n \in [\![-N, N]\!] \mid \langle \alpha n + \beta \rangle \in I^\alpha(S)\})}{(2N+1)} = \frac{\lfloor \frac{2N+1}{q} \rfloor}{2N+1} + \frac{k_0}{2N+1}$$

   where $k_0 \leq \text{card}(H_N') < q$.
   So, $\text{overfreq}_{\alpha,\beta}(S) = \lim_{N \to +\infty} \frac{\lfloor \frac{2N+1}{q} \rfloor}{2N+1} = \frac{1}{q} = \mu(I^\alpha(S)) = \text{freq}_\alpha(S)$.
   (b) Suppose now that α is irrational. By results of density of the sequences $(\langle \alpha n + \rho \rangle)_{n \in \mathbb{Z}}$ for an irrational number α [6], we have

   $$\text{overfreq}_{\alpha,\beta}(S) = \lim_{N \to +\infty} \frac{\text{card}(\{i \in E_N \mid \langle \alpha i + \beta \rangle \in I^\alpha(S)\})}{2N+1}$$
   $$= \mu(I^\alpha(S))$$

So, for all $\alpha, \beta \in [0,1]$, $\text{overfreq}_{\alpha,\beta}(S) = \text{freq}_\alpha(S)$. $\square$

Remarks

− The function $\alpha \mapsto \text{overfreq}_{\alpha,\beta}(S)$ does not depend on $\beta \in [0,1]$.
− If $\alpha = \frac{p}{q}$, then $\text{card}(\mathcal{C}(m,\alpha)) = q$ and for all $S \in \mathcal{C}(m,\alpha)$, $\text{overfreq}_{\alpha,\beta}(S) = \frac{1}{q}$ which does not depend on $S \in \mathcal{C}(m,\alpha)$.

Definition 17. *A function $f : \mathbb{R} \to \mathbb{R}$ is called a piecewise affine function if there exists a finite collection $(C_i)_{i \in I}$ of open interval and affine functions $f_i :$ $\mathbb{R} \to \mathbb{R}$ for $i \in I$, such that :*

- $C_i \cap C_{i'} = \emptyset$ *for* $i, i' \in I$ *and* $i \neq i'$,
- $\bigcup_{i \in I} \overline{C_i} = \mathbb{R}$ *and*
- *The restriction of f to C_i is f_i for all $i \in I$ (for all $i \in I, f(x) = f_i(x)$ for all $x \in C_i$).*

Property 9. *Let $f, g : \mathbb{R} \to \mathbb{R}$ be two piecewise affine functions. Then $-f, f + g, f - g, \max(f, g)$ and $\min(f, g)$ are also piecewise affine functions.*

Theorem 2. *For any digital segment S, the function $\alpha \mapsto \mathrm{freq}_\alpha(S)$ is a continuous function which is piecewise affine.*

Proof. $\mathcal{I}m(S) = \{(\alpha, \beta) \in \mathbb{R}^2 \mid S(k) \leq \alpha k + \beta < S(k) + 1 \text{ for all } k \in [\![0, m]\!]\}$.
Then $I^\alpha(S) = [\max_{k \in [\![0,m]\!]}(S(k) - \alpha k), \min_{k \in [\![0,m]\!]}(S(k) + 1 - \alpha k))$.

So, $\mathrm{freq}_\alpha(S) = \max(0, \min_{k \in [\![0,m]\!]}(S(k) + 1 - \alpha k) - \max_{k \in [\![0,m]\!]}(S(k) - \alpha k))$.
Affine functions, max and min are continuous functions. Then $\alpha \mapsto \mathrm{freq}_\alpha(S)$
is a continuous function which is piecewise affine because it is composition of
continuous functions and by Property 9 it is piecewise affine function.   □

Proposition 2. *Let $\alpha_1, \alpha_2 \in [0, 1]$ such that $\alpha_1 < \alpha_2$. Suppose that the function $\alpha \mapsto \mathrm{freq}_\alpha(S)$ is affine on $[\alpha_1, \alpha_2]$ for any digital segment S. Let $\alpha \in (\alpha_1, \alpha_2)$. Then*

$$\mathcal{C}(m, \alpha) = \mathcal{C}(m, \alpha_1) \bigcup \mathcal{C}(m, \alpha_2)$$

Proof. Consider $\lambda_1, \lambda_2 > 0$ such that $\alpha = \lambda_1 \alpha_1 + \lambda_2 \alpha_2$ and $\lambda_1 + \lambda_1 = 1$ $(\lambda_1, \lambda_2$
are barycentric coordinates of α relatively to α_1, α_2). By affinity of $\alpha \mapsto \mathrm{freq}_\alpha(S)$
on $[\alpha_1, \alpha_2]$ we have:

$$\mathrm{freq}_\alpha(S) = \lambda_1 \mathrm{freq}_{\alpha_1}(S) + \lambda_2 \mathrm{freq}_{\alpha_2}(S)$$

If $S \notin \mathcal{C}(m, \alpha)$ then by Proposition 1, $\mathrm{freq}_\alpha(S) = 0$ and so for any i, $\mathrm{freq}_{\alpha_i}(S) = 0$
because $\lambda_1, \lambda_2 > 0$, which implies that for any i, $S \notin \mathcal{C}(m, \alpha_i)$. Conversely as
$\lambda_1, \lambda_2 > 0$ and $\lambda_1 + \lambda_2 = 1$, if $S \in \mathcal{C}(m, \alpha)$, then by Proposition 1, $\mathrm{freq}_\alpha(S) > 0$
and thus, there must exist a $i \in \{1, 2\}$ such that $\mathrm{freq}_{\alpha_i}(S) > 0$.   □

Theorem 3. *Let f, f' be two consecutive m-Farey numbers such that $f < f'$.
Then the function $\alpha \mapsto \mathrm{freq}_\alpha(S)$ is affine in $[f, f']$ for all $S \in \mathfrak{S}_m$. Moreover
for any $\alpha, \alpha' \in [0, 1] \setminus \mathcal{F}_m$ $\mathcal{C}(m, \alpha) = \mathcal{C}(m, \alpha')$ if and only if there exists two
consecutive m-Farey numbers f, f' such that $\alpha, \alpha' \in (f, f')$.*

Proof. 1. Let $i \in [\![0, m]\!]$ and $\alpha, \alpha' \in (f, f')$, such that $\alpha \neq \alpha'$ and $k = \lfloor \alpha i \rfloor <$
$k' = \lfloor \alpha' i \rfloor$. Then $\alpha \in [\frac{k}{i}, \frac{k+1}{i})$ and $\alpha' \in [\frac{k'}{i}, \frac{k'+1}{i})$. Which is absurd because
f, f' are two consecutive m-Farey numbers. So, for all $i \in [\![0, m]\!]$ there exists
$k_i \in \mathbb{N}$ such that $k_i = \lfloor \alpha i \rfloor$ for all $\alpha \in (f, f')$.

Let $i, j \in [\![0, m]\!]$ such that $i \neq j$ and suppose that there exists $\alpha', \alpha'' \in$
(f, f') such that $\langle \alpha' i \rangle < \langle \alpha' j \rangle$ and $\langle \alpha'' i \rangle \geq \langle \alpha'' j \rangle$. So, we have $k_i - \alpha' i <$

$k_j - \alpha' j$ and $k_i - \alpha'' i \geq k_j - \alpha'' j$. Then there exists $\alpha \in]\alpha', \alpha''] \subset (f, f')$ such that $k_i - \alpha i = k_j - \alpha j$, which imply that α is a m-Farey number. Which is absurd because f, f' are consecutive m-Farey numbers.

Let $k \in [\![-1, m-1]\!]$ and $S \in \mathfrak{S}_m$. Then two cases are possible: $I^\alpha(S) = [B_k^\alpha, B_{k+1}^\alpha) \cap (f, f') = \emptyset$ or $(f, f') \subset I^\alpha(S)$. Then, in the first case we have $\mathrm{freq}_\alpha(S) = 0$ and in the second case we have for all $\alpha \in (f, f')$, $\mathrm{freq}_\alpha(S) = (1 - \langle \alpha i \rangle) - (1 - \langle \alpha j \rangle) = k_i - k_j + \alpha(i - j)$ where $B_k^\alpha = 1 - \langle \alpha j \rangle$ and $B_{k+1}^\alpha = 1 - \langle \alpha i \rangle$ for all $\alpha \in (f, f')$.

2. Let $\alpha, \alpha' \in [0, 1] \setminus \mathcal{F}_m$ such that $\mathcal{C}(m, \alpha) = \mathcal{C}(m, \alpha')$ and suppose that there exists $f = \frac{p}{q} \in \mathcal{F}_m$ such that $\alpha < f < \alpha'$ where $p, q \in \mathbb{N}$ and $q \leq m$. As for all $S \in \mathfrak{S}_m, Im(S)$ is a convex polygon of $\mathbb{R}^2$, then $\{\alpha | (\alpha, \beta) \in Im(S)\} = \{\alpha \mid \mathrm{freq}_\alpha(S) > 0\}$ is an interval. So, for all $S \in \mathcal{C}(m, \alpha)$, as $\mathrm{freq}_\alpha(S) > 0$ and $\mathrm{freq}_{\alpha'}(S) > 0$ then $\mathrm{freq}_{\frac{p}{q}}(S) > 0$ and thus $S \in \mathcal{C}(m, \frac{p}{q})$. Which is absurd because $\mathrm{card}(\mathcal{C}(m, \alpha)) = m + 1$ and $\mathrm{card}(\mathcal{C}(m, \frac{p}{q})) \leq m$. $\square$

Several combinatorial properties of the set of digital segments (which corresponds to the factors of Sturmian words) [3,2,8] is a consequence of the affinity of the function $\alpha \mapsto \mathrm{freq}_\alpha(S)$ on $[f, f']$ for all $S \in \mathfrak{S}_m$ where $f < f'$ are a consecutive m-Farey numbers: Theorem 3.

Corollary 3. *Let $f = \frac{p}{q}, f' = \frac{p'}{q'}$ be two consecutive m-Farey numbers such that $f < f'$. Then for any $\alpha \in (f, f')$, $\mathcal{C}(m, \alpha) = \mathcal{C}(m, f) \cup \mathcal{C}(m, f')$. Moreover, for $\alpha \in [f, f']$,*

- *if $S \in \mathcal{C}(m, f) \setminus \mathcal{C}(m, f')$, then $\alpha \mapsto \mathrm{freq}_\alpha(S) = p' - q'\alpha$,*
- *if $S \in \mathcal{C}(m, f') \setminus \mathcal{C}(m, f)$, then $\alpha \mapsto \mathrm{freq}_\alpha(S) = q\alpha - p$,*
- *if $S \in \mathcal{C}(m, f) \cap \mathcal{C}(m, f')$, then $\alpha \mapsto \mathrm{freq}_\alpha(S) = (p' - q'\alpha) + (q\alpha - p) = (q - q')\alpha + p' - p$.*

Proof. By Theorem 3, for all $S \in \mathfrak{S}_m$, the function $\alpha \mapsto \mathrm{freq}_\alpha(S)$ is affine on $[f, f']$ and by using Proposition 2 we have, for any $\alpha \in (f, f'), \mathcal{C}(m, \alpha) = \mathcal{C}(m, f) \cup \mathcal{C}(m, f')$.

As the function $\alpha \mapsto \mathrm{freq}_\alpha(S)$ is affine on $[f, f']$, then if $\alpha \in [f, f']$ then $\alpha = \lambda f' + (1 - \lambda)f$ with $\lambda = qq'\alpha - pq' \in [0, 1]$ and thus, $\mathrm{freq}_\alpha(S) = \lambda \mathrm{freq}_{f'}(S) + (1 - \lambda)\mathrm{freq}_f(S)$. Then

- *if $S \in \mathcal{C}(m, f) \setminus \mathcal{C}(m, f')$, then $\mathrm{freq}_\alpha(S) = \lambda 0 + (1 - \lambda)\frac{1}{q} = p' - q'\alpha$,*
- *if $S \in \mathcal{C}(m, f') \setminus \mathcal{C}(m, f)$, then $\mathrm{freq}_\alpha(S) = \lambda \frac{1}{q'} + (1 - \lambda)0 = q\alpha - p$,*
- *if $S \in \mathcal{C}(m, f) \cap \mathcal{C}(m, f')$, then $\mathrm{freq}_\alpha(S) = \lambda \frac{1}{q'} + (1 - \lambda)\frac{1}{q} = (q - q')\alpha + p' - p$.* $\square$

The results of Corollary 3 are proved in [3] by using other techniques.

Property 10. *Let f, f', f'' be three consecutive m-Farey numbers and α, γ such that $\alpha < f' = \frac{p}{q} < \gamma < f''$ where p, q is positive coprime numbers. Then*

1. $(\mathcal{C}(m, \gamma) \setminus \mathcal{C}(m, f')) \cap \mathcal{C}(m, \alpha) = \emptyset$.
2. *If $f < \alpha < f'$, then $\mathrm{card}(\mathcal{C}(m, \gamma) \setminus \mathcal{C}(m, \alpha)) = m - q + 1$.*

Proof. By Property 8, the set $\mathcal{I}m(S)$ is a convex polygon of $\mathbb{R}^2$. So, by Property 7 $(\alpha, \beta) \in \mathcal{I}m(S)$ if and only if $\text{freq}_\alpha(S) > 0$. Thus, the set $\{\alpha \mid \text{freq}_\alpha(S) > 0\} = \{\alpha \mid (\alpha, \beta) \in \mathcal{I}m(S)\}$ is a convex set of $\mathbb{R}$ *(i.e. it is an interval)* because it is a projection of $\mathcal{I}m(S)$.

1. So, $S \in \mathcal{C}(m, \alpha) \cap \mathcal{C}(m, \gamma)$ if and only if $S \in \mathcal{C}(m, f')$. Thus, $(\mathcal{C}(m, \gamma) \setminus \mathcal{C}(m, f')) \cap \mathcal{C}(m, \alpha) = \emptyset$.
2. If $f < \alpha < f'$, then $S \in \mathcal{C}(m, \alpha) \cap \mathcal{C}(m, \gamma)$ if and only if $S \in \mathcal{C}(m, f')$. But $\text{card}(\mathcal{C}(m, f')) = q$ and $\text{card}(\mathcal{C}(m, \alpha)) = \text{card}(\mathcal{C}(m, \gamma)) = m + 1$ because $f' = \frac{p}{q}$, $0 \leq p \leq q \leq m$ and $\alpha, \gamma \in ([0, 1] \setminus \mathcal{F}_m)$. Then $\text{card}(\mathcal{C}(m, \gamma) \setminus \mathcal{C}(m, \alpha)) = m - q + 1$ $\square$

Property 11. We have three possibilities for the function $\alpha \mapsto \text{freq}_\alpha(S)$:

- If S_0 is the horizontal segment, then $\text{freq}_\alpha(S_0) = -m\alpha + 1$ for $\alpha \in [0, \frac{1}{m}]$ and 0 elsewhere.
- If S_1 is the diagonal segment, then $\text{freq}_\alpha(S_1) = m\alpha + 1 - m$ for $\alpha \in [\frac{m-1}{m}, 1]$ and 0 elsewhere.
- If $S \in \mathfrak{S}_m \setminus \{S_0, S_1\}$, then there exists three m-Farey numbers f', f, f'' such that $f' < f < f''$, $f = \frac{p}{q}$ and $\text{freq}_\alpha(S) = \frac{1}{q(f-f')}(\alpha - f) + \frac{1}{q}$ for $\alpha \in [f', f]$ and $\text{freq}_\alpha(S) = \frac{1}{q(f-f'')}(\alpha - f) + \frac{1}{q}$ for $\alpha \in [f, f'']$ and 0 elsewhere.

Property 11 is is illustrated in Fig. 1.

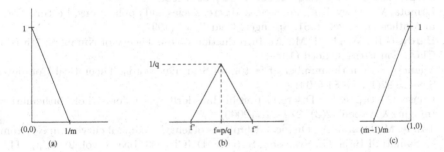

Fig. 1. The function $\alpha \mapsto \text{freq}_\alpha(S)$, for a digital segment S: horizontal segment (a), diagonal segment (c), nor horizontal and nor diagonal segment (b)

Corollary 4. *Let $m \in \mathbb{N}^*$. Then $\text{card}(\mathfrak{S}_m) = 1 + \sum_{i=1}^{m}(m-i+1)\varphi(i)$ where φ is the Euler's totient function (i.e. $\varphi(n) = \text{card}(\{i \mid 1 < i < n \text{ and } \gcd(i, n) = 1\})$ for $n \in \mathbb{N}^*$).*

Corollary 4 is proved initially in [2,8]. The proof proposed here use new arguments.

Corollary 5. *Let $m \in \mathbb{N}^*$. Then $\text{card}(\mathfrak{S}_{H,m}) = m + 1(1 + \sum_{i=1}^{m}\varphi(i))$.*

Proof. $\text{card}(\mathfrak{S}_{H,m}) = \text{card}(\mathfrak{S}_m) + \sum_{i=1}^{m}i\varphi(i)$. But $\text{card}(\mathfrak{S}_m) = 1 + \sum_{i=1}^{m}(m - i + 1)\varphi(i)$. Thus, $\text{card}(\mathfrak{S}_{H,m}) = m + 1(1 + \sum_{i=1}^{m}\varphi(i))$. $\square$

5 Conclusion

In this paper we have proved that the frequency of a digital segment of size m in a digital straight line is a continuous piecewise affine function in the slopes of the straight line. This has consequences on the combinatorics of the set of digital segment of size m, in particular its give a new proofs of several results on the factors of Sturmian words [3,2,8]. We obtain also some properties of digital segments for Hausdorff discretization.

A generalization, to digital plan, of some notions of this paper is studied in [5].

References

1. Barnsley, M.F.: Fractals Everywhere, 2nd edn. Academic Press, London (1993)
2. Berenstein, C.A., Lavine, D.: On the Number of Digital Straight Line Segments. IEEE Trans. on Pat. Anal. And Mach. Intell. 10(6), 880–887 (1988)
3. Berthé, V.: Fréquences des facteurs des suites sturmiennes. Theoretical Computer Science 165, 295–309 (1996)
4. Brimkov, V.E., Coeurjolly, D., Klette, R.: Digital planarity - a review. Discrete Appl. Math. 155(4), 468–495 (2007)
5. Daurat, A., Tajine, M., Zouaoui, M.: About the frequencies of some patterns in digital planes. Application to area estimators - extended version (Preprint, 2007), http://hal.archives-ouvertes.fr/hal-00174960
6. Drmota, M., Tichy, R.F.: Sequences, discrepancies and applications. Lecture Notes in Mathematics, vol. 1651. Springer, Heidelberg (1997)
7. Hardy, G.H., Wright, E.M.: An Introduction to the Theory of Numbers, 5th edn. Clarendon Press, Oxford (1978)
8. Mignosi, F.: On the number of factors of Sturmian words. Theoretical Computer Science 82(1), 71–84 (1991)
9. Ronse, C., Tajine, M.: Discretization in Hausdorff Space. Journal of Mathematical Imaging & Vision 12(3), 219–242 (2000)
10. Tajine, M., Daurat, A.: On local definitions of length of digital curves. In: Nyström, I., Sanniti di Baja, G., Svensson, S. (eds.) DGCI 2003. LNCS, vol. 2886, pp. 114–123. Springer, Heidelberg (2003)
11. Tajine, M., Ronse, C.: Topological Properties of Hausdorff discretizations. In: ISMM 2000 (2000)
12. Tajine, M., Ronse, C.: Topological properties of Hausdorff discretization, and comparison to other discretization schemes. Theoretical Computer Science 283, 243–268 (2002)

Scaling of Plane Figures That Assures Faithful Digitization

Valentin E. Brimkov

Mathematics Department, SUNY Buffalo State College
Buffalo, NY 14222, USA
brimkove@buffalostate.edu

Abstract. In this paper we propose a method for obtaining a faithful digitization of certain broad classes of plane figures, so that the original continuous object and its digitization feature analogous geometric properties. The approach is based on an appropriate scaling of a given figure so that the obtained one admits digitization satisfying some desirable conditions. Informally speaking, we show that from certain point on, a continuous object and its digitization are in a sense equivalent. In terms of computational complexity, the scaling factor is easily computable. As a corollary of the presented theory we prove the strong NP-hardness of the problem of obtaining a polyhedron reconstruction in which the facets are trapezoids or triangles.

Keywords: digital geometry, lattice polygon, scaling factor, polyhedral reconstruction, NP-hard problem.

1 Introduction

Digitizing a real object in a way preserving some of its basic properties is of significant importance for several areas of visual computing [8,13]. Independently of the particular method used, the quality of the obtained digitization depends on the *grid resolution* h, that is the inverse of the *grid constant* defined as the number of grid elements per unit of distance. Obviously, if the grid resolution is very low, e.g, if the size of the digitized object S is comparable with the one of the grid cell, than the digitization would carry very little or no information about S. The higher the grid resolution the better (more useful) representation of the real object is provided by its digitization. A reasonable digitization (obtained, e.g., by a tomography scanner) may assure conditions for a faithful reconstruction of the original real object. High grid resolution may also make possible to compute with sufficiently high precision various properties of that real object whose characteristics are usually unknown. A lot of work been devoted to characterizing the asymptotic behavior of estimators of various properties (such as curve length, perimeter, normal, curvature, etc.) as the grid resolution tends to infinity (see the extensive bibliography at the end of Ch. 10 of [7] as well as [11] for theoretical foundations). Intuitively, investigations of this kind can also indicate how continuous advances in technology would impact the quality of object analysis.

V.E. Brimkov, R.P. Barneva, H.A. Hauptman (Eds.): IWCIA 2008, LNCS 4958, pp. 87–98, 2008.

Of course, infinite grid refinement is impossible in practice. In this paper we will show that for certain purposes it is also unnecessary. Roughly speaking, we will demonstrate that for any plane figure S from a certain class there is a number, called *scaling factor*, such that if S is magnified by that factor (that is proportional to the squared diameter of S), the obtained (larger) figure will admit digitization that faithfully represents the shape and some properties of the original S.

The paper is organized as follows. Scaling of plane polygons (possibly non-convex and possessing holes) is considered in Section 2. It is proved that the obtained polygon is minimally enclosing, i.e., has minimal number of sides over all possible enclosing polygons containing the same set of integer points. This last set admits efficient reconstruction with minimal number of sides. Moreover, it is shown that between any two minimally enclosing polygons there is a 1-to-1 correspondence, such that corresponding vertices are "close" to each other. In Section 3 the investigation goes deep into study of scaling that preserves intrinsic geometric features of the object, assuring, e.g., minimal decomposition with respect to the enclosed set of integer points. Using the obtained results, as a bi-product it is shown in Section 3.1 that the problem of optimal reconstruction by trapezoids of a 3D integer set is strongly NP-hard. In Section 4 the considerations are extended to more general objects whose border is composed by straight line segments and convex curves. The paper concludes with some final remarks in Section 5. Due to space constraints, the proofs of some theorems and other details are differed to the full length journal version of the paper.

Basic Definitions and Notations

Let S be a *body* in $\mathbb{R}^n$, i.e., a subset of $\mathbb{R}^n$ of full topological dimension $dim(S) = n$. Denote $S_\mathbb{Z} = S \cap \mathbb{Z}^n$. For a set $A \subseteq \mathbb{R}^n$, $d(A) = \max_{x,y \in A} \|x - y\|$ is its *diameter* where $\|.\|$ is the Euclidean norm. By kA we denote the homothetic image of A under a homothety with center $\mathbb{O}$ and a constant of proportionality $k \in \mathbb{R}_+$. By $conv(A)$ we denote the convex hull of A. Given two sets $A, B \subset \mathbb{R}^n$, $\rho(A, B) = \inf_{x,y}\{\rho\{x,y\} : x \in A, y \in B\}$ is the Euclidean distance between them. Given a polyhedron $P \subset \mathbb{R}^n$, the number of its i-facets is denoted by $f_i(P)$, $0 \le i \le n$.

2   Scaling I - Preserving Geometry for Optimal Polygonal Reconstruction

Given a simple polygon P, we want to determine an appropriate grid resolution or, equivalently, appropriate scaling of P, i.e., a homothetic polygon $Q = kP$, $k \in \mathbb{R}_+$, that assures certain properties. In this section we pursue scaling that satisfies the following conditions (to be formalized later).

1. The obtained polygon Q has the minimal possible number of sides over all polygons that contain the same set $M = Q_\mathbb{Z}$ of integer points;
2. M well approximates the shape of the original polygon P.

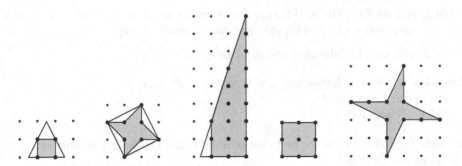

Fig. 1. *From left to right:* The first three polygons are non-minimally enclosing while the last two are minimally enclosing for the integer points contained in them

Moreover, it will be shown that the digitization M admits efficient polygonal reconstruction with a minimal number of sides.

Recall that the *polygonal reconstruction* problem is the following.[1] Given a nonempty set $M \subset \mathbb{Z}^2$, one looks for a (possibly non-convex) polygon P, such that the set $P_\mathbb{Z}$ of integer points contained in P is precisely M. Such a polygon will be called *enclosing* for M. An enclosing polygon with a minimal number of sides will be called *minimally enclosing* for M.

Obviously, any triangle T is minimally enclosing for $T_\mathbb{Z}$, provided that $T_\mathbb{Z}$ contains at least three non-collinear points. An arbitrary polygon P is enclosing for the set $P_\mathbb{Z}$ but not necessarily minimally enclosing (see Fig. 1). In what follows, we will assume that $P_\mathbb{Z}$ encloses at least three non-collinear integer points as w.l.o.g. the origin $\mathbb{O}$ is among these.

Now let $P \subset \mathbb{R}^2$ be a polygon in $\mathbb{R}^2$. Let $W = \{v^1, v^2, \dots, v^m\}$ be the set of its vertices where $v^i = (v_1^i, v_2^i) \in \mathbb{Z}^2$, $1 \le i \le m$.

Let $d(P) = \max_{v^j, v^k \in W} \{\|v^i - v^j\|\}$, $1 \le j, k \le m$ be the diameter of P.

For any $v^i \in W$, define $\rho_i(P) = \min_{j,k \neq i} \rho(v^i, \overline{v^j v^k})$, where $v^i \notin \overline{v^j v^k}$ and $\overline{v^j v^k}$ is the straight line through v^j and v^k. Let $\rho(P) = \min_i \rho_i(P)$.

Now we define a *scaling element* of P as

$$\delta_P = \max\{3, d(P), 1/\rho(P)\} \tag{1}$$

To technically simplify some considerations, in the rest of this section we will assume that P is a lattice polygon, i.e., one with integer vertices. All results hold for arbitrary polygons as well. (Note: any rational polygon can be represented as an integer one through a proper increase of the grid resolution.)

For any lattice polygon P we have that $\delta_P \ge 3$, as the bound 3 is reached, e.g., for a lattice unit square. We also have

Fact 1. δ_P *can be computed with $O(m \log m)$ algebraic operations.*

---

[1] The 3D version of the problem, called polyhedral reconstruction, will be considered in Section 3.1.

Proof. Follows from the well-known fact that the diameter of a set of m points can be computed with $O(m \log m)$ algebraic operations [10].   □

We will also use the following technical fact.

Lemma 1. *Given a lattice polygon P with $\mathbb{O} \in P$, then*
  (a) $\rho(P) \leq d(P)$, and
  (b) $1/\rho(P) \leq d(P)$.

Proof. (a) Follows from the fact that in a right triangle the hypotenuse is longer than each of the legs.

(b) Let $d(P) = \|u^* - v^*\| = \sqrt{v_2^* - u_2^*}$ for some points $u^*, v^* \in W$, $u^* = (u_1^*, u_2^*)$, $v^* = (v_1^*, v_2^*)$. It suffices to show that $1/\rho(P) \leq d(P)$, i.e., that $\frac{1}{\rho(w, \overline{uv})} \leq d(P)$ for any triple of points $u = (u_1, u_2), v = (v_1, v_2), w = (w_1, w_2) \in W$ with $w \notin \overline{uv}$.

$\overline{uv}$ has equation $ax_1 + bx_2 + c = 0$, where $a = v_2 - u_2, b = -(v_1 - u_1)$, and $c = (v_1 - u_1)u_2 - (v_2 - u_2)u_1$. Then we consecutively obtain

$$\rho(w, \overline{uv}) = \frac{|aw_1 + bw_2 + c|}{\sqrt{a^2 + b^2}},$$

$$1/\rho(w, \overline{uv}) = \frac{\sqrt{a^2 + b^2}}{|aw_1 + bw_2 + c|} =$$

$$\frac{\sqrt{(v_2 - u_2)^2 + (v_1 - u_1)^2}}{|(v_2 - v_1)w_1 - (v_1 - u_1)w_2 + (v_1 - u_1)u_2 - (v_2 - u_2)u_1|}.$$

By the definition of a point set diameter, the numerator of the last expression above is always less than or equal to $d(P)$. Since $w \notin \overline{uv}$, the denominator is nonzero. Moreover, since u, v, and w are integer points, it is greater than or equal to 1. Hence, $1/\rho(w, \overline{uv}) \leq d(P)$, which completes the proof.   □

The first implication of the above lemma is that for lattice polygons the scaling element δ_P can be written as

$$\delta_P = \max\{3, d(P)\} \qquad (2)$$

We have the following theorem.

Theorem 1. *Let P be a lattice polygon. For any integer $k \geq 2\delta_P^2$, the polygon kP is minimally enclosing for $(kP)_{\mathbb{Z}}$.*

Proof. Given a polygon Q, for sufficiently small real number $\epsilon > 0$ one can define an *ϵ-extension* Q_ϵ of Q as follows. To each side a of Q there corresponds a side a' of Q_ϵ; it lies outside Q and is a portion of a straight line l_a parallel to a and at a distance ϵ from it. Clearly, if ϵ is small enough, Q_ϵ is a polygon with the same number of vertices and sides as Q, so that there is a one-to-one correspondence between vertices/sides of Q and Q_ϵ. See Fig. 2 (left). (Obviously, if ϵ is not enough small, such a correspondence may not exist.)

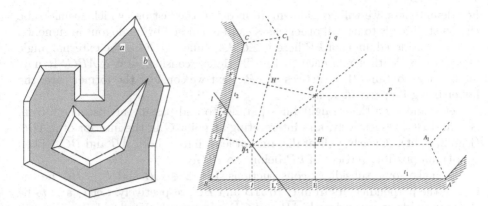

Fig. 2. *Left:* A polygon (in gray) and its ϵ-extension. *Right:* Illustration to the proof of Theorem 1.

It is not hard to realize that if $k \geq 2\delta_P^2$, then the above-mentioned correspondence between kP and $(kP)_\epsilon$ is present for $\epsilon = \sqrt{2}$. In fact, since $\delta_P \geq 3$, the following conditions hold:

$$All\ sides\ of\ kP\ are\ longer\ than\ 18 \tag{3}$$

and

$$\rho(kP) \geq 6 \tag{4}$$

(To obtain (4), recall that by Lemma 1 (b), $d(kP)\rho(kP) \geq 1$. Then $\rho(kP) = k\rho(P) \geq 2\delta_P^2\rho(P) = 2\delta_P(\delta_P\rho(P)) \geq 2\delta_P(d(P)\rho(P)) \geq 2\delta_P \geq 6$.)

Under these conditions, for any two consecutive sides a and b of kP that intersect at point M, the corresponding lines l_a and l_b will intersect at a point M' that lies outside kP and appears to be a vertex of $(kP)_\epsilon$. Connecting corresponding vertices such as M and M' by segments partitions the set $(kP)_\epsilon \setminus kP$ into a set T of trapezoids (Fig. 2, left). Their number equals the one of the sides of P and kP.

Now let P' be an arbitrary enclosing polygon for $(kP)_\mathbb{Z}$ different from kP. Consider the set of sides of P' that are not identical to sides of kP and are not portions of sides of kP. Among these there must be ones that intersect some trapezoids of the partition of $(kP)_\epsilon \setminus kP$ described above. Let side u of P' intersects trapezoid $t_1 = ABB_1A_1$, where $\overline{AB}$ is a side of kP. Let side $\overline{BC}$ is a side of kP that is adjacent to $\overline{AB}$ and $t_2 = BCC_1B_1$ be its corresponding trapezoid. If $\overline{BC}$ is also a side of P' or contains a side v of P' as a subsegment, then there are at least two different sides—u and $\overline{BC}$ (resp. v)—of P' intersecting the two consecutive trapezoids t_1 and t_2. Assume that the above condition is not the case. We will show that nevertheless there is a side v of P', $v \neq u$, that intersects t_2.

One can distinguish four cases depending on the measure of $\angle ABC$ (resp. $\angle CBA$), i.e., if the latter belongs to the interval $(0, \frac{\pi}{2})$, $[\frac{\pi}{2}, \pi)$, $(\pi, \frac{3}{2}\pi]$, or $[\frac{3}{2}\pi, 2\pi)$.

For definiteness, we will consider more in detail the last one which seems to be the most difficult to us. All other cases can be handled by analogous arguments.

The considered angle of kP here is $\angle CBA$ while $\angle ABC$ is the external angle at vertex B. With a reference to Fig. 2 (right), consider first $\triangle ABC$. It may or may not contain other vertices of kP. Next we consider the former case, the latter being its particular case.

Let p and q be the straight lines parallel to and passing at distance 6 from $\overline{AB}$ and $\overline{BC}$, respectively, and intersecting at point G in the interior of $\angle ABC$ (Fig. 2, right). (Clearly, G is at distance at least 6 from both $\overline{AB}$ and $\overline{BC}$.) Then by (4), all possible vertices of kP belong to $\angle pGq$.

We make some subsidiary constructions. Let $E \in \overline{AB}$ and $F \in \overline{BC}$ be the feet of the perpendiculars from G to $\overline{AB}$ and $\overline{BC}$, respectively. Let t_1 and t_2 be the trapezoids associated with $\overline{AB}$ and $\overline{CD}$, respectively. In what follows we will evaluate their size, or more precisely, the size of the maximal ($\sqrt{2} \times s$)-rectangle they may contain. For our purposes it will suffice if $s \geq \sqrt{2}$.

It is clear that if, say, C and F are close to each other (or coincide), then, by (3), the length of $\overline{CF}$ will be close (or equal) to 18. If C is between B and F, then the size of t_1 will be even larger and the needed bound easily obtainable. Therefore we consider the less trivial case when F is between B and C and the distance between F and C is at least $\sqrt{2}$. Analogous hypothesis is assumed for A, E, and B.

Let $BEH'B_1$ and $BFH''B_1$ be the respective portions of trapezoids t_1 and t_2, cut of by the perpendiculars $\overline{GE}$ and $\overline{GF}$, and let $\overline{B_1I}$ and $\overline{B_1J}$ be the perpendiculars from B_1 to $\overline{BE}$ and $\overline{BF}$, respectively. Note that triangles $\triangle BEG$ and $\triangle BFG$ are congruent and all corresponding considerations are symmetric. Since $\angle ABC < \pi/2$ and using (4), $|BF| > |FG| > 6$. Keeping this last inequalities in mind, from the similar triangles $\triangle BJF$ and $\triangle B_1GH'$ we consecutively obtain

$$\frac{|FJ|}{|BF|} = \frac{|GH''|}{|GF|} > \frac{6 - \sqrt{2}}{|GF|},$$

i.e.,

$$|FJ| > (6 - \sqrt{2})\frac{|BF|}{|GF|} > 6 - \sqrt{2}.$$

Finally, construct a straight line l passing through point B_1 and perpendicular to $\overline{BG}$. Let $l \cap \overline{BE} = L'$ and $l \cap \overline{BF} = L''$. Consider again the side u of P' that intersects the trapezoid $t_1 = ABB_1A_1$. By the geometric constructions we have that u either does not intersect any of the trapezoids $L'EH'B_1$ or $L''FH''B_1$, or if it intersects one of these, it does not intersect the other. Since $\angle ABC < \pi/2$, from the congruent right triangles $\triangle IL'B_1$ and $\triangle JL''B_1$ we have $IL' < \sqrt{2}$ and $JL'' < \sqrt{2}$. Then either $L'EH'B_1$ or $L''FH''B_1$ or both contain a rectangle of size $(6 - 2\sqrt{2}) \times \sqrt{2}$, that is larger than $3.16 \times \sqrt{2}$. Then at least one of $L'EH'B_1$ or $L''FH''B_1$ will contain a disc of diameter $\sqrt{2}$. Such a disc always contains at least one integer point in its interior or four points on its border circle. Then this point or points must be cut off by a side v of P', as $v \neq u$.

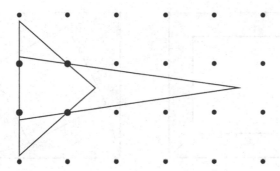

Fig. 3. Corresponding vertices of two minimally enclosing polygons may be arbitrarily far from each other

Thus it follows that the number of sides of P' is not less than the number of trapezoids, that is the number of sides of kP. Hence, kP is minimally enclosing for $(kP)_{\mathbb{Z}}$. □

An integer k satisfying $k \geq 2\delta_P^2$ is called *enclosure scaling factor* for P. The procedure of computing kP is called *enclosure scaling* of P.

Fig. 3 demonstrates that a set of integer points may admit different minimally enclosing reconstructions that are "far" from each other in terms of distance between corresponding vertices. The following theorem reveals the importance of an appropriate scaling of a polygon P for assuring that (i) any other minimally enclosing reconstruction of M is "close" to the digitized polygon P, and (ii) its digitization $M = P_{\mathbb{Z}}$ adequately approximates P.

Theorem 2. *Let k be an integer with $k \geq 2\delta_P^3$. By Theorem 1, kP is minimally enclosing for $(kP)_{\mathbb{Z}}$. Let P' be another minimally enclosing polygon for $(kP)_{\mathbb{Z}}$. Then there is a one-to-one correspondence between the vertices of kP and P', for which there exists a positive constant c, such that a vertex v of kP is corresponding to a vertex v' of P' if and only if $||v - v'|| \leq c\delta_P^2$.*

We conclude this section by one more remark. There are several algorithms that decompose a digital curve into digital straight segments [4,5]. These algorithms are straightforwardly adaptable to algorithms that linearize the border of digital objects and thus represent them as digital polygons. Theorems 1 and 2 can imply that the digital polygon $(kP)_{\mathbb{Z}}$ has the same number of "digital sides" as kP. The above-mentioned algorithms can also be used to efficiently obtain linear reconstruction of $(kP)_{\mathbb{Z}}$ that is minimally enclosing for $(kP)_{\mathbb{Z}}$.

3   Scaling II - Preserving Geometry for Partitioning

In this section our investigation goes deep into study of scaling that preserves intrinsic geometric features of the object, assuring, e.g., minimal decomposition with respect to the enclosed set of integer points. Using the obtained results, it

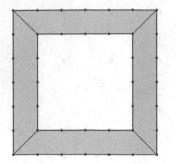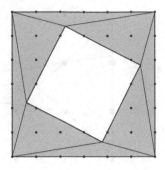

Fig. 4. Two minimally enclosing polygons for the same set of points. The one on the left is minimally decomposable while the one on the right is not.

is shown in Section 3.1 that the problem of optimal reconstruction by trapezoids of a 3D integer set is strongly NP-hard.

Let $P \subset \mathbb{R}^2$ be a polygon. By $\nu(P)$ we denote the minimal number of convex polygons (with vertices among those of P) into which P can be decomposed. We will say that P is *minimally decomposable* with respect to $M = P_{\mathbb{Z}}$ if $\nu(P)$ is minimal over all enclosing polygons for P. Fig. 4 shows that a minimally enclosing polygon may be non-minimally decomposable.

Finding $\nu(P)$ is a strongly NP-hard problem [9], so testing if a polygon is minimally decomposable or not is hard, in general. The following theorem states that any lattice polygon P can be scaled by a sufficiently large integer k, so that the obtained polygon kP is minimally decomposable.

Theorem 3. *For any integer $k \geq 2\delta_P^3$, $\nu(kP) \leq \nu(P')$ for any polygon P' (not necessarily a lattice one) that is enclosing for $(kP)_{\mathbb{Z}}$.*

The proof of the above theorem is based on a number of lemmas that admit lengthy technical proofs. An integer k satisfying $k \geq 2\delta_P^3$ is called *decomposition scaling factor* for P. The procedure of computing kP is called *decomposition scaling* of P. We will conclude this section by showing that the results of the last two sections can be used for studying the complexity of a 3D polyhedral reconstruction problem.

3.1 Corollary: The Optimal Polyhedral Reconstruction by Trapezoids Is Strongly NP-Hard

The *polyhedral reconstruction* problem is the 3D extension of the polygonal reconstruction considered earlier. Given a set $M \subset \mathbb{Z}^3$, one looks for a polyhedron P that is enclosing for M.[2] An enclosing polyhedron with a minimal number of facets is *minimally enclosing* for M. Usually, the 2-dimensional facets of P (2-facets, for short) are required to be convex polygons, as two adjacent polygons may be co-planar. Their number $f_2(P)$ is desired to be as small as possible.

---

[2] In practice, M is often obtained through "digitization" of some (usually unknown) set $S \subset \mathbb{R}^2$ of full dimension.

The main motivation for the above problem comes from medical imaging and other visualization problems where discrete volumes of voxels result from scanning and MRI techniques. Since digital medical images involve a huge number of points, it is quite problematic to apply traditional rendering or texture algorithms to obtain satisfactory visualization. Moreover, one can face difficulties in storing or transmitting data of that size. There are multiple sources of data being transmitted for many diverse uses, such as telemedicine, mine detection, tele-maintenance, ATR, visual display, cueing, and others. In all these applications the coding compression methodology used is paramount. For this, one can try to transform a discrete data set to a polyhedron P such that the number of its 2-facets is as small as possible. Such polyhedrizations are also searched for the purposes of geometric approximation of surfaces as well as for surface area and volume estimation.

It was recently shown that the optimization PR problem is strongly NP-hard [2]. Here we show that the following special variant of PR is NP-hard as well.

Optimal Polyhedral Reconstruction by Trapezoids (OptPRT)
Instance: A set $M \subset \mathbb{Z}^3$ and a bound $\beta \in \mathbb{Z}_+$.
Problem: Decide if there is a polyhedron P, such that $M = P_\mathbb{Z}$ and with no more than β facets that are either trapezoids or triangles, some of which may be co-planar.

We prove that OptPRT is strongly NP-hard by exhibiting a pseudopolynomial reduction to it from the following problem known to be strongly NP-complete [1].

Minimal Number Trapezoidal Partition (MNTP)
Instance: A simple polygon P (given by a sequence of pairs of integer-coordinate points in the plane) with non-rectilinear holes and a bound $\alpha \in \mathbb{Z}_+$.
Problem: Decide if P can be decomposed into no more than α trapezoids or triangles with vertices among those of P.

For example, the polygon in Fig. 4 (left) admits a minimal partition into four trapezoids, while the one in Fig. 4 (right) can be partitioned into eight triangles but does not admit partition involving trapezoids.

Note that MNTP is in P if holes are not allowed [3].

For getting acquainted with the notion of strong NP-hardness, pseudopolynomial reduction, and related matters the reader is referred to [6]. Here we only recall some basic points. Let $\Pi = (D_\Pi, Y_\Pi)$ and $\Pi' = (D_{\Pi'}, Y_{\Pi'})$ be decision problems with instance sets D_Π and $D_{\Pi'}$, respectively, and sets of instances with answer "yes" Y_Π and $Y_{\Pi'}$, respectively. Denote by $Max[I]$, $Length[I]$, $Max'[I']$, $Length'[I']$ the maximal number and the input length of the instances $I \in D_\Pi$ and $I' \in D_{\Pi'}$, respectively. A *pseudopolynomial reduction* from Π to Π' is a function $f : D_\Pi \to D_{\Pi'}$ such that:

(a) for all $I \in D_\Pi$, $I \in Y_\Pi$ iff $f(I) \in Y_{\Pi'}$,
(b) f can be computed in time polynomial in two variables: $Max[I]$ and $Length[I]$,

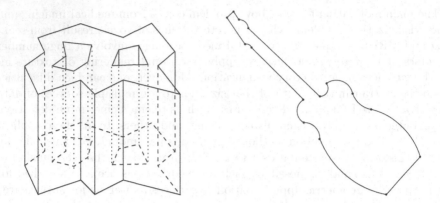

Fig. 5. *Left:* Illustration to the proof of Theorem 4. *Right:* An object whose boundary consists of convex curve and line segments. The joint point of two such segments is a "vertex" of the object boundary.

(c) there exists a polynomial q_1 such that, for all $I \in D_\Pi$, $q_1(Length'[f(I)]) \geq Length[I]$,

(d) there exists a two-variable polynomial q_2 such that, for all $I \in D_\Pi$, $Max'[f(I)] \leq q_2(Max[I], Length[I])$.

It is well-known [6] that if Π is strongly NP-hard and there is a pseudopolynomial reduction from Π to Π', then Π' is strongly NP-hard. We now sketch a proof of the following theorem.

Theorem 4. *OptPRT is strongly NP-hard.*

Proof. It is not hard to see that by Fact 1 and Lemma 1 the pseudopolynomial reduction conditions (b),(c), and (d) are all satisfied. We show next that condition (a) is met as well, i.e., given an instance $I = (X, \alpha)$ of MNTP, it is possible to construct an instance $I' = (M, \beta)$ of OptPRT such that the solution of I' is "yes" if and only if the solution of I is "yes."

We construct a polytope $P = X \times \tau \subset \mathbb{R}^2 \times \mathbb{R}^1$, i.e., $P \subset \mathbb{R}^3$ is the Cartesian product of X and an interval $\tau = [0, t]$ where t is an integer constant greater than two (see Fig. 5, left). Then let $M = P \cap \mathbb{Z}^3$. Finally, we set $\beta = 2\alpha + k$.

The so-constructed instance $I' = (P, 2\alpha + k)$ of OptPRT has solution "yes" iff the instance $I = (X, \alpha)$ of MNTP has solution "yes." In the one direction the proof is trivial: if the solution for I is "yes" then obviously so is the solution for I' since P can be decomposed into no more than $2\alpha + k$ trapezoids. To demonstrate the other direction, we have to make sure that if the set M can be represented as $M = P' \cap \mathbb{Z}^3$ for some polyhedron P' with no more than $2\alpha + k$ facets, then X can be partitioned into no more than α trapezoids. For this, it is enough to show that any such polyhedron $P' \neq P$ cannot have smaller number of convex facets than P, which follows from theorems 1 through 3. $\qquad \square$

4   Scaling of More General Objects

Consider a plane set $S \subset \mathbb{R}^2$ whose border $bd(S)$ consists of straight line segments and smooth strictly convex curve segments (line segments and curve segments, for short). A common point of two adjacent curve/line segments will be called a *vertex* of $bd(S)$. See Fig. 5 (right).

In addition to the scaling conditions for polygons that here apply to the line segments of S, we introduce the following. Given the set S, define a number γ_S (depending on S) as follows.

Let

$$\gamma_1^S = \max_{u,s}\{\rho(u,s)\}$$

where u denotes an arbitrary vertex and s an arbitrary curve/line segment of $bd(S)$. Let

$$\gamma_2^S = \max_{s_1,s_2}\{\rho(s_1,s_2)\}$$

where s_1, s_2 are arbitrary curve/line segments of $bd(S)$. Finally, let γ_3^S be the minimal number for which

$$r(\gamma_3^S s) \geq 3$$

for all curve segments s of $bd(S)$, where $r(s)$ is the minimal radius of curvature of s over all points of s. Then we set a scaling factor for S to be equal to

$$\gamma_S = \max\{\delta_S, \gamma_1^S, \gamma_2^S, \gamma_3^S, 3\}$$

Similar to the case of polygons, it can be shown that scaling by an integer k, that is a suitable low-degree polynomial of γ_S, assures a digitization $(kS)_{\mathbb{Z}}$ of kS that adequately approximates the shape of the original set S.

5   Concluding Remarks

In this paper we proposed scaling of plane figures (polygons or ones having curve segments) that leads to obtaining faithful digitizations. The scaling factors δ_P is efficiently computable. Computation issues regarding γ_S will be considered in a future work.

Clearly, both δ_P and γ_S are not necessarily the minimal possible factors that assure the properties stated in theorems 1,2, and 3. An open problem is to study the computational complexity of finding these minima.

Another interesting problem is the following. Given a polygon P, decide if it is minimally enclosing and/or whether it admits minimal decomposition with respect to $P_{\mathbb{Z}}$.

One can also look for other scaling procedures that assure digitizations preserving other desirable geometric properties of real objects.

Acknowledgements

The author thanks the three anonymous referees for their useful remarks and suggestions.

References

1. Asano, Ta., Asano, Te., Imai, H.: Partitioning a polygonal region into trapezoids. J. ACM 33, 290–312 (1986)
2. Brimkov, V.E.: Discrete volume polyhedrization: Complexity and bounds on performance. In: Tavares,, et al. (eds.) Computational Modelling of Objects Represented in Images: Fundamentals, Methods and Applications. Proceedings of the International Symposium CompIMAGE 2006, Coimbra (Portugal), October 21-22, pp. 117–122. Taylor & Francis, Abington (2006)
3. Chaselle, B., Dobkin, D.P.: Decomposing a polygon into its convex parts. In: Proc. 11th Annual ACM Sympos. on Theory Comput., pp. 38–48 (1979)
4. Debled-Renesson, I., Reveillès, J.-P.: A linear algorithm for segmentation of digital curves. International Journal of Pattern Recognition and Artificial Intelligence 9(4), 635–662 (1995)
5. Feschet, F., Tougne, L.: On the min dss problem of closed discrete curves. Discrete Applied Mathematics 151(1-3), 138–153 (2005)
6. Garey, M., Johnson, D.: Computers and Intractability. W.H. Freeman & Company, San Francisco (1979)
7. Klette, R., Rosenfeld, A.: Digital Geometry - Geometric Methods for Digital Picture Analysis. Morgan Kaufmann, San Francisco (2004)
8. Latecki, L.J., Rosenfeld, A.: Recovering a polygon from noisy data. Computer Vision and Image Understanding 86, 1–20 (2002)
9. Lingas, A.: The power of non-rectilinear holes. In: Nielsen, M., Schmidt, E.M. (eds.) ICALP 1982. LNCS, vol. 140, pp. 369–383. Springer, Heidelberg (1982)
10. Preparata, F.P., Shamos, M.I.: Computational Geometry: An Introduction. Springer, New York (1985)
11. Serra, J.: Image Analysis and Mathematical Morphology. Academic Press, London (1982)
12. Sivignon, I., Dupont, F., Chassery, J.-M.: Decomposition of three-dimensional discrete objects surface into discrete plane pieces. Algorithmica 38, 25–43 (2004)
13. Stelldinger, P., Latecki, L.J., Siqueira, M.: 3D object digitization: Topological equivalence between a 3D object and the reconstruction of its digital image. IEEE Transactions on Pattern Analysis and Machine Intelligence 29(1), 126–140 (2007)

Computing Admissible Rotation Angles from Rotated Digital Images

Yohan Thibault[1,2], Yukiko Kenmochi[1], and Akihiro Sugimoto[1,2]

[1] Université Paris-Est, LABINFO-IGM, UMR CNRS 8049, A2SI-ESIEE, France
[2] National Institute of Informatics, Japan
thibauly@esiee.fr, y.kenmochi@esiee.fr, sugimoto@nii.ac.jp

Abstract. Rotations in the discrete plane are important for many applications such as image matching or construction of mosaic images. In this paper, we propose a method for estimating a rotation angle such that the rotation transforms a digital image A into another digital image B. In the discrete plane, there are many angles that can give the rotation from A to B, called admissible angles for the rotation from A to B. For such a set of admissible angles, there exist two angles α_1, α_2 that are its upper and lower bounds. To find those upper and lower bounds, we use hinge angles as used in Nouvel and Rémila [5]. Hinge angles are particular angles determined by a digital image, such that any angle between two consecutive hinge angles gives the identical digital image after the rotation with the angle. Our proposed method obtains the upper and lower bounds of hinge angles from a given Euclidean angle and from a pair of digital images.

1 Introduction

Rotations in the discrete plane are required in many applications for image computation such as image matching or construction of mosaic images [4]. For the moment, the method to estimate the rotation angle is to approximate the rotation matrix by minimizing errors [4]. In the continuous plane, the Euclidean rotation is well defined and possesses the property of bijectivity. This implies that for two angles γ_1, γ_2 and a set of points A, if the Euclidean rotation of angle γ_1 applied to A gives the same result as the Euclidean rotation of angle γ_2 applied to A, then we have $\gamma_1 = \gamma_2$.

In the discrete plane, however, the property of bijectvity does not hold. To understand this reason, we have to first define the discretized Euclidean rotation, abbreviated to DER hereafter. DER is the discretization of the Euclidean rotation, namely, the application of the rounding function after applying the rotation matrix to a set of points. Thus two points in the Euclidean plane may give the same point in the discrete plane after the discretization. Because of this reason, two angles γ_1, γ_2 give the same result for a set of points A even if $\gamma_1 \neq \gamma_2$. In other words, we can define the admissible rotation angles S such that any angle in S gives the same rotation result for a set of points A. Note

V.E. Brimkov, R.P. Barneva, H.A. Hauptman (Eds.): IWCIA 2008, LNCS 4958, pp. 99–111, 2008.
© Springer-Verlag Berlin Heidelberg 2008

that S depends on A. Another way to define the admissible rotation angles is for two corresponding sets of points A and B, where B is the rotation of A by an unknown angle, to find the set S of angles which transforms A into B. The two most interesting angles in S are the upper and the lower bounds because with only these two angles we can deduce the other angles in S. Therefore, the aim of this paper is to find these two angles from a given rotation angle or from two given corresponding sets of grid points. Because we identify the exact bounds, we have to avoid computation with real numbers. Thus, in this paper, we only work with discrete geometry tools which guarantee to avoid computation errors. Moreover, because we assume that our data are discretized from continuous images of an object, the discrete rotation between two different sets of points has to give the same result as DER.

Some work on discrete rotations already exists. The first discrete rotation is the CORDIC algorithm [6]. Estimation of the rotation angle is done by addition or subtraction using pre-computed values to achieve the needed precision. It gives almost the same result as DER but an approximation of the angle. Andres described in [1],[2] some discrete rotations such as the rotation by discrete circles, the rotation by Pythagorean lines or the quasi-shear rotation. Computation done during these rotations are exact, but they are bijective. Thus they cannot give the same results as DER.

On the other hand, Nouvel and Rémila proposed in [5] another discrete rotation based on hinge angles which gives the same results as DER. It is known that hinge angles are particular angles determined by a digital image, such that any angle between two consecutive hinge angles gives the identical rotated digital image. This means that hinge angles correspond to the discontinuity of DER. Nouvel and Rémila showed that each hinge angle is represented by an integer triplet, so that any discrete rotation of a digital image is realized only with integer calculation. Because their algorithm gives the same results as DER, we see that hinge angles represented by integer triplets give sufficient information for executing any digital image rotation.

In this paper, we propose a discrete method for finding the lower and upper bounds of admissible rotation angles. Our method uses hinge angles, because we can obtain the same result as DER and they allow exact computations. The input data of our method is two sets of points A and B, where point correspondences across the two sets are known. The output of the algorithm is two hinge angles that give the lower and upper bounds of the admissible rotation angles for A and B.

In the following of this paper, we first introduce the notion of hinge angles and their properties. Then, we show how to obtain such a hinge angle from a given angle so that we can efficiently obtain a rotated digital image from the integer triplet. We then present a method for obtaining from a pair of digital images, two hinge angles which constitute the upper and lower bounds of the admissible rotation angles from the pair of digital images.

2   Hinge Angles

Let us consider points of $\mathbb{Z}^2$ as centers of pixels and rotate them such that the rotation center has integer coordinates. Hinge angles are particular angles which make some points of $\mathbb{Z}^2$ rotated to points on the frontier between adjacent pixels. In this section, we give the definition of hinge angles and their properties related to Pythagorean angles.

2.1   Definition of Hinge Angles

Let $\boldsymbol{x}$ be a point in $\mathbb{R}^2$ such that $\boldsymbol{x} = (x, y)$. We say that $\boldsymbol{x}$ has a semi-integer coordinate if $x - \frac{1}{2} \in \mathbb{Z}$ or $y - \frac{1}{2} \in \mathbb{Z}$. The set of points each of which has a semi-integer coordinate is denoted by $\mathscr{H}$, and is called the half grid. Thus, $\mathscr{H}$ represents the set of points on the frontiers of all pixels whose centroids are points in $\mathbb{Z}^2$.

Definition 1. *An angle α is called a hinge angle if at least one grid point in $\mathbb{Z}^2$ exists such that its image by the Euclidean rotation with α belongs to $\mathscr{H}$.*

Because $\mathscr{H}$ can be seen as the discontinuity of the rounding functions, hinge angles can be regarded as the discontinuity of the discretized Euclidean rotation. More simply, hinge angles determine a transit of a grid point from a pixel to its adjacent pixel during the rotation.

The following proposition is important because it shows that we can represent every hinge angle with three integers.

Proposition 1. *An angle α is a hinge angle if there is an integer triple (P, Q, K) such that*

$$2Q \cos\alpha + 2P \sin\alpha = 2K + 1. \tag{1}$$

The proof is given in [5].

Geometrically, a hinge angle α is formed by two rays that go through (P, Q) and a half-grid point such as $(K + \frac{1}{2}, \lambda)$ respectively sharing the origin as their endpoints as shown in Figure 1 (left). From this proposition, all calculations related to hinge angles can be done only with integers. Hereafter, α indicates a hinge angle.

We denote by $\alpha(P, Q, K)$ the hinge angle generated by an integer triple (P, Q, K). Setting $\lambda = \sqrt{P^2 + Q^2 - (K + \frac{1}{2})^2}$, the following equations can be easily derived from (1) and Figure 1 (left).

$$\cos\alpha = \frac{P\lambda + Q(K + \frac{1}{2})}{P^2 + Q^2}, \qquad \sin\alpha = \frac{P(K + \frac{1}{2}) - Q\lambda}{P^2 + Q^2}. \tag{2}$$

Note that we can have a half grid point $(\lambda, K + \frac{1}{2})$ instead of $(K + \frac{1}{2}, \lambda)$. In such a case, the above equations become

$$\cos\alpha = \frac{Q\lambda + P(K + \frac{1}{2})}{P^2 + Q^2}, \qquad \sin\alpha = \frac{P\lambda - Q(K + \frac{1}{2})}{P^2 + Q^2}. \tag{3}$$

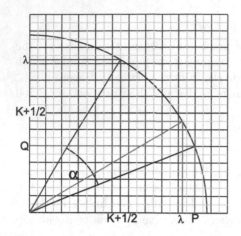
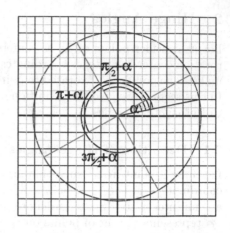

Fig. 1. A hinge angle $\alpha(P, Q, K)$ (left) and four symmetrical hinge angles (right)

The symmetries on hinge angles are important, because it allows us to restrict rotations in the first quadrant of the circle such that $\alpha \in [0, \frac{\pi}{2}]$.

Corollary 1. *Each triple (P, Q, K) corresponds to four symmetrical hinge angles such as $\alpha + \frac{\pi k}{2}$ where $k = 0, 1, 2, 3$.*

Figure 1(right) gives an example of Corollary 1. In order to distinguish the case $(K + \frac{1}{2}, \lambda)$ from the case $(\lambda, K + \frac{1}{2})$, we change the sign of K; we use $\alpha(P, Q, K)$ for the case of $(K + \frac{1}{2}, \lambda)$, and $\alpha(P, Q, -K)$ for the case of $(\lambda, K + \frac{1}{2})$. Note that the symmetries allow us to restrict α to the range $[0, \frac{\pi}{2}]$. Thus we know that K is always positive.

2.2 Properties Related to Pythagorean Angle

Because hinge angles are strongly related to Pythagorean angles, properties of Pythagorean angles are needed to prove some properties of hinge angles. Thus, we first give the definition of Pythagorean angles and their properties.

Definition 2. *An angle θ is called Pythagorean if both its cosine and sine belong to the set of rational numbers $\mathbb{Q}$.*

We can deduce from Definition 2 that each Pythagorean angle θ is defined by an integer triplet (a, b, c) such that

$$\cos\theta = \frac{a}{c}, \qquad \sin\theta = \frac{b}{c}. \tag{4}$$

In the following, θ indicates a Pythagorean angle. The following lemma is needed for the proof of the next proposition.

Lemma 1. *Let (a, b, c) be an integer triplet generating a Pythagorean angle with $|a| < |b| < |c|$. If $\gcd(a, b, c) = 1$, then c is odd.*

Proof. We assume that c is even such that $c = 2d$ where d in $\mathbb{Z}$. Then we obtain a and b are both odd because of $a^2 + b^2 = c^2 = (2d)^2$. Otherwise, we would have $\gcd(a, b, c) = 2n$ for $n \in \mathbb{Z}$. Setting $a = 2e + 1$ and $b = 2f + 1$ where $e, f \in \mathbb{Z}$, we obtain $(2e + 1)^2 + (2f + 1)^2 = 4d^2$, which can be rewritten by $e^2 + e + f^2 + f + \frac{1}{2} = d^2$. This indicates that d does not belong to $\mathbb{Z}$. We therefore conclude that c is odd.

If $\gcd(a, b, c) = i$, then $\gcd(\frac{a}{i}, \frac{b}{i}, \frac{c}{i}) = 1$ and the triple of integers $(\frac{a}{i}, \frac{b}{i}, \frac{c}{i})$ generates the same Pythagorean angle as (a, b, c).

Proposition 2. *Let E_h be the set of hinge angles and E_p be the set of Pythagorean angles. Then we have $E_h \cap E_p = \emptyset$.*

Proof. Assume that there exists an angle α such that $\alpha \in E_h$ and $\alpha \in E_p$. Since α in E_p, we can find an integer triplet (a, b, c), generating α such that $\gcd(a, b, c) = 1$. By substitution of (4) in (2), we obtain

$$2\frac{Qa + Pb}{c} = 2K + 1, \tag{5}$$

from which we derive $2\frac{Qa+Pb}{c} \in \mathbb{Z}$. Because we know that c is odd according to Lemma 1, we obtain $\frac{Qa+Pb}{c} \in \mathbb{Z}$. However, this contradicts the fact that for any pair n, m in $\mathbb{Z}$, we never have $2n = 2m + 1$. Therefore α cannot belong to both E_h and E_p simultaneously.

This proposition shows that it is not possible to rotate from a point (i, j) in $\mathbb{Z}^2$ to a point (x, y) such as $x = i + \frac{1}{2}$, $y = j + \frac{1}{2}$, where $(i, j) \in \mathbb{Z}^2$, if the angle of the rotation is a hinge angle.

3 Computing the Lower Bound Hinge Angle from a Pythagorean Angle

In this section, we propose a method for computing a lower bound hinge angle α_1 from a given angle for rotating a given digital image. Note that with minor modifications, this method can also find the upper bound hinge angle α_2, and thus, by applying twice this method, we can obtain two hinge angles that enclose the given angle. The set S of angles γ such that $\alpha_1 \leq \gamma \leq \alpha_2$ is called admissible rotation angles, denoted by ARA. All rotations of the given digital image done by an angle in S give the same result. Nouvel and Rémila presented a method to compute all possible hinge angles for a grid point or a pixel in a digital image [5]. Their method can be used for finding our interesting hinge angle which is the lower bound of the admissible rotation angles. Its time complexity is $O(n \log(n))$ where n is the number of all hinge angles for a given grid point. Note that n depends on the coordinates of the grid point. In subsection 3.1, we improve their method by using a tree structure for hinge angles, so that our method brings the complexity $O(\log(n))$. In subsection 3.2, we present a method for finding the lower bound hinge angle for a given digital image, namely, for all pixels in the image.

3.1   Computing the Lower Bound Hinge Angle for a Grid Point

For each grid point $p = (P, Q)$ in $\mathbb{Z}^2$, there are less than $n = \lfloor \sqrt{P^2 + Q^2} + \frac{1}{2} \rfloor$ different hinge angles [5]. This means that we have a sequence of $K_i, i = 0, 1, .., n - 1$ in $\mathbb{Z}$, where $0 \leq K_i < n$ for each p. Because we can compare any pair of associated hinge angles $\alpha_i(P, Q, K_i)$, we obtain a totally ordered set $\{\alpha_1(P, Q, K_1), \alpha_2(P, Q, K_2), ..., \alpha_{max}(P, Q, K_{max})\}$ in the ascending order such that $\alpha_1 < \alpha_2 < ... < \alpha_{max}$. Given a Pythagorean angle θ, in order to find the lower bound hinge angle α_i such that $\alpha_i < \theta < \alpha_{i+1}$, we use a tree structure. Binary search allows us to find α_i in $O(\log(n))$, providing that we can compare a hinge angle with a Pythagorean angle in a constant time. The algorithm is described in Figure 2.

```
Function: Find a hinge angle
    Input (Point p(P,Q), Pythagorean angle θ)
    Output (α(P,Q,K))
    var K_max = ⌊√(P² + Q²) − 1⌋;
    var K_min = 0;
    var K = ⌊(K_max+K_min)/2⌋;
    While (K_max − K_min ≠ 1)
        if (α(P,Q,K) > θ)
            K_max = K;
        else
            K_min = K;
        K = ⌊(K_max+K_min)/2⌋;
    end while
    return α(P,Q,K);
```

Fig. 2. Function for finding a hinge angle

The following proposition shows that the comparison between a hinge angle and a Pythagorean angle is executed in a constant time.

Proposition 3. *Let α be a hinge angle and θ be a Pythagorean angle. We can check whether $\alpha > \theta$ in a constant time with integer calculation.*

Proof. Let $\alpha(P, Q, K)$ be a hinge angle in $[0, \frac{\pi}{2}]$ and $\theta(a, b, c)$ be a Pythagorean angle in $[0, \frac{\pi}{2}]$. From (2) and (4), we obtain

$$\cos \alpha - \cos \theta = \frac{P(K + \frac{1}{2}) + Q\lambda}{P^2 + Q^2} - \frac{a}{c}.$$

If θ is greater than α, $\cos \alpha - \cos \theta > 0$. Thus

$$cP(2K + 1) - 2a(P^2 + Q^2) > -2cQ\lambda. \tag{6}$$

Since we know that c, Q, λ are positive, the right-hand side of (6) is always negative. Thus, if the left-hand side of (6) is not negative, then $\theta > \alpha$. Otherwise, we take squares of (6), so that we only have to check whether the following inequality holds:

$$\left[cP(2K+1) - 2a(P^2 + Q^2)\right]^2 < 4c^2 Q^2 \lambda^2. \tag{7}$$

Note that because $\lambda = \sqrt{P^2 + Q^2 - (K + \frac{1}{2})^2}$, we see that $4\lambda^2$ in the right-hand side of (7) contains only integer values. Therefore, we can also verify (7) with integer calculation. If it is true, $\theta > \alpha$; otherwise $\alpha > \theta$. Note that because of Proposition 2, it is impossible to obtain $\theta = \alpha$.

We mention the importance of the rotation with angle $\frac{\pi}{2}$ and its multiples. In fact, if the angle of a rotation is equal to $\frac{\pi}{2}, \pi, \frac{3\pi}{2}$, we just have to flip x and/or y-coordinates by changing their signs. It gives the reason that we can restrict the input angle θ to $0 < \theta < \frac{\pi}{2}$.

3.2 Computing the Lower Bound Hinge Angle for a Set of Points

In this subsection, we present an algorithm, based on the previous one, for computing the lower bound hinge angle from a given Pythagorean angle θ for a digital image consisting of m grid points such that $A = \{p_1, p_2, ..., p_m\}$. The output is a triplet of integers that represents a hinge angle. The algorithm computes all hinge angles for all points in A, and sorts them to keep the largest one. More precisely, we first compute the lower bound hinge angle for the first point of A, and store it as the reference. Then, we compute the hinge angle for the second point in A and compare it with the reference to keep the larger one. After repeating this procedure for all points in A, our algorithm returns the lower bound hinge angle α such that $\alpha < \theta$. The time complexity of this algorithm is $O(m \log(n))$ because we call m times the function of binary search (Figure 2) whose time complexity is $O(\log(n))$. Figure 3 illustrates our algorithm. As shown in the following proposition, the comparison between two hinge angles is realized in a constant time, so that it does not change the global complexity.

Proposition 4. *Let α_1, α_2 be two hinge angles. We can check if $\alpha_1 > \alpha_2$ in a constant time and with full integer calculation.*

The proof is similar to that of Proposition 3.

Note that our input is a Pythagorean angle, as the one in [5], in this paper. However, we can replace it by an Euclidean angle because there exists a method in linear time complexity $O(m)$ to approximate a given Euclidean angle with a Pythagorean angle with a precision of $\frac{1}{10^m}$ [3].

4   Digital Image Rotation by a Hinge Angle

In this section, we present an algorithm for rotating a digital image with a given lower bound hinge angle, which is obtained by the algorithm described in Subsection 3.2. It is already proved in [5] that we can obtain the same result as the DER with respect to the original angle. Note that our input is a hinge angle

```
Function: Find hinge angle for a digital image
    Input (Digital image A, Pythagorean angle θ)
    Output (hinge angle)
    var HA, HA_temps \* hinge angle *\;
    HA = Find hinge angle(first point of A, angle
of rotation);
    for each p ∈ A\{p₁}
        HA_temps = Find hinge angle(p, θ);
        if (HA < HA_temps) HA = HA_temps;
    end for
    Return (HA);
```

Fig. 3. Function for finding the lower bound hinge angle for a digital image

```
Function: Discrete rotation
    Input (a digital image A, a hinge angle α)
    Output (Rotated image A')
    var HA : hinge angle;
    for each p ∈ A
        HA = Find hinge angle(p, α);
        move p to (K, ⌊λ + ½⌋) or (⌊λ + ½⌋, K),
depending on the sign of K and put it to A' ;
    end for
    Return (A');
```

Fig. 4. Discrete rotation algorithm by a hinge angle

and the input of the algorithm presented in Figure 2 is a Pythagorean angle. In spite of this difference, we can apply the same algorithm thanks to Proposition 4. The algorithm is presented in Figure 4. It supposes that the center of rotation is the origin. For each point, it calls the function of binary search (Figure 2) to find the corresponding hinge angle, which designates its new position. If we consider n as the biggest coordinate of all points in A, we can assume that there are less than $4n^2$ points in A. Thus we can conclude that the complexity of our algorithm is $O(n^2 \log(n))$. The first advantage of our method is that it does not require any float number calculation. The second advantage is that the exact rotation of the digital image is obtained with only an integer triplet. We need neither matrices nor angles for realizing the rotation.

5   Obtaining Admissible Rotation Angles from Two Digital Images

Let us assume that a set of grid points in the first image and its corresponding set in the second image are given: $A = \{p_1, p_2, ..., p_l\}$ and $B = \{q_1, q_2, ..., q_l\}$ are given where p_i corresponds to q_i. Given A and B, we obtain a hinge angle pair α_1, α_2, such that $\alpha_1 \leq \gamma < \alpha_2$ where γ is an admissible rotation angle consistent with the point correspondences between A and B. Hereafter, we assume that A is the original point set and B is the rotated point set by angle γ. In this section, we show how to obtain the ARA from these two digital images.

To simplify the notation, we denote by $ARA(p_i, q_i) = (\alpha_{i1}, \alpha_{i2})$ the pair of angles, which gives the lower and the upper bounds of possible angles of the rotation for the pair of points (p_i, q_i). Note that the angles α_{i1}, α_{i2} are hinge angles. $ARA(A_{n+1}, B_{n+1})$ denotes the two most restrictive angles for all point i such as $i \leq n + 1$. We formally define it by $ARA(A_{n+1}, B_{n+1}) = ARA(A_n, B_n) \cap ARA(p_{n+1}, q_{n+1})$.

5.1   Setting Rotation Centers

For any rotation, we need to set a rotation center. In this paper, we choose one of the grid points in a digital image for the rotation center. Assuming centers are p_1 and q_1 for A and B respectively, we define two functions $\mathscr{T}_A$ and $\mathscr{T}_B$ such that

$$\mathscr{T}_A(p_i) = p_i - p_1,$$
$$\mathscr{T}_B(q_i) = q_i - q_1,$$

for all $p_i \subset A, q_i \in B$, so that we can consider the rotation centers to be the origin after the translations. Hereafter, we will use new sets of points $A' = \{\mathscr{T}_A(p_1), \mathscr{T}_A(p_2), ..., \mathscr{T}_A(p_l)\}$ and $B' = \{\mathscr{T}_B(q_1), \mathscr{T}_B(q_2), ..., \mathscr{T}_B(q_l)\}$ instead of A and B. However, for simplicity, we will denote them by $A = \{p_1, p_2, ..., p_l\}$ and $B = \{q_1, q_2, ..., q_l\}$.

5.2   Computing Hinge Angles from Two Corresponding Point Pairs

In this subsection, we consider the case with $A = \{p_1, p_2\}$ and $B = \{q_1, q_2\}$ where $p_i = (P_i, Q_i)$ and $q_i = (R_i, S_i)$. Let us first define a circle $\mathscr{C}(p_2)$ with center p_1 that goes through p_2. Thus the radius of $\mathscr{C}(p_2)$ is $r = d(p_1, p_2)$ where $d(p_1, p_2)$ is the Euclidean distance between p_1 and p_2. Let us consider the half grid around q_2 such that

$$\mathscr{H}(q_2) = \{(x, y) \in \mathscr{H} : S_2 - \frac{1}{2} \leq y \leq S_2 + \frac{1}{2} \text{ when } x = R_2 + \frac{1}{2},$$
$$R_2 - \frac{1}{2} \leq x \leq R_2 + \frac{1}{2} \text{ when } y = S_2 + \frac{1}{2}\}.$$

Setting p_1 and q_1 to be the rotation centers, for finding a hinge angle pair, we need to detect intersections between $\mathscr{C}(p_2)$ and $\mathscr{H}(q_2)$. In other words, we study corners of $\mathscr{H}(q_2)$ in the interior of $\mathscr{C}(p_2)$. Setting four corners of $\mathscr{H}(q_2)$ such that $C_1(q_2) = (R_2 - \frac{1}{2}, S_2 - \frac{1}{2}), C_2(q_2) = (R_2 - \frac{1}{2}, S_2 + \frac{1}{2}), C_3(q_2) = (R_2 + \frac{1}{2}, S_2 + \frac{1}{2}), C_4(q_2) = (R_2 + \frac{1}{2}, S_2 - \frac{1}{2})$ as shown in Figure 5, we define a binary function $\mathscr{F}$ such as

$$\mathscr{F}(C_i(q_2)) = \begin{cases} 0 & \text{if } C_i(q_2) \text{ is outside of } \mathscr{C}(p_2), \\ 1 & \text{otherwise.} \end{cases}$$

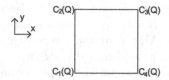

Fig. 5. The half grid $\mathscr{H}(q)$, namely a pixel around q and its four corners

In order to obtain $\mathscr{F}(C_i(q_2))$ with integer calculation, we compare each of $4((R_2 \pm \frac{1}{2})^2 + (S_2 \pm \frac{1}{2})^2)$ with $4r^2$. Note that we may assume that the intersection between $\mathscr{C}(p_2)$ and $\mathscr{H}(q_2)$ is not null. If no intersection between $\mathscr{C}(p_2)$ and $\mathscr{H}(q_2)$ exists, then p_2 and q_2 are not corresponding.

Proposition 5. *If two points p_2 and q_2 are corresponding, a circle $\mathscr{C}(p_2)$ and a pixel boundary $\mathscr{H}(q_2)$ always have two distinct intersections.*

The mathematically rigorous proof is omitted in this paper because of the page limitation. The proof is accomplished by distinguishing the following two cases; in the other cases, we have always two intersections.

The first case is that $\mathscr{C}(p_2)$ goes through one of the four corners of $\mathscr{H}(q_2)$. Because any angle between p_2 and $C_i(q_2)$ at the origin is a Pythagorean angle it cannot be a hinge angle from Proposition 2. Thus, this case never happens.

The second case is that $\mathscr{C}(p_2)$ and $\mathscr{H}(q_2)$ have the unique intersection on one of edges of $\mathscr{H}(q_2)$. This case may happen only when one coordinate of q_2 is zero. A circle centered at the origin can cross twice a half grid parallel to one of the axes if and only if the circle arc between those intersections cuts another axis. Therefore, if the intersection is single, it should be on an axis, so that λ should be null. However, it is impossible by the definition of hinge angles.

From Proposition 5, we always have two intersections between $\mathscr{C}(p_2)$ and $\mathscr{H}(q_2)$, and see that there are four cases corresponding to different possibilities to have 0,1,2 or 3 corners in the interior of $\mathscr{C}(p_2)$, as illustrated in Figure 6.

- Case A: $\mathscr{C}(p_2)$ includes no corner. Thus we have $\mathscr{F}(C_i(q_2)) = 0$ for all $i = 1, 2, 3, 4$, similarly to the above second impossible case. This case can only happen when $R_2 = 0$ or $S_2 = 0$. Supposing that R_2 and S_2 are not null, we assume that they are positive. In the first quadrant, the y-coordinate (respectively x-coordinate) of points in $\mathscr{C}(p_2)$ is strictly decreasing with respect to x (respectively y). Thus it cannot intersect twice a line parallel to the x-axis (respectively y-axis). Therefore, if $S_2 = 0$, $ARA(p_2, q_2) = (\alpha_{21}(P_2, Q_2, R_2 - 1), \alpha_{22}(P_2, Q_2, R_2 - 1))$. In this case the two hinge angles are symmetrical with respect to the y-axis.
- Case B: $\mathscr{C}(p_2)$ includes only one corner. For example, if $C_1(q_2)$ is in the circle, we obtain $ARA(p_2, q_2) = (\alpha_{21}(P_2, Q_2, R_2 - 1), \alpha_{22}(P_2, Q_2, -S_2 + 1))$.
- Case C: $\mathscr{C}(p_2)$ includes two corners that should have one common coordinate. For example, if $C_1(q_2)$ and $C_2(q_2)$ are in the circle, we obtain $ARA(p_2, q_2) = (\alpha_{21}(P_2, Q_2, -S_2 + 1), \alpha_{22}(P_2, Q_2, -S_2))$.

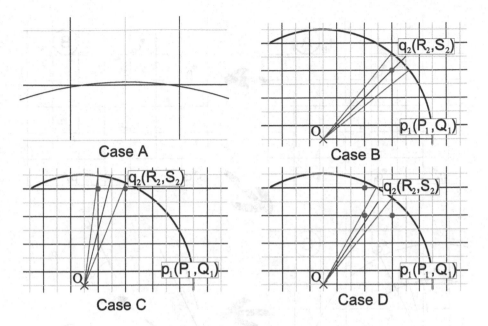

Fig. 6. Illustration of cases A,B,C and D

- Case D: $\mathscr{C}(p_2)$ includes three corners. For example, if $C_1(q_2), C_2(q_2)$ and $C_4(q_2)$ are in $\mathscr{C}(q_2)$, then we obtain $ARA(p_2, q_2) = (\alpha_{21}(P_2, Q_2, R_2), \alpha_{22}(P_2, Q_2, -S_2))$.

The main function of our algorithm for finding the two hinge angles consist of three steps. The first step sets the rotation center at p_1 and q_1, as described in the previous subsection. The second step computes which corners are in the interior of $\mathscr{C}(q_2)$ and then stocks the result as an index. The index is calculated by $index = \sum_i 2^i \times \mathscr{F}(C_i(q_2))$. Therefore we can easily identify which corners are in the interior of $\mathscr{C}(p_2)$ from the index. The third step calls a function that returns hinge angles corresponding to the index. There exist fourteen possible values for the index. Note that geometrically the index can be neither 5 nor 10. The index value 15 implies an error such that all corners are in the interior of $\mathscr{C}(q_2)$. Since the index value 0 corresponds to the case A, we should verify whether $\mathscr{H}(q_2)$ really intersects with $\mathscr{C}(q_2)$. Note that for all other index values, we can make a pair (d, e) such that $d + e = 15$. The two indices of such a pair design the same pair of hinge angles. Each step of this algorithm has the constant time complexity. Thus the global complexity of this algorithm is also O(1).

5.3   Incremental Hinge Angle Computing

In general, the corresponding point sets contain more than two points. Therefore, in this section, we extend our algorithm in the previous section to two sets of corresponding point pairs, A and B, each of which has l points where $l > 2$.

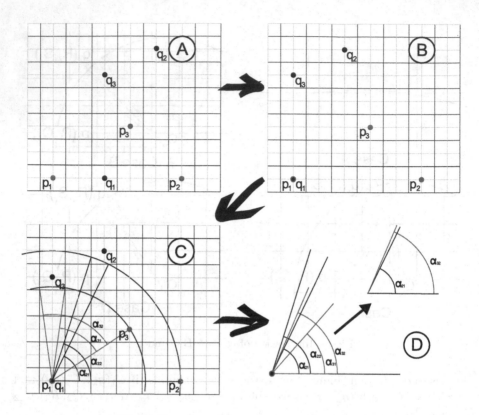

Fig. 7. Running of the incremental algorithm

A new algorithm handles all points incrementally. This algorithm is divided into two parts. The first part is to initialize the algorithm by computing ARA (p_2, q_2). Note that $ARA(p_1, q_1)$ cannot be computed because p_1 and q_1 are the centers of the rotation. The second part computes $ARA(A_{i+1}, B_{i+1})$ for $i = 2, \ldots, n-1$. The time complexity of this algorithm is O(l). As explained in the previous subsection, the function is realized in a constant time O(1). Moreover, as explained in Section 3, we can compare two hinge angles in a constant time O(1). Therefore, the full computation of this algorithm for l points takes the time complexity of $l \times (O(1) + O(1)) = O(l)$.

5.4   Example of the Running of the Algorithm

Figure 7 gives an example of the incremental algorithm for two sets of three points. Given input data of the algorithm as shown in Figure 7 (A), we first obtain the result of the translation described in subsection 5.1 as illustrated in (B). We then compare, for each pair of points (p_i, q_i) with $i \geq 2$, the distance of p_i from the origin with the distance of each corner from $\mathcal{H}(q_i)$ to deduce the corresponding hinge angle as explained in subsection 5.2. Finally, we obtain (D) which shows the intersection of all $ARA(p_i, q_i)$ obtained in (C).

6   Conclusion

In this paper, we have shown how to obtain a hinge angle which is the lower bound approximation to a given Euclidean angle. We have also shown that we can efficiently obtain a rotated digital image from the integer triplet identically to that from the Pythagorean angle. We then have presented a method for obtaining the upper and lower bounds of the ARA from a pair of digital images.

The future work will extend our proposed method into two directions. The first direction is to extend this algorithm to the 3D case. The second direction is to create a 2D matching algorithm based on hinge angles. Current methods for matching can be improved by restricting the searching area. The admissible rotation angles obtained by our method will be useful for the restriction of the searching area.

Acknowledgment

This work was supported by grants from Région Ile-de-France.

References

1. Andres, E.: Cercles Discrets et Rotations Discrétes. Thesis, Université Louis Pasteur Strasbourg (1992)
2. Andres, E.: The Quasi-Shear Rotation. In: Miguet, S., Ubéda, S., Montanvert, A. (eds.) DGCI 1996. LNCS, vol. 1176, pp. 307–314. Springer, Heidelberg (1996)
3. Anglin, W.S.: Using Pythagorean triangles to approximate angles. American Mathematical Monthly 95, 540–541 (1988)
4. Hansen, M., Anandan, P., Dana, K., van der Wal, G., Burt, P.: Real-time scene stabilization and mosaic construction. In: 2nd IEEE workshop on Applications of Computer Vision, pp. 54–64 (1994)
5. Nouvel, B., Rémila, E.: Incremental and transitive discretized rotations. In: Reulke, R., Eckardt, U., Flach, B., Knauer, U., Polthier, K. (eds.) IWCIA 2006. LNCS, vol. 4040, pp. 199–213. Springer, Heidelberg (2006)
6. Volder, J.E.: The CORDIC Trigonometric Computing Technique. IRE Transactions on Electronic Computers EC-8, 330–334 (1959)

On the Number of hv-Convex Discrete Sets

Péter Balázs*

Department of Image Processing and Computer Graphics
University of Szeged
Árpád tér 2, H-6720 Szeged, Hungary
pbalazs@inf.u-szeged.hu

Abstract. One of the basic problems in discrete tomography is the reconstruction of discrete sets from few projections. Assuming that the set to be reconstructed fulfills some geometrical properties is a commonly used technique to reduce the number of possibly many different solutions of the same reconstruction problem. The class of hv-convex discrete sets and its subclasses have a well-developed theory. Several reconstruction algorithms as well as some complexity results are known for those classes. The key to achieve polynomial-time reconstruction of an hv-convex discrete set is to have the additional assumption that the set is connected as well. This paper collects several statistics on hv-convex discrete sets, which are of great importance in the analysis of algorithms for reconstructing such kind of discrete sets.

Keywords: discrete tomography, hv-convex discrete set, connectedness, analysis of algorithms.

1 Introduction

Discrete tomography (DT) [15,16] aims to reconstruct a discrete set (a finite subset of the two-dimensional integer lattice defined up to translation) from the number of its elements lying on the same line along several (usually horizontal, vertical, diagonal, and antidiagonal) directions, called projections. It has several applications in pattern recognition, image processing, electron microscopy, angiography, non-destructive testing, and so on. The main challenge in DT is that practical limitations every time reduce the number of available projections to at most about four – which results in a possibly extremely large number of solutions of the same reconstruction task. This can cause the reconstructed discrete set to be quite different from the original one. In addition, the reconstruction problem can be NP-hard, depending on the number and directions of the projections. One way of eliminating these problems is to suppose that the set to be reconstructed has some geometrical properties. In this way we can reduce the search space of the possible solutions and we can achieve fast and rare ambiguous reconstructions.

---

* This work was supported by OTKA grant T48476.

V.E. Brimkov, R.P. Barneva, H.A. Hauptman (Eds.): IWCIA 2008, LNCS 4958, pp. 112–123, 2008.

The class of *hv*-convex discrete sets and its subclasses are very frequently studied in DT. The reconstruction in those classes has a well-developed theory including heuristic and exact reconstruction algorithms as well as some important complexity results. It turned out, that the key to achieve polynomial-time reconstruction of an *hv*-convex discrete set is to have the additional assumption that the set is connected as well.

In this paper we describe a method for counting elements of several subclasses of the *hv*-convex class. The paper is structured as follows. First, the necessary definitions are introduced in Section 2. In Section 3 we describe recursive formulas for counting *hv*-convex discrete sets, possibly with certain additional properties. After that, in Section 4 we collect some statistics that can affect the complexity of several reconstruction algorithms developed for the *hv*-convex class. Section 5 is for the conclusion.

2 Definitions

The finite subsets of the 2D integer lattice are called *discrete sets*. The *size* of a discrete set is defined by the size of its minimal bounding discrete rectangle (i.e. not the number of its elements). A discrete set F of size $m \times n$ is defined up to a translation and it is usually represented by a binary picture formed from unitary cells (see Fig. 1). We refer to the topmost row of the discrete set as the first row, and to the leftmost column of the set as the first column. Thus, the upper left corner of the minimal bounding rectangle of a discrete set is always the $(1, 1)$ position, and the remaining positions of the minimal bounding rectangle (and of the discrete set as well) are addressed consequently.

A discrete set F is *4-connected* (*8-connected*), if for any two positions $P \in F$ and $Q \in F$ of the set there exist a sequence of distinct positions $(i_0, j_0) = P, \ldots, (i_k, j_k) = Q$ such that $(i_l, j_l) \in F$ and $|i_l - i_{l+1}| + |j_l - j_{l+1}| = 1$ $(\max\{|i_l - i_{l+1}|, |j_l - j_{l+1}|\} = 1)$ for each $l = 0, \ldots, k - 1$ (see Figs. 2a and 2b). The 4-connected sets are also called polyominoes [14]. If the discrete set is not 4-connected then it consists of several polyominoes. The maximal 4-connected subsets of a discrete set F are called the *components of F*. For, example the discrete set in Fig. 1 has three components: $\{(1, 4), (2, 4)\}$, $\{(2, 2), (3, 2), (4, 1), (4, 2)\}$, and $\{(3, 5), (4, 5), (4, 6), (5, 6), (6, 5), (6, 6)\}$. A discrete set is called *horizontally and vertically convex* (shortly, *hv*-convex) if all the rows and columns of the

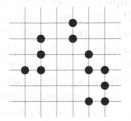
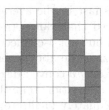

Fig. 1. A discrete set represented by its elements (*left*) and a binary picture (*right*)

set are 4-connected (see Figs. 2c, 2d, and 2e). Let us introduce the following notations for some classes of hv-convex sets:

- $\mathcal{P}$ for the class of hv-convex polyominoes;
- $\mathcal{Q}$ for the class of hv-convex 8-connected discrete sets;
- $\mathcal{HV}'$ for the class of hv-convex discrete sets with nonempty rows and columns;
- $\mathcal{HV}$ for the class of hv-convex discrete sets which possibly can have empty rows and columns.

The following inclusions are trivial,

$$\mathcal{P} \subset \mathcal{Q} \subset \mathcal{HV}' \subset \mathcal{HV} . \tag{1}$$

A polyomino F is *northeast directed* (NE-directed for short) if there is a particular point $P \in F$ such that for each point $Q \in F$ there is a sequence $P_0 = P, \ldots, P_t = Q$ of distinct points of F such that each point P_l of the sequence is north or east of P_{l-1} for each $l = 1, \ldots, t$ (see Fig. 2f). Similar definitions can be given for SW-, SE-, and NW-directedness. An hv-convex polyomino is called *NW/NE-parallelogram polyomino* if it is both NW- and SE-directed or both NE- and SW-directed, respectively (see Fig. 2g).

3   Enumeration of hv-Convex Discrete Sets

The class of hv-convex discrete sets is one of the most important classes in discrete tomography. Although the reconstruction from two projections in this class is NP-hard [20] several methods can solve this problem by applying some heuristic [17], metaheuristic [8] or optimization [11] technique. Besides, for hv-convex polyominoes and hv-convex 8-connected sets different polynomial-time reconstruction algorithms have been developed. One of them approximates the solution iteratively by a nondecreasing sequence of so-called kernel sets and by a nonincreasing sequence of so-called shell sets (see [6,5,9]). This algorithm has a worst case time complexity of $O(mn \cdot \log mn \cdot \min\{m^2, n^2\})$. An other approach is based on an observation that the reconstruction task can be transformed into a 2SAT task that is solvable in polynomial time [10,18]. This latter algorithm has a worst case time complexity of $O(mn \cdot \min\{m^2, n^2\})$. In [4] the two algorithms were compared, and the observations concerning the average execution times of the two reconstruction approaches led to the design of a hybrid reconstruction algorithm that has the same worst case time complexity of $O(mn \cdot \min\{m^2, n^2\})$ and remains fast in the average case as well. Recently, an algorithm has been also published that can perform the reconstruction in the class of hv-convex 8-connected but not 4-connected discrete sets in $O(mn \cdot \min\{m, n\})$ time [3].

Summarizing the above-mentioned contributions we can say that the reconstruction of an hv-convex discrete set is in general a difficult problem but the additional information that the set satisfies some connectedness properties as well can adequately facilitate the reconstruction. Now, one can naturally pose the question whether an hv-convex discrete set – chosen randomly using a uniform distribution – often fulfills some connectedness properties as well. To answer

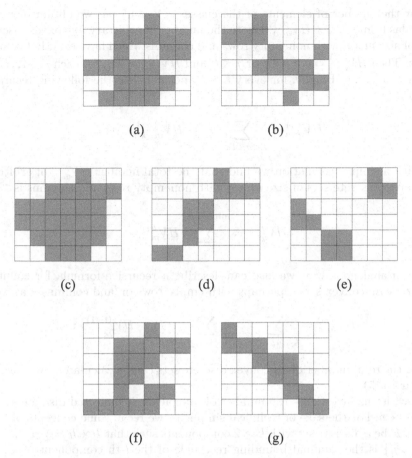

(a)   (b)

(c)   (d)   (e)

(f)   (g)

Fig. 2. (a) a polyomino, (b) an 8-connected discrete set, (c) an hv-convex polyomino, (d) an hv-convex 8-connected discrete set, (e) a general hv-convex discrete set, (f) an NE-directed polyomino, and (g) an NW-parallelogram polyomino

this question we have to identify the cardinality of the class of hv-convex discrete sets and its subclasses.

Regarding the class $\mathcal{P}$ we already have nice closed formulas for describing the number of hv-convex polyominoes according to several parameters. In [12] it was proved that the number P_{n+4} of hv-convex polyominoes with a semiperimeter value of $n+4$ is

$$P_{n+4} = (2n + 11)4^n - 4(2n + 1) \binom{2n}{n}. \tag{2}$$

Later, based on the above result in [13] it was shown that the number $P_{m+1,n+1}$ of hv-convex polyominoes of size $(m + 1) \times (n + 1)$ is

$$P_{m+1,n+1} = \frac{m + n + mn}{m + n} \binom{2m + 2n}{2m} - \frac{2mn}{m + n} \binom{m + n}{m}^2. \tag{3}$$

For the number of elements of the classes $\mathcal{HV}'$ and $\mathcal{HV}$ we obtain recursive formulas from [1]. Let $HV_{m,n}^{\prime(t)}$ denote the number of arbitrary hv-convex discrete sets of size $m \times n$ with nonempty rows and columns which have exactly t components. Then $HV_{i,j}^{\prime(t)} = 0$ if $i < t$ or $j < t$, and $HV_{i,j}^{\prime(1)} = P_{i,j}$ for each $i = 1, \ldots, m$ and $j = 1, \ldots, n$. Finally, for every $t > 1$ and $m, n \geq 1$ the following recursion holds

$$HV_{m,n}^{\prime(t)} = \sum_{k<m,\ l<n} P_{k,l} \cdot HV_{m-k,n-l}^{\prime(t-1)} \cdot t \ . \tag{4}$$

With a simple calculation we find that the total number $HV_{m,n}'$ of arbitrary hv-convex discrete sets of size $m \times n$ with nonempty rows and columns is

$$HV_{m,n}' = \sum_{t=1}^{\min\{m,n\}} HV_{m,n}^{\prime(t)} \ . \tag{5}$$

In an analogous way, we also can describe a recursive formula for counting arbitrary hv-convex sets – perhaps with empty rows or/and columns – as well.

$$HV_{m,n}^{(t)} = \sum_{k<m,\ l<n} \ \sum_{i\leq m-k,\ j\leq n-l} P_{k,l} \cdot HV_{i,j}^{(t-1)} \cdot t \ . \tag{6}$$

Then, the total number of hv-convex discrete sets can be calculated by a formula similar to (5).

Now, let us investigate the number of hv-convex 8-connected discrete sets of size $m \times n$. For the sake of technical simplicity we recall some concepts of [1].

Let F be a discrete set with $k \geq 2$ components such that $I_l \times J_l = \{i_l, \ldots, i_l'\} \times \{j_l, \ldots j_l'\}$ is the minimal bounding rectangle of the l-th component of F. We say that the components of F are *disjoint* if for any $1 \leq l, l' \leq k$ $l \neq l'$ implies that $I_l \cap I_{l'} = \emptyset$ and $J_l \cap J_{l'} = \emptyset$. Now, without loss of generality we can assume that $i_l < i_{l+1}$ for each $l = 1, \ldots, k-1$. F is called *canonical* if $j_l < j_{l+1}$ for each $l = 1, \ldots, k-1$. F is called *anticanonical* if $j_l > j_{l+1}$ for each $l = 1, \ldots, k-1$. That is, the discrete set is canonical (anticanonical) if - omitting empty rows and columns - the minimal bounding rectangles of the components are connected to each other with their bottom right hand and upper left hand (bottom left hand and upper right hand) corners (see Fig. 3).

Let us introduce the notations $D_{m,n}$, $L_{m,n}$, and $Q_{m,n}$ for the number of NW-directed polyominoes, NW-parallelogram polyominoes, and hv-convex 8-connected discrete sets of size $m \times n$, respectively. Moreover let $T_{m,n}$ denote the number of canonical 8-connected discrete sets whose components are all NW-parallelogram polyominoes. With these notations we obtain

Theorem 1. *For each* $m, n > 1$

$$T_{m,n} = L_{m,n} + \sum_{k<m,\ l<n} L_{k,l} \cdot T_{m-k,n-l} \tag{7}$$

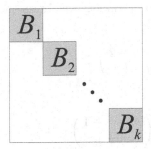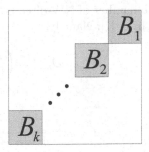

Fig. 3. The relative position of the minimal bounding rectangles of the components $B_1, \ldots, B_k$ of a canonical (*left*) and an anticanonical (*right*) discrete set

and

$$Q_{m,n} = P_{m,n} + 2 \sum_{i<m,\ j<n} \sum_{k\leq m-i,\ l\leq n-j} D_{i,j} \cdot D_{k,l} \cdot T_{m-k-i,n-j-l} \ . \qquad (8)$$

Proof. Let $\mathcal{T}$ denote the class of canonical 8-connected discrete sets which components are all NW-parallelogram polyominoes. With this notation Equation (7) can be proven in the following way. A discrete set $F \in \mathcal{T}$ of size $m \times n$ is either a NW-parallelogram polyomino (i.e. it has just one component) or it contains a NW-parallelogram polyomino of size $k \times l$ (where $k < m$ and $l < n$) as a subset in the upper left hand corner and the remaining part of F is a discrete set of size $(m-k) \times (n-l)$ which also belongs to the $\mathcal{T}$ class (see Fig. 3 again). This observation can be concisely expressed by the recursive formula (7).

To prove Equation (8) we recall the following observations from [3]. A set of $\mathcal{Q}$ is either an hv-convex polyomino or it consists of several hv-convex components. Let $F \in \mathcal{Q}$ be a discrete set having components $F_1, \ldots, F_k$ such that $\{i_l, \ldots, i'_l\} \times \{j_l, \ldots, j'_l\}$ is the minimal bounding rectangle of the l-th ($l = 1, \ldots, k$) component of F. Without loss of generality we can assume that $1 = i_1 \leq i'_1 < i_2 \leq i'_2 < \ldots \leq i'_k = m$. Then, either $1 = j_1 \leq j'_1 < j_2 \leq j'_2 < \ldots \leq j'_k = n$, or $n = j_1 \geq j'_1 > j_2 \geq j'_2 > \ldots \geq j'_k = 1$. Consequently, such a set of $\mathcal{Q}$ is always canonical or anticanonical.

Due to symmetry the number of canonical and anticanonical sets of $\mathcal{Q}$ which are not polyominoes are the same. Therefore, it is sufficient to calculate the number of canonical discrete sets of $\mathcal{Q}$ and multiply the result by 2. For a canonical set of $\mathcal{Q}$ it is always true that $F_1, \ldots, F_{k-1}$ are NW-directed and $F_2, \ldots, F_k$ are SE-directed (that is, $F_2, \ldots, F_{k-1}$ are NW-parallelogram polyominoes). In particular, we also get that there are hv-convex 8-connected sets which have just two components and with no parallelogram polyominoes between them. Additionally, the structure of a canonical set of $\mathcal{Q}$ is the following. It contains an NW-directed polyomino of size $i \times j$ in the upper-left corner (where $i < m$ and $j < n$), an SE-directed polyomino of size $k \times l$ in the bottom right corner (where $k \leq m-i$ and $l \leq n-j$) and the remaining part (if exists) is a canonical discrete set of size $(m-i-k) \times (n-j-l)$ which components are all NW-parallelogram polyominoes. Thus, we get the formula (8). $\qquad \square$

For the number of NW-directed (parallelogram) polyominoes we obtain the formulas from [7]. Namely,

$$D_{m,n} = \binom{m+n-2}{m-1} \binom{m+n-2}{n-1}$$ (9)

and

$$L_{m,n} = \frac{1}{m+n-1} \binom{m+n-1}{m-1} \binom{m+n-1}{n-1} .$$ (10)

Setting $D_{1,j} = D_{i,1} = L_{0,j} = L_{1,j} = L_{i,0} = L_{i,1} = T_{1,j} = T_{i,1} = 1$ and $T_{0,j} = T_{i,0} = 0$ for each $i = 1, \ldots, m$ and $j = 1, \ldots, n$ (for technical reasons we set $T_{0,0} = 1$) we are now able to determine the number of hv-convex 8-connected discrete sets of size $m \times n$ for an arbitrary m and n.

4   Statistics on hv-Convex Discrete Sets

The recursive formulas of Section 3 allow us to examine some important properties of hv-convex discrete sets that can affect the reconstruction complexity. In order to get such statistics we first calculated the number of hv-convex discrete sets in the classes studied. Table 1 shows the number of elements in the classes $\mathcal{P}$, $\mathcal{Q}$, $\mathcal{HV}'$, and $\mathcal{HV}$ with minimal bounding rectangles of semi-perimeter n for the first 15 values of n – represented by P_n, Q_n, HV'_n, and HV_n, respectively (the first column can also be calculated via formula (2) and it enumerates the first 15 elements of Sequence A005436 in [19]). For $n = 5$ the corresponding hv-convex binary images are shown in Fig. 4.

Knowing the relations of (1) and with the aid of the statistics presented in Table 1, we can describe the relative cardinality of the classes examined. With this information we can, for example, address questions concerning the relative occurrence of certain hv-convex discrete sets and calculate the probability that an hv-convex discrete set chosen from a uniform random distribution has some special properties which can facilitate the reconstruction task.

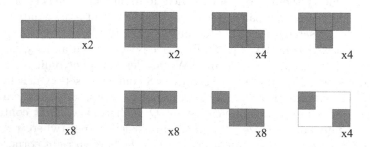

Fig. 4. Some hv-convex binary pictures with a perimeter value of 10. The numbers tell us that there are other solutions that can be obtained by mirroring or/and rotating the given polyomino.

Table 1. The values of P_n, Q_n, HV_n', and HV_n

n	P_n	Q_n	HV_n'	HV_n
2	1	1	1	1
3	2	2	2	2
4	7	9	9	9
5	28	36	36	40
6	120	154	162	184
7	528	668	732	860
8	2344	2916	3368	4058
9	10416	12741	15520	19240
10	46160	55570	71618	91440
11	203680	241692	329988	435136
12	894312	1047604	1518090	2072672
13	3907056	4524464	6971112	9883264
14	16986352	19470660	3196392	47193776
15	73512288	83500968	146390016	225779728

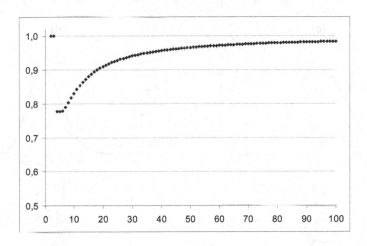

Fig. 5. The ratio P_n/Q_n (*vertical axis*) depending on the semiperimeter value n (*horizontal axis*)

Example 1. Using the entries of Table 1 we can calculate the probability that an *hv*-convex discrete set with semi-perimeter value of 6 chosen from a uniform random distribution is an *hv*-convex polyomino (i.e. it consists of one component), which turns out to be $120/184 \approx 0.65$. If we increase the semi-perimeter value to 10, say, then this probability decreases to $46160/91440 \approx 0.50$. Such information is especially useful in the reconstruction task as *hv*-convex polyominoes can be reconstructed from their horizontal and vertical projections in polynomial time. In contrast, if the *hv*-convex set has at least two components then the reconstruction is NP-hard (see the introduction here). Hence with this method we can calculate the probability that the reconstruction of the randomly chosen

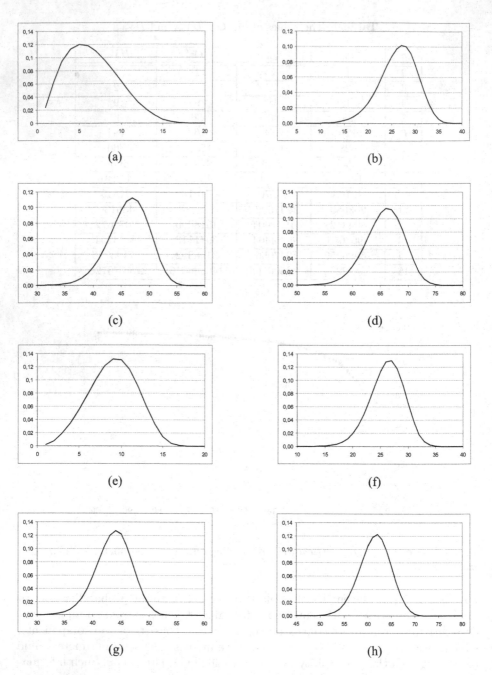

Fig. 6. The distributions of the number of components – which depend on the size of the test data – in the $\mathcal{HV}'$ ((a)-(d)) and $\mathcal{HV}$ ((e)-(h)) classes

Table 2. The expectation value $E_{\mathcal{HV'}}(n)$ $(E_{\mathcal{HV}}(n))$ and the variance $D^2_{\mathcal{HV'}}(n)$ $(D^2_{\mathcal{HV}}(n))$ of the components of a set with a minimal bounding rectangle of size $n \times n$ in the $\mathcal{HV'}$ $(\mathcal{HV})$ class. The values have been rounded to 5 digits

n	$E_{HV'}(n)$	$D^2_{HV'}(n)$	$E_{HV}(n)$	$D^2_{HV}(n)$
20	6.53981	9.84446	9.03570	8.12406
40	26.33821	16.00766	26.11090	9.54114
60	46.30283	12.92260	43.68220	10.00145
80	65.70631	12.05665	61.49588	10.72577

hv-convex set can be performed using a polynomial-time algorithm to reconstruct an hv-convex polyomino.

Example 2. In [3] the authors presented a very fast algorithm for the reconstruction of hv-convex 8-connected but not 4-connected discrete sets. From the first few entries of Table 1 we have the suggestion that the number of such kind of sets rapidly decreases as the semiperimeter value increases. To verify this, we calculated the first 100 values of P_n/Q_n (see Fig. 5). From this figure it is evident, that – unfortunately – even for sets of relatively small sizes there is almost no chance to apply this fast reconstruction algorithm in practice (assuming that the sets to be reconstructed are from a uniform random distribution), and things get worse if we want to reconstruct sets of bigger sizes.

With the aid of the formulas (4) and (6) it is also possible to describe the true distribution of the number of components of the generated hv-convex discrete set of the $\mathcal{HV'}$ class since, in this case, we can enumerate the discrete sets of a given class that have k components. This piece of information is also very useful when reconstructing images like these. For example, as was discussed in the introduction of Section 3, if the hv-convex set consists of a single component then the reconstruction from two projections can be solved in polynomial time, otherwise it is NP-hard. Furthermore, the number of components of an hv-convex set also affects the accuracy of the reconstruction heuristic that was presented in [2]. Namely, the more components the hv-convex discrete set has, it is more likely that ambiguity will occur in the reconstruction.

Table 2 lists the expectation values and the variances of the variables which represent the number of components of a discrete set generated using a uniform random distribution from the $\mathcal{HV'}$ and $\mathcal{HV}$ classes when the size of the minimal bounding rectangle is $n \times n$ for some fixed positive integer n. In addition, the corresponding distributions are depicted in Fig. 6.

Statistics about the expected number of components can be especially useful in the reconstruction task. It tells us something about the discrete set to be reconstructed before we attempt to reconstruct it. Thus, such statistics opens the way to the design of reconstruction algorithms that exploit information known beforehand about the expected number of components. The author believes that such algorithms could be more effective in practice than the previously developed ones which do not make use of such prior knowledge.

5   Conclusions

In this paper we have presented recursive formulas to count hv-convex discrete sets (which possibly have certain connectedness properties as well). With the aid of these formulas we have collected some statistics on several subclasses of hv-convex discrete sets. We can use these statistics to analyze the performance of certain reconstruction algorithms developed for the classes studied. In addition, it turned out that it is also possible to say something about the number of the components of an hv-convex discrete set before we attempt to reconstruct it (if the set arises from a uniform random distribution). Incorporating this prior knowledge into the reconstruction process can hopefully yield more effective reconstruction algorithms in the future.

References

1. Balázs, P.: A framework for generating some discrete sets with disjoint components using uniform distributions. Theor. Comput. Sci. (submitted)
2. Balázs, P.: On the ambiguity of reconstructing hv-convex binary matrices with decomposable configurations. Acta Cybernetica (accepted)
3. Balázs, P., Balogh, E., Kuba, A.: Reconstruction of 8-connected but not 4-connected hv-convex discrete sets. Disc. Appl. Math. 147, 149–168 (2005)
4. Balogh, E., Kuba, A., Dévényi, C., Del Lungo, A.: Comparison of algorithms for reconstructing hv-convex discrete sets. Lin. Alg. Appl. 339, 23–35 (2001)
5. Barcucci, E., Del Lungo, A., Nivat, M., Pinzani, R.: Medians of polyominoes: A property for the reconstruction. Int. J. Imaging Systems and Techn. 9, 69–77 (1998)
6. Barcucci, E., Del Lungo, A., Nivat, M., Pinzani, R.: Reconstructing convex polyominoes from horizontal and vertical projections. Theor. Comput. Sci. 155, 321–347 (1996)
7. Barcucci, E., Frosini, A., Rinaldi, S.: On directed-convex polyominoes in a rectangle. Discrete Math. 298, 62–78 (2005)
8. Batenburg, K.J.: An evolutionary algorithm for discrete tomography. Disc. Appl. Math. 151, 36–54 (2005)
9. Brunetti, S., Del Lungo, A., Del Ristoro, F., Kuba, A., Nivat, M.: Reconstruction of 4- and 8-connected convex discrete sets from row and column projections. Lin. Alg. Appl. 339, 37–57 (2001)
10. Chrobak, M., Dürr, C.: Reconstructing hv-convex polyominoes from orthogonal projections. Inform. Process. Lett. 69(6), 283–289 (1999)
11. Dahl, G., Flatberg, T.: Optimization and reconstruction of hv-convex $(0,1)$-matrices. Disc. Appl. Math. 151, 93–105 (2005)
12. Delest, M.P., Viennot, G.: Algebraic languages and polyominoes enumeration. Theor. Comput. Sci. 34, 169–206 (1984)
13. Gessel, I.: On the number of convex polyominoes. Ann. Sci. Math. Québec 24, 63–66 (2000)
14. Golomb, S.W.: Polyominoes. Charles Scriber's Sons, New York (1965)
15. Herman, G.T., Kuba, A. (eds.): Discrete Tomography: Foundations, Algorithms and Applications. Birkhäuser, Boston (1999)
16. Herman, G.T., Kuba, A. (eds.): Advances in Discrete Tomography and its Applications. Birkhäuser, Boston (2007)

17. Kuba, A.: The reconstruction of two-directionally connected binary patterns from their two orthogonal projections. Comp. Vision, Graphics, and Image Proc. 27, 249–265 (1984)
18. Kuba, A.: Reconstruction in different classes of 2D discrete sets, Lecture Notes in Comput. In: Bertrand, G., Couprie, M., Perroton, L. (eds.) DGCI 1999. LNCS, vol. 1568, pp. 153–163. Springer, Heidelberg (1999)
19. Sloane, N.J.A.: The on-line encyclopedia of integer sequences, http://www.research.att.com/~njas/sequences/
20. Woeginger, G.W.: The reconstruction of polyominoes from their orthogonal projections. Inform. Process. Lett. 77, 225–229 (2001)

Finding the Orthogonal Hull of a Digital Object: A Combinatorial Approach

Arindam Biswas[1], Partha Bhowmick[1], Moumita Sarkar[1],
and Bhargab B. Bhattacharya[2]

[1] Computer Science and Technology Department
Bengal Engineering and Science University, Shibpur, Howrah, India
{abiswas,partha}@cs.becs.ac.in, moumita_becit@yahoo.co.in
[2] Advanced Computing and Microelectronics Unit
Indian Statistical Institute, Kolkata, India
bhargab@isical.ac.in

Abstract. A combinatorial algorithm to compute the orthogonal hull of a digital object imposed on a background grid is presented in this paper. The resolution and complexity of the orthogonal hull can be controlled by varying the grid spacing, which may be used for a multiresolution analysis of a given object. Existing algorithms on finding the convex hull are based on divide and conquer strategy, sweepline approach, etc., whereas the proposed algorithm is combinatorial in nature whose time complexity depends on the object perimeter instead of the object area. For a larger grid spacing, the perimeter of an object decreases in length in terms of grid units, and hence the runtime of the algorithm reduces significantly. The algorithm uses only comparison and addition in the integer domain, thereby making it amenable to usage in real-world applications where speed is a prime factor. Experimental results including the CPU time demonstrate the elegance and efficacy of the proposed algorithm.

1 Introduction

The convex hull of an object A, denoted by $CH(A)$, is the smallest convex set that contains A. There exist a number of algorithms [3,9,17] to find the convex hull of a point set or a polygonal object A having arbitrary shape on the real plane. The time complexities of some of the referred ones are of order $O(n^3)$ (brute force), $O(n \log n)$ (Graham scan [12]), $O(nh)$ (Jarvis march [13]), and $O(n \log h)$ (Kirkpatrick-Siedel's algorithm [15]), where, n is the number of points/vertices constituting A, and h is the number of vertices of $CH(A)$. Also, there are other algorithms on finding the convex hull, e.g., [2,7,23]. A detailed analytical study of the convex hull algorithms may be seen in [1].

Apart from the concept of convex hull, other types of hulls, such as pseudo-hull, near-hull [16], digital convex hull [8], relative convex hull [20], and α-hull [11], can be also found in literature, which are designed for specific applications. However, the execution of these algorithms for sufficiently large digital object is not as fast as required in a practical application.

V.E. Brimkov, R.P. Barneva, H.A. Hauptman (Eds.): IWCIA 2008, LNCS 4958, pp. 124–135, 2008.
© Springer-Verlag Berlin Heidelberg 2008

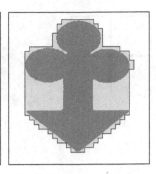

Fig. 1. A sample 2D object, its convex hull (left), and its orthogonal hulls for grid size $g = 22$ (middle) and $g = 8$ (right)

Some of the typical applications involving convex hulls are analysis of landmark data, shape analysis and classification, measuring the polygonal entropy, and many such areas of computer vision and pattern recognition [5,10,14,18]. A properly defined convex polygon describing a real or a digital object is often considered to be the domain of interest of the underlying object. As a result, the subject has received a considerable attention amongst researchers [6,21,22,24].

This paper presents a novel algorithm for finding the orthogonal hull (Sec. 2) of a given digital object, the hull edges lying on a set of equally spaced horizontal and vertical grid lines. The ordered list of hull vertices is obtained by an analysis of the object occupation of the four neighboring quadrants corresponding to a grid point. The orthogonal hull consists of fewer vertices with an increase of the grid size (spacing between two consecutive horizontal/vertical grid lines), enabling a multiresolution analysis of the object. The algorithm is based on the fact that a polygon is orthogonally convex if and only if a counterclockwise traversal of its boundary never makes two consecutive right turns. The algorithm involves only comparison and addition/subtraction in the integer domain, and hence runs very fast, as demonstrated by the CPU time in our experiments.

The convex hull and the orthogonal hulls for $g = 22$ and $g = 8$, corresponding to a digital object, are shown in Fig. 1. The convex hull algorithm (Graham Scan) on a digital object (22404 pixels) shown in this figure takes a few seconds, whereas the proposed algorithm on finding the orthogonal hull takes only a few milliseconds ($g = 22 : 0.94$ msec., $g = 8 : 3.16$ msec.). The Graham Scan, however, takes less time for the object contour; but finding the object contour needs an edge extraction algorithm. On the contrary, given an object, the proposed algorithm runs on the object contour without resorting to edge extraction.

Apart from the speed, the proposed algorithm has the ability to capture the shape information of the object. For example, for $g = 22$, the orthogonal hull of the object shown in Fig. 1 is vertically symmetrical, which conforms to the vertical symmetry of the object; for $g = 8$, the orthogonal hull is also almost symmetrical. The vertices of the orthogonal hull are reported in order in terms of their types (90^0 and 270^0), from which the symmetry can be ascertained. The non-convex regions — detected and removed to derive the corresponding convex

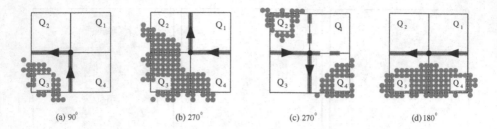

Fig. 2. Different vertex types

regions using certain semantic rules based on a combinatorial analysis — also captures the shape complexity of the concerned object, which can be used in subsequent applications.

2   Definitions and Preliminaries

A subset of $\mathbb{Z}^2$ in which every pair of points is k-connected[1] is called a k-**connected set.** (In this paper, we define a digital object A to be an 8-connected subset $A \subset \mathbb{Z}^2$ whose complement $\mathbb{Z}^2 \setminus A$ is a 4-connected set [19].) The **background grid** $(\mathcal{G})$ is defined as a set of uniformly spaced horizontal and vertical grid lines, $\mathcal{G} = (\mathcal{H}, \mathcal{V})$, where $\mathcal{H}$ and $\mathcal{V}$ represent two sets of equi-spaced horizontal and vertical grid lines respectively. The **grid size** g is defined as the distance between two consecutive horizontal/vertical grid lines. A **grid point** is the point of intersection of a horizontal and a vertical grid line. P is an **orthogonal polygon** if and only if each of its vertices is a grid point of $\mathcal{G}$ and each of its edges is axis-parallel. The **orthogonal convex hull**, or simply **orthogonal hull**, of a digital object A, denoted by $OH(A)$, is the smallest area orthogonal polygon such that (i) no point $p \in A$ lies on or outside $OH(A)$ and (ii) intersection of $OH(A)$ with any horizontal or vertical line is either empty or a line segment.

2.1   Orthogonal Traversal of the Object Contour

In order to detect and remove the concavities, we traverse around the object contour, orthogonally along the grid lines. The nature of traversal is such that we visit the vertices (in order) of the smallest-area orthogonal cover of the digital object using an efficient combinatorial technique based on object containments of the four cells incident at a particular grid point [4]. The characteristics of a grid point p in $\mathcal{G}$ is determined by object containments of the four neighboring cells of size $g \times g$ incident at p.

During traversal, a grid point is determined either as a vertex or as a non-vertex point. Since we traverse orthogonally, a grid point, p, if detected as a

---

[1] Two points p and q are said to be k-connected ($k = 4$ or 8) in a set S if and only if there exists a sequence $\langle p = p_0, p_1, \ldots, p_n = q \rangle \subseteq S$ such that $p_i \in N_k(p_{i-1})$ for $1 \leq i \leq n$.

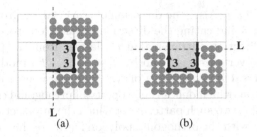

Fig. 3. Concave regions (require reduction)

vertex, can be a 90^0 vertex or a 270^0 vertex. Otherwise, p is simply a point on the edge of the orthogonal cover, or a grid point lying inside/outside the object and also inside/outside the orthogonal cover. The classification of p is based on the object containment of the four cells $(Q_1 - Q_4)$ incident at p and their combinatorial arrangement (Fig. 2). If number of object-containing cells incident at p is i, then p is classified to class C_i as follows.

C_0 : p is not a vertex, since none of Q_is has object containment.

C_1 : Exactly one of the Q_is is intersected by the object. Hence p is classified as a 90^0 vertex, as shown in Fig. 2(a).

C_2 : Two cells occupied by the object can have two different arrangements:
(i) if adjacent cells are occupied, then p is an edge point (Fig. 2(d));
(ii) if diagonally opposite cells are occupied, then p is a 270^0 vertex (Fig. 2(c)).
Out of the four edges incident at p, exactly two edges will be traversed so that p becomes a 270^0 vertex. For example, if p is visited from the left/right, then the outgoing edge from p will be directed downwards/upwards (shown in firm/dotted lines).

C_3 : Three cells are occupied by the object and hence p is classified as a 270^0 vertex (Fig. 2(b)).

C_4 : All four cells are occupied by the object, and so p is not a vertex.

Henceforth in this paper, a 90^0 vertex is referred to as a type '**1**' vertex, and a 270^0 vertex as a type '**3**' vertex. The start point of the traversal is determined by a row-wise scan of the grid points and start vertex is the first grid point classified as a 90^0 vertex (we assume, w.l.o.g., that the object lies left during traversal).

3   Finding the Orthogonal Hull

The concavities present in the orthogonal cover are detected and removed when the object boundary is traversed along the grid lines as mentioned in Sec. 2.1. Hence the proposed algorithm finds the orthogonal hull of a digital object without any prior knowledge about its orthogonal cover. Deriving the orthogonal hull is, therefore, very fast.

During the traversal, each vertex, v_i, is represented by a two-tuple $\langle t_i, l_i \rangle$, where, t_i ($= 1$ or 3) is the type of the vertex and l_i is the length of the line segment from v_i to v_{i+1}. Also, the direction of traversal, d_i, from v_i, can assume the value $0, 1, 2$, or 3, indicating the direction towards left, top, right, or bottom respectively. The direction of traversal at v_i, derived from the previous direction and the type of the vertex v_i, is given by $d_i = (d_{i-1} + t_i) \bmod 4$.

During the traversal, if two vertices of type **3** appear consecutively, then it implies a concave region, which defies the property of orthogonal convexity (Sec. 2). Illustrated in Fig. 3 is two such patterns for which the intersection of a vertical or a horizontal line, L, with the orthogonal polygon has more than one segment. The goal is to identify such regions and derive the edges of the orthogonal hull such that the properties of orthogonal convexity are maintained. In this incremental algorithm, the part of the orthogonal hull obtained upto a point does not contain two consecutive vertices of type **3**, which acts as the invariant of the algorithm. Whenever such an occurrence appears, we apply necessary reduction rules to maintain the algorithm invariant and to ensure the orthogonal convexity, thereof. However, rest of the patterns, **13**, **31**, and **11**, are in conformance with the algorithm invariant and hence do not violate the properties of orthogonal convexity.

3.1   Setting the Rules

Let v_1, v_2, v_3, and v_4 be four consecutive vertices for which the rule has to be applied in order to remove the concavity, if any. Let the vertex preceding v_1 be v_0, if any. If v_0 exists, then the rule is applied on the vertex tuple formed by $v_0, v_1, \ldots, v_4$; otherwise, the rule is applied on the tuple formed by $v_1, v_2, \ldots, v_4$.

Since a reduction rule is applied only when two consecutive vertices have type **3**, there can be two sets of rules depending on whether the type of the vertex following the pattern **33** is **1** or **3**. We consider that the two consecutive vertices of type **3** are designated by v_2 and v_3, and the vertex v_1 preceding v_2 is of type **1**. For, in our algorithm, the traversal always starts from a vertex of type **1**, which is verified from the combinatorial arrangement of its four neighboring cells (Sec. 2.1). This policy of starting the traversal always ensures that there will be at least one type-**1** vertex preceding two consecutive vertices of type **3**.

Pattern 1331: This signifies a type-**1** vertex followed by two consecutive type-**3** vertices and another type-**1** vertex. Occurrence of two consecutive **3**s essentially signifies a concavity in the object, as explained earlier. The rules for removal of the associated concavities are stated in Fig. 4. The concave regions are detected and coalesced to their corresponding convex products using the related edge lengths in the reduction mechanism. There can arise three cases depending on the relation between l_1 and l_3, which are as follows:

Rule **R11***:* Applied when $l_1 = l_3$.
If v_0 exists, then v_1, v_2, v_3, and v_4 are removed, and the length of v_0 is modified to $l_0 + l_2 + l_4$. If v_0 does not exist, then v_2, v_3, and v_4 are removed, and l_1 is modified to $l_2 + l_4$.

Rule **R12***:* Applied when $l_1 > l_3$.
v_2 is modified to v_2' such that l_1 becomes $l_1 - l_3$, l_2 is modified to $l_2 + l_4$, and v_4 is removed. This rule is irrespective of whether or not v_0 exists.

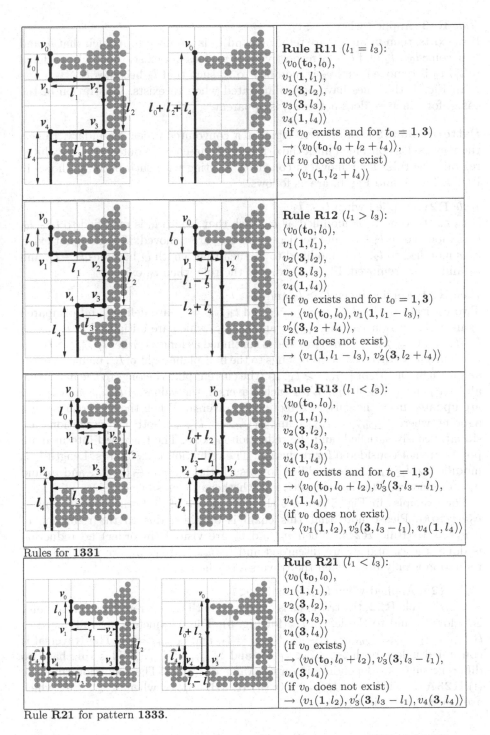

Rule R11 $(l_1 = l_3)$:
$\langle v_0(\mathbf{t_0}, l_0),$
$v_1(\mathbf{1}, l_1),$
$v_2(\mathbf{3}, l_2),$
$v_3(\mathbf{3}, l_3),$
$v_4(\mathbf{1}, l_4)\rangle$
(if v_0 exists and for $t_0 = \mathbf{1, 3}$)
$\rightarrow \langle v_0(\mathbf{t_0}, l_0 + l_2 + l_4)\rangle,$
(if v_0 does not exist)
$\rightarrow \langle v_1(\mathbf{1}, l_2 + l_4)\rangle$

Rule R12 $(l_1 > l_3)$:
$\langle v_0(\mathbf{t_0}, l_0),$
$v_1(\mathbf{1}, l_1),$
$v_2(\mathbf{3}, l_2),$
$v_3(\mathbf{3}, l_3),$
$v_4(\mathbf{1}, l_4)\rangle$
(if v_0 exists and for $t_0 = \mathbf{1, 3}$)
$\rightarrow \langle v_0(\mathbf{t_0}, l_0), v_1(\mathbf{1}, l_1 - l_3),$
$v_2'(\mathbf{3}, l_2 + l_4)\rangle,$
(if v_0 does not exist)
$\rightarrow \langle v_1(\mathbf{1}, l_1 - l_3), v_2'(\mathbf{3}, l_2 + l_4)\rangle$

Rule R13 $(l_1 < l_3)$:
$\langle v_0(\mathbf{t_0}, l_0),$
$v_1(\mathbf{1}, l_1),$
$v_2(\mathbf{3}, l_2),$
$v_3(\mathbf{3}, l_3),$
$v_4(\mathbf{1}, l_4)\rangle$
(if v_0 exists and for $t_0 = \mathbf{1, 3}$)
$\rightarrow \langle v_0(\mathbf{t_0}, l_0 + l_2), v_3'(\mathbf{3}, l_3 - l_1),$
$v_4(\mathbf{1}, l_4)\rangle$
(if v_0 does not exist)
$\rightarrow \langle v_1(\mathbf{1}, l_2), v_3'(\mathbf{3}, l_3 - l_1), v_4(\mathbf{1}, l_4)\rangle$

Rules for **1331**

Rule **R21** $(l_1 < l_3)$:
$\langle v_0(\mathbf{t_0}, l_0),$
$v_1(\mathbf{1}, l_1),$
$v_2(\mathbf{3}, l_2),$
$v_3(\mathbf{3}, l_3),$
$v_4(\mathbf{3}, l_4)\rangle$
(if v_0 exists)
$\rightarrow \langle v_0(\mathbf{t_0}, l_0 + l_2), v_3'(\mathbf{3}, l_3 - l_1),$
$v_4(\mathbf{3}, l_4)\rangle$
(if v_0 does not exist)
$\rightarrow \langle v_1(\mathbf{1}, l_2), v_3'(\mathbf{3}, l_3 - l_1), v_4(\mathbf{3}, l_4)\rangle$

Rule **R21** for pattern **1333**.

Fig. 4. Concavity detection and removal rules for patterns **1331** and **1333**

Rule **R13**: Applied when $l_1 < l_3$.
If v_0 exists, then v_1 and v_2 are removed, and v_3 is modified to v_3' such that l_0 and l_3 become $l_0 + l_2$ and $l_3 - l_1$ respectively. If v_0 does not exist, then l_1 is modified to l_2, v_2 is removed, and v_3 is modified to v_3' such that l_3 becomes $l_3 - l_1$.

In Fig. 4, the rules have been illustrated when v_0 exists. Illustration of the cases, for which v_0 does not exist, is apparent.

Pattern 1333: Such a pattern signifies a convoluted object boundary. Hence, the traversal is continued until the orthogonal chain comes out of the convoluted region. The rules for removal of concavity starting with such pattern, shown in Fig. 4, Fig. 5, and Fig. 6, are as follows.

Rule **R21**: Applied when $l_1 < l_3$.
If v_0 exists, then v_3 is modified as v_3', such that length l_0 is modified to $l_0 + l_2$, l_3 is modified to $l_3 - l_1$, and vertices v_1 and v_2 are removed. If v_0 does not exist, l_1 is modified to l_2, v_3 is modified to v_3' such that length l_3 becomes $l_3 - l_1$, and v_1 and v_2 are removed. Fig. 4 illustrates the rule when v_0 exists.

Rule **R22**: Applied when $l_1 \geqslant l_3$ and $l_4 \geqslant l_2$.
Two critical lengths, $l_{1\_crit} = l_1 - l_3$ and $l_{4\_crit} = l_2$, are defined and compared against the current lengths, l_{1_curr} and l_{4_curr}, which are initialized as $l_{1\_curr} = l_1 - l_3$ and $l_{4\_curr} = l_4$. The traversal is continued as long as either $l_{1\_curr} \geqslant l_{1\_crit}$ (the currently traversed point belongs to the half-plane right of l_V, inclusive of l_V, as shown in Fig. 5), or $l_{4\_curr} \geqslant l_{4\_crit}$ (the current point belongs to the half-plane above l_H, inclusive of l_H). With each traversal, the values of l_{1_curr} and l_{4_curr} are updated from the current direction of traversal. When the traversal reaches a point where $l_{1\_curr} < l_{1\_crit}$ and $l_{4\_curr} < l_{4\_crit}$, both the conditions are simultaneously satisfied, and the reduction is done. The traversed intermediate points are not considered for reduction. The reduction is as follows. Length l_1 is modified to l_{1_curr}, v_2 is modified to v_2' such that $l_2 = l_2 - l_{4\_curr}$, and v_3 and v_4 are removed. This is irrespective of whether v_0 exists or not.

For example, in Fig. 5, we get the current set of vertices for reduction as $v_0 v_1 v_2 v_3 v_4$. Please note that this chain is obtained due to the reduction of $v_1 v_x v_y v_z v_3$ (Rule **R21**). When v_p and v_q are visited (in order) no reduction is done, the v_p and v_q are discarded and l_{1_curr} and l_{4_curr} are updated. The reduction is only done when the traversal reaches v_r.

Rule **R23**: Applied when $l_1 \geqslant l_3$ and $l_4 < l_2$.
Similar to rule **R22**, the traversal is continued until the currently traversed point is below l_H and to the left of l_V (for Fig. 6). The parameters are initialized as $l_{1\_crit} = l_1 - l_3$, $l_{1\_curr} = l_1 - l_3$, $l_{4\_crit} = l_2$, and $l_{4\_curr} = l_4$. The traversal is continued unless both $l_{1\_curr} < l_{1\_crit}$ and $l_{4\_curr} < l_{4\_crit}$. This rule has two different subcases depending upon the value of l_{1_curr}. The reduction rules are (i) **R23A**: applied when $l_{1\_curr} \geqslant 0$; (ii) **R23B**: applied when $l_{1\_curr} < 0$, which are stated clearly in Fig. 6.

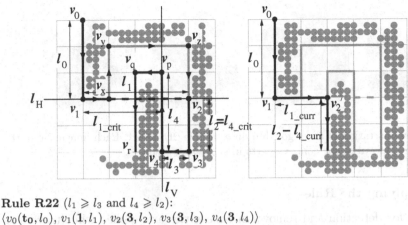

Rule R22 ($l_1 \geqslant l_3$ and $l_4 \geqslant l_2$):

$\langle v_0(\mathbf{t_0}, l_0), v_1(\mathbf{1}, l_1), v_2(\mathbf{3}, l_2), v_3(\mathbf{3}, l_3), v_4(\mathbf{3}, l_4)\rangle$
$\rightarrow \langle v_0(\mathbf{t_0}, l_0), v_1(\mathbf{1}, l_{1\_curr}), v_2'(\mathbf{3}, l_2 - l_{4\_curr})\rangle$ (if v_0 exists)
$\rightarrow \langle v_1(\mathbf{1}, l_{1\_curr}), v_2'(\mathbf{3}, l_2 - l_{4\_curr})\rangle$ (if v_0 does not exist)

Fig. 5. Rule **R22** for pattern **1333**

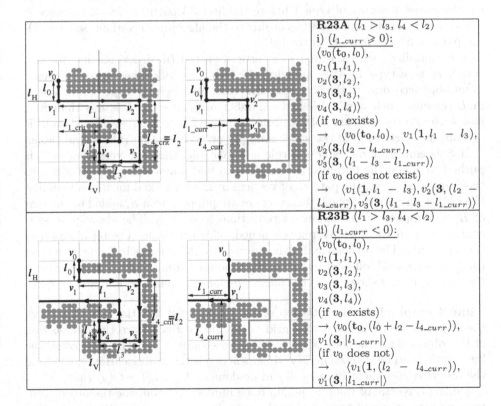

R23A ($l_1 > l_3$, $l_4 < l_2$)
i) ($l_{1\_curr} \geqslant 0$):
$\langle v_0(\mathbf{t_0}, l_0),$
$v_1(\mathbf{1}, l_1),$
$v_2(\mathbf{3}, l_2),$
$v_3(\mathbf{3}, l_3),$
$v_4(\mathbf{3}, l_4)\rangle$
(if v_0 exists)
$\rightarrow \langle v_0(\mathbf{t_0}, l_0), v_1(\mathbf{1}, l_1 - l_3),$
$v_2'(\mathbf{3}, (l_2 - l_{4\_curr}),$
$v_3'(\mathbf{3}, (l_1 - l_3 - l_{1\_curr}))\rangle$
(if v_0 does not exist)
$\rightarrow \langle v_1(\mathbf{1}, l_1 - l_3), v_2'(\mathbf{3}, (l_2 - l_{4\_curr}), v_3'(\mathbf{3}, (l_1 - l_3 - l_{1\_curr}))\rangle$

R23B ($l_1 > l_3$, $l_4 < l_2$)
ii) ($l_{1\_curr} < 0$):
$\langle v_0(\mathbf{t_0}, l_0),$
$v_1(\mathbf{1}, l_1),$
$v_2(\mathbf{3}, l_2),$
$v_3(\mathbf{3}, l_3),$
$v_4(\mathbf{3}, l_4)\rangle$
(if v_0 exists)
$\rightarrow \langle v_0(\mathbf{t_0}, (l_0 + l_2 - l_{4\_curr})),$
$v_1'(\mathbf{3}, |l_{1\_curr}|)\rangle$
(if v_0 does not)
$\rightarrow \langle v_1(\mathbf{1}, (l_2 - l_{4\_curr})),$
$v_1'(\mathbf{3}, |l_{1\_curr}|)\rangle$

Fig. 6. Rules **R23A** and **R23B** for pattern **1333**. Note that illustrations on the right only consists of the cases where v_0 exists.

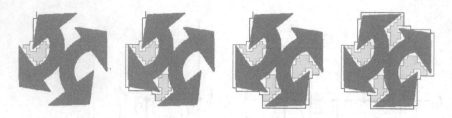

Fig. 7. Demonstration of the algorithm on a sample 2D object. Each image shows the result after removal of the concave parts in successive steps.

3.2   Applying the Rules

The rules for detection and removal of concavities are applied while traversing along the object contour in an orthogonal path. A stack S contains the vertices after applying the reduction rules so far. The vertices, currently under inspection, are stored in a variable-length link list, L, whose maximum length is five. The head of the list L has the least recent vertex. According to our traversal strategy, we always get a vertex of type **1** before the first **33** pattern (Sec. 3.1). Since S does not contain a pattern **33** (according to the algorithm invariant, see Sec. 3), the pattern **33** occurs only in the list L.

L is initialized with the first five vertices visited (in order) so that its first vertex v_0 is of type **1**. After applying the reduction rule, the vertices decrease in number according to the corresponding rule, although the type of first vertex in L remains unchanged. When the number of vertices in L is less than five, new vertex/vertices is/are popped from S and added to the rear of L to make it contain five vertices until the stack is empty or no reduction rule is applicable.

If S is empty or no reduction rule is applicable, then the vertex v_0 of L is pushed to S and the traversal is continued to find a new vertex to be added to the front of L. Next, the pattern of vertices in L are checked for its reducibility. If a reduction rule is applied, then vertices are popped from S, added to the rear of L, and applied with subsequent reductions iteratively. The above process is continued until the start vertex is reached, which indicates the termination of the algorithm. The orthogonal hull is reported by the vertices in L, starting from the front of L and followed by the vertices popped from S. Fig 7 demonstrates the algorithm and shows how the concave regions are removed in different steps.

Time Complexity: Since the object is a connected set, the containment of the object in a cell incident at a grid vertex p is verified from the intersection of the object with the four edges of the corresponding cell. For each edge, the intersection can be checked in $O(g)$ time, where g is the grid size. Hence, checking the object containment in any cell can be done in $4 \times O(g) = O(g)$ time.

During traversal of the grid points lying immediately outside the object contour, we visit each grid point p_i from the preceding one p_{i-1} using the information on intersection of the object with the edges incident at p_{i-1}. For example, in Fig. 2(a), if p_{i-1} is of type **1** which has been visited along the vertical edge (from its predecessor, p_{i-2}), then p_i is visited along the horizontal edge from p_{i-1}. Thus, the number of grid points visited while traversing orthogonally along the object contour is bounded by $O(n/g)$, where n is the number of points constituting the object contour. The resultant time complexity for visiting all the vertices is, therefore, given by $O(n/g) \cdot O(g) = O(n)$. Note that, each grid point lying on the orthogonal path of traversal is visited either once (Fig. 2(a, b, d)) or twice (Fig. 2(c)). The time spent over detection and removal of concavities is associated with checking a pattern in the list L and applying the reduction rule, whenever necessary. If a pattern does not contain two consecutive **3**s, then the vertex v_0 is pushed to the stack S in $O(1)$ time, and a new vertex is visited in $O(g)$ time. If the pattern undergoes a reduction, then the reduced L needs (at least) one or (at most) two pops from S. For, in a particular iteration, popping is done from S until we get a vertex type **1**, and no two consecutive vertices in S are of type **3**. Maximum number of reductions is bounded by $O(n/g) - 4$, since at most $O(n/g)$ vertices are visited and the orthogonal hull consists of at least four vertices. Thus, total number of stack operations (push and pop) is given by $(O(n/g) - 4) \cdot O(1) = O(n/g)$. Hence, the total time complexity for finding the orthogonal hull of a digital object is given by $O(n) + O(n/g) = O(n)$.

4   Experimental Results and Discussions

The proposed algorithm is implemented in C on a Sun_Ultra 5_10, Sparc, 233 MHz, the OS being the SunOS Release 5.7 Generic, and has been tested on (i) database D1 containing 1034 logo images (received on request, from Prof. Anil K. Jain and Aditya Vailya of Michigan State Univ., USA); (ii) a collected database D2 having 520 shape images; (iii) 100 test curves; (iv) a selected database of optical characters.

The algorithm can be easily extended to an image having more than one connected set and depending upon the grid size, the outcome may be one orthogonal

Table 1. Number of object pixels, area of convex hull, areas of orthogonal hull and the CPU time consumed in milliseconds (for different grid sizes), CPU time required for convex hull computation using Graham Scan, number of convex hull vertices and number of vertices of orthogonal hull (for different grid sizes) are shown above

image name	image size	#pixels	$|CH|$	$A(CH)$	$T(CH)$		$|OH|$	$A(OH)$	$T(OH)$
logo245	288 × 288	9868	82	39509	987	$g = 4$	86	35392	9
						$g = 16$	32	41216	2
logo247	288 × 288	10096	67	32419	1221	$g=4$	58	29824	10
						$g = 16$	30	35328	1
logo353	288 × 288	18122	46	32241	1812	$g=4$	116	30792	14
						$g = 16$	48	35840	3

$|CH|$ and $|OH|$ denote the number of convex hull vertices and the number of orthogonal hull vertices; $A(CH)$ and $A(OH)$ are their respective areas; and $T(\cdot)$ indicates the CPU time in milliseconds.

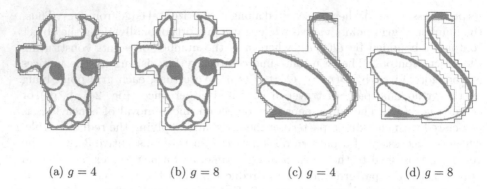

(a) $g = 4$ (b) $g = 8$ (c) $g = 4$ (d) $g = 8$

Fig. 8. The orthogonal hulls of two images for grid spacings, $g = 4$ and $g = 8$

hull (comprising more than one connected set). Some of the results for sample images are shown in Fig. 8 for $g = 4$ and $g = 8$. The orthogonal hull is shown in blue color whereas the gray color polygon indicates the smallest-area orthogonal polygon describing the object which is traversed during the derivation of the orthogonal hull. In Table 3.2, the area of the convex hull and the areas of the orthogonal hull for different grid sizes are presented. The number of convex hull vertices, the CPU time required for computation of the convex hull, the number of vertices of the orthogonal hull, and the CPU times at different grid sizes are also presented in the table. It can be seen from the data that the number of vertices of $OH(A)$ decreases and its area increases with the increase of grid size. Also, the computation time drops drastically for higher grid sizes.

5   Conclusion

A combinatorial algorithm is presented to construct the orthogonal convex hulls of a digital object for various grid resolutions. The worst-case time complexity of the algorithm is linear in the size of the contour. The actual runtimes on different images reinforce its speedy execution. The algorithm is a single-pass algorithm, and outputs the (types and outgoing edge-lengths of) vertices of the orthogonal hull in order. Hence, we can utilize the orthogonal hull description in some suitable application like shape analysis, shape-based image retrieval, etc. For example, the number of times the reduction rules are applied during the traversal for removal of a concavity can be used to measure the distribution of shape complexity over a large and complex object. Presently, we are working on shape analysis of objects using their orthogonal hulls, which will be reported in future.

References

1. Avis, D., Bremner, D.: How good are convex hull algorithms? In: Symp. Computational Geometry, pp. 20–28 (1995)
2. Barber, B., Dobkin, D., Huhdanpaa, H.: The quickhull algorithm for convex hull. The Geometry Center, University of Minnesota, TR-GCG53 (July 1993)

3. Berg, M.D., Kreveld, M.V., Overmars, M., Schwarzkopf, O.: Computational Geo. Algo. & Appl. Springer, Heidelberg (2000)
4. Biswas, A., Bhowmick, P., Bhattacharya, B.B.: TIPS: On Finding a Tight Isothetic Polygonal Shape Covering a 2D Object. In: Kalviainen, H., Parkkinen, J., Kaarna, A. (eds.) SCIA 2005. LNCS, vol. 3540, pp. 796–799. Springer, Heidelberg (2005)
5. Bookstein, F.: Morphometric Tools for Landmark Data: Geometry and Biology. Cambridge Univ. Press, Cambridge (1991)
6. Boxer, L.: Computing deviations from convexity in polygons. PRL 14, 163–167 (1993)
7. Chazelle, B.: An optimal convex hull algorithm in any fixed dimension. Discrete Comput. Geom. 10, 377–409 (1993)
8. Chaudhuri, B.B., Rosenfeld, A.: On the computation of digital convex hull and circular hull of a digital region. Patt. Rec. 31, 2007–2016 (1998)
9. Cormen, T.H., Leiserson, C.E., Rivest, R.L.: Introduction to Algorithms. PHI (2000)
10. Costa, L.daF., Cesar, J.R.M.: Shape Analysis and Classification. CRC Press, Boca Raton (2001)
11. Edelsbrunner, H.: Weighted alpha shapes. TR-UIUCDCS-R-92-1760, Dept. Comput. Sci., Univ. Illinois, Urbana, IL (1992)
12. Graham, R.: An efficient algorithm for determining the convex hull of a finite point set. Info. Proc. Letters 1, 132–133 (1972)
13. Jarvis, R.: On the identification of the convex hull of a finite set of points in the plane. Info. Proc. Letters 2, 18–21 (1973)
14. Hyde, S., Andersson, S., Blum, Z., Lidin, S., Larsson, K., Landh, T., Ninham, B.: The Language of Shape. Elsevier, Amsterdam (1997)
15. Kirkpatrick, D., Seidel, R.: The ultimate planar convex hull algorithm? SIAM Jour. Comput. 15, 287–299 (1986)
16. Klette, R., Rosenfeld, A.: Digital Geometry: Geometric Methods for Digital Image Analysis. In: Morgan Kaufmann Series in Computer Graphics and Geometric Modeling, Morgan Kaufmann, San Francisco (2004)
17. Preparata, F.P., Shamos, M.I.: Computational Geometry – An Introduction. Springer, New York (1985)
18. Pitty, A.: Geomorphology. Blackwell, Malden (1984)
19. Rosenfeld, A., Kak, A.C.: Digital Picture Processing. Academic Press, London (1976) (Second Edition, 1982)
20. Sklansky, J., Kibler, D.F.: A theory of nonuniformly digitized binary pictures. IEEE Transactions on Systems, Man, and Cybernetics SMC-6(9), 637–647 (1976)
21. Sonka, M., Hlavac, V., Boyle, R.: Image Processing, Analysis, and Machine Vision. Chapman and Hall, Boca Raton (1993)
22. Stern, H.: Polygonal entropy: A convexity measure. PRL 10, 229–235 (1989)
23. Swart, G.: Finding the convex hull facet by facet. Journal of Algorithms, 17–48 (1985)
24. Zunic, J., Rosin, P.L.: A new convexity measure for polygons. IEEE Trans. PAMI 26(7), 923–934 (2004)

A Discrete Approach for
Supervised Pattern Recognition

João P. Papa[1], Alexandre X. Falcão[1], Celso. T.N. Suzuki[1],
and Nelson D.A. Mascarenhas[2]

[1] Institute of Computing, State University of Campinas,
Av. Albert Einstein 1216, Campinas, São Paulo, Brazil
{papa.joaopaulo,alexandre.falcao,celso.suzuki}@gmail.com
[2] Department of Computing, Federal University of São Carlos,
Rod. Washington Luis, Km 235, São Carlos, São Paulo, Brazil
nelson@dc.ufscar.br

Abstract. We present an approach for supervised pattern recognition
based on combinatorial analysis of optimum paths from key samples (prototypes), which creates a discrete optimal partition of the feature space
such that any unknown sample can be classified according to this partition. A training set is interpreted as a complete graph with at least one
prototype in each class. They compete among themselves and each prototype defines an optimum-path tree, whose nodes are the samples more
strongly connected to it than to any other. The result is an optimum-
path forest in the training set. A test sample is assigned to the class
of the prototype which offers it the optimum path in the forest. The
classifier is designed to achieve zero classification errors in the training
set, without over-fitting, and to learn from its errors. A comparison with
several datasets shows the advantages of the method in accuracy and
efficiency with respect to support vector machines.

Keywords: supervised learning, optimum-path forest, image foresting
transform, pattern recognition, graph-search algorithms.

1 Introduction

Graph-based approaches for pattern recognition are usually unsupervised (data
clustering). They mostly follow a same principle of creating a neighborhood
graph for the data samples and then removing inconsistent arcs based on some
criterion [14]. Other approaches interpret data clustering as a graph-cut problem [26]. Problems in these approaches are the overlap between different clusters
and no guarantee of success, because it is hard to assign samples to their correct class without any prior knowledge. Such problems are better treated by
supervised approaches.

Artificial neural networks (ANN) [13] and support vector machines (SVM) [4]
are supervised approaches actively pursued in the last years. An ANN multi-
layer perceptron (ANN-MLP) trained by backpropagation, for example, is an

V.E. Brimkov, R.P. Barneva, H.A. Hauptman (Eds.): IWCIA 2008, LNCS 4958, pp. 136–147, 2008.

unstable classifier. Its accuracy may be improved at the computational cost of using multiple classifiers and algorithms (e.g., bagging and boosting) for training classifier collections [16]. However, it seems that there is an unknown limit in the number of classifiers to avoid accuracy degradation [23]. ANN-MLP assumes that the classes can be separated by hyperplanes in the feature space. Such assumption is unfortunately not valid in practice. SVM was proposed to overcome the problem by assuming it is possible to separate the classes in a higher dimensional space by optimum hyperplanes.

Although SVM usually provides reasonable accuracies, its computational cost rapidly increases with the training set size and the number of support vectors. As a binary classifier, multiple SVMs are required to solve a multi-class problem [11]. Two main approaches are one-versus-all (OVA) and one-versus-one (OVO). OVA projects c SVMs to separate each class from the others. The decision is taken for the class with highest confidence value. OVO requires $\frac{c(c-1)}{2}$ SVMs by taking into account all binary combinations between classes. The decision is usually taken by majority vote. Tang and Mazzoni [28] proposed a method to reduce the number of support vectors in the multi-class problem. Their approach suffers from slow convergence and high computational complexity, because they first minimize the number of support vectors in several binary SVMs, and then share these vectors among the machines. Panda et al. [20] presented a method to reduce the training set size before computing the SVM algorithm. Their approach aims to identify and remove samples likely related to non-support vectors. However, in all SVM approaches, the assumption of separability may also not be valid in any space of finite dimension [8].

We present a supervised pattern classifier which exploits the *strength of connectedness* between samples in the feature space. Each sample in the training set is a node of a complete graph (i.e., the arcs connect all pairs of nodes) and each arc is weighted by the distance between the feature vectors of its corresponding nodes (Figure 1a). A path in the graph is a sequence of nodes connecting two terminal samples, each path has a cost given by a *path-cost function* (e.g., function f_{max} which assigns the maximum arc weight along the path), and a path is optimum when its cost is minimum. The strength of connectedness between two samples is inversely proportional to the cost of an optimum path between them. The method identifies the nodes (prototypes) that best represent all classes and computes the minimum-cost paths from the prototypes to the remaining nodes, such that each sample becomes part of one optimum path tree rooted at its most strongly connected prototype (Figure 1b). This procedure partitions the graph (training set) into an optimum path forest (OPF), concluding the training step.

The classification of a test sample evaluates the optimum paths from the prototypes to this sample incrementally, as though it were part of the graph (Figure 1c). The optimum path from the most strongly connected prototype, its label and path cost (classification cost) are assigned to the test sample (Figure 1d). Note the difference between an OPF classifier with f_{max} and the nearest neighbor approach [10]. The test sample is assigned to a given class, even when its closest labeled sample is from another class.

The OPF classifier extends the image foresting transform (IFT) — a tool for the design of image processing operators based on connectivity [12] — from the image domain to the feature space. The optimum path forest with f_{max} produces in the feature space similar result to the image segmentation based on relative-fuzzy connectedness [25] and watershed transform by markers (prototypes) [3,30,17]. These aspects represent important theoretical contributions by relating distinct approaches based on the same underlying concepts. Besides, we compute prototypes by exploiting the relation between minimum spanning trees and minimum-cost path trees for f_{max} [1]. This property holds that the label of a prototype will be the same of the training samples in its optimum path tree (i.e., the training samples are always correctly classified by the prototypes).

Other contribution of this work concerns learning algorithms, which can teach a classifier from its errors on a third evaluation set without increasing the training set size. As the samples in the test set can not be seen during the project, the evaluation set is necessary for this purpose. The basic idea is to randomly interchange samples of the training set with misclassified samples of the evaluation set, retrain the classifier and evaluate it again, repeating this procedure for a few iterations. The effectiveness is measured by comparing the results on the unseen test set before and after the learning algorithm. It should be expected an improvement in performance for any stable classifier.

The OPF framework has also been investigated for unsupervised pattern recognition with applications to image segmentation [24]. The supervised OPF classifier was first presented in [21] and it has been successfully used for texture object recognition [18]. The present work differs from the previous works in the learning methodology, which is simpler and can achieve better performance in most cases. It also uses more datasets to compare the classifiers.

Section 2 introduces the OPF classifier and its novel learning procedure. In Section 3, we compare the OPF classifier with support vector machines [4]. This comparison uses databases with outliers and non-separable multiple classes. Conclusions are further discussed in Section 4.

2   Optimum Path Forest Classifier

This section aims to present the new and fast approach to pattern recognition called Optimum Path Forest (OPF) [21]. The OPF approach works by modeling the patterns as being nodes of a graph in the feature space, where every pair of nodes are connected by an arc (complete graph). This classifier creates a discrete optimal partition of the feature space such that any unknown sample can be classified according to this partition. This partition is an optimum path forest computed in $\Re^n$ by the image foresting transform (IFT) algorithm [12].

Let Z_1, Z_2, and Z_3 be respectively the training, evaluation, and test sets with $|Z_1|$, $|Z_2|$, and $|Z_3|$ samples such as feature vectors. Let $\lambda(s)$ be the function that assigns the correct label i, $i = 1, 2, \ldots, c$, from class i to any sample $s \in Z_1 \cup Z_2 \cup Z_3$. Z_1 and Z_2 are labeled sets used to the design of the classifier and the unseen set Z_3 is used to compute the final accuracy of the classifier. Let $S \subset Z_1$

be a set of prototypes of all classes (i.e., key samples that best represent the classes). Let v be an algorithm which extracts n attributes (texture properties) from any sample $s \in Z_1 \cup Z_2 \cup Z_3$ and returns a vector $v(s) \in \Re^n$. The distance $d(s,t)$ between two samples, s and t, is the one between their feature vectors $v(s)$ and $v(t)$ (e.g., Euclidean or any other valid metric).

Let (Z_1, A) be a complete graph whose the nodes are the samples in Z_1. We define a path as being a sequence of distinct samples $\pi = \langle s_1, s_2, \ldots, s_k \rangle$, where $(s_i, s_{i+1}) \in A$ for $1 \leq i \leq k - 1$. A path is said trivial if $\pi = \langle s_1 \rangle$. We assign to each path π a cost $f(\pi)$ given by a path-cost function f. A path π is said optimum if $f(\pi) \leq f(\pi')$ for any other path π', where π and π' end at a same sample s_k. We also denote by $\pi \cdot \langle s, t \rangle$ the concatenation of a path π with terminus at s and an arc (s, t). The OPF algorithm uses the path-cost function f_{max}, because of its theoretical properties for estimating optimum prototypes:

$$f_{max}(\langle s \rangle) = \begin{cases} 0 & \text{if } s \in S, \\ +\infty & \text{otherwise} \end{cases}$$
$$f_{max}(\pi \cdot \langle s, t \rangle) = \max\{f_{max}(\pi), d(s,t)\}. \tag{1}$$

We can observe that $f_{max}(\pi)$ computes the maximum distance between adjacent samples in π, when π is not a trivial path. The OPF algorithm assigns one optimum path $P^*(s)$ from S to every sample $s \in Z_1$, forming an optimum path forest P (a function with no cycles which assigns to each $s \in Z_1 \backslash S$ its predecessor $P(s)$ in $P^*(s)$ or a marker nil when $s \in S$. Let $R(s) \in S$ be the root of $P^*(s)$ which can be reached from $P(s)$. The OPF algorithm computes for each $s \in Z_1$, the cost $C(s)$ of $P^*(s)$, the label $L(s) = \lambda(R(s))$, and the predecessor $P(s)$, as follows.

Algorithm 1. – OPF ALGORITHM

INPUT: A λ-labeled training set Z_1, prototypes $S \subset Z_1$ and the pair (v, d) for feature vector and distance computations.
OUTPUT: Optimum path forest P, cost map C and label map L.
AUXILIARY: Priority queue Q and cost variable cst.

1.   For each $s \in Z_1 \backslash S$, set $C(s) \leftarrow +\infty$.
2.   For each $s \in S$, do
3.   └ $C(s) \leftarrow 0$, $P(s) \leftarrow nil$, $L(s) \leftarrow \lambda(s)$, and insert s in Q.
4.   While Q is not empty, do
5.   Remove from Q a sample s such that $C(s)$ is minimum
6.   For each $t \in Z_1$ such that $t \neq s$ and
7. $C(t) > C(s)$, do
8.   Compute $cst \leftarrow \max\{C(s), d(s,t)\}$.
9.   If $cst < C(t)$, then
10.   If $C(t) \neq +\infty$, then remove t from Q.
11. $P(t) \leftarrow s$, $L(t) \leftarrow L(s)$, $C(t) \leftarrow cst$ and insert t in Q.

Lines $1-3$ initialize maps and insert prototypes in Q. The main loop computes an optimum path from S to every sample s in a non-decreasing order of cost

(Lines $4 - 11$). At each iteration, a path of minimum cost $C(s)$ is obtained in P when we remove its last node s from Q (Lines 5). Lines $9 - 11$ evaluate if the path that reaches an adjacent node t through s is cheaper than the current path with terminus t and update the position of t in Q, $C(t)$, $L(t)$ and $P(t)$ accordingly. The label $L(s)$ may be different from $\lambda(s)$, leading to classification errors in Z_1. The training finds prototypes with none classification errors in Z_1. The OPF algorithm works with two phases: training and classification (test), as described in the following two sections.

2.1   Training Phase

We say that S^* is an optimum set of prototypes when Algorithm 1 propagates the labels $L(s) = \lambda(s)$ for every $s \in Z_1$. Set S^* can be found by exploiting the theoretical relation between *Minimum Spanning Tree* (MST) [1] and optimum path tree for f_{max}. The training essentially consists of finding S^* and an OPF classifier rooted at S^*.

By computing an MST in the complete graph (Z_1, A), we obtain a connected acyclic graph whose nodes are all samples of Z_1 and the arcs are undirected and weighted by the distances d between adjacent samples. The spanning tree is optimum in the sense that the sum of its arc weights is minimum as compared to any other spanning tree in the complete graph. In the MST, every pair of samples is connected by a single path which is optimum according to f_{max}. That is, the minimum-spanning tree contains one optimum-path tree for any selected root node [1].

The optimum prototypes are the closest elements of the MST with different labels in Z_1. By removing the arcs between different classes, their adjacent samples are inserted in S^* and Algorithm 1 can compute an optimum-path forest with zero classification errors in Z_1. It is not difficult to see that the optimum paths between classes should pass through the same removed arcs of the minimum-spanning tree. The choice of prototypes as just described blocks these passages, avoiding samples of any given class be reached by optimum paths from prototypes of other classes.

Note that, a given class may be represented by multiple prototypes (i.e., optimum-path trees) and there must exist at least one prototype per class.

2.2   Classification

For any sample $t \in Z_3$, the OPF consider all arcs connecting t with samples $s \in Z_1$, as though t were part of the graph (Figure 1c). Considering all possible paths from S^* to t, we wish to find the optimum path $P^*(t)$ from S^* and label t with the class $\lambda(R(t))$ of its most strongly connected prototype $R(t) \in S^*$. This path can be identified incrementally, by evaluating the optimum cost $C(t)$ as

$$C(t) = \min\{\max\{C(s), d(s, t)\}\}, \ \forall s \in Z_1. \tag{2}$$

Let the node $s^* \in Z_1$ be the one that satisfies the above equation (i.e., the predecessor $P(t)$ in the optimum path $P^*(t)$). Given that $L(s^*) = \lambda(R(t))$, the

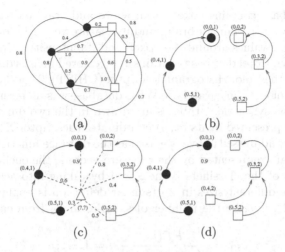

Fig. 1. (a) Complete weighted graph for a simple training set. (b) Resulting optimum path forest for f_{max} and two given prototypes (circled nodes). The entries (x, y) over the nodes are, respectively, the cost and the label of the samples. The arrow indicates the predecessor node in the optimum path. (c) Test sample (gray square) and its connections (dashed lines) with the training nodes. (d) The optimum path from the most strongly connected prototype, its label 2, and classification cost 0.4 are assigned to the test sample.

classification simply assigns $L(s^*)$ to t. An error occurs when $L(s^*) \neq \lambda(t)$. Note the difference between an OPF classifier with f_{max} and the nearest neighbor approach [10]. The test sample is assigned to a given class, even when its closest labeled sample is from another class (Figure 1d).

3 Experimental Results

Two experiments were conducted to demonstrate the discriminative power of the OPF classifier for pattern recognition as compared to support vector machines (SVM).

In the first experiment (Section 3.1), we compare the accuracy of the OPF and SVM classifiers, by training them on set Z_1 and testing them on set Z_3 for several randomly selected instances of Z_1 and Z_3.

In the second experiment (Section 3.2), we evaluate the ability of OPF and SVM to learn from their errors on a third evaluation set Z_2.

Several datasets have been used for comparative analysis: MPEG-7 [19], Brain [7], Corel [9], Wisconsin Breast Cancer (WBC) [2] and four synthetic databases [15]: Cone-torus, Saturn, Petals and Boat. The samples in these datasets represent shapes, images, pixels (voxels).

The MPEG-7 is a database with 1400 shapes equally distributed in 70 classes. We used the Fourier Shape descriptor (FD) [22] and the Multiscale Fractal Dimension descriptor (MS) [29] as the feature vectors that will encode the shapes.

The Brain database contains voxels from white and gray matters in magnetic resonance images of the human brain, and its feature vector is composed by the lowest and highest values around the voxel, and its intensity value.

The well known Corel database contains 1607 colored images distributed in 49 classes. We used the Color Histogram descriptor (CHIST) [27] as been the feature vector to represent each image and the WBC database was represented by its own descriptor. As the synthetic databases are drawn in the two dimensional space, each sample is represented by its (x, y) coordinates (descriptor XY). Recall that for all descriptors addressed above, we used as the distance function between two samples, which are represented by their feature vectors, the euclidean distance.

The accuracy of the classifiers is measured by taking into account that the classes may have different sizes in Z_2 (similar definition is applied for Z_3). Let $NZ_2(i)$, $i = 1, 2, \ldots, c$, be the number of samples in Z_2 from each class i. We define

$$e_{i,1} = \frac{FP(i)}{|Z_2| - |NZ_2(i)|} \quad i = 1, \ldots, c \tag{3}$$

and

$$e_{i,2} = \frac{FN(i)}{|NZ_2(i)|}, \quad i = 1, \ldots, c \tag{4}$$

where $FP(i)$ and $FN(i)$ are the false positives and false negatives, respectively. That is, $FP(i)$ is the number of samples from other classes that were classified as being from the class i in Z_2, and $FN(i)$ is the number of samples from the class i that were incorrectly classified as being from other classes in Z_2. The errors $e_{i,1}$ and $e_{i,2}$ are used to define

$$E(i) = e_{i,1} + e_{i,2}, \tag{5}$$

where $E(i)$ is the partial sum error of class i. Finally, the accuracy Acc of the classification is written as

$$Acc = \frac{2c - \sum_{i=1}^{c} E(i)}{2c} = 1 - \frac{\sum_{i=1}^{c} E(i)}{2c}. \tag{6}$$

3.1   Effectiveness of the Classifiers Before Learning with the Errors

We used the well known LibSVM [5] software as the SVM implementation. This package implements optimization procedures to increase accuracy, at the price of a higher computational time, and uses the Radial Bases Function (RBF) as the kernel to map the samples in the feature space to a higher dimensional space.

We executed the OPF and SVM algorithms 10 times to compute their accuracies (Equation 6), using different randomly selected training Z_1 and test Z_3 sets. The results are displayed in the following format: $x \pm y(z)$, where x, y and z are, respectively, the mean accuracy and its standard deviation and the mean Kappa coefficient [6]. The training (Z_1) and test (Z_3) sets percentage were, respectively, 50% and 50% for all datasets used. Recall that the evaluation set (Z_2) is only used in the next experiment. The Table 1 presents the results.

Table 1. Results obtained before learning with the errors. The word **OWN** remains to the own database descriptor.

Database (descriptor)	OPF	SVM
MPEG-7 (FD)	**0.67±0.01(0.33)**	0.60±0.01(0.20)
MPEG-7 (MS)	**0.83±0.01(0.66)**	**0.83±0.01(0.66)**
Corel (CHIST)	**0.84±0.01(0.71)**	0.83±0.01(0.70)
WBC (OWN)	**0.94±0.01(0.96)**	0.84±0.17(0.69)
Cone-torus (XY)	**0.84±0.01(0.68)**	0.83±0.02(0.67)
Saturn (XY)	0.74±0.01(0.64)	**0.84±0.04(0.67)**
Petals (XY)	**0.98±0.01(0.97)**	**0.98±0.01(0.97)**
Boat (XY)	**1.0±0.0(1.0)**	**1.0±0.0(1.0)**

Note that the OPF classifier achieved 5 wins, 2 ties and just 1 lose. Note also that the OPF outperformed SVM in about 10% on the MPEG-7 dataset using the FD descriptor and on the WBC dataset.

3.2 Effectiveness of the Classifiers by Learning with the Errors

The idea of this section is to evaluate the ability of learning with the errors in a third evaluation set Z_2. For randomly selected instances of Z_1, Z_2 and Z_3, the classifiers are trained on Z_1 and tested on Z_2 during a few iterations that replace samples of Z_1 by misclassified samples of Z_2. After 10 iterations, the classifiers are finally tested on Z_3. This process is also repeated 10 times to measure the average accuracies of the classifiers. The learning procedure aims to capture from Z_2 the most informative samples.

The algorithm outputs a *learning curve* over T iterations, which reports the accuracy values of each instance of the classifier during learning (Figure 2), and the final OPF/SVM classifier. It is shown in Algorithm 2.

Algorithm 2. – LEARNING ALGORITHM

INPUT: Training and evaluation sets labeled by λ, Z_1 and Z_2, number T of iterations, and the pair (v, d) for feature vector and distance computations.

OUTPUT: Learning curve $\mathcal{L}$ and the best OPF/SVM classifier.

AUXILIARY: False positive and false negative arrays, FP and FN, of sizes c, list LM of misclassified samples and variable Acc.

1. *For each iteration $I = 1, 2, \ldots, T$, do*
2. | $LM \leftarrow \emptyset$
3. | *Train OPF/SVM with Z_1.*
4. | *For each class $i = 1, 2, \ldots, c$, do*
5. | └ $FP(i) \leftarrow 0$ *and* $FN(i) \leftarrow 0$.
6. | *For each sample $t \in Z_2$, do*
7. | | *Use the classifier obtained in Line 3 to classify t with a label $L(t)$.*
8. | | *If $L(t) \neq \lambda(t)$, then*
9. | | | $FP(L(t)) \leftarrow FP(L(t)) + 1$.

10. $\qquad \Big\lfloor \quad \Big\lfloor \quad FN(\lambda(t)) \leftarrow FN(\lambda(t)) + 1.$
11. $\qquad \Big\lfloor \quad \Big\lfloor \quad LM \leftarrow LM \cup t.$
12. $\qquad$ *Compute Acc by Equation 6 and save the current instance of the classifier.*
13. $\qquad \mathcal{L}(I) \leftarrow Acc$
14. $\qquad$ *While* $LM \neq \emptyset$
15. $\qquad \Big\lfloor \quad LM \leftarrow LM \backslash t$
16. $\qquad \Big\lfloor \quad$ *Replace t by a randomly object of the same class*
17. $\qquad \Big\lfloor \quad \Big\lfloor \quad$ *in* Z_1, *under some hard constraints.*
18. *Select the instance of the classifier with highest accuracy* $\mathcal{L}(\mathcal{I})$

The Learning algorithm can be used for both OPF and SVM classifier, by changing Lines 3 and 16 − 17. For OPF, Line 3 is performed by computing $S^* \subset Z_1$ as in Section 2.1 and the predecessor map P, label map L and cost map C by Algorithm 1. The classification is done by finding $s^* \in Z_1$ that satisfies Equation 2, i.e., $P(t) = s^*$ and $L(t) \leftarrow L(s^*)$. The constraints in Lines 16 − 17 refer to not use the prototypes in the sample interchanging process between Z_1 and Z_2. For SVM implementation we use the LibSVM package [5] with Radial Basis Function (RBF) kernel and parameter optimization in Line 3. The hard constraints in Lines 16 − 17 refer to not use the support vectors in the sample interchanging process.

A different version of the learning algorithm for OPF was presented in [21]. The algorithm computed a relevance degree for each sample of Z_1, based on the numbers of right and wrong classifications on Z_2 involving that sample in the optimum path. The relevance degree of a sample was computed by subtracting the number of wrong classifications from the number of right classifications involving that sample. Samples with negative degrees were considered irrelevant, and so selected for the interchanging process. The idea was to eliminate outliers from Z_1, since these samples are usually irrelevant. The present version is simpler and faster than the previous one — i.e., it does not need to compute the relevance degrees of all nodes in Z_1. Nevertheless, it has shown higher accuracies in all tested datasets except sometimes for Brain [7].

The Z_1, Z_2 and Z_3 sets percentage were, respectively, 20%, 30% and 50%, for all databases used. Table 2 presents the results. Note that these accuracies were obtained over the unseen test set Z_3. Again, we executed the OPF and SVM algorithms 10 times to compute their accuracies (Equation 6). The results were displayed in the same format used in the previous section.

After the learning procedure, the OPF classifier achieved 6 wins and 2 ties. Note that all accuracies were increased, both for OPF and SVM. Recall that the proposed learning approaches can improve the performance of the classifiers over the Z_3 set without increasing the training set size.

We also computed the execution time of the methodologies. Table 3 displays these values in seconds. The OPF classifier was, on average, 118.315 times faster than SVM. Note that the SVM algorithm had a slow performance due to the fact of the optimization procedure implemented in the LibSVM [5]. However, by removing the optimization procedures, this processing time could be decreased. In turn, this could produce lower classification rates.

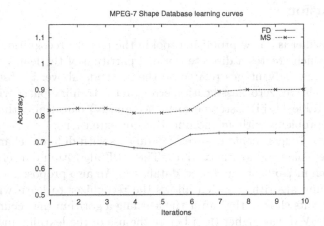

Fig. 2. OPF learning curves for MPEG-7 Shape Database using the Fourier Shape descriptor (FD) and the Multiscale Fractal Dimension (MS). These accuracies were obtained over the evaluation set Z_2.

Table 2. Results obtained by learning with the errors. The word **OWN** remains to the own database descriptor.

Database (descriptor)	OPF	SVM
MPEG-7 (FD)	**0.75±0.03(0.51)**	0.62±0.01(0.27)
MPEG-7 (MS)	**0.88±0.02(0.76)**	0.84±0.01(0.69)
Corel (CHIST)	**0.86+0.03(0.73)**	0.84±0.01(0.71)
WBC (OWN)	**0.98±0.01(0.96)**	0.96±0.01(0.92)
Cone-torus (XY)	**0.92±0.01(0.84)**	0.85±0.02(0.68)
Saturn (XY)	**0.93±0.02(0.87)**	0.84±0.03(0.67)
Petals (XY)	**1.0±0.0(1.0)**	**1.0±0.0(1.0)**
Boat (XY)	**1.0±0.0(1.0)**	**1.0±0.0(1.0)**

Table 3. Classifiers execution times

Database (descriptor)	OPF	SVM
MPEG-7 (FD)	**1.17**	260.77
MPEG-7 (MS)	**1.15**	250.76
Corel (CHIST)	**2.09**	270.73
WBC (OWN)	**0.73**	127.38
Cone-torus (XY)	**0.5**	28.01
Saturn (XY)	**0.31**	15.02
Petals (XY)	**0.29**	14.11
Boat (XY)	**0.29**	14.11

4   Conclusion

The OPF classifier is a new promising tool in the pattern recognition and correlated fields, which creates a discrete optimal partition of the feature space and presents four new advantages related to the methods above: i) absence of parameters, ii) achieves zero classification errors in the training phase without data over-fitting (Table 1), iii) faster performance and iv) can deal with multi-class classification problems without modifications or extensions.

We conducted experiments in several databases with different characteristics, such as shape, color and synthetic data. The OPF algorithm outperformed the SVM approach in 7 out of the 8 used databases. We also proposed a simple but efficient learning algorithm, which allows the recognizer to learn with its own errors. The results showed that after the learning algorithm the accuracy rate of the OPF and SVM was higher than before the use of the learning methodology. We also showed that OPF algorithm is much faster than the support vector machines approach, which is a very interesting advantage in large databases.

References

1. Allène, C., Audibert, J.Y., Couprie, M., Cousty, J., Keriven, R.: Some links between min-cuts, optimal spanning forests and watersheds. In: Mathematical Morphology and its Applications to Image and Signal Processing (ISMM'07), MCT/INPE, pp. 253–264 (2007)
2. Asuncion, A., Newman, D.: UCI machine learning repository (2007)
3. Beucher, S., Meyer, F.: The morphological approach to segmentation: The watershed. Mathematical Morphology in Image Processing, 433–481 (1993)
4. Boser, B., Guyon, I., Vapnik, V.: A training algorithm for optimal margin classifiers. In: Proc. 5th Workshop on Comp. Learning Theory, pp. 144–152 (1992)
5. Chang, C.C., Lin, C.J.: LIBSVM: A library for support vector machines (2001), http://www.csie.ntu.edu.tw/~cjlin/libsvm.
6. Cohen, J.: A coefficient of agreement for nominal scales. Educational and Psychological Measurement 20, 37–46 (1960)
7. Collins, D., Zijdenbos, A., Kollokian, V., Sled, J., Kabani, N., Holmes, C., Evans, A.: Design and construction of a realistic digital brain phantom. IEEE Trans. on Medical Imaging 17(3), 463–468 (1998)
8. Collobert, R., Bengio, S.: Links between perceptrons, mlps and svms. In: Proceedings of the 21th Int. Conf. on Machine learning, p. 23 (2004)
9. Corel: Corel stock photo images, http://www.corel.com
10. Cover, T., Hart, P.: Nearest neighbor pattern classification. IEEE Transactions on Information Theory 13(1), 21–27 (1967)
11. Duan, K., Keerthi, S.S.: Which is the best multiclass svm method? an empirical study. Multiple Classifier Systems, 278–285 (2005)
12. Falcão, A., Stolfi, J., Lotufon, R.: The image foresting transform: Theory, algorithms, and applications. IEEE Trans. Pattern Anal. Mach. Intell. 26(1), 19–29 (2004)
13. Haykin, S.: Neural networks: A comprehensive foundation. Prentice Hall, Englewood Cliffs (1994)

14. Jain, A.K., Dubes, R.C.: Algorithms for clustering data. Prentice-Hall, Inc., Upper Saddle River, NJ, USA (1988)
15. Kuncheva, L.: Fuzzy classifier design. Physica-Verlag and Springer (2000)
16. Kuncheva, L.I.: Combining Pattern Classifiers: Methods and Algorithms. Wiley-Interscience, Chichester (2004)
17. Lotufo, R.A., Falcão, A.X., Zampirolli, F.A.: Fast euclidean distance transform using a graph-search algorithm. In: Proc. of the 13th Braz. Symposium on Computer Graphics and Image Processing, pp. 269–275 (2000)
18. Montoya-Zegarra, J., Papa, J., Leite, N., Torres, R., Falcão, A.: Rotation-invariant texture recognition. In: Bebis, G., Boyle, R., Parvin, B., Koracin, D., Paragios, N., Tanveer, S.-M., Ju, T., Liu, Z., Coquillart, S., Cruz-Neira, C., Müller, T., Malzbender, T. (eds.) ISVC 2007, Part II. LNCS, vol. 4842, pp. 193–204. Springer, Heidelberg (2007)
19. MPEG-7: Mpeg-7: The generic multimedia content description standard, part 1. IEEE MultiMedia 09(2), 78–87 (2002)
20. Panda, N., Chang, E.Y., Wu, G.: Concept boundary detection for speeding up svms. In: Proc. of the 23rd Int. Conf. on Machine learning, pp. 681–688 (2006)
21. Papa, J.P., Falcão, A.X., Miranda, P.A.V., Suzuki, C.T.N., Mascarenhas, N.D.A.: Design of robust pattern classifiers based on optimum-path forests. In: Mathematical Morphology and its Applications to Image and Signal Processing (ISMM 2007), MCT/INPE, pp. 337–348 (2007)
22. Persoon, E., Fu, K.: Shape Discrimination Using Fourier Descriptors. IEEE Transanctions on Systems, Man, and Cybernetics 7(3), 170–178 (1977)
23. Reyzin, L., Schapire, R.E.: How boosting the margin can also boost classifier complexity. In: Proc. of the 23rd Int. Conf. on Machine learning, pp. 753–760 (2006)
24. Rocha, L.M., Falcão, A.X., Meloni, L.G.P.: A robust extension of the mean shift algorithm using optimum path forest. In: 8th International Workshop on Combinatorial Image Analysis (accepted, 2008)
25. Saha, P.K., Udupa, J.K.: Relative fuzzy connectedness among multiple objects: theory, algorithms, and applications in image segmentation. Comput. Vis. Image Underst. 82(1), 42–56 (2001)
26. Shi, J., Malik, J.: Normalized cuts and image segmentation. IEEE Trans. on Pattern Analysis and Machine Intelligence 22(8), 888–905 (2000)
27. Swain, M., Ballard, D.: Color Indexing. International Journal of Computer Vision 7(1), 11–32 (1991)
28. Tang, B., Mazzoni, D.: Multiclass reduced-set support vector machines. In: Proc. of the 23rd Int. Conf. on Machine learning, pp. 921–928 (2006)
29. Torres, R., Falcão, A.X., Costa, L.: A graph-based approach for multiscale shape analysis. Pattern Recognition 37(6), 1163–1174 (2004)
30. Vincent, L., Soille, P.: Watersheds in digital spaces: An efficient algorithm based on immersion simulations. IEEE TPAMI 13(6), 583–598 (1991)

Robust Decomposition of Thick Digital Shapes

Alexandre Faure and Fabien Feschet*

Univ. Clermont 1, LAIC, IUT, Campus des Cézeaux, F-63172 Aubière, France
{afaure,feschet}@laic.u-clermont1.fr

Abstract. This paper deals with the widely studied problem of decomposition of digital shapes. The Tangential Cover [7] is a powerful tool that computes the set of all maximal segments of a digital curve. In previous works [4], the Tangential Cover has been extended to a new class of "thick digital curves". This extension brought up some major issues. In the present paper, we generalize even more the notion of Tangential Cover, in order to fix those issues. We propose a new relevant way of representing thick digital curves, as sets of consecutive triangles. Then, we study the use of this representation to define a generalized Tangential Cover, and we show some results produced by our technique.

1 Introduction

The decomposition of digital shapes into specific subparts is an important task, often among the first stages in image processing, pattern recognition or shape analysis. One of the main goals is to obtain methods as generic as possible, while maintaining low time complexities and providing robustness to noise. Classically, the primitives used are digital straight segments (DSS). Those are the object of many publications, among which the Tangential Cover [5]. The task we are occupied with is the generalization of the Tangential Cover to primitives other than DSS. In particular, we extend it to a new class of "thick digital curves". Those allow the handling of irregularities inherent to digital shapes. In the present article, we exhibit our new method to manage this extension, using triangulations of thick digital curves. Our generalized Tangential cover can be used to manage other primitives than straight or approximately straight ones.

We start off this article by recalling the context of this study, providing the original definition of the Tangential Cover and its direct application to DSS. Then, we generalize the notion of Tangential Cover. We furnish a definition of thick digital curves, and adapt the generalized Tangential Cover to such curves. We then justify our use of triangulations, and the issues emerging from it. We present the different elements obtained when triangulating thick digital curves. Then, we propose an algorithm to obtain the requisite total ordering on the set of triangles. Finally, we exhibit results of our technique, conclude and give some perspectives regarding the cleaning of noisy curves.

---

* This work was supported by the French National Agency of Research (ANR) under contract GEODIB.

V.E. Brimkov, R.P. Barneva, H.A. Hauptman (Eds.): IWCIA 2008, LNCS 4958, pp. 148–159, 2008.

2   Definitions And Context

2.1   The Tangential Cover

Definition 1. *A discrete curve C is an ordered set of integer points $(p_1, \ldots, p_n)$ such that the real polygonal line passing through them, in order, is a simple polygonal line.*

The curve is closed when the next point of p_n is p_1. For a discrete curve C, we will denote by $\mathcal{C}$ the real polygonal line passing through the points of C.

The notion of maximal digital straight segment (maximal DSS) plays a particular role in our study. We refer to the book [8] for basic properties of DSS. For any i and j in $\mathbb{Z}$, let us denote by $[i, j]$ the integer interval corresponding to $\{k \in \mathbb{Z}, \ i \le k \le j\}$. We consider, in the sequel, a discrete curve $C = (p_1, \ldots, p_n)$.

Definition 2. *A set $P = (p_i, \ldots, p_j)$ is called a maximal digital straight segment (DSS) if and only if P is a DSS with $\forall k \in [i, j]$, $p_k \in P$ and maximal with respect to inclusion of subsets of C.*

Obviously, using the order of the points in C, a DSS P is maximal if and only if both sets $P \cup \{p_{j+1}\}$ and $P \cup \{p_{i-1}\}$ are not DSS.

Definition 3. *The tangential cover $T(C)$ of the curve C is the set of all maximal DSS of C.*

The tangential cover has proven itself a powerful tool for studying digital shapes [5]. Intrinsic properties of the maximal DSS have been used in [7] and [6] to incrementally build the tangential cover with a linear complexity. The incremental construction is quite simple. Points of C are added in order to the front end of a current DSS. According to definition 2, when the test fails, the current DSS is maximal. Points are then deleted from the rear end of the DSS, until the addition of a new point is valid (Fig. 1, left). The process stops when all maximal DSS have been constructed. The resulting tangential cover $T(C)$ is a canonical representation of closed digital curves. In order to graphically represent this construction, $T(C)$ is mapped into the class of circular arcs graph, as seen in [5].

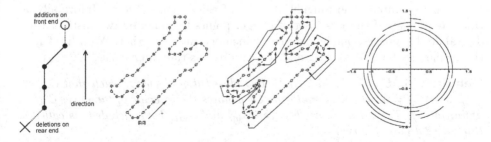

Fig. 1. Left: additions and deletions performed on a DSS. Right: a chromosome digital shape; the set of its maximal segments; its tangential cover, where to set apart overlapping arcs, the radius is increased or decreased accordingly.

The tangential cover has been used to determine the polygonalization of discrete curves with the least number of vertices. This problem is known as min-DSS [8]. It is solved in [6], using the circular arcs graph rather than the original curve. There, the tangential cover either uses standard or naive discrete segments based on the arithmetical definition of Reveillès [11].

2.2   The Generalized Tangential Cover

In order to apply the notion of tangential cover in other contexts than the one of DSS, we now extend it in the most general possible way.

Let S be a finite set composed of the elements $(e_1, \ldots, e_n)$ that we suppose to be totally ordered and thus numbered by integer indices. We first define the notion of subpart. To do this, we introduce a predicate $\mathcal{P}$ with value in $\{\text{true}, \text{false}\}$.

Definition 4. *A valid subpart T of the set S regarding a given predicate $\mathcal{P}$ is a set $(e_l, \ldots, e_r)$ of elements of S, such that $l \leq r$, $\forall k \in [l, r]$, $e_k \in T$ and $\mathcal{P}(T)$ is true.*

The subparts of S are ordered by inclusion which is however a partial order. But using the order of the elements of S, we can notice that a subpart T will be maximal if and only if $\mathcal{P}(T) = \text{true}$ and neither $T \cup \{e_{r+1}\}$ nor $T \cup \{e_{l-1}\}$ are valid for the predicate $\mathcal{P}$. This is exactly the same property as for the DSS case.

Definition 5. *The (generalized) Tangential Cover of S, with respect to $\mathcal{P}$, is the set of all valid maximal subparts of S.*

When the elements correspond to the integer points of C and when $\mathcal{P}$ is the predicate *"to be a DSS"*, our new definition leads to the usual notion of tangential cover. The latter was used to solve the min-DSS [8] problem. Here, the generalized Tangential Cover may be used to solve a more general so-called *min-subpart* problem, the generalization being the nature of the subparts and the predicate of maximality.

2.3   Thick Digital Curves

We wish to compute the tangential cover of thick digital curves. Informally, a thick digital closed curve is a set of integer points bounded by two distinct thin digital Jordan curves, one being strictly included into the other. We will refer to the former as the *inner curve*, and to the latter as the *outer curve*.

Definition 6. *Let C_{int} and C_{ext} be two closed digital curves such that $C_{int}^{\circ} \subset C_{ext}^{\circ}$, and $C_{int}^{\circ} \neq \emptyset$, where the notation C° denotes the interior of the curve C (see definition 1). The set of points between C_{int} and C_{ext} (both included) is called a thick digital closed curve.*

Thin digital curves are particular thick digital curves where the two sets C_{int} and C_{ext} are equal. Figure 2 represents a thick digital curve. Our definition 6 is related to the notion of *"simple polygons with one hole"*.

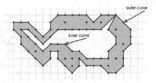

Fig. 2. A thick digital closed curve

3   Extension to Thickness

Let us now extend the tangential cover to thick digital curves, according to definition 6. First, we notice that a total ordering on the elements of the input curve is needed in order to compute its tangential cover. This means that notions of *previous* and *next* must make sense for each atomic element. Second, a predicate must be defined to obtain valid subparts of the thick curve. We currently postpone the problem of a total ordering and focus on the predicate.

As a predicate, we use the notion of "α-thickness" [1], which is equivalent to the notion of "blurred segments", introduced in [3].

Definition 7. *A set of points is a subset of an α-thick digital line if and only if its convex hull has an isothetic thickness inferior to α.*

The isothetic thickness of a convex set is the minimum between its vertical and its horizontal thicknesses (see figure 3, left). The vertex for which the isothetic thickness is reached gives us an important information regarding the convex hull's direction. The facing edge is the slope of the convex set (see figure 3, right). An algorithm that computes the isothetic thickness of a given convex hull in logarithmic time at worst is given in [1]. If the isothetic thickness of the convex hull of a set of elements is inferior to a fixed α value, then the set is a valid subpart of the input thick digital curve. Now if this same set is maximal (cannot be extended left nor right), then it belongs to the thick tangential cover. This implies of course that convex hull of elements should be defined.

In order to build the set of all maximal subparts of a thick digital curve, we need to compute its successive convex hulls. According to the construction process of the tangential cover which we described in section 2.1, this means that we need to add *and* subtract points from convex hulls. Managing dynamic convex hulls is a difficult problem, and known algorithms (such as [10]) remain

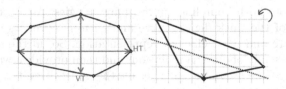

Fig. 3. The isothetic thickness of this convex hull is the minimum between its vertical (VT) and horizontal (HT) thicknesses

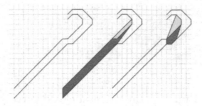

Fig. 4. A problem occurring when trying to compute maximal segments on a thick digital curve

too costly in time complexity for our usage. We observe that our construction process only requires deletions on the rear end and insertions on the front end of convex hulls. Thus, we need an incremental and decremental method rather than an actual dynamic one, where insertions and deletions may be performed in the middle of the convex hull. Buzer proposed such an approach in [2], based on Melkman's famous algorithm [9]. Its time complexity is linear in the size of the input simple polygonal chain.

The recent work [4] allow us to plug the isothetic thickness computation algorithm to this incremental/decremental management of convex hulls. Such an assembly is not trivial at all since the two methods were not designed to work together. The resulting time complexity is $O(n \log n)$. Successive convex hulls of the curves subparts are built, and then their isothetic thicknesses are tested to determine if their validity. Hence, the desired thick tangential cover is built. However, in [4], a thick digital curve was only represented by the points of the inner and the outer bounding curves. A strategy was proposed to add points alternatively from one of the two curves but this leads to the generation of subparts that do not belong to the thick curve (see Fig. 4 right). A solution based on the control of the speed on the inner and the outer curve was proposed in [4] but this does not solve all problems and cannot be used to provide a generalized approach. The problem of this study was the inexistence of a total ordering as well as an imprecise definition of the elements of computation.

One fact was missing in [4]: the management of the interior (in the weak sense where boundaries are included) of the thick digital curve. The only way to overcome this serious drawback is to look at the curve as a series of subparts based on the two bounding curves and only representing its interior. To achieve that, a triangulation of the thick curve (seen as a polygon with one hole) appears to us as the best candidate. Then, the objects added to the tangential cover would not be points from both bounding curves, but triangles. As the triangles lie on the boundaries of the thick digital curve and entirely cover its interior, this is the most accurate and most efficient representation. So, in the sequel, we first present a precise study of our triangulations and then provide a way to construct a total ordering of the triangles. As a result, the generalized tangential cover is perfectly defined on thick digital curves.

4   Triangulation

In order to better handle the interior of thick digital curves, we want to triangulate those. There exist several ways to compute triangulations of such polygons. We experimented some of them, and finally chose a method based on constrained Delaunay triangulation [12]. Its main benefit regarding our usage is to avoid the creation of narrow triangles. Moreover, no new points (Steiner points) are added and all triangles in the resulting triangulation have all their vertices on the inner, on the outer or on both curves. Given an input of an inner and an outer curve, the algorithm returns a set of adjacent triangles. Let us have a look at all possible elements of such a triangulation (see Fig. 5).

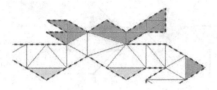

Fig. 5. A triangulation for a given thick shape. White triangles are "heterogeneous", grey ones are "homogeneous". Dark grey triangles lie strictly on the outer border, light grey ones lie on the inner border.

We wish to take care of the interior of the thick curve, and not only of both of its bounding curves. Intuitively, to achieve that goal, the subparts that we create must not entirely lie on only one bounding curve. However, in the general case, this is not possible. Triangles whose vertices all belong to the same curve (inner or outer) are called *homogeneous*. Other triangles are called *heterogeneous*. For our usage, which is to find maximal subparts of the curve regarding a predefined predicate, we will restrain the validity of the subparts based on the triangles and their homogeneity.

Definition 8. *Let a thick digital curve C, and its associated triangulation $T = \{T_1, T_2, ...T_n\}$. A valid subpart of C is a set of adjacent triangles $\{T_i, ...T_j\}$ which contains at least one heterogeneous triangle.*

A set of adjacent homogeneous triangles only lie by definition on one of the bounding curves. We consider that such a set is not a valid subpart of the curve. Indeed, it only represents an excrescence. The adjacency relation between triangle is the edge adjacency.

Definition 9. *Two triangles are considered neighbors if and only if they share one edge.*

For each connected non trivial subset of a triangulation, each triangle has at least one neighbor. It is obvious that there can be at most three. Any heterogeneous triangle has at least two neighbors. It may also have three neighbors. In such a case, we prove that exactly one of its neighbor is homogeneous.

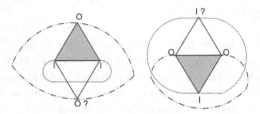

Fig. 6. Illustration for proof 4

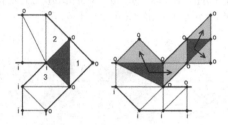

Fig. 7. Triangulation between the inner (i) and outer (o) curves. Left, an heterogeneous 3-N triangle. Right, two consecutive examples of homogeneous 3-N triangles (dark grey), which lead to two *pockets* (light grey) apiece. Finally, there are three *dead-end* triangles.

Lemma 1. *An heterogeneous triangle with three neighbors has exactly one homogeneous neighbor.*

Proof. We recall that the curves interior is not empty: $\mathcal{C}_{int}^{\circ} \neq \emptyset$. Let us have a look at figure 6 (left). If the neighbor emerging from the "I-I" edge was heterogeneous, this would mean that the inner curves interior would cross a triangle, which is impossible since, in our method, we only triangulate the interior of the thick curve. The same goes for the symmetric situation (figure 6, right) where the inner curve would not strictly be included into the outer one. Thus the triangle has one homogeneous neighbor and this is necessarily the only one.   □

In the sequel, we call k-N triangle, a triangle with exactly k neighbors. The classification of homogeneous triangles is easy : 3-N homogeneous triangles have at least two homogeneous neighbors, 2-N homogeneous triangles have at least one homogeneous neighbor, 1-N homogeneous triangles can have either a homogeneous or a heterogeneous neighbor.

An homogeneous triangle is always part of an homogeneous subset of the curve, which we will call a *pocket* of the curve. 3-N homogeneous triangles are shared between several pockets. Such pockets lead inevitably to one 1-N homogeneous triangle, which we will call a *dead-end* triangle. We refer to figure 7 for all definitions. Dead-end triangles and pockets are related to heterogeneous 3-N triangles. Indeed, if it was not the case, then we could make a path in the triangulation only composed of homogeneous triangles. As the thick curve is finite, we would obtain that all triangles of the triangulation are homogeneous, which is impossible. Thus, we alway reach a heterogeneous triangle by following the adjacency from an homogeneous triangle.

As stated in section 3, the computation of a curves tangential cover requires a total ordering on the elements of the curve. In order to use the set of triangles as an input to the tangential cover algorithm, it is mandatory to build such a total ordering between the triangles produced by the triangulation.

5 Total Ordering

We gave all possible configurations of the obtained triangles in section 4. We will denote by HET (resp. HOM) heterogeneous (resp. homogeneous) triangles. In the case of closed curves, the resulting ordering will be circular, thus the choice of the starting triangle is not primordial. We first determine an orientation (let us for example choose the counterclockwise orientation, but the result would still be valid for the clockwise one). Then we create the ordered list of edge-neighbors for each triangle, regarding the chosen orientation. For 2-N triangles, the choice of direction is trivial. Indeed, the next triangle is the first neighbor according to the counterclockwise orientation. The difficult part is to determine an ordering when cases of 1- or 3-neighboring appear. We proved in section 4 that 1-N HOM triangles, the dead-end, may only be encountered after the traversal of a 3-N HET triangle.

Let us first recall that there always exists an HET triangle in a thick digital closed curve, since all vertices of the triangulation are points from the inner and outer curves (see section 4). If there exist 2-N HET triangles, we choose any of those as the starting element. Otherwise, we start with any 3-N HET triangle, getting rid of one of its associated pocket, which we process afterwards. This way, we can always start with a 2-N HET triangle, for which the direction of traversal is obvious. We follow the adjacency path until a 3-N HET triangle is encountered. Then, exactly one of its neighbors is homogeneous (lemma 1). We stated in section 4 that such a triangle is always part of one or two "pocket(s)" of the curve, which lead to dead-end triangles. Let us notice that there is no possibility of constructing a total ordering of adjacent triangles without duplicating some triangles. Moreover, with the α-thick predicate, since convex hull are invariant when duplicating triangles we will not introduce new subparts. In the general case, only the triangles of some pockets are inserted at most three times in the list. This is a logical behavior since pockets correspond to irregular branches of the curve that must be described. Thus, the strategy is to explore the pockets in depth, and to go back the same way to the 3-N HET that generated the pocket. Meanwhile, we duplicate all encountered triangles.

As for 3-N HOM triangles, the strategy is not different. We previously stated that such triangles conduct to two pockets. Using counterclockwise orientation, we determine an order between those two pockets. We then explore each one forth and back. This strategy can be seen as a Depth-First Search in a tree, that would

```
T_0 ← Find or Create a 2-neighbored triangle;
T ← FirstNeighbor(T_0);
while T ≠ T_0 do
    if T is Heterogeneous then
        if T is 2-neighbored then
            | T ← FirstNeighbor(T);
        else
            | T' ← HomogeneousNeighbor(T);
            | if Link(T,T') is marked then
            |     | T ← FirstHeterogeneousNeighbor(T);
            | else
            |     | Mark(Link(T,T')); T ← T';

    else
        if T is 2-neighbored or 3-neighbored then
            | T' ← FirstFreeLinkedNeighbor (T);
            | Mark(Link(T,T')); T ← T';
        else
            | T' ← Neighbor(T);
            | Mark(Link(T,T')); T ← T';

end
Process last pocket if necessary
```

Algorithm 1. Total Ordering

go back after encountering leaves. When all pockets emerging from a 3-N HET have been explored, the path continues with the heterogeneous neighbor that has not been visited yet. Our algorithm is detailed in pseudo-code in algorithm 1.

Preliminary to the execution of this algorithm, a list of every neighbor T' of each triangle T is created. This list is ordered using counterclockwise direction. Then, during the execution, when a neighbor T' is chosen from T, we put a mark on it. Thus, we can not enter in a previously seen pocket.

6   Results

First, let us exhibit the results of our total ordering on a discrete shape (see Fig. 8, right). The obtained path seems really natural, no conflicts are encountered and all difficult cases are solved using algorithm 1.

We now plug the triangulation onto the tangential cover algorithm. Let us take a look at Fig. 9. The α parameter is fixed to 2 pixels. Fig. 9$a)$ represents the ordering between the triangles. Then, successive maximal subparts are built. Crossed triangles are the ones that have been suppressed from the previous

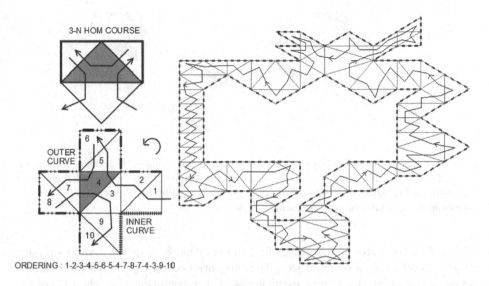

Fig. 8. Left : the course for homogeneous 3-neighbored triangles. Right : Total ordering obtained for a triangulation of a closed thick digital curve.

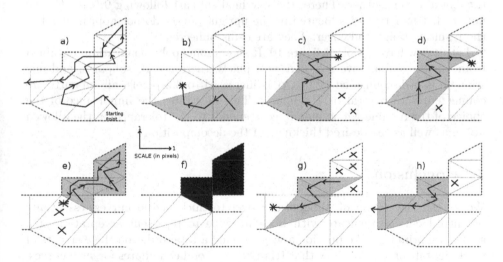

Fig. 9. The building of the tangential cover on a small part of a closed thick digital curve. Successive maximal subparts are represented in grey. An invalid maximal subpart is drawn in black.

subpart in order to add a new triangle, while maintaining an isothetic thickness inferior to α (our generalized tangential cover predicate of validity). Starred triangles represent elements which can not be added to the current subpart, in order to keep up with the validity of the predicate. An interesting case occurs on Fig.9e). The obtained maximal subpart only consists in homogeneous triangles,

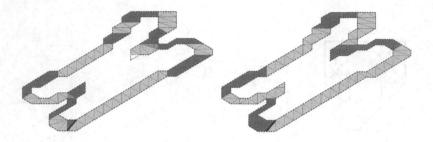

Fig. 10. Some polygonalizations of the thick chromosome shape. Thickness parameter is fixed to $\alpha = 2$ pixels (left) and $\alpha = 3.2$ pixels (right). Light grey and dark grey represent α-thick subparts. Bi-colored triangles are shared by two consecutive subparts. α-disconnected triangles are represented in white.

all lying on the outer curve. According to definition 8, it is not a valid subpart (9*f*)). Therefore, it is rejected from the tangential cover as an irrelevant excrescence of the curve, regarding α-thickness. The recognition may then carry on with the next subparts. Let us also remark that subpart 9*g*), although valid, is not maximal since it is included in subpart 9*d*). Thus, it does not belong to the tangential cover neither. There, the maximal subpart following 9*d*) is 9*h*). We notice that four triangles located in the upright pocket do not appear at all in the resulting tangential cover. They are α-disconnected.

Let us now have a look at figure 10. It shows some polygonalizations obtained from our tangential covers. For $\alpha = 2$ pixels, there are 16 subparts, and one triangle is α-disconnected. When α is increased to 3.2 pixels, all subparts are connected, with a total of 7 subparts. This illustrates the importance of the choice of the α value which merely corresponds to the tolerance on the allowed noise, as well as the desired thickness of the decomposition.

7   Conclusion

We extend the notion of Tangential Cover to thick digital curves, essentially by introducing a predicate with boolean value to represent acceptation of a geometric primitive. We provide a way to build a total order on the elements of thick digital curves. We show that triangles are good candidates for such curves. Finally, we show some results produced by this technique. The computation of the generalized Tangential Cover has a time complexity of $O(n \log n)$, n being the size of both bounding lines of the curve. This time complexity is very low regarding the total size of the input shape, when considering points in the interior of the thick digital curve. In future works, we plan to study the size and the behavior of the obtained tangential covers, by varying the α parameter. Also, the fact that we automatically eliminate non-relevant parts of the curve could bring along a new method for cleaning noisy curves.

References

1. Buzer, L.: Digital line recognition, convex hull, thickness, a unified and logarithmic technique. In: Reulke, R., Eckardt, U., Flach, B., Knauer, U., Polthier, K. (eds.) IWCIA 2006. LNCS, vol. 4040, pp. 189–198. Springer, Heidelberg (2006)
2. Buzer, L.: Computing multiple convex hulls of a simple polygonal chain in linear time. In: European Workshop on Computational Geometry (2007)
3. Debled-Rennesson, I., Feschet, F., Rouyer-Degli, J.: Optimal blurred segments decomposition in linear time. In: Andrès, É., Damiand, G., Lienhardt, P. (eds.) DGCI 2005. LNCS, vol. 3429, pp. 371–382. Springer, Heidelberg (2005)
4. Faure, A., Feschet, F.: Tangential cover for thick digital curves. In: 14th Intl. Conference on Discrete Geometry for Computer Imagery (submitted, 2008)
5. Feschet, F.: Canonical representations of discrete curves. Pattern Anal. Appl. 8(1-2), 84–94 (2005)
6. Feschet, F., Tougne, L.: On the min dss problem of closed discrete curves. Discrete Applied Mathematics 151(1-3), 138–153 (2005)
7. Feschet, F., Tougne, L.: Optimal time computation of the tangent of a discrete curve: Application to the curvature. In: Bertrand, G., Couprie, M., Perroton, L. (eds.) Discrete Geometry for Computer Imagery, 8th International Conference, DCGI 1999, Marne-la-Vallee, France, March 17-19, 1999, pp. 31–40 (1999)
8. Klette, R., Rosenfeld, A.: Digital Geometry: Geometric Methods for Digital Picture Analysis. In: Computer Graphics and Geometric Modeling, Morgan Kaufman, San Francisco (2004)
9. Melkman, A.A.: On-line construction of the convex hull of a simple polyline. Inf. Process. Lett. 25(1), 11–12 (1987)
10. Overmars, M., van Leeuwen, J.: Maintenance of configurations in the plane. J. Comput. Syst. Sci. 23(2), 166–204 (1981)
11. Réveillès, J.P.: Géométrie discrète, calcul en nombres entiers et algorithmique. PhD thesis, Université Louis Pasteur (1991)
12. Seidel, R.: Constrained delaunay triangulations and voronoi diagrams with obstacles. Technical Report 260, Institute for Information Processing, Graz, Austria (1988)

Segmentation of Noisy Discrete Surfaces*

Laurent Provot and Isabelle Debled-Rennesson

LORIA Nancy
Campus Scientifique - BP 239
54506 Vandœuvre-lès-Nancy Cedex, France
{provot,debled}@loria.fr

Abstract. We propose in this paper a segmentation process that can deal with noisy discrete objects. A flexible approach considering arithmetic discrete planes with a variable width is used to avoid the over-segmentation that might happen when classical segmentation algorithms based on regular discrete planes are used to decompose the surface of the object. A method to choose a seed and different segmentation strategies according to the shape of the surface are also proposed.

1 Introduction

Three-dimensional discrete objects are widely used in medical area. Due to their internal structure and their huge size, the manipulation of such objects is not an easy task. Rendering algorithms for instance, cannot apply usual techniques to obtain a nice visualization of the objects. A general idea to address this problem is to transform the discrete volume into a Euclidean polyhedra. The segmentation of the border of such objects into discrete primitives is thus a natural first step and several studies have been led on the subject.

In [6], L. Papier and J. Françon propose a segmentation of the surface of a discrete object into pieces of standard arithmetic discrete planes. These standard planes are recognized using a Fourier-Motskin elimination algorithm [7] and are forced to be homeomorphic to a topological disk in order to be used in a polyhedrization process. They state that the resulting segmentation heavily depends on the choice of the seed to start the recognition of a new face and the tracking order of the points chosen for enlarging the current face, but do not address these problems.

In [2], J. Burguet and R. Malgouyres have developed a polyhedrization algorithm based on the computation of a topological Voronoï diagram. Seeds are distributed on the surface of the discrete object according to its curvature [9] and a thinning algorithm is used to generate the skeleton of this surface without the seeds. The resulting Voronoï regions can then be seen as a segmentation of the surface.

In the framework of surface area estimation [8], R. Klette and H.J. Sun have proposed a segmentation of the surface into digital planar segments (DPS) – which

---

* This work is supported by the ANR in the framework of the GEODIB project.

V.E. Brimkov, R.P. Barneva, H.A. Hauptman (Eds.): IWCIA 2008, LNCS 4958, pp. 160–171, 2008.

are actually pieces of standard discrete planes. To incrementally check whether a set of points is a DPS, they compute the convex hull of this set to retrieve a specific pair of parallel planes for which the main diagonal distance has to be less than $\sqrt{3}$. A breadth-first search of the surfel graph representing the surface is used to incrementally add points into the DPS.

In [13], I. Sivignon et al. have compared different tracking processes to decompose the surface of a discrete object into naive discrete planes. A dual-space approach is used to incrementally recognize the pieces of planes [15]. In [14] the authors have also studied the relation between a segmentation into naive and standard discrete planes depending on the considered surface definition.

Although these methods are reversible and behave well with regular discrete objects, they might lead to an over-segmentation with noisy ones. In this paper we try to be more flexible and address the problem by considering a segmentation into pieces of discrete planes with a variable width. A first pre-processing step is done to compute geometric features of the surface of a possibly noisy discrete object. We then use these goemetric features to take into account the shape of the object to choose the seeds and to guide the incremental growth of the segments.

In section 2, after recalling the definition of blurred pieces of discrete planes, we summarize results from [11] about geometric features for noisy discrete surfaces. The different steps of the segmentation process are then described in section 3, followed by some results on noisy and non-noisy objects. The paper ends up with a conclusion and some perspectives in section 4.

2   Background

We recall in this section the definition of a width-ν blurred piece of discrete plane, an arithmetical discrete primitive introduced in [10], that allows to deal with noisy discrete data. Relying on this primitive, we present the notion of a width-ν patch centered at a border point of a discrete objet and some features of the border obtained from this patch. More details about the construction of the patch and the study of different features of the border can be found in [11].

2.1   Blurred Pieces of Discrete Planes

One can see a blurred piece of discrete plane as an arithmetic discrete plane for which some points are missing. More formally:

Definition 1. *Let N be a norm on $\mathbb{R}^3$ and $\mathcal{E}$ a set of points in $\mathbb{Z}^3$. We say that the discrete plane $\mathcal{P}(a, b, c, \mu, \omega)^1$ is a **bounding plane of** $\mathcal{E}$ if all the points of $\mathcal{E}$ belong to $\mathcal{P}$, and we call **width of** $\mathcal{P}(a, b, c, \mu, \omega)$, the value $\frac{\omega - 1}{N(a,b,c)}$.*

*A bounding plane of $\mathcal{E}$ is said **optimal** if its width is minimal.*

---

[1] An arithmetic discrete plane $\mathcal{P}(a, b, c, \mu, \omega)$ is the set of integer points (x, y, z) verifying $\mu \leq ax + by + cz < \mu + \omega$, where $(a, b, c) \in \mathbb{Z}^3$ is the normal vector of the plane. $\mu \in \mathbb{Z}$ is named the translation constant and $\omega \in \mathbb{Z}$ the arithmetical thickness.

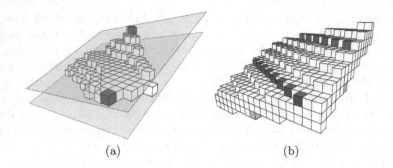

(a)   (b)

Fig. 1. (a) A width-3 blurred piece of discrete plane and (b) a piece of its optimal bounding plane $\mathcal{P}(4, 8, 19, -80, 49)$, using the Euclidean norm

Definition 2. *A set $\mathcal{E}$ of points in $\mathbb{Z}^3$ is a **width-ν blurred piece of discrete plane** if and only if the width of its optimal bounding plane is less than or equal to ν.*

Two recognition algorithms of blurred pieces of discrete planes have been proposed in [10]. The first one considers the Euclidean norm and, for a set of points P in $\mathbb{Z}^3$, it solves the recognition problem by using the geometry of the convex hull of P. The second one considers the infinity norm and uses methods from linear programming to solve the recognition problem.

Thereafter, we denote by $\mathcal{O}_b$ a possibly noisy 6-connected discrete object. We call *surface* or *border* of $\mathcal{O}_b$ the set of points $\mathcal{B}_b$ which have a 6-neighbor that does not belong to $\mathcal{O}_b$. All the results we present on this type of objects have been obtained by considering the geometrical approach which uses the Euclidean norm.

2.2 Width-ν Discrete Patches

If we are working on a noisy discrete surface and need to extract some of its local geometric features, such as the normal vector or the curvature, it is wise to use estimators that take into account the irregularity of this surface to compute these kinds of features. A way to achieve this task at a point p of the surface is to gather the information of points lying in an extended neighborhood of p. The notion of patch we present hereafter takes place in this framework, considering an adaptative neighborhood around p.

Definition 3. *Let $\mathcal{B}_b$ be the border of a discrete object, p a point in $\mathcal{B}_b$ and ν the greatest real value allowed. Let d be a distance. At each point $q \in \mathcal{B}_b$ we associate the weighting factor $d_p(q) = d(p, q)$. We call **width-ν patch centered at** p, and denote by $\Gamma_\nu(p)$, a width-ν blurred piece of discrete plane incrementally recognized from p by adding points q of $\mathcal{B}_b$ following the increasing values of $d_p(q)$.*

About the Incremental Recognition: We construct a *width-ν patch centered at* p using the incremental recognition algorithm of blurred pieces of discrete

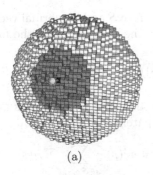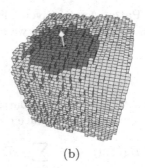

(a)   (b)

Fig. 2. An example of width-2 patches spread on the surface of different noisy objects. (a) A sphere of radius 20 and (b) a cube of edge 25.

planes introduced in [10]. We add the points following the increasing values of d_p and, as soon as the width of the blurred piece of discrete plane becomes greater than ν, we stop the recognition process.

About the Distance d: To uniformly spread the patch in all directions, the best solution would be to use a geodesic distance. Nevertheless, for efficiency, we have chosen to rely on a distance based on a chamfer mask $\langle 3, 4, 5 \rangle$ which is a good approximation of the geodesic distance [1]. The aim is to have a well-balanced patch around p which looks almost circular. With this method we obtain patches like those in Fig. 2.

2.3   Patch Features

A patch $\Gamma_\nu(p)$, as previously defined, characterizes the planarity of the surface around p (with respect to the width ν). Thus, the more the patch is spread, the less the surface around p is bent.

In addition, if the growth of $\Gamma_\nu(p)$ stopped, it means that the close neighboring points outside $\Gamma_\nu(p)$ would bend the patch too much if they were added. In that case the patch could no longer be regarded as flat. Therefore, it is possible to deduce a conformation of the discrete surface around p by studying the patches centered along the points of the outline of $\Gamma_\nu(p)$.

The following definitions give a formal quantization of all these observations.

Width-ν Normal. With the previous intuition we can see that the normal vector of $\Gamma_\nu(p)$ is a good estimation of the normal at p. Thus, assimilating the normal vector of $\Gamma_\nu(p)$ to the normal of the surface at p, we define a normal vector estimator for each point of the surface of a possibly noisy discrete object.

Definition 4. *Let $\mathcal{B}_b$ be the border of a discrete object and p a point of $\mathcal{B}_b$. We call **width-ν normal at** p the normal vector*

$$\overrightarrow{n_\nu}(p) = \overrightarrow{n}(\Gamma_\nu(p))$$

where $\overrightarrow{n}(\Gamma_\nu(p))$ is the normal vector of the patch $\Gamma_\nu(p)$.

Width-ν Patch Area. Given a Euclidean surface $\mathcal{S}$ and its normal vector field $\{\vec{n}\}$, in the continuous space we can compute the area of $\mathcal{S}$ with the formula:

$$\mathcal{A}(\mathcal{S}) = \int_{\mathcal{S}} \vec{n}(s)\, ds$$

The discrete version of this equation given in [4] has been adapted as follows to compute the area of the surface of a width-ν patch:

$$E_{\mathcal{A}}(\Gamma_{\nu}(p)) = \sum_{s \in \mathcal{S}_{\Gamma_{\nu}(p)}} \vec{n_{\nu}}(p).\vec{n}_{el}(s) = \vec{n_{\nu}}(p). \sum_{s \in \mathcal{S}_{\Gamma_{\nu}(p)}} \vec{n}_{el}(s)$$

where $\mathcal{S}_{\Gamma_{\nu}(p)}$ is the set of surfels[2] of the patch surface, and $\vec{n}_{el}(s)$ the elementary normal vector of s.

Shape Estimator. An estimator that enables the characterization of the shape (concave, convex or flat) of the surface around a border point of a possibly noisy discrete object has been developed. It is based on the study of the conformation of the patches which are centered on points belonging to the outline of $\Gamma_{\nu}(p)$.

Definition 5 (Patch Outline). *Let $\mathcal{B}_b$ be the border of a discrete object $\mathcal{O}_b$. We denote by $\mathcal{S}_b$ the set of surfels of $\mathcal{B}_b$ which are incident to a point that does not belong to $\mathcal{O}_b$, and $\mathcal{S}_{\Gamma_{\nu}(p)}$ the subset of $\mathcal{S}_b$ that belongs to $\Gamma_{\nu}(p)$. A point q belongs to the outline of $\Gamma_{\nu}(p)$ if the voxel representation of q has a surfel $s \in \mathcal{S}_{\Gamma_{\nu}(p)}$ and if there exists a surfel $s' \in \mathcal{S}_b \setminus \mathcal{S}_{\Gamma_{\nu}(p)}$ such that s and s' are adjacent by edge.*

Let C be the set of points that belong to the outline of $\Gamma_{\nu}(p)$. Our shape estimator of the surface around a point p is then given by the formula :

$$\mathcal{F}_{\nu}(p) = \frac{1}{|C|} \sum_{\forall q \in C} (\widehat{\vec{n_{\nu}}(p), \vec{n_{\nu}}(q)}) \cdot \frac{E_{\mathcal{A}}(\Gamma_{\nu}(q))}{E_{\mathcal{A}}(\Gamma_{\nu}(p))}$$

where $(\widehat{\vec{n_{\nu}}(p), \vec{n_{\nu}}(q)})$ is the oriented angle value between the two normal vectors. So, the estimator $\mathcal{F}_{\nu}(p)$ is a weighted mean of the angle values between $\vec{n_{\nu}}(p)$ and the $\vec{n_{\nu}}(q_i)_{1 \leq i \leq |C|}$.

$\mathcal{F}_{\nu}(p)$ is positive when the surface around p is rather convex and $\mathcal{F}_{\nu}(p)$ is negative when the surface around p is rather concave. An increasing value of $|\mathcal{F}_{\nu}(p)|$ means that the surface around p is more strongly concave or convex. Moreover, if $\Gamma_{\nu}(p)$ is big, a value $\mathcal{F}_{\nu}(p)$ close to zero means that the area around p is almost flat (according to the width ν we chose). If $\Gamma_{\nu}(p)$ is small, then the area around p is strongly distorted, but in a way we can neither qualify concave, nor qualify convex (a saddle point for instance).

---

[2] Faces of a voxel are called *surfels*.

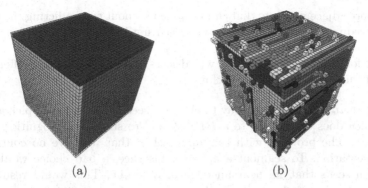

(a)   (b)

Fig. 3. Segmentation of (a) a regular cube of edge 25 and (b) a noisy counterpart by using the DSD algorithm (http://liris.cnrs.fr/isabelle.sivignon/DSD.html) proposed by I. Sivignon [13,14].

3   Segmentation

3.1   Introduction

The segmentation of a tridimensional discrete object we will describe in this section consists in partitioning the border of the object into pieces of discrete planes. Some studies have been led on the subject [8,13,14,2,6] but they all consider regular planes with a fixed width (mainly naive or standard arithmetic discrete planes). Although these methods give good results with regular discrete objects (Fig. 3(a)), it is not always the case when we have to deal with irregular or noisy discrete objects (Fig. 3(b)). In particular, irregularities force to create lots of small segments. The approach we present hereafter is more flexible and considers a segmentation into pieces of planes with a variable width, width-ν blurred pieces of discrete planes (denoted $BPDP_\nu$ in the sequel) to be specific, to deal with noisy data.

3.2   Segmentation into Blurred Pieces of Discrete Planes

Firstly, a pre-processing step is done on the borber $\mathcal{B}_b$ of the discrete object we want to segment. Given a real ν, for each point $p \in \mathcal{B}_b$ we compute a width-ν patch centered at p as explained in section 2.2. At each point p we can thus associate the features presented in section 2.3, that is:

- the normal vector $\overrightarrow{n_\nu}(p)$,
- the area factor $E_\mathcal{A}(\Gamma_\nu(p))$,
- and the shape factor $\mathcal{F}_\nu(p)$.

Our segmentation process can be summed up to the following steps: a *seed* is chosen among points of $\mathcal{B}_b$ to start a first $BPDP_\nu$ *recognition* that grows through a process of accretion. An adjacent point is selected and added to $BPDP_\nu$ if it satisfies some *required criteria*. The $BPDP_\nu$ eventually stops growing when there are no more adjacent points that can be added without contradicting the criteria. This procedure is repeated from a new seed until all points of $\mathcal{B}_b$ belong to a $BPDP_\nu$.

In the following paragraphs we will discuss more in details the different key points of this segmentation algorithm.

Seed Selection. The easiest way to choose a seed is to randomly pick a border point which does not belong to a $BPDP_\nu$ and to start the recognition process from there. The problem with this approach is that we have no control over the segmentation. To segment a cube for instance, a bad choice would be to start from seeds that lie near an edge of the cube. This would result in an over-segmentation as shown in Fig 4.

A better choice is to start from seeds that are lying in flat areas. It is indeed more meaningful to give a higher priority to flat areas than to bent areas since the underlying primitive of a $BPDP_\nu$ is an arithmetical discrete plane. Chances to have a better approximation are thus higher. Therefore we have chosen to rely on the area estimator $E_\mathcal{A}(\Gamma_\nu(p))$ to find the seeds. The idea is to pick the border point p (not yet processed) which has the highest $E_\mathcal{A}(\Gamma_\nu(p))$ value as the next seed.

$BPDP_\nu$ Recognition. The algorithm used to incrementally recognize width-ν blurred pieces of discrete planes is the geometrical one proposed in [10] by considering the Euclidean norm.

The spreading of a $BPDP_\nu$ heavily depends on the way the neighborhood of the seed is visited as explained in [13]. For the same reasons as before, the value

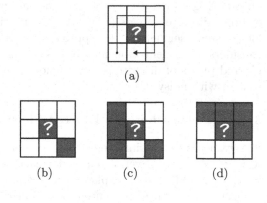

Fig. 4. Over-segmentation due to randomly chosen seeds

Fig. 5. The points that belong to the $BPDP_\nu$ are in grey (a) Processing order of the neighborhood. (b-d) Some possible configurations when we try to add the point with the question mark: (b) it cannot be added because it is not 4-connected to another grey point; (c) it cannot be added because it creates a hole; and (d) it can be added.

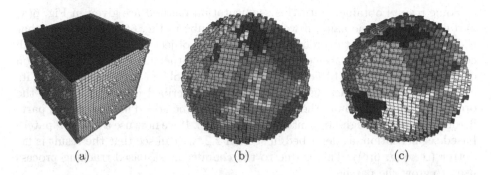

Fig. 6. Width-2 segmentation of a weakly (a) noisy cube of edge 30; (b) noisy sphere of radius 20 with the method presented in section 3.2. (c) Segmentation of the same sphere using only width-2 patches.

$E_{\mathcal{A}}(\Gamma_{\nu}(p))$ is used in the accretion process. Points p adjacent to the evolving $BPDP_{\nu}$ are added according to their decreasing $E_{\mathcal{A}}(\Gamma_{\nu}(p))$ values.

To implement this behaviour we use a priority queue Q. We start by pushing the seed into the queue with a weight equals to its area factor and mark this seed as visited. Then, while Q is not empty, we pop out of Q the point p with the highest weight w and we add p to the evolving $BPDP_{\nu}$ if it satisfies the requiered criteria presented in the following paragraph. We then add the non-visited 26-neighbours of p which belong to the border and their associated area factor into the priority queue Q and mark them as visited.

Using this technique the $BPDP_{\nu}$ does not stop growing if a point cannot be added.

Required Criteria. The first criterion that has to be satisfied is that the width of the evolving $BPDP_{\nu}$ must not exceed ν when a point p is added. But this is implicitly checked in the recognition algorithm.

Moreover, as we plan to use the segmentation in a future work to develop a polyhedrization algorithm for noisy discrete objects, the $BPDP_{\nu}$ segments have to satisfy some constraints of good formation. In particular we want a $BPDP_{\nu}$ segment to be 4-connected and without holes, i.e. homeomorphic to a topological disk, according to the main direction of the normal vector of its seed. To check these constraints we use a simplified version of a method proposed in [12] (p.153). We work in the projection plane associated to the normal vector of the seed. We consider the 8-neighborhood of the point we are trying to add in the evolving $BPDP_{\nu}$ and process the 8-neighbors in the order shown in Fig. 5(a). During the processing a zero-initialized counter is incremented at each time we go from a point which belongs to $BPDP_{\nu}$ to a point which does not, and vice-versa. At the end, if the counter value is greater than two it means that the point cannot be added whitout creating a hole. In the same time we check that at least one 4-neighbor belongs to $BPDP_{\nu}$. If a point does not pass these checks it is marked as non-visited to give the tracking process the opportunity to visit it later on.

Some results obtained with this segmentation method are given in Fig. 6(a) and 6(b). On the one hand, if we look at the cube in Fig. 6(a), we can see that the segmentation is rather good and partition the object into six segments which correspond to the six faces of the cube. On the other hand, the segmentation of the sphere in Fig. 6(b) could be better. The problem with the sphere is its curved border. The tracking process of points described in section 3.2 has the opportunity to skirt round the points that cannot be added and on curved parts it tends to create rough-crescent-shaped $BPDP_\nu$. If we now use a width-ν patch-based segmentation, as described in section 2.2, we can see that the result is far better (see Fig. 6(c)). This is due to the chamfer-mask-based tracking process used to grow the patches.

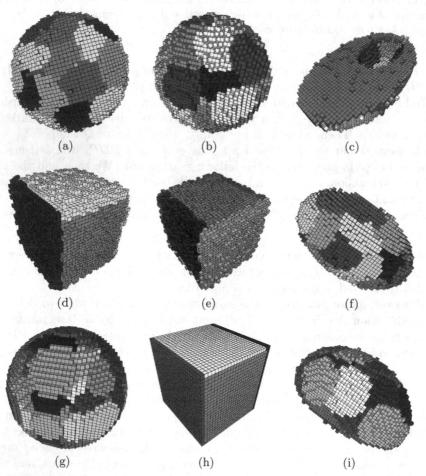

(a)   (b)   (c)

(d)   (e)   (f)

(g)   (h)   (i)

Fig. 7. Results of different segmentations by using the hybrid method on synthetical objects: (a, b) a width-2 segmentation of a weakly noisy sphere of radius 20; (d, e) a width-3 segmentation of a strongly noisy cube of edge 25; (g, h) a width-1 segmentation of non-noisy objects; and (c, f, i) a width-2 segmentation of an half hallowed ellipsoid.

3.3   Hybrid Method

Relying on the previous observations, we have developed an hybrid segmentation method. For seeds that lie in a flat part of the object we develop a $BPDP_\nu$ segment as described in 3.2 and for seeds that lie in curved parts we develop a patch $\Gamma_\nu(p)$ (see section 2.2).

To distinguish between flat parts and curved parts we use the shape factor $\mathcal{F}_\nu(p)$. As previously explained, an increasing value of $|\mathcal{F}_\nu(p)|$ means that the surface around p is more strongly bent. Thus, given a threshold value σ and a seed s, if $|\mathcal{F}_\nu(s)| < \sigma$ we develop a $BPDP_\nu$ segment, otherwise we develop a patch $\Gamma_\nu(s)$.

Results obtained with this method are shown in Fig. 7 and 8. In Fig. 7 synthetical objects with different shapes have been segmented at different widths. We can see in Fig. 7(g) and 7(h) that the method still works for non-noisy objects.

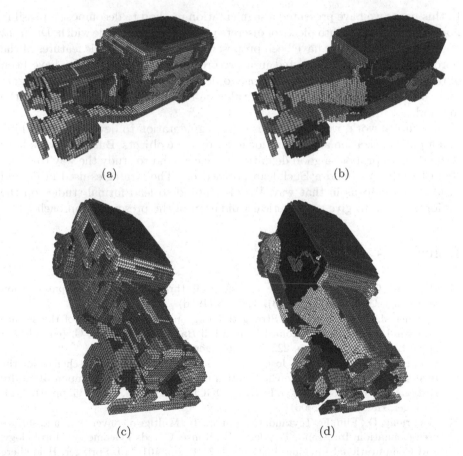

(a)

(b)

(c)

(d)

Fig. 8. The segmentation of a car (a, c) by using the DSD algorithm of I. Sivignon; and (b, d) by using the width-2 hybrid method

Furthermore, due to the hybrid approach, both the flat and the curved areas of the noisy half hallowed ellipsoid in Fig 7(c), 7(f) and 7(i), are well segemented. In Fig. 8 a real-life object – an old Dodge car, available on the TC18 website[3] – has been segmented with both, the DSD algorithm proposed by I. Sivignon and the hybrid approach. We can notice that, with the hybrid approach, the flat areas, i.e the roof, the parts of the hood and the windshield, are well segmented, with respect to the shape of the car and with a little number of segments. In curved areas the two segmentations are close, but it is difficult to decide what is a "good" segmentation in these areas.

Note that, at this time, the threshold have to be set manualy, but we would like to investigate more to find a way to automaticaly choose a good value for σ.

4   Conclusion

In this paper we have presented a segmentation method to decompose a possibly noisy discrete object into pieces of discrete planes with a variable width. Different segmentation strategies have been proposed, guided by geometric features of the border of the object, computed in a pre-process step. Good results have been obtained for both noisy and non-noisy objects, but we still have to investigate more to automaticaly choose good values for the different parameters of the method.

In a future work, we intend to use this segmentation to develop a smoothing and a polyhedrization algorithms for noisy discrete objects. But some work have to be done to propose a good definition for facets and to study the way to group them together to build a Euclidean polyhedron. The strategies used in [5] and in [3] could help us in that way. We also intend to lead formal studies on the notion of *noise* to give a theoretical validation of the presented approach.

References

1. Borgefors, G.: On digital distance transforms in three dimensions. Computer Vision and Image Understanding 64(3), 368–376 (1996)
2. Burguet, J., Malgouyres, R.: Strong thinning and polyhedrization of the surface of a voxel object. In: Nyström, I., Sanniti di Baja, G., Borgefors, G. (eds.) DGCI 2000. LNCS, vol. 1953, pp. 222–234. Springer, Heidelberg (2000)
3. Coeurjolly, D., Dupont, F., Jospin, L., Sivignon, I.: Optimization schemes for the reversible discrete volume polyhedrization using marching cubes simplification. In: Kuba, A., Nyúl, L.G., Palágyi, K. (eds.) DGCI 2006. LNCS, vol. 4245, pp. 413–424. Springer, Heidelberg (2006)
4. Coeurjolly, D., Flin, F., Teytaud, O., Tougne, L.: Multigrid convergence and surface area estimation. In: Asano, T., Klette, R., Ronse, C. (eds.) Geometry, Morphology, and Computational Imaging. LNCS, vol. 2616, pp. 101–119. Springer, Heidelberg (2003)

---

[3] http://www.cb.uu.se/~tc18/code_data_set/3D_images.html

5. Dexet, M., Coeurjolly, D., Andres, E.: Invertible polygonalization of 3d planar digital curves and application to volume data reconstruction. In: Bebis, G., Boyle, R., Parvin, B., Koracin, D., Remagnino, P., Nefian, A., Meenakshisundaram, G., Pascucci, V., Zara, J., Molineros, J., Theisel, H., Malzbender, T. (eds.) ISVC 2006. LNCS, vol. 4292, pp. 514–523. Springer, Heidelberg (2006)
6. Françon, J., Papier, L.: Polyhedrization of the boundary of a voxel object. In: Bertrand, G., Couprie, M., Perroton, L. (eds.) DGCI 1999. LNCS, vol. 1568, pp. 425–434. Springer, Heidelberg (1999)
7. Françon, J., Schramm, J.M., Tajine, M.: Recognizing arithmetic straight lines and planes. In: Miguet, S., Ubéda, S., Montanvert, A. (eds.) DGCI 1996. LNCS, vol. 1176, pp. 141–150. Springer, Heidelberg (1996)
8. Klette, R., Sun, H.J.: Digital planar segment based polyhedrization for surface area estimation. In: Arcelli, C., Cordella, L.P., Sanniti di Baja, G. (eds.) IWVF 2001. LNCS, vol. 2059, pp. 356–366. Springer, Heidelberg (2001)
9. Lenoir, A.: Fast estimation of mean curvature on the surface of a 3d discrete object. In: Ahronovitz, E. (ed.) DGCI 1997. LNCS, vol. 1347, pp. 175–186. Springer, Heidelberg (1997)
10. Provot, L., Buzer, L., Debled-Rennesson, I.: Recognition of blurred pieces of discrete planes. In: Kuba, A., Nyúl, L.G., Palágyi, K. (eds.) DGCI 2006. LNCS, vol. 4245, pp. 65–76. Springer, Heidelberg (2006)
11. Provot, L., Debled-Rennesson, I.: Geometric feature estimators for noisy discrete surfaces. Technical report, Loria (2007),
    http://www.loria.fr/~provot/doc/geom_feature_estimators.pdf
12. Sivignon, I.: De la caractérisation des primitives à la reconstruction polyédrique de surfaces en géométrie discrète. Thèse, I.N.P. de Grenoble (2004)
13. Sivignon, I., Dupont, F., Chassery, J.M.: Decomposition of a three-dimensional discrete object surface into discrete plane pieces. Algorithmica 38(1), 25–43 (2004)
14. Sivignon, I., Dupont, F., Chassery, J.M.: Discrete surface segmentation into discrete planes. In: Klette, R., Žunić, J. (eds.) IWCIA 2004. LNCS, vol. 3322, pp. 458–473. Springer, Heidelberg (2004)
15. Vittone, J., Chassery, J.M.: Recognition of digital naive planes and polyhedrization. In: Nyström, I., Sanniti di Baja, G., Borgefors, G. (eds.) DGCI 2000. LNCS, vol. 1953, pp. 296–307. Springer, Heidelberg (2000)

MRF Labeling with a Graph-Shifts Algorithm

Jason J. Corso[1], Zhuowen Tu[2], and Alan Yuille[3]

[1] Computer Science and Engineering, University at Buffalo, State University of New York
[2] Center for Computational Biology, Laboratory of Neuro Imaging
University of California, Los Angeles
[3] Department of Statistics, University of California, Los Angeles
jcorso@cse.buffalo.edu

Abstract. We present an adaptation of the recently proposed graph-shifts algorithm for labeling MRF problems from low-level vision. Graph-shifts is an energy minimization algorithm that does labeling by dynamically manipulating, or shifting, the parent-child relationships in a hierarchical decomposition of the image. Graph-shifts was originally proposed for labeling using relatively small label sets (e.g., 9) for problems in high-level vision. In the low-level vision problems we consider, there are much larger label sets (e.g., 256). However, the original graph-shifts algorithm does not scale well with the number of labels; for example, the memory requirement is quadratic in the number of labels. We propose four improvements to the graph-shifts representation and algorithm that make it suitable for doing labeling on these large label sets. We implement and test the algorithm on two low-level vision problems: image restoration and stereo. Our results demonstrate the potential for such a hierarchical energy minimization algorithm on low-level vision problems with large label sets.

1   Introduction

Markov random field (MRF) models [2] play a key role in both low- and high-level vision problems [13]. Example low-level vision problems are image restoration and stereo disparity calculation. Fast and accurate labeling of MRF models remains a fundamental problem in Bayesian vision. The configuration space is combinatorial in the labels and the energy landscape is rife with local minima. This point is underscored by the recent comparative survey of methods for low-level labeling by Szeliski et al. [15].

In recent years, multiple new algorithms have been proposed for solving the energy minimization problem associated with MRF labeling. For example, graph cuts [3] is one such algorithm that guarantees achieving a *strong* local minimum for two-class energy functions. However, processing times for the graph cuts remain in the order of several minutes on modern hardware. Max Product Belief propagation [8] computes local maxima of the posterior, but it is not guaranteed to converge for the loopy graphs present in low-level vision. Efficient implementations can lead to running times in the order of seconds [6]. However, despite its high-regard and widespread use, it performed poorly in the recent benchmark study [15].

In this paper, we work with a recently proposed approach called graph-shifts [4,5]. Graph-shifts is a class of algorithms that do energy minimization on dynamic, adaptive graph hierarchies. The graph hierarchy represents an adaptive decomposition of the

V.E. Brimkov, R.P. Barneva, H.A. Hauptman (Eds.): IWCIA 2008, LNCS 4958, pp. 172–184, 2008.
© Springer-Verlag Berlin Heidelberg 2008

input image; they are adaptive in the sense that the graph hierarchy and neighborhood structure is data-dependent in contrast to conventional pyramidal schemes [1] in which the hierarchical neighborhood structure is fixed. They are dynamic in the sense that the algorithm iteratively reduces the energy by changing the parent-child relationships, or *shifting*, in the graph, which results in a change in the underlying labels at the pixels. Graph-shifts stores a representation of the combinatorial energy landscape, and is able to efficiently compute the optimal energy reducing move at each iteration.

The original graph-shifts algorithm [4,5] was defined on a conditional random field (CRF) [11,12] with comparatively few labels (e.g., 8) and applied to high-level labeling problems in medical imaging. Recall that a CRF is a MRF with a broader conditioning on the observed data than is typical in MRF and MAP-MRF [9] formulations. But, in the low-level labeling problems considered in this paper, the label sets are much larger (e.g. 32, 256). The original graph-shifts algorithms scales linearly in pixels; however a factor linear in labels is incurred at each iteration. The memory requirement is quadratic in the labels. In practice, these complications lead to slower convergence as the number of labels grow. The main focus and contribution of this paper is how we adapt graph-shifts to work efficiently with large, ordered label sets (e.g., 256). This requires four improvements on the original graph-shifts algorithm: 1) an improved way of caching the binary energy terms, 2) efficient sorting of the potential shift list and 3) an improved spawn shift, and 4) new, efficient rules for keeping the hierarchy in synch with the energy landscape after shifting. We demonstrate this algorithm on standard benchmark data in image restoration and stereo calculation.

We consider labeling problems of the following form. Define a pixel lattice Λ with pixels $i \in \Lambda$, and associate a label (random variable) y_i with each pixel. The labels take values in a label set $\mathcal{L} = \{1, \ldots, k\}$ and represent various problem specific physical quantities we want to estimate, like intensities, disparities, etc. The joint probability over the labels on the lattice, $y \doteq \{y_i\}_{i \in \Lambda}$, is given by a Gibbs distribution:

$$P(y) = \frac{1}{Z} \exp \left[-\sum_i U(y_i, i) - \sum_{\langle i,j \rangle} V(y_i, y_j) \right] \tag{1}$$

where Z is the partition function, and $\langle i, j \rangle$ represents the set of neighbors on Λ.

Equation 1 is the standard MRF. The clique potential functions, U and V, encode the local energy associated with various labelings at pixel sites. U is conditioned on the pixel location to permit the incorporation of some external data in the one-clique potential computation; for example, the input intensity field in image restoration (see section 5.1). The goal is to find a labeling y that maximizes $P(y)$, or equivalently minimizes the energy given by the sum over all clique potentials U and V. To simplify notation, we will consider problem as energy function minimization in the remainder of the paper.

The remainder of the paper is as follows: a short literature survey is in the next section. In section 3 we review the graph-shifts approach to energy minimization. Then, in section 4, we present our adaptations that make it possible to apply the algorithm on large label sets. In section 5 we analyze the proposed algorithm and give comparative results to efficient belief propagation [6] for the image restoration problem.

2   Related Work

Many algorithms have been proposed to solve the energy minimization problem associated with labeling the MRFs. The iterated conditional modes [2] is an early algorithm developed near the beginning of MRF research. It iteratively updates the pixel labels in a greedy fashion choosing the new labeling that gives the steepest decrease in the energy function. This algorithm converges slowly since it flips a single label at a time, and is very sensitive to local minima. Simulated annealing [9], on the other hand, is a stochastic global optimizer that given a *slow enough* cooling rate will always converge to the global minimum. However, in practice, the *slow enough* is a burden.

Some more recent algorithms are able to consistently approach the global minimum in compute times in the order of minutes. Graph cuts [3,10] can guarantee a so-called *strong* local minimum for a defined class of (metric or semi-metric) energy functions. Max-product loopy belief propagation (BP) [8] computes a low energy solution by passing messages (effectively, max of conditional distributions) between neighbors in a graph. When (if) convergence is reached, BP will have computed a local max to the label posterior at each node. Although not guaranteed to converge for loopy graphs, it has performed well in a variety of low-level vision tasks [7], and can be effectively implemented for low-level vision tasks to run in just a few seconds [6]. Tree reweighted belief propagation (TRW) [16] is a similar message-passing algorithm that has the goal of computing the upper bound on the log partition function of any undirected graph. Although the recent comparative analysis [15] did not find a single best method, a modified version of the TRW approach did consistently outperform the other methods.

3   Graph-Shifts

Following the notation from [4], define a graph G to be a set of nodes $\mu \in \mathcal{U}$ and a set of edges. The graph is hierarchical and composed of multiple layers with the nodes at the lowest layer representing the image pixels. Call two connected nodes on the same layer neighbors using the predicate $N(\mu, \nu) = 1$ and $N(\mu, \nu) = 0$ otherwise. Two connected nodes on different (adjacent) layers are called parent-child nodes. Each node has a single parent (except for the nodes at the top layer, which have no parent) and has the same label as its parent. Every node has at least one child (except for the nodes at the bottom layer). Let $C(\mu)$ be the set of children of node μ and $A(\mu)$ be the parent of node μ. A node μ on the bottom layer (i.e. on the lattice) has no children, and hence $C(\mu) = \emptyset$. At the top of the graph is a special *root* layer with a single node $\bar{\mu}$ for each of the k labels. The label of the root nodes is fixed to a single value. Since all non-root nodes in the hierarchy can trace their ancestry back to a single root node, an instance of the graph G is equivalent to a labeling of the image.

The coarser layers are computed recursively by an iterative bottom-up coarsening procedure. We use the coarsening method defined in [4] without modification. The basic idea is that edges in the graph are randomly turned on or off based on the local intensity similarity. The *on* edges induce a connected components clustering, and each component defines a new node in the next coarse layer in the hierarchy. Thus, nodes at coarser layers in the hierarchy represent (roughly) homogeneous regions in the images.

The procedure is adaptive and the resulting hierarchy is data-dependent. This is in contrast to traditional pyramidal schemes [1] which fix the coarse level nodes independent of the data. A manually tuned *reduction* parameter governs the amount of coarsening that happens at each layer in the bottom up procedure. In section 5, we give advice based on empirical experimentation on how to choose this parameter.

3.1 Energy in the Hierarchy

The original energy function (1) is defined at the pixel level only. Now, we extend this definition to propagate the energies up the hierarchy by recursing on the potentials:

$$\hat{U}(y_\mu, \mu) = \begin{cases} U(y_\mu, \mu) & \text{if } C(\mu) = \emptyset \\ \displaystyle\sum_{\nu \in C(\mu)} \hat{U}(y_\mu, \mu) & \text{otherwise} \end{cases} \tag{2}$$

$$\hat{V}(y_{\mu_1}, y_{\mu_2}) = \begin{cases} V(y_{\mu_1}, y_{\mu_2}) & \text{if } C(\mu_1) = C(\mu_2) = \emptyset \\ \displaystyle\sum_{\substack{\nu_1 \in C(\mu_1), \\ \nu_2 \in C(\mu_2): \\ N(\nu_1, \nu_2)=1}} \hat{V}(y_{\nu_1}, y_{\nu_2}) & \text{otherwise} \end{cases} \tag{3}$$

By defining the recursive energies in this form, we are able to explore the full label set $\mathcal{L}$ at each layer in the hierarchy rather than work on a reduced label set at each layer, which is typical of pyramidal coarse-to-fine approaches. By operating with the complete label set in the whole hierarchy, graph-shifts is able to quickly switch between scales at each iteration when selecting the next steepest shift to take (further discussion in section 5).

By using (2) and (3), we can compute the exact energy caused by any node in the hierarchy. Furthermore, the complete energy (1) can be rewritten in terms of the roots:

$$E(y) \doteq \sum_{i \in \mathcal{L}} \hat{U}(y_{\overline{\mu}_i}, \overline{\mu}_i) + \sum_{\substack{i,j \in \mathcal{L} \\ N(\overline{\mu}_i, \overline{\mu}_j)=1}} \hat{V}(y_{\overline{\mu}_i}, y_{\overline{\mu}_j}) \tag{4}$$

3.2 The Graph Shift and Minimizing the Energy

A *graph shift* is defined as an operation that changes the label of a node by dynamically manipulating the connectivity structure of the graph hierarchy. There are two types of graph shifts: a split-merge shift and a spawn shift. During a split-merge shift (figure 1(a)), a node μ detaches itself from its current parent $A(\mu)$ and takes the parent of a neighbor $A(\nu)$. By construction, the shift also relabels the entire sub-tree rooted at μ such that $y_\mu = y_\nu$. During a spawn shift (figure 1(b)), a node μ creates (or spawns) a new top-level root node $\overline{\mu}$ and dynamically creates a chain connecting μ to $\overline{\mu}$ with one new node per layer. The new tree is assigned whatever label (one that none of μ's neighbors already had) was associated with the spawn shift. After making either shift, the hierarchy must be resynchronized with the changed energy landscape (section 4.3).

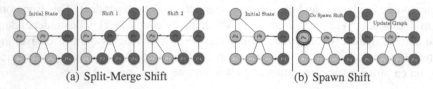

| (a) Split-Merge Shift | (b) Spawn Shift |

Fig. 1. Toy examples of the split-merge and spawn shifts with two classes, light and dark gray

The basic idea behind the graph-shifts algorithms is to select shift that would most reduce the energy at each iteration. Using (2) and (3), the exact energy gradient, or the *shift gradient*, can be computed as

$$\Delta E(\mu, y_\mu \to \hat{y}_\mu) = \hat{U}(\hat{y}_\mu, \mu) - \hat{U}(y_\mu, \mu) \;+\; \sum_{\nu:N(\mu,\nu)=1} \left[\hat{V}(\hat{y}_\mu, y_\nu) - \hat{V}(y_\mu, y_\nu) \right] .$$

(5)

This directly leads to the graph-shifts algorithm. After initialization the graph hierarchy, iterate the following steps until convergence:

1. Compute and select the graph shift that most reduces the energy.
2. Apply this shift to the graph.
3. Update the graph hierarchy to resynchronize with the new energy landscape.

4   Adapting Graph-Shifts for Large Label Sets

This section describes the adaptations we make to the original graph-shifts algorithms to increase its efficiency when dealing with large label sets. The first two adaptations (section 4.1) consider how the shifts are stored in various caches up the hierarchy. The third one considers a reduced spawning label set (section 4.2). The fourth one, in section 4.3, discusses how to update the hierarchy and potential shift list after executing a shift.

4.1   Computing and Selecting Shifts

Here, we discuss two representational details that reduce the amount of computation required when computing potential shifts. First, though the energy recursion formulas (2) and (3) provide a mathematically convenient way of computing the energy at a given node, repeatedly recursing down the entire tree to the leaves to compute the energy at a node is often redundant. So, an *energy cache* is stored at each node in the hierarchy. The cache directly stores the unary energy $\hat{U}(y_\mu, \mu)$ for a node μ in a vector of k dimension. The unary cache can be efficiently evaluated in a bottom-up fashion at the leaves first and them *pushing* them up the hierarchy. The memory cost is $O(kn \log n)$ for n pixels.

[4] suggests such a caching scheme for both the unary and the binary terms of the energy. However, storing a similar complete cache for the binary term is not plausible for large label sets because its cost is quadratic in the labels, $O(k^2 cn \log n)$ with c being the average cardinality of the nodes. However, recall the binary term is a function of only the labels at the two nodes, and the sub-trees have the same labels as their

parents. So, for the binary energy, the recursion formulas serve to count the length of the boundary between two nodes, which is then multiplied by the cost on the two labels:

$$\hat{V}(y_{\mu_1}, y_{\mu_2}) = V(y_{\mu_1}, y_{\mu_2})B(\mu_1, \mu_2) \tag{6}$$

$$B(\mu_1, \mu_2) = \begin{cases} N(\mu_1, \mu_2) & \text{if } C(\mu_1) = C(\mu_2) = \emptyset \\ \displaystyle\sum_{\substack{\nu_1 \in C(\mu_1) \\ \nu_2 \in C(\mu_2)}} B(\nu_1, \nu_2) & \text{otherwise.} \end{cases} \tag{7}$$

This form of the binary energy suggests caching the boundary length B between nodes in the graph. By doing so, we save on the k^2 factor for the cache and the resulting memory cost is $O(cn \log n)$. We discuss how to update this cache after a shift in 4.3.

Second, a complete list of potential shifts is maintained by the algorithm. After initializing the hierarchy, the full list is computed, and all those shifts with a negative gradient (5) are stored. To further reduce the list size, only a single shift is stored for any given node. The list is updated after each shift (section 4.3). Computing the best shift at each iteration results searching this list. [4] choose to store an unsorted list to save the $O(s \log s)$, for s shifts, cost of initially sorting the list at the expense of $O(s)$ search every iteration. However, an $O(s)$ is already paid during the initial computation. Hence, the "sorting cost" is only an additional $O(\log s)$ cost if we sort while computing the list. Searching every iterations is then only $O(1)$. Thus, we choose to sort the list.

4.2   An Improved Spawn Shift

The original spawn shift [5] requires the evaluation of a shift gradient when switching to potentially any new label. The cost of this evaluation grows with the number of labels, $O(k)$, but the cost of computing the best split-merge shift for a node is $O(c)$ using the caches. In the problems we consider $k \gg c$. We exploit the label ordering and search only a sub-range of the labels based on the current label. For node μ with label y_μ, we search the range $\{y_\mu - \kappa, y_\mu + \kappa\}$ where κ is a user selected parameter (we use 3).

4.3   Updating the Hierarchy After a Shift

A factor of crucial importance to the graph-shifts algorithm is dynamically keeping the hierarchy in synch with the energy landscape it represents. Since a shift is a very local change to the labeling, updating the hierarchy can be done quickly. After each shift, the following steps must be performed to ensure synchrony. Assume the shift occurs at level l, pixels correspond to level 0 and there are T levels in the hierarchy.

1. Update the unary caches at levels $t = \{l + 1, \ldots, T\}$.
2. Update the boundary length caches at levels $t = \{l + 1, \ldots, T\}$.
3. Recompute the shift for all affected nodes and update their entries in the potential shift list (removing them if necessary). Affected nodes will be present on all graph levels. For nodes below l, any node in the subtree or an immediate neighbor of the subtree of the shifted node must be updated. For nodes above only the parents and their neighbors must be updated. This an $O(c \log n)$ number of updates.

Updating the unary caches is straightforward. For a split-merge shift where node μ shifts to ν and takes $A(\nu)$ as a new parent, the update equations are

$$\hat{U}\left(y, A(\mu)\right)' = \hat{U}\left(y, A(\mu)\right) - \hat{U}\left(y, \mu\right) \qquad \forall y \in \mathcal{L} \tag{8}$$

$$\hat{U}\left(y, A(\nu)\right)' = \hat{U}\left(y, A(\nu)\right) + \hat{U}\left(y, \mu\right) \qquad \forall y \in \mathcal{L} . \tag{9}$$

Consider a spawn shift where μ spawns a new sub-tree to the root level. Equation (8) applies to the old parent $A(\mu)$. The new parent $A^*(\mu)$ is updated by

$$\hat{U}\left(y, A^*(\mu)\right) = \hat{U}\left(y, \mu\right) \qquad \forall y \in \mathcal{L} . \tag{10}$$

Each of these equations must be applied recursively to the root level T in the hierarchy.

Since the boundary length terms involve two nodes, they result in more complicated update equations. For a shift from node μ to ν the update equations for level $l + 1$ are

$$B\left(A(\mu), A(\nu)\right)' = B\left(A(\mu), A(\nu)\right) - \sum_{\substack{\eta:\, A(\eta)=A(\nu), \\ N(\mu,\eta)=1}} B(\mu, \eta) \tag{11}$$

$$B\left(A(\nu), A(\omega)\right)' = B\left(A(\nu), A(\omega)\right) + B(\mu, \omega)$$
$$B\left(A(\mu), A(\omega)\right)' = B\left(A(\mu), A(\omega)\right) - B(\mu, \omega)$$
$$\forall \omega:\, A(\omega) \neq A(\nu), N(\mu, \omega) = 1 \tag{12}$$

where $A(\mu)$ is the ancestor of μ before the shift takes place. The second term on the righthand side of (11) arises because μ can be a neighbor to more than one child of $A(\nu)$. When, μ shifts to become a child of $A(\nu)$, then it will a become sibling of such a node and the boundary length from $A(\mu)$ to $A(\nu)$ must account for it. Again, the updates must be applied recursively to the top of the hierarchy. These equations are also applicable in the case of a spawn shift with the additional knowledge, that if $B(A(\nu), \cdot)$ is 0, then a new edge must be created in the graph connecting these two nodes.

5   Experiments

We consider two low-level labeling problems in this section: image restoration and stereo. We also present a number of evaluative results on the efficiency and robustness of the graph-shifts algorithm for low-level MRF labeling. In all of the results, unless otherwise stated, a truncated linear binary potential function was used. It is defined on two labels and is fixed by two parameters, β_1, β_2:

$$V(y_i, y_j) = \min(\beta_1 \|y_i - y_j\|, \beta_2) . \tag{13}$$

5.1   Image Restoration

Image restoration is the problem of removing the noise and other artifacts of an acquired image to restore it back to its original, or ideal state. The label set comprises the 256 gray-levels. To analyze this problem, we work with the well-known penguin image; it's ideal image is given on the right in figure 2. In figure 3, we present some restoration results for various possible potential functions on images that have been perturbed by independent Gaussian noise of three increasing variances (in each row). In the following potentials, let x_i be the inputted intensity in the corrupted image. The second column shows a Potts model on both the unary and binary potential functions. The third column shows a truncated linear unary potential and a Potts binary potential with the fourth column showing both truncated linear potentials. Truncated quadratic results are shown in figure 4 in comparison with EBP. In these results, we can see that the graph-shifts algorithm is able to find a good minimum to the energy function, and it's clear that the stronger potential functions are giving better results. In the truncated linear terms here, $\alpha_1 = \beta_1 = 1$, $\alpha_2 = 100$, and $\beta_2 = 20$. The two terms were equally weighted.

Fig. 2. The ideal image

$$\text{Potts} \qquad U_P(y_i, i) = \delta(x_i, y_i) \qquad (14)$$

$$\text{Truncated Linear} \qquad U_L(y_i, i) = \min(\alpha_1 ||x_i - y_i||, \alpha_2) \qquad (15)$$

$$\text{Truncated Quadratic} \qquad U_Q(y_i, i) = \min(\alpha_1 ||x_i - y_i||^2, \alpha_2) \qquad (16)$$

Figure 4 and table 1 present a comparison of the image restoration graph-shifts with the efficient belief propagation (EBP) [6] algorithm. Here, we use a truncated quadratic energy function (16) with exactly the same energy function and parameters: $\alpha_1 = \beta_1 = 1$, $\alpha_2 = 100$ and $\beta_2 = 20$. In these restoration results, we use the sum of squared differences error between the original (ideal) image and the restored image that is outputted by each algorithm to measure the accuracy. Note, however, that the energy minimization algorithms are not directly minimizing the sum-of-squared differences error function. When computing the SSD error on these images we disregarded the outside row and column since the EBP implementation cannot do labeling near the image borders.

From inspecting the scores in table 1, we find the two algorithms are both able to find good minima for the two inputs with smaller noise. The graph-shifts minimum achieves lower SSD error scores for these two images. This could be due to the extra high-frequency information near the bottom of the image that it was able to retain, but the EBP implementation smoothed it out. However, for the higher noise variance, EBP is able to converge to a similar minimum and its SSD error is much low than graph-shifts in this case. We also see that the two algorithms run in the order of seconds (they are run on the same hardware). However, the speed comparison is not completely fair: EBP is implemented in C++ while graph-shifts is implemented in Java (one expects at

Table 1. Quantitative comparison of time and SSD error between efficient belief propagation and graph-shifts. Speed is not directly comparable as BP is in C++ but graph-shifts is in Java.

	Time (ms)		SSD Error	
Variance	Graph-Shifts	Efficient BP	Graph-Shifts	Efficient BP
10	18942	2497	2088447	2524957
20	24031	2511	2855951	3105418
30	24067	2531	5927968	3026555

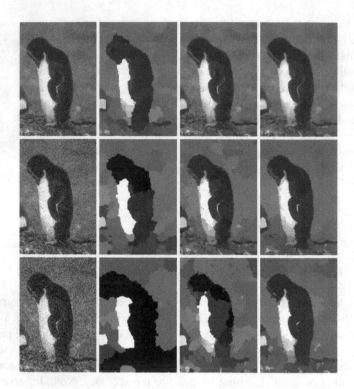

Fig. 3. Visual comparison of the performance of different energy functions with the same graph-shifts parameters. The images on the left are the input noisy images (with variances of 10, 20, and 30 in the rows). The remaining columns are Potts+Potts,Truncated Linear+Potts,Truncated Linear+Truncated Linear in the unary + binary terms, respectively.

least a factor of two speedup). We note that some of the optimizations suggested in the EBP algorithm [6] can also apply in the computation and evaluation of the graph shifts to further increase efficiency. The clear message from the time scores is that the graph-shifts approach is of the same order of magnitude as current state of the art inference algorithms (seconds).

5.2   Stereo

We present results on two images from the Middlebury Stereo Benchmark [14]: the sawtooth image has 32 disparities with 8 levels of subpixel accuracy or a total of 256 labels, and the Tsukuba image has 16 labels. The energy functions we use here to model the stereo problem remain very simple MRFs. The unary potential is a truncated linear function on the pixel-wise difference between the left image $I_L(u, v)$ and the right image $I_R(u - y_i, v)$, where $i = (u, v)$:

$$U_S(y_i, i) = \min(\alpha_1 \|I_L(u, v) - I_R(u - y_i, v)\|, \alpha_2) \ . \tag{17}$$

Figure 5 shows the two results. The parameters are $\alpha_1 = \beta_1 = 1$, $\alpha_2 = 100$ and $\beta_2 = 20$ for the tsukuba image and $\alpha_1 = \beta_1 = 1$, $\alpha_2 = \beta_2 = 10$. The inferred

Fig. 4. Visual comparison between the proposed graph-shifts algorithms and the efficient belief propagation [6]. The two pairs of images have noise with variance 10, 20, and 30. Images in each pair are the EBP restoration followed by the graph-shifts restoration. See Fig. 3 for input images.

disparity, on the right, is very close the ground truth nearly everywhere in the image. These results are in the range of the other related algorithms [3,6]. However, graph-shifts can compute them in only seconds. Even without a specific edge/boundary model, which many methods in the benchmark use, the graph-shifts minimizer is able to maintain good boundaries. For lack of space, we cannot discuss the stereo results in more detail.

5.3   Evaluation

We use the truncated linear unary and binary potentials on the penguin image (for restoration) with a variance of 20 for the noise in all results in this section unless otherwise noted. The parameters on the potentials are $(1, 100)$ and $(1, 20)$ for unary and binary respectively.

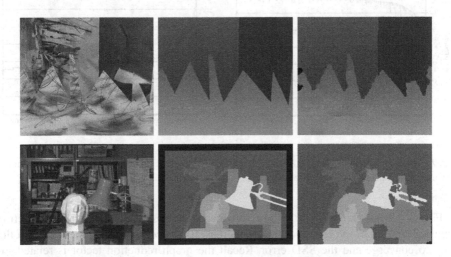

Fig. 5. Results on computing stereo with the graph-shifts algorithm on an MRF with truncated linear potentials. Left column is the left image of the stereo pair, middle column is the ground truth disparity, and the right column is the inferred disparity.

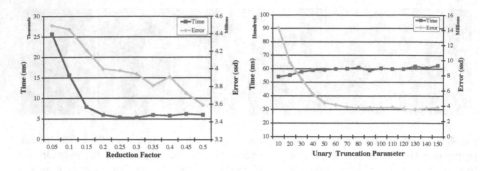

Fig. 6. Evaluation plots. (left) How does the convergence time vary with the height of the hierarchy (reduction parameter)? (right) How robust is the convergence speed when varying parameters of the potential functions?

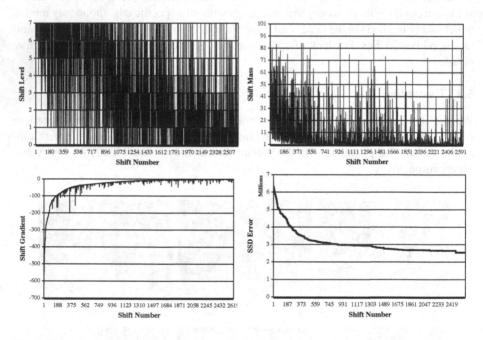

Fig. 7. These four graphs show statistics about the graph-shift process during one of the image restoration runs. See text for full explanation.

Figure 5.3-left shows how the time to converge varies with changing the reduction criterion. As the graph reduction factor increases, we see an improvement in both the time to converge and the SSD error. Recall the graph reduction factor is related inversely to the amount of coarsening that occurs in each layer. So, with a larger factor, the taller the hierarchy we find and the stronger the homogeneity properties in each graph node. Thus, the shifts that are taken by the graph-shifts with larger graph reduction are more *targetted* and result in fewer total necessary shifts. Figure 5.3-right demonstrates robustness to variation in the parameters of the energy functions. As we vary the truncation parameter on the unary potential the time to converge stays roughly constant, but the SSD error changes (as expected).

Figure 7 shows four different graphs that explore the actual graph-shift process. Each graph shows the shift number on the horizontal access and the vertical axis shows one of the following measurements (left-to-right) the level at which each shift occurs, the mass of the shift (number of pixels whose label changes), the shift gradient (5) and the SSD error of the current labeling. We show the SSD error to demonstrate that it is being reduce by minimizing the MRF; the SSD error is not the objective function directly being minimized. The upper-left plot highlights a significant difference in the energy minimization procedure created by the graph-shifts algorithm and the traditional multi-level coarse-to-fine approach. We see that as the algorithm proceeds it is greatly varies in which level to select the current shift; recall that graph-shifts will select the shift (at any level) with the current steepest negative shift gradient. This *up-and-down* action contrasts the coarse-to-fine approach which would complete all shifts at the top level and the proceeds down until the pixel level.

6   Conclusion

In this paper we present an adaptation of the recently proposed graph-shifts algorithm to the case of MRF labeling with large label sets. Graph-shifts does energy minimization by dynamically changing the parent-child relationships in a hierarchical decomposition of the image, which encodes the underlying pixel labeling. Graph-shifts is able to efficiently compute the optimal shift at every iteration. However, this efficiency comes from keeping the graph in synch with the underlying energy. The large label sets make ensuring this synchrony difficult. We made four suggestions for adapting the original graph-shifts algorithm to maintain its computational efficiency and quick run-times (order of seconds) for MRF labeling with large label sets. The results on image restoration and stereo are an indication of the potential in such a hierarchical energy minimization algorithm. The results also indicate that the quality of the minimization depends on the properties of the hierarchy, like height and homogeneity of nodes. In future work, we plan to develop a methodology to systematically optimize these for a given problem.

References

1. Anandan, P.: A Computational Framework and an Algorithm for the Measurement of Visual Motion. International Journal of Computer Vision 2(3), 283–310 (1989)
2. Besag, J.: Spatial interaction and the statistical analysis of lattice systems (with discussion). J. Royal Stat. Soc., B 36, 192–236 (1974)
3. Boykov, Y., Veksler, O., Zabih, R.: Fast Approximate Energy Minimization via Graph Cuts. IEEE Transactions on Pattern Analysis and Machine Intelligence 23(11), 1222–1239 (2001)
4. Corso, J.J., Tu, Z., Yuille, A., Toga, A.W.: Segmentation of Sub-Cortical Structures by the Graph-Shifts Algorithm. In: Karssemeijer, N., Lelieveldt, B. (eds.) Proceedings of Information Processing in Medical Imaging, pp. 183–197 (2007)
5. Corso, J.J., Yuille, A.L., Sicotte, N.L., Toga, A.W.: Detection and Segmentation of Pathological Structures by the Extended Graph-Shifts Algorithm. In: Ayache, N., Ourselin, S., Maeder, A. (eds.) MICCAI 2007, Part I. LNCS, vol. 4791, pp. 985–994. Springer, Heidelberg (2007)

6. Felzenszwalb, P.F., Huttenlocher, D.P.: Efficient Belief Propagation for Early Vision. International Conference on Computer Vision 70(1) (2006)
7. Freeman, W.T., Pasztor, E.C., Carmichael, O.T.: Learning Low-Level Vision. International Journal of Computer Vision (2000)
8. Frey, B.J., MacKay, D.: A Revolution: Belief Propagation in Graphs with Cycles. In: Proceedings of Neural Information Processing Systems (NIPS) (1997)
9. Geman, S., Geman, D.: Stochastic Relaxation, Gibbs Distributions, and Bayesian Restoration of Images. IEEE Transactions on Pattern Analysis and Machine Intelligence 6, 721–741 (1984)
10. Kolmogorov, V., Zabih, R.: What Energy Functions Can Be Minimized via Graph Cuts? In: European Conference on Computer Vision, vol. 3, pp. 65–81 (2002)
11. Kumar, S., Hebert, M.: Discriminative Random Fields: A Discriminative Framework for Contextual Interaction in Classification. In: International Conference on Computer Vision (2003)
12. Lafferty, J., McCallum, A., Pereira, F.: Conditional Random Fields: Probabilistic Models for Segmenting and Labeling Sequence Data. In: Proceedings of International Conference on Machine Learning (2001)
13. Li, S.Z.: Markov Random Field Modeling in Image Analysis, 2nd edn. Springer, Heidelberg (2001)
14. Scharstein, D., Szeliski, R.: A Taxonomy and Evaluation of Dense Two-Frame Stereo Correspondence Algorithms. International Journal of Computer Vision 47(2), 7–42 (2002)
15. Szeliski, R., Zabih, R., Scharstein, D., Veksler, O., Kolmogorov, V., Agarwala, A., Tappen, M., Rother, C.: A Comparative Study of Energy Minimization Methods for Markov Random Fields. In: Leonardis, A., Bischof, H., Pinz, A. (eds.) ECCV 2006, Part II. LNCS, vol. 3952, pp. 16–29. Springer, Heidelberg (2006)
16. Wainwright, M.J., Jaakkola, T.S., Willsky, A.S.: Tree-Reweighted Belief Propagation Algorithms and Approximate ML Estimation by Pseudo-Moment Matching. In: AISTATS (2003)

Label Space:
A Multi-object Shape Representation

James Malcolm[1], Yogesh Rathi[2], and Allen Tannenbaum[1]

[1] Georgia Institute of Technology, Atlanta, GA
{malcolm,tannenba}@ece.gatech.edu
[2] Brigham and Women's Hospital, Boston, MA
yogesh@bwh.harvard.edu

Abstract. Two key aspects of coupled multi-object shape analysis are the choice of representation and subsequent registration to align the sample set. Current techniques for such analysis tend to trade off performance between the two tasks, performing well for one task but developing problems when used for the other.

This article proposes $\mathcal{L}^n$ *label space*, a representation that is both flexible and well suited for both tasks. We propose to map object labels to vertices of a regular simplex, *e.g.* the unit interval for two labels, a triangle for three labels, a tetrahedron for four labels, etc. This forms a linear space with the property that all labels are equally separated.

On examination, this representation has several desirable properties: algebraic operations may be done directly, label uncertainty is expressed as a weighted mixture of labels, interpolation is unbiased toward any label or the background, and registration may be performed directly.

To demonstrate these properties, we describe variational registration directly in this space. Many registration methods fix one of the maps and align the rest of the set to this fixed map. To remove the bias induced by arbitrary selection of the fixed map, we align a set of label maps to their intrinsic mean map.

1 Introduction

Multi-object shape analysis is an important task in the medical imaging community. When studying the neuroanatomy of patients, clinical researchers often develop statistical models of important structures which are then useful for population studies or as segmentation priors [7,9,10,11,12]. The first step for this problem consists in choosing an appropriate shape descriptor capable of representing its statistical variability.

A common starting point for shape representation is a simple scalar label map, each pixel indicating the object present at that pixel, *e.g.* a one indicating object #1, a two indicating object #2, etc. Many techniques go on to map this entire volume to another space, the value of each pixel contributing to describe the shape. In this new space, arbitrary topologies may be represented, correspondences are naturally formed between pixels, and there are no control points to distribute.

The simplest implicit representation is a binary map where each pixel indicates the presence or absence of the object. Signed distance maps (SDM's) are another example of an implicit representation, each pixel having the distance to the nearest object boundary, a negative distance for points inside the object [7,12].

V.E. Brimkov, R.P. Barneva, H.A. Hauptman (Eds.): IWCIA 2008, LNCS 4958, pp. 185–196, 2008.

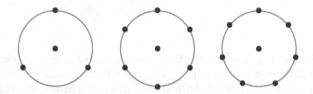

Fig. 1. Tsai et al. [11] proposed mapping each pixel from object label to a point in a space shaped as a non-regular simplex, each vertex corresponding to an object label. Visualized here for the case of two objects and background, the bottom left background (0,0) is a distance of 1 from both labels top (0,1) and right (1,0), while labels are separated from each other by a distance of $\sqrt{2}$.

Fig. 2. Example configurations for the S^1 hypersphere representation of [2]: three, six, and seven labels *(left to right)* with background at the center

For the multi-object setting, binary maps may be extended to scalar label maps, each pixel holding a scalar value corresponding to the presence of a particular object; however, this representation is not well suited for algebraic manipulation. For example, if labels are left as scalar values, the arithmetic average of labels with values #1 and #3 would incorrectly indicate the label of value #2, not a mixture of labels #1 and #3.

To address this, mappings of object labels to linear vector spaces were proposed, an approach to which our method is most closely related. The work of Tsai et al. [11] introduced two such representations, each for a particular task. For registration, the authors proposed mapping scalar labels to binary vectors with entries corresponding to labels; a one in an entry indicates the presence of the corresponding label at that pixel location. As an example for the case of two labels and background, Figure 1 visualizes the spatial configuration each pixel is mapped onto. Here the background is at the bottom left origin (0,0) with one label at (1,0) and the other at (0,1). It is also important to note that he goes on to perform registration considering each entry of these vectors separately. For shape analysis, Tsai et al. [11] proposed mapping scalar labels to layered SDM's, in this case each layer giving the signed distance to the corresponding object's interface.

Note that in both vector valued representations described in Tsai et al. [11], each label lies on its own axis and so the dimension of the representation grows linearly with the number of labels, *e.g.* two objects require two dimensions, three objects require three dimensions. To address this spatial complexity, Babalola and Cootes [2,3] propose a lower dimension approximation to replace the binary vectors in registration. By mapping labels to the unit hypersphere S^n, they demonstrate that even configurations involving dozens of labels can be efficiently represented with label locations distributed uniformly on a hypersphere. Figure 2 gives examples for S^1.

Finally, Pohl et al. [10] indirectly embeds label maps in the logarithm-of-odds space using as intermediate mappings either the binary or SDM representations of [11].

Fig. 3. The first three $\mathcal{L}^n$ label space configurations: a unit interval $\mathcal{L}^2$ in $\mathbb{R}$ for two labels, a triangle $\mathcal{L}^3$ in $\mathbb{R}^2$ for three labels, and a tetrahedron $\mathcal{L}^4$ in $\mathbb{R}^3$ for four labels *(left to right)*

Particularly well suited for probabilistic computations, the logarithm-of-odds space is also a field providing closed operations for addition and scalar multiplication. As with the representations of Tsai et al. [11], the dimensionality of the logarithm-of-odds space increases with each additional object. We should also note that the work of [10] did not address registration, but instead assumed an already registered atlas via [8].

Once the representation is settled upon, registration must be performed to eliminate variation due to differences in pose. A common approach is to register the set to a reference image; however, this then introduces a bias to the shape of the chosen reference. Joshi et al. [6] propose unbiased registration with respect the mean sample as a template reference. Assuming a general metric space of transformations, they describe registering a sample set with respect to its intrinsic mean and use the L_2 distance for demonstration. A similar approach uses the minimum description length to measure distance from the intrinsic mean [13]. Instead of registering to a mean template, an alternative approach is to minimize per-pixel entropy. Using binary maps Miller et al. [8] demonstrate that this has a similar tendency toward the mean sample. This approach has also been demonstrated on intensity images [14,15]. Among these energy-based registration techniques, iterative solutions include those that are variational [11,6] and those that use sampling techniques [15].

1.1 Our Contributions

This paper proposes a multi-object implicit representation that maps object labels to the vertices of a regular simplex, going from a scalar label value to a vertex coordinate position in a high dimensional space which we term *label space* and denote by $\mathcal{L}^n$ for n labels. Visualized in Figure 3, this regular simplex is a hyper-dimensional analogue of an equilateral triangle, n vertices capable of being represented in $n - 1$ dimensions ($\mathcal{L}^n \subset \mathbb{R}^{n-1}$). Lying in a linear vector space, this space has several desirable properties: all labels are equally separated in space, addition and scalar multiplication are natural, label uncertainty is expressed as a weighted combination of label vertices, and interpolation is unbiased toward any label including the background.

The proposed method addresses several problems with current implicit mappings. For example, while the binary vector representation of Tsai et al. [11] was proposed for registration, we will demonstrate that it induces a bias sometimes leading to misalignment, and since our $\mathcal{L}^n$ label space representation equally spaces labels, there is no such bias. Additionally, compared to the SDM representation, the proposed method introduces no inherent per-pixel variation across equally labeled regions making it more robust for statistical analysis. Hence, the proposed method better encapsulates the functionality of both representations. Further, the registration energy of Tsai et al. [11] is designed to consider each label independent of the others. In contrast, $\mathcal{L}^n$ label space

Fig. 4. For the S^1 hypersphere configurations of [2], cases such as these yield erroneous results during interpolation. Judged by nearest neighbor, interpolating between two labels resolves to background, ambiguously either background or another label, and finally another label (*left to right*).

jointly considers all labels. We will also demonstrate that, while lowering the spatial demands of the mapping, the hypersphere representation of Babalola and Cootes [2] biases interpolation and can easily lead to erroneous results. The arrangement of our proposed label space incurs no such bias allowing linear combinations of arbitrary labels.

The rest of this paper is organized as follows. Section 2 explores several problems that can develop with the implicit representations described above [2,10,11]. Section 3 then describes the proposed $\mathcal{L}^n$ label space representation documenting several of its properties. Section 4 demonstrates variational registration directly within this representation, and finally in Section 5 we summarize our work.

2   Related Work

2.1   Shape Representation

The signed distance map (SDM) has been used as a representation in several studies [1,7,10,11,12]; however, it may produce artifacts during statistical analysis [4]. For example, small deviations at the interface cause large variations in the surface far away, thus it inherently contains significant per-pixel variation. Additionally, ambiguities arise when using layered signed distance function to represent multiple objects: what happens if more than one of the distance functions indicates the presence of an object? Such ambiguities and distortions stem from the fact SDM's lie in a manifold where these linear operations introduce artifacts [4,5].

Label maps have inherently little per-pixel variation, pixels far from the interface having the same label as those just off the interface. For statistical analysis in the case of one object, Dambreville et al. [4] demonstrated that binary label maps have higher fidelity compared to SDM's. However, for the multi-object setting, the question then becomes one of how to represent multiple shapes using binary maps? What is needed is a richer feature space suitable for a uniform pair-wise separation of labels.

An example of such a richer feature space is that of Babalola and Cootes [2] where labels are mapped to points on the surface of a unit hypersphere S^n placing the background at the center. This is similar to the binary vector representation described by Tsai et al. [11] to spread labels out; however, Babalola and Cootes [2] argue that lower dimensional approximations can be made. They demonstrate that configurations involving dozens of labels can be efficiently represented by distributing label locations uniformly on the unit hypersphere using as few as three dimensions. Since any label may neighbor the background, the background must be placed at the hypersphere center, equally spaced from all other labels. The fundamental assumption is that pixels only vary between labels that are located near to each other on the hypersphere, so the placement of labels is crucial to avoid erroneous label mixtures. For example, Figure 4 demonstrates that if two labels far from each other are mixed, the result may be attributed erroneously to other labels. Notice in particular that the central placement of the background gets in the way when interpolating across the sphere. Smoothing in Figure 7 also demonstrates these inherent effects of the lower dimensional approximation, effects that cannot be avoided unless the dimension approaches label cardinality.

The logarithm-of-odds representation of Pohl et al. [10] provides the third and final shape representation we compare against. Aside from the normalization requirement for closed algebraic manipulation, the main concern when using this representation is the choice of intermediate mapping, a choice that directly impacts the resulting probabilities. The authors explore the use of both representations from [11]; however, both choices have inherent drawbacks.

For the layered SDM intermediate mapping, Pohl et al. [10] notes that SDM's are a subspace of the logarithm-of-odds space. This means that, while the layered SDM's are exactly the logarithm-of-odds representation, results after algebraic manipulation in the logarithm-of-odds space often yield invalid SDM's (but still valid logarithm-of-odds representations). Using such results, computing probabilities as described in [10] may yield erroneous likelihoods. Notice also, that the generalized logistic function is used to compute probabilities. This introduces additional problems as the use of the exponential ensures that these probabilities will always have substantial nonzero character across the entire domain, even in areas never indicated by the sample set.

Using smoothed binary maps as intermediates also leads to problems. To begin, using binary maps directly would mean probabilities of either zero or one, which in the log domain produce singularities. Smoothing lessens such effects yet results in a loss of fine detail along the interface. Also, Pohl et al. [10] shows examples where after normalization the logarithm-of-odds representation develops artifacts at the interface between objects, an effect which is magnified in the logarithm domain.

2.2 Registration

Tsai et al. [11] propose a binary vector representation specifically for registration. As Figure 1 shows, this representation places labels at the corners of a right-triangular simplex; however, unlike this present work, it is not a regular simplex but has a bias with respect to the background. The background, located at the origin, is a unit distance from any other label, while any two labels, located along a positive axis, are separated by a distance of $\sqrt{2}$. The effect may be seen in registration where there is a bias to misalign labels over the background (penalty 1) rather than over other labels (penalty $\sqrt{2}$).

To demonstrate the effect of this induced bias, consider the example in Figure 5 with black background and two rectangles of label #1, one with strip of label #2 along its top. Using the representation and registration energy of Tsai et al. [11], there are two global minima: the image overlapping and the image shifted up. In the first case, label

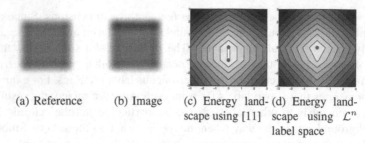

(a) Reference   (b) Image   (c) Energy land- (d) Energy land-
                                          scape using [11] scape using $\mathcal{L}^n$
                                                            label space

Fig. 5. Alignment of an image with a reference template using the representation of Tsai et al. [11] results in two possible alignments, the shifted one misaligning along both the top and bottom with respect to the reference *(red dots indicate minima)*. For just x- and y-translation, isocontours of the energy landscape show the non-unique energy minima in (c).

#1 is misaligned over label #2, while in the second case that a strip of pixels at both the top and bottom are misaligned over the background; that is, because of this bias, there can be twice as many pixels misaligned in the shifted case than in the unshifted. These global minima (indicated by red dots in the energy landscapes) are shown only for translation; considering additional pose parameters further increases the number of local minima in the energy landscape representing misalignments. Also, this is not inherent in the energy, as the same phenomena is observed using the energy in (1). Since all labels are equidistant in the proposed representation, there are fewer minima and hence less chance of misalignment.

3   Label Space

Our goal is to create a robust representation where algebraic operations are natural, label uncertainty is captured, and interpolation is unbiased toward any label. To this end we propose mapping each label to a vertex of a regular simplex; given n labels, including the background, we use a regular simplex which lies in $n - 1$ dimensions and denote this by $\mathcal{L}^n$ (see Figure 3). A regular simplex is an n-dimensional analogue of an equilateral triangle.

In this space, algebraic operations are as natural as vector addition, scalar multiplication, inner products, and norms; hence, there is no need for normalization as in [10]. Label uncertainty is realized as the weighted mixture of vertices. For example, a pixel representing labels #1, #2, and #3 with equal characteristic would simply be the point $p = \frac{1}{3}v_1 + \frac{1}{3}v_2 + \frac{1}{3}v_3$, a point equidistant from those three vertices (see Figure 6). Also, we have that such algebraic operations are unbiased toward any label since all labels

Fig. 6. Proposed $\mathcal{L}^3$ label space for the case of three labels: a point indicating the equal presence of all three labels *(left)*, and a point indicating the unequal mixed presence of just the left and top labels *(right)*

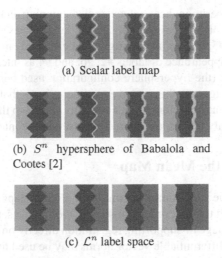

(a) Scalar label map

(b) S^n hypersphere of Babalola and Cootes [2]

(c) $\mathcal{L}^n$ label space

Fig. 7. Progressive smoothing directly on scalar label maps, the hypersphere representation of Babalola and Cootes [2], and $\mathcal{L}^n$ label space. Both the scalar label maps and hypersphere representations develop intervening strips of erroneous labels. Only label space is able to correctly capture the label mixtures during smoothing. The rightmost hypersphere in Figure 4 depicts the S^1 configuration used here in (b).

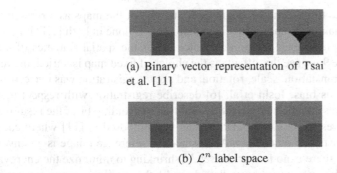

(a) Binary vector representation of Tsai et al. [11]

(b) $\mathcal{L}^n$ label space

Fig. 8. Progressive smoothing directly on binary vector representation of Tsai et al. [11] and $\mathcal{L}^n$ label space. Smoothing among several labels in the binary vector representation yields points closer to background *(black)* than any of the original labels. Label space is able to correctly begin to smooth out the sharp corners of the bottom two regions without erroneous introduction of the black background label.

are equally spaced; hence, there is no bias with respect to the background as is found in both [2,11]. Label space is robust to statistical analysis much like binary label maps, a specific case of label space. Additionally, problems encountered in the intermediate representations of [10] are avoided. Specifically, smoothing is unnecessary and so fine detail is retained, and interfaces are correctly maintained.

To demonstrate some of these properties, we performed progressive smoothing using the various representations described: scalar label values, the binary vector representation of Tsai et al. [11], the S^n representation of Babalola et al. [2], and $\mathcal{L}^n$ label space.

In Figure 7, the first experiment has each example beginning on the left with the jagged stripes of labels #5, #7, and #3, respectively. Scalar label values show the appearance of intervening labels #4, #5, and #6 as the original labels blend, and the hypersphere representation shows the appearance of labels #2, #6, and #4 as interpolation is performed across the hypersphere (the hypersphere configuration used here is the rightmost depicted in Figure 4). In Figure 8, the second experiment shows that the smoothing among multiple labels using binary vectors produces points closest to the background (black). In both experiments, only label space correctly preserves the interfaces.

4   Registering to the Mean Map

We demonstrate here the variational registration of a set of maps to their intrinsic mean map, thereby respecting the first order statistics of the sample set. The proposed representation has the advantage of supporting registration directly on the representation. By directly we mean that differentiable vector norms may be used to compare labels.

In this section, we begin with a review of reference-based approaches for rigid registration borrowing the notation of [11]. After demonstrating how a bias can be induced by the choice of reference template, we demonstrate unbiased registration using the mean map as the reference template in the manner of [6]. We conclude with experiments on synthetic maps, the 2D slices from [11] with three labels, and 2D slices with eight labels.

Common approaches to registration begin by fixing one of the maps as a reference and registering the remaining maps to this fixed map. This is done in both [2,11]; however, as Joshi et al. [6] describes, this initial choice biases the spatial statistics of the aligned maps. In Figure 9 we see this effect: as the choice of fixed map is varied, the resulting atlas varies in translation, scale, rotation, and skew (registration was performed as in [11]). To avoid this bias, Joshi et al. [6] describe registration with respect to a reference that best represents the sample set. In addition to avoiding bias, the resulting gradient descent involves far less computation than that proposed in [11] where each map is compared against each other map. Also, since the reference image is a convex combination of the set, there is no fear of the set $\tilde{M}$ shrinking to minimize the energy.

Before presenting the energy used, we first describe the problem borrowing notation from [11]. For the set of label maps $M = \{m_i\}_{i=1}^N$, our goal is to estimate the set of corresponding pose parameters $P = \{\mathbf{p}_i\}_{i=1}^N$ for optimal alignment. We denote as $\tilde{m}$ the label map m transformed by its pose parameters. An advantage of implicit representations over explicit ones is that, once the label maps have undergone

Fig. 9. Label maps from patient MRI data after registration where a different label map has been fixed in each run. The choice of which map to fix can subtly distort measurements and hence the statistical model constructed from the registered set.

this transformation, we can assume direct per-pixel correspondence between maps and use a vector norm to perform comparison. We model pose using an affine model, and so for 2D, the pose parameter is the vector $\mathbf{p} = [x \; y \; s_x \; s_y \; \theta \; k]^T$ corresponding to x-,y- translation, x-,y-scale, in-plane rotation, and shear. Note that this is a fully affine model as compared to the rigid transformation model used in [11]. The transformed map is defined as $\tilde{m}(\tilde{x}, \tilde{y}) = m(x, y)$ where coordinates are mapped according to $[\tilde{x} \; \tilde{y} \; 1]^T = T(\mathbf{p}) [x \; y \; 1]^T$, where $T(\mathbf{p})$ is the decomposable transformation matrix

$$
T(\mathbf{p}) = \underbrace{\begin{bmatrix} 1 & 0 & x \\ 0 & 1 & y \\ 0 & 0 & 1 \end{bmatrix}}_{M(x,y)} \underbrace{\begin{bmatrix} \cos(\theta) & -\sin(\theta) & 0 \\ \sin(\theta) & \cos(\theta) & 0 \\ 0 & 0 & 1 \end{bmatrix}}_{R(\theta)} \underbrace{\begin{bmatrix} s_x & 0 & 0 \\ 0 & s_y & 0 \\ 0 & 0 & 1 \end{bmatrix}}_{H(s_x, s_y)} \underbrace{\begin{bmatrix} 1 & k & 0 \\ k & 1 & 0 \\ 0 & 0 & 1 \end{bmatrix}}_{K(k)}
$$

for a translation matrix $M(x, y)$, rotation matrix $R(\theta)$, anisotropic scale matrix H (s_x, s_y), and shear matrix $K(k)$, all for the parameters taken from $\mathbf{p}$.

As in [6,15], we assume the intrinsic mean map $\tilde{\mu}$ of the sample set to best represent the population. We then attempt to minimize the energy defined as the squared distance between each transformed label map $\tilde{m}$ and this mean map $\tilde{\mu}$ of the set $\tilde{M}$ as it converges:

$$
d^2 = \sum_{i=1}^{N} \|\tilde{m}_i - \tilde{\mu}\|^2, \tag{1}
$$

where $\tilde{\mu} = \frac{1}{N} \sum_{i=1}^{N} \tilde{m}_i$, and while $\| \cdot \|$ may be any differentiable norm, we take it to be the elemental L_2 inner product $\|x\| = \langle x, x \rangle^{1/2} = \int x^2 dx$. Notice how using a vector norm here jointly considers all labels in contrast to the energy proposed by Tsai et al. [11]. Further, since the reference map $\tilde{\mu}$ is intrinsic, there is no concern of the set $\tilde{M}$ shrinking to minimize (1). Hence, there is no need for the normalizing term introduced in [11] which allows for a reduced complexity energy here.

This work uses a variational approach to registration. Specifically we perform gradient descent to solve for the pose parameters minimizing this distance. We find the gradient of this distance, taken with respect to the pose $\mathbf{p}_j$, to be:

$$
\nabla_{\mathbf{p}_j} d^2 = 2 \langle \nabla_{\mathbf{p}_j} \tilde{m}_j, \tilde{m}_j - \tilde{\mu} \rangle. \tag{2}
$$

Notice that terms involving other label maps ($\tilde{m}_i$ for $i \neq j$) fall out and that the gradient of the mean contributes nothing. It remains to define $\nabla_{\mathbf{p}_j} \tilde{m}_j$. For the kth element of the pose parameter vector $\mathbf{p}_j$, using the chain rule produces $\nabla_{\mathbf{p}_j^k} \tilde{m}_j =$

$\begin{bmatrix} \frac{\partial \tilde{m}_j}{\partial \tilde{x}} & \frac{\partial \tilde{m}_j}{\partial \tilde{y}} & 0 \end{bmatrix} \frac{\partial T(\mathbf{p}_j)}{\partial \mathbf{p}_j^k} \begin{bmatrix} x \\ y \\ 1 \end{bmatrix}$, where $\frac{\partial T(\mathbf{p}_j)}{\partial \mathbf{p}_j^k}$ is computed for each pose parameter where

matrix derivatives are taken componentwise. Finally, gradient descent proceeds by repeated calculation of $\nabla_{\mathbf{p}_j} d^2$ and adjustment of $\mathbf{p}_j$ for each map in the set until convergence.

To illustrate this technique, we first performed alignment of a synthetic 2D set. The unaligned set consists of 15 maps of three labels and background. Figure 10 shows

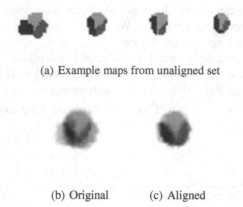

(a) Example maps from unaligned set

(b) Original   (c) Aligned

Fig. 10. Alignment of a set of 15 synthetic maps with three labels and background. The original and aligned sets are superimposed for visualization.

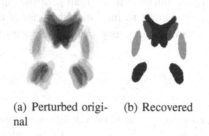

(a) Perturbed origi-   (b) Recovered
nal

Fig. 11. From the dataset used by Tsai et al. [11], one map is chosen and perturbed under several transformations, yet registration is able to recover the pose parameters to bring the perturbed versions back to the original chosen map. The perturbations ranged up to translations of 5% of the image, rotational differences of 20°, and scale changes +/- 5% of the image. The original and aligned sets are superimposed for visualization.

examples from this set as well as the original and aligned sets. For visualization, we created a superimposed map for both the original unaligned set and the aligned set by summing the scalar label values pixelwise and dividing by the number of maps, hence this is the mean scalar map.

We then turned to verifying our method using the 2D data from the study by Tsai et al. [11]. Taking one map from this set, we formed a new set by transforming this map arbitrarily. Restricting ourselves to the rigid rotation pose model used in that study, we formed transformations involving translations of 5% of the image size, rotational differences of 20°, and scale changes of +/- 5% of the image. Figure 11 shows that the technique successfully recovered the initial map. Figure 12 shows alignment on the entire data set.

Lastly, we performed registration using 2D maps obtained from expert manual segmentation of 33 patient MRI scans involving eight labels and background. Figure 13 shows examples from the original unaligned set as well as the superimposed maps after alignment.

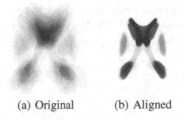

(a) Original   (b) Aligned

Fig. 12. Alignment of a set of 30 maps used in the study by Tsai et al. [11]. The original and aligned sets are superimposed for visualization.

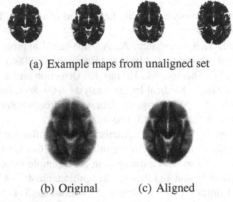

(a) Example maps from unaligned set

(b) Original   (c) Aligned

Fig. 13. Alignment of a set of 33 maps with eight labels and background obtained from manual MRI segmentations. The original and aligned sets are superimposed for visualization.

5   Conclusion

This paper describes a new implicit multi-object shape representation. After detailing several drawbacks to current representations, we demonstrated several of its properties. In particular, we demonstrated that algebraic operations may be done directly, label uncertainty is expressed naturally as a mixture of labels, interpolation is unbiased toward any label or the background, and registration may be performed directly.

Modeling shapes in label space does have its limitations. One key drawback to label space is the spatial demand. To address this we are examining lower dimensional approximations much like Babalola and Cootes [2]. Some interpolation issues such as those noted in Figure 4 might be avoided by taking into consideration the empirical presence of neighbor pairings when determining label distribution.

Acknowledgements

This work was supported in part by grants from NSF, AFOSR, ARO, MURI, as well as by a grant from NIH (NAC P41 RR-13218) through Brigham and Women's Hospital. This work is part of the National Alliance for Medical Image Computing

(NAMIC), funded by the National Institutes of Health through the NIH Roadmap for Medical Research, Grant U54 EB005149. Information on the National Centers for Biomedical Computing can be obtained from `http://nihroadmap.nih.gov/bioinformatics`

References

1. Abd, H., Farag, A.: Shape representation and registration using vector distance functions. In: Computer Vision and Pattern Recognition (2007)
2. Babalola, K., Cootes, T.: Groupwise registration of richly labelled images. In: Medical Image Analysis and Understanding (2006)
3. Babalola, K., Cootes, T.: Registering richly labelled 3d images. In: Proc. of the Int. Symp. on Biomedical Images (2006)
4. Dambreville, S., Rathi, Y., Tannenbaum, A.: A shape-based approach to robust image segmentation. In: Int. Conf. on Image Analysis and Recognition (2006)
5. Golland, P., Grimson, W., Shenton, M., Kikinis, R.: Detection and analysis of statistical differences in anatomical shape. Medical Image Analysis 9, 69–86 (2005)
6. Joshi, S., Davis, B., Jomier, M., Gerig, G.: Unbiased diffeomorphic atlas construction for computational anatomy. NeuroImage 23, 150–161 (2004)
7. Leventon, M., Grimson, E., Faugeras, O.: Statistical shape influence in geodesic active contours. In: Computer Vision and Pattern Recognition, pp. 1316–1324 (2000)
8. Miller, E., Matsakis, N., Viola, P.: Learning from one example through shared densities on transforms. In: Computer Vision and Pattern Recognition, pp. 464–471 (2000)
9. Nain, D., Haker, S., Bobick, A., Tannenbaum, A.: Multiscale 3-d shape representation and segmentation using spherical wavelets. Trans. on Medical Imaging 26(4), 598–618 (2007)
10. Pohl, K., Fisher, J., Bouix, S., Shenton, M., McCarley, R., Grimson, W., Kikinis, R., Wells, W.: Using the logarithm of odds to define a vector space on probabilistic atlases. Medical Image Analysis (to appear, 2007)
11. Tsai, A., Wells, W., Tempany, C., Grimson, E., Willsky, A.: Mutual information in coupled multi-shape model for medical image segmentation. Medical Image Analysis 8(4), 429–445 (2003)
12. Tsai, A., Yezzi, A., Wells, W., Tempany, C., Tucker, D., Fan, A., Grimson, W., Willsky, A.: A shape-based approach to the segmentation of medical imagery using level sets. Trans. on Medical Imaging 22(2), 137–154 (2003)
13. Twining, C., Marsland, C., Taylor, S.: Groupwise non-rigid registration: The minimum description length approach. In: British Machine Vision Conf. (2004)
14. Warfield, S., Rexillius, J., Huppi, R., Inder, T., Miller, E., Wells, W., Zientara, G., Jolesz, F., Kikinis, R.: A binary entropy measure to assess nonrigid registration algorithms. In: Niessen, W.J., Viergever, M.A. (eds.) MICCAI 2001. LNCS, vol. 2208, pp. 266–274. Springer, Heidelberg (2001)
15. Zöllei, L., Learned-Miller, E., Grimson, E., Wells, W.: Efficient population registration of 3d data. In: Workshop on Comp. Vision for Biomedical Image Applications (ICCV) (2005)

A New Image Segmentation Technique Using Maximum Spanning Tree

Qiang He[1] and Chee-Hung Henry Chu[2]

[1] Department of Mathematics, Computer and Information Sciences
Mississippi Valley State University
Itta Bena, MS 38941
QiangHe@mvsu.edu
[2] The Center for Advanced Computer Studies
University of Louisiana at Lafayette
Lafayette, LA 70504-4330
cice@cacs.louisiana.edu

Abstract. An alternative to the gradient-based image segmentation methods are those methods that use eigenvectors based on an affinity matrix built from pairwise pixel similarity. In this paper, we describe a new image segmentation algorithm using the maximum spanning tree. Our method works on the affinity matrix; however, instead of computing eigenvalues and eigenvectors, we show that image segmentation could be transformed into an optimization problem: finding the maximum spanning tree of the graph with image pixels as vertices and pairwise similarities as weights. The experimental results on synthetic and real data show good performance of this algorithm.

Keywords: image segmentation, affinity matrix, maximum spanning tree.

1 Introduction

The objective of image segmentation and clustering is to extract meaningful regions out of an image. A graph theoretic approach, as an alternative to the gradient-based methods, is usually based on the eigenvectors of an affinity matrix [3,4,5,6,7]. The theoretical foundation of this development is the Spectral Graph Theory [1], through which the combinatorial features of a graph can be revealed by its spectra. This characteristic can be applied into graph partitioning and preconditioning. The typical eigendecompostion based segmentation work is called the normalized cuts [6]. The normalized cut measure incorporates the local similarity within cluster and total dissimilarity between clusters. The minimization of this measure is to solve the Rayleigh quotient, a generalized eigenvalue solving problem. However, solving a standard eigenvalue problem for all eigenvectors has exponential complexity and is very time consuming. Shi [6] made use of the sparsity of the affinity matrix and introduced the Lanczos method to simplify the computation of eigenvalues.

In this paper, we give a new image segmentation algorithm using the maximum spanning tree [2]. Our method works on affinity matrix and addresses the physical meanings of an affinity matrix. Instead of computations of eigenvalues and

V.E. Brimkov, R.P. Barneva, H.A. Hauptman (Eds.): IWCIA 2008, LNCS 4958, pp. 197–204, 2008.

eigenvectors, we proved that the image segmentation could be transformed into an optimization problem: finding the maximum spanning tree of the graph with image pixels as vertices and pairwise similarities as weights. Section 2 describes the related theory and Section 3 gives the experimental results on synthetic and real data to illustrate the performance of this algorithm. Finally, we draw a conclusion and discuss future work in Section 4.

2 Method Descriptions

In this section, we first discuss the characteristics of affinity matrix and then define an optimization measure based on the weighted graph associated with an image. The solution to the optimization problem satisfies the clustering standard with maximal within-class similarity and minimum between-class similarity.

2.1 Affinity Matrix

The affinity matrix is a symmetric matrix and describes the pairwise pixel similarity. Every element $W_{i,j}$ of an affinity matrix $\mathbf{W}$ represents the similarity between pixels i and j. There are various definitions for the similarity measures. In general, $W_{i,j}$ can be defined as

$$W_{i,j} = e^{-\|x_i - x_j\|^2 / 2\sigma^2} \tag{1}$$

where $\|.\|$ is Euclidean distance and σ is a free parameter. This is somewhat similar to the definition to Gaussian distribution.

The characteristics of an affinity matrix (or similarity measures) are listed as follows.

1.   Symmetric property
    The affinity matrix is symmetric, that is, $W_{i,j} = W_{j,i}$. So it can be diagonalized.
2.   Unit normalization
    That is, $0 \le W_{i,j} \le 1$. The similarity $W_{i,j}$ between pixels i and j increases as $W_{i,j}$ goes from 0 to 1 while the dissimilarity decreases.
3.   Transitive property
    If pixels i and j are similar and pixels j and k are similar, then pixels i and k are similar.
4.   Coherence property
    That is, $W_{i,j} \ge W_{l,k}$ holds for $\forall i, j, l, k$ if pixels i and j are in the same cluster while pixels l and k are in different clusters.
5.   If similarity $W_{i,j}$ is greater than some threshold, then we say that pixels i and j are similar.

2.2 Similarity Measure for Cluster and Whole Image

Now, we define a similarity measure for one cluster and the whole image. If we consider the affinity matrix represents a weight matrix of a complete graph with all pixels as vertices. Then, there is a maximum spanning tree for this complete graph. Obviously, there is a subaffinity matrix for every cluster. We define the cluster similarity measure as the product of those weights in its maximum spanning tree. That is,

$$S_h = \prod_{i=1}^{N_h-1} P_i^h \tag{2}$$

where h represents cluster number, N_h is the number of entities (pixels) in cluster h, and P_i^h are weights in the maximum spanning tree of cluster h. Because of the symmetric, transitive, and coherence properties of affinity matrix, we can understand this as follows. Given a pixel p in cluster h, in order to find all pixels in cluster h, we first find the pixel q with maximum similarity to pixel p. Then we find another pixel not in set $\{p, q\}$, but with maximum similarity either with p or with q. Repeatedly, until all pixels of cluster h are found. We can see that this measure is reasonable to represent the maximum within-cluster similarity for cluster h.

After we define cluster similarity measure, we further define a similarity $\mathbf{S}$ for the whole image, as follows

$$\mathbf{S} = \prod_{h=1}^{c} S_h = \prod_{h=1}^{c} \prod_{i=1}^{N_h-1} P_i^h \tag{3}$$

where c is the number of clusters of image or number of segmentation components.

For convenience, sometimes, we use log on $\mathbf{S}$. We have

$$\log \mathbf{S} = \sum_{h=1}^{c} \log S_h = \sum_{h=1}^{c} \sum_{i=1}^{N_h-1} \log P_i^h \tag{4}$$

Next, we will show that to maximize the similarity measure $\mathbf{S}$ is to maximize the within-cluster similarity and minimize the between-cluster similarity, which is preferred by the clustering and image segmentation.

Proposition 1. The following optimization problem

$$\underset{h,i}{\operatorname{argmax}} \, \mathbf{S} = \underset{h,i}{\operatorname{argmax}} \prod_{h=1}^{c} \prod_{i=1}^{N_h-1} P_i^h \tag{5}$$

guarantees that the within-cluster similarity is maximum and the between-cluster similarity is minimal.

Proof. By contradiction.
Assume that there is a pixel p in cluster m is misclassified into cluster n. In the maximum spanning tree of cluster m, pixel p either connects two edges in the

middle of the tree or connects one edge as a leaf node. Consider that the pixel p is removed from the cluster m. If p is a leaf node, then one its associated edge (also the weight) is removed from the maximum spanning tree. If p is in the middle of the tree, then two its associated edges (also the weights) are removed from the maximum spanning tree. But a new edge must be added to connect the two separate parts into a new maximum spanning tree. When pixel p is added into cluster n, it is either in the middle of the tree or exists as a leaf node. However, because of the coherence property of affinity matrix, the pairwise similarity between p and any pixel in cluster n is the smallest in the maximum spanning tree of cluster n, then p can not be added in the middle of the maximum spanning tree of cluster n. So p is added as a leaf node and one more edge (also the weight) is added onto the new maximum spanning tree of cluster n.

If p is removed as a leaf node from the cluster m, its cluster similarity measure S_m becomes

$$S'_m = \frac{S_m}{w_r} \qquad (6)$$

where w_r is the removed weight.

If p is removed as a node in the middle of the maximum spanning tree of cluster m, its cluster similarity measure S_m becomes

$$S'_m = \frac{w_a S_m}{w_b w_c} \qquad (7)$$

where w_b and w_c are the removed weights from the maximum spanning tree and w_a is the added new weight. From Prim's algorithm [2] (for minimal spanning tree, but inverse weights of maximum spanning tree, we can use it), $w_a \le w_b$ or $w_a \le w_c$. Or else, $w_a > w_b$ and $w_a > w_c$, w_a will be in the maximum spanning tree.

After p is added as a leaf node into the cluster n, its cluster similarity measure S_n becomes

$$S'_n = w_d S_n \qquad (8)$$

where w_d is the added weight. Because of coherence property, $w_d < w_r, w_a, w_b, w_c$.

Then either,

$$S'_m \times S'_n = \frac{S_m}{w_r} \times w_d S_n = \frac{w_d}{w_r} S_m S_n < S_m S_n \qquad (9)$$

or

$$S'_m \times S'_n = \frac{w_a S_m}{w_b w_c} \times w_d S_n = \frac{w_a w_d}{w_b w_c} S_m S_n < S_m S_n \qquad (10)$$

Therefore, the maximum within-cluster similarity and the minimal between-cluster similarity are guaranteed under the above similarity measure.

2.3 Maximum Spanning Tree

After we define the above optimization problem, we want to solve it. We show that the above optimization problem can be solved by finding a maximum spanning tree for the complete weighted graph of an image. First, we give a brief introduction to Prim's algorithm, which, in graph theory, is used to find a minimum spanning tree for a connected weighted graph. If we inverse all weights of affinity matrix, to find a maximum spanning tree of the original graph is equavalent to finding the minimum spanning tree of the graph with new weights. So Prim's algorithm can be used to find a maximum spanning tree.

Prim's algorithm is an algorithm that finds a minimum spanning tree for a connected weighted graph, where the sum of weights of all the edges in the tree is minimized. If the graph is not connected, then it only finds a minimum spanning tree for one of the connected components.

The time complexity of the Prim's algorithm is $O(|E|\log|V|)$, where $|E|$ is number of edges in the graph and $|V|$ is number of nodes. For a complete graph, the number of edges $|E| = \binom{|V|}{2} = \dfrac{|V|(|V|-1)}{2}$. This is also the maximum number of edges that a graph can have. So the time complexity of the Prim's algorithm is also $O(|V|^2 \log|V|)$. If we only use the transitive property of affinity matrix and compute the local similarity for a pixel, then there are only eight similarities for a pixel between it and its eight neighboring pixels. The number of edges becomes $|E| = 8|V|$ and the complexity becomes $O(|V|\log|V|)$. This will reduce the complexity considerably. On the other hand, the complexity to a standard eigenfunction problems takes $O(|V|^3)$, where $|V|$ is the number of nodes in the graph.

Next, we show that the optimization problem can be solved by finding the maximum spanning tree and then removing the $c-1$ minimal weights.

Proposition 2. The following optimization problem

$$\underset{h,i}{\operatorname{argmax}} \mathbf{S} = \underset{h,i}{\operatorname{argmax}} \prod_{h=1}^{c} \prod_{i=1}^{N_h-1} P_i^h$$

can be solved by finding the maximum spanning tree of the graph associated with the image and then removing the $c-1$ minimal weights.

Proof. Prim's algorithm is used to find the maximum spanning tree.

Searching the maximum spanning tree of the whole image starts from a pixel x in some cluster m. According to coherence property of affinity matrix, the maximum spanning tree of cluster m must be put into the maximum spanning tree of the image first. Then the maximum spanning tree of cluster m will connect some pixel y in another cluster n through a between-cluster edge. From pixel y, the maximum spanning tree of cluster n is put into the maximum spanning tree of the image next. Repeatedly, all maximum spanning trees of clusters will be put in the maximum spanning tree of the image. The $c-1$ minimal weights connect those maximum spanning trees of c clusters into the maximum spanning tree of the whole image. Therefore, we obtain the solution

$$\max \mathbf{S} = \frac{T}{\prod\limits_{l=1}^{c-1} w_l} = \frac{\prod\limits_{k=1}^{N-1} T_k}{\prod\limits_{l=1}^{c-1} w_l} \tag{11}$$

where

$T = \prod\limits_{k=1}^{N-1} T_k$ is the product of all weights in the maximum spanning tree of the whole image. w_l are the $c-1$ minimal weights.

3 Experimental Results

We test our algorithm using synthetic data and real data. The synthetic data are binary data, generated by drawing white squares on the black background. The real data is a kid picture. The number of clusters (components) is give as prior here. The selection of number of clusters is a model selection problem that depends on the application, and is beyond the scope of our discussion here.

Test on synthetic data

The synthetic data are a picture with two white squares on the black background. The similarity is computed using Equation 1 based on intensity values. We chose $\sigma = 0.1$. Then the within-class similarity is

$$e^{-\|x_i - x_j\|^2/2\sigma^2} = e^{-\|0-0\|^2/2 \cdot 0.1^2} = e^{-\|1-1\|^2/2 \cdot 0.1^2} = e^0 = 1 \tag{12}$$

and the between-class similarity is

$$e^{-\|x_i - x_j\|^2/2\sigma^2} = e^{-\|1-0\|^2/2 \cdot 0.1^2} = e^{-50} \approx 0 \tag{13}$$

For the data with two white squares, the maximum spanning tree of the image has two edges with zero weights. All other edges have weight one. The zero weight edges are two minimal weights and separate the maximum spanning tree of the image into three spanning trees. One of the new spanning trees represents the background and the other two represent two foreground squares.

The segmentation results are shown in Figure 1. We can see that we obtain perfect results.

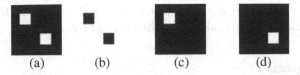

Fig. 1. Image segmentation on synthetic data. (a). the original image (b). the segmented background (white part) (c). the first segmented square (white part) (d). the second segmented square (white part).

Test on real data

The real data is a kid picture. The similarity is computed using Equation 1 based on average values of three channel values. We still chose $\sigma = 0.1$. The segmentation results are shown in Figure 2. The results are reasonable. Since boundary contours of kid face and shirt are not consistent with the background and foreground (kid face or shirt), they can be clearly separated as we chose 5 components.

In real images, because of noise and outliers in cluster, some within-cluster similarities are very small. Correspondingly, some very small weights do not represent the between-cluster separation weights. In practice, we follow the order of edges from the Prim's algorithm and pick those edge weights with considerable differences from (viz. smaller than) its previous and afterwards edge weights as the between-cluster separation edges. This method works robustly for real images. In practice, the χ^2-statistic [4] for histograms may give a better similarity measure for color and texture.

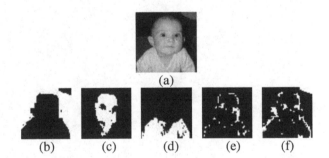

Fig. 2. Image segmentation on kid picture. (a). original image (b). the segmented background (white part) (c). the segmented kid face (white part) (d). the segmented shirt (white part) (e,f). the segmented contours (white part).

4 Conclusions

In this paper, we presented a new graph-based image segmentation algorithm. This algorithm finds the maximum spanning tree of the graph associated with the image affinity matrix. Instead of solving eigenvalues and eigenvector, we proved that the image segmentation could be transformed into an optimization problem: finding the maximum spanning tree of a graph with image pixels as vertices and pairwise similarities as weights. In our future work, we will explore different similarity measures and test the segmentation algorithm on more data.

Acknowledgments

This work was supported in part by the State of Louisiana Governor's Information Technology Initiative.

References

1. Chung, F.R.K.: Spectral Graph Theory. American Mathematical Society (1997)
2. Cormen, T.H., Leiserson, C.E., Rivest, R.L., Stein, C.: Introduction to Algorithms, 2nd edn. MIT Press and McGraw-Hill (2001)
3. Perona, P., Freeman, W.T.: A factorization approach to grouping. In: Burkardt, H., Neumann, B. (eds.) Proc ECCV, pp. 655–670 (1998)
4. Puzicha, J., Rubner, Y., Tomasi, C., Buhmann, J.M.: Empirical evaluation of dissimilarity measures for color and texture. In: Proceedings of the IEEE International Conference on Computer Vision, pp. 1165–1173 (1999)
5. Scott, G.L., Longuet-Higgins, H.C.: Feature grouping by relocalisation of eigenvectors of the proximity matrix. In: Proc. British Machine Vision Conference, pp. 103–108 (1990)
6. Shi, J., Malik, J.: Normalized cuts and image segmentation. In: Proc. IEEE Conf. Computer Vision and Pattern Recognition, pp. 731–737 (1997)
7. Weiss, Y.: Segmentation using eigenvectors: a unifying view. In: Proceedings IEEE International Conference on Computer Vision, pp. 975–982 (1999)

Reducing the Coefficients of a Two-Dimensional Integer Linear Constraint

Emilie Charrier[1,2,3] and Lilian Buzer[1,2]

[1] Laboratory CNRS-UMLV-ESIEE, UMR 8049
[2] ESIEE, 2, boulevard Blaise Pascal, Cité DESCARTES, BP 99
93160 Noisy le Grand CEDEX, France
[3] DGA/D4S/MRIS
charriee@esiee.fr, buzerl@esiee.fr

Abstract. Let us consider a two-dimensional linear constraint C of the form $ax + by \leq c$ with integer coefficients and such that $|a| \leq |b|$. A constraint C' of the form $a'x + b'y \leq c'$ is equivalent to C relative to a domain iff all the integer points in the domain satisfying C satisfy C' and all the integer points in the domain not satisfying C do not satisfy C'. This paper introduces a new method to transform a constraint C into an equivalent constraint C' relative to a domain defined by $\{(x, y) | h \leq x \leq h + D\}$ such that the absolute values of a' and b' do not exceed D. Our method achieves a $O(\log(D))$ time complexity and it can operate when the constraints coefficients are real values with the same time complexity. This transformation can be used to compute the convex hull of the integer points which satisfy a system of n two-dimensional linear constraints in $O(n \log(D))$ time where D represents the size of the solution space. Our algorithm uses elementary statements from number theory and leads to a simple and efficient implementation.

Keywords: Linear constraint, integer convex hull, Bezout identity, logarithmic time complexity, continued fraction.

1 Introduction

Given a two-dimensional linear constraint C of the form $ax + by \leq c$ with integer coefficients, the *size* of the constraint corresponds to the maximum of the absolute value of its coefficients. We suppose that we have $|a| \leq |b|$, so the size of the constraint is equal to $|b|$. Considering the domain defined by $\{(x, y) | h \leq x \leq h + D\}$, the two linear constraints C and C' are *equivalent* relative to this domain iff all the integer points in the domain satisfying C satisfy C' and all the integer points in the domain not satisfying C do not satisfy C'. If we focus on the integer points included in the domain and satisfying the constraint C, we can *reduce* this constraint relative to the domain. This reduction consists in transforming the constraint C into an equivalent constraint C' such that the size of C' does not exceed the size D of the domain.

Such a reduction can be useful in integer linear programming. Indeed, given a set of two-dimensional linear integer constraints, the solution space is defined

V.E. Brimkov, R.P. Barneva, H.A. Hauptman (Eds.): IWCIA 2008, LNCS 4958, pp. 205–216, 2008.

by the set of points satisfying these constraints. If the solution space is bounded, it is included in a domain defined by $\{(x,y)|h \leq x \leq h + D, \ h' \leq y \leq h' + D\}$. We can first reduce the constraints relative to the domain so that the size of the reduced constraints does not exceed D and then compute the convex hull of the integer points included in their solution space. Computing the convex hull of the integer solutions of a set of linear constraints is a well-studied problem. The first algorithm was proposed by Schrijver([10]). The algorithm successively approximates the integer convex hull and it is guaranteed to terminate in a finite number of iterations. However, the time complexity is exponential in the size of the input. Polynomiality of two-variable integer programming problem was established by Hirschberd and Wong in [3] and by Kannan in [4] for special cases. Scarf established it in the general case in [8] and [9]. Lenstra proved in [6] that integer programming in arbitrary fixed dimension can be solved in polynomial time. The best worst-case time complexity algorithm we know to compute the integer convex hull is proposed by Harvey (see [2]). His algorithm has a $O(n \log(A_{max}))$ time complexity in the worst case where A_{max} denotes the maximal size of the constraints and where n denotes the number of constraints in the set. Thus, we can reduce each constraint relative to the domain in $O(\log(D))$ time and then compute the integer convex hull of the solution space in $O(n \log(D))$ time complexity in the worst case using Harvey's algorithm.

In Sec. 2, we recall some notions from number theory. Then, in Sec. 3, we introduce some definitions. We describe Harvey's algorithm and our approach to compute the integer convex hull between two constraints in Sec. 4. In Sec. 5, we compare the time complexity of the two methods when they are used to reduce a constraint.

2   Number Theory

2.1   Bezout's Identity

In number theory, the *Bezout's identity* is a linear Diophantine equation. It states that if a and b are non-zero integers, then there exist two integers x and y, called *Bezout numbers*, such that $ax + by = \gcd(a, b)$ where $\gcd(a, b)$ denotes the greatest common divisor of a and b. More generally, the linear Diophantine equation $ax + by = c$ admits integer solutions iff the gcd of a and b divides c. The Bezout numbers x and y of the equation $ax + by = c$ can be determined with the extended Euclidean algorithm. The set of solutions is given by:

$$\left\{ \left(x_0 + \frac{kb}{\gcd(a, b)}, \ y_0 - \frac{ka}{\gcd(a, b)} \right) \mid k \in \mathbb{Z} \right\} \tag{1}$$

You can consult [7] for more details on Bezout's identity and the extended Euclidean algorithm. A vector $u = (u_x, u_y)$ is called *irreducible* iff $\gcd(u_x, u_y)$ is equal to one.

Definition 1. *Let $u = (u_1, u_2)$ denote an irreducible vector. A Bezout vector of u is a vector $v = (v_1, v_2)$ with integer coordinates such that $u \wedge v = \pm 1$, where*

$u \wedge v$ is equal $u_1v_2 - u_2v_1$. The value $u \wedge v$ corresponds to the z component of the cross product of u and v and it is the signed area of a parallelogram generated by u and v (this means that v is an irreducible vector for which the parallelogram $(0, u, u + v, v)$ contains no integer points in its interior).

We can easily find such a vector by using the extended Euclidean algorithm.

Example 1. Let u denote the vector $(3, 2)$, let V denote the set of the Bezout vectors of u, then: $V = \{(1, 1) + ku | k \in \mathbb{Z}\}$

2.2 Continued Fractions

Definitions and Properties We present the definition of *continued fractions* and some of their properties (see [1] for a more detailed introduction).

Definition 2. *The simple continued fraction decomposition of a real number x corresponds to:*

$$x = a_0 + \cfrac{1}{a_1 + \cfrac{1}{a_2 + \cfrac{1}{a_3 + \frac{1}{\ddots}}}}$$

where a_0 denotes an integer value and where a_i denotes a strictly positive integer value for all $i \geq 1$. We usually use the notation $x = [a_0, a_1, a_2, \cdots]$ where the a_i are called the partial quotients of x.

The *principal convergents* of x correspond to its rational approximations p_k/q_k. The two convergents p_0/q_0 and p_1/q_1 are respectively set to $0/1$ and $1/0$ whereas the others are obtained by truncating the continued fraction decomposition after the k-th partial quotient. The numerator and the denominator of the principal convergents are computed as follows:

$$\begin{cases} p_0 = 0 & p_1 = 1 & p_{k+2} = p_k + a_k p_{k+1} \\ q_0 = 1 & q_1 = 0 & q_{k+2} = q_k + a_k q_{k+1} \end{cases} \tag{2}$$

Each convergent of odd order is less than the whole continued fraction and each convergent of even order is greater than the whole continued fraction. Each convergent is closer in value to the whole continued fraction than the preceding. The *intermediate convergents* of two principal convergents p_k/q_k and p_{k+2}/q_{k+2} are defined as:

$$\frac{p_k + ip_{k+1}}{q_k + iq_{k+1}}, \quad i = 1 \cdots a_k - 1 \tag{3}$$

Let S_O (resp. S_E) denote the series defined by all the principal and the intermediate convergents of odd (resp. even) order. If x is a rational number then the last term of one of the two series is not equal to x. The rational number x is added to the end of this series. Let us enunciate a useful proposition (see [1] for the proof).

Proposition 1. *Let A_I denote the rational number which is less (resp. not less) than a real number B, which most closely approximates B and such that its denominator does not exceed an integer value d. The rational number A_I is the term of the series S_O (resp. S_E) of B with the greatest denominator which does not exceed d.*

Example 2. We want to find the rational number whose denominator does not exceed 60 and which is the best approximation of $779/207$. We have $779/207 = [3, 1, 3, 4, 2, 5]$. The odd convergents are $0/1, 3/1, 15/4, 143/38$ and the even convergents are $1/0, 4/1, 64/17, 779/207$. It follows the two series:

$$S_O = \frac{0}{1}, \frac{1}{1}, \frac{2}{1}, \frac{3}{1}, \frac{7}{2}, \frac{11}{3}, \frac{15}{4}, \frac{79}{21}, \frac{143}{38}, \frac{779}{207}$$

$$S_E = \frac{1}{0}, \frac{4}{1}, \frac{19}{5}, \frac{34}{9}, \frac{49}{13}, \frac{64}{17}, \frac{207}{55}, \frac{350}{93}, \frac{493}{131}, \frac{636}{169}, \frac{779}{207}$$

We deduce that the rational number $207/55$ in S_E is the best approximation of $779/207$, which is greater than $779/207$ and whose denominator does not exceed 60. The rational number $143/38$ in S_O is the best approximation of $779/207$, which is less than $779/207$ and whose denominator does not exceed 60.

Geometrical Interpretation. We can establish a correspondence between a rational number a/b and an integer vector of coordinates (b, a) in the Euclidean plane. As a result, the two series S_O and S_E approximating the rational number p/q corresponds to two series of integer vectors in the Euclidean plane approximating the integer vector (q, p). We can interpret Prop. 1 in the Euclidean plane. The rational number which most closely approximates the rational number p/q such that its denominator does not exceed d corresponds in the Euclidean plane to the integer vector which most closely approximates the vector (q, p) such that its abscissa does not exceed d. The two series S_O and S_E interpreted in the Euclidean plane are called *Klein's polygonal lines* (see [5]). Figure 1 shows an example of the series S_O and S_E where $p/q = 4/11$ and $d = 6$ interpreted in the Euclidean plane. In this example, $P = (5, 2)$ and $Q = (3, 1)$ correspond to the best approximations of the vector $(11, 4)$ and whose abscissa does not exceed 6.

3   Reducing the Coefficients of a Linear Constraint

3.1   Introduction

First, we introduce some notations. The notation $\lfloor \cdot \rfloor$ denote the floor function and the notation $\lceil \cdot \rceil$ denote the ceiling function. Let C denote a two-dimensional linear constraint of the form $ax + by \leq c$ where a, b and c are integer values. The *support line* of the constraint C is the straight line of the form $ax + by = c$. Let A denote the *size* of the constraint C, A is defined as the value $\max\{|a|, |b|\}$. We suppose that we have $|a| \leq |b|$ and so the size of C is equal to $|b|$. Let K denote a domain defined by $\{(x, y) | h \leq x \leq h + D\}$. The value D corresponds to the size of the domain.

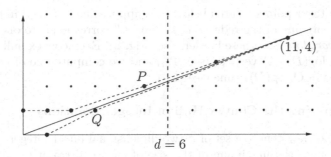

Fig. 1. Example of Klein's polygonal lines for $p/q = \frac{4}{11}$ and $d = 6$

3.2 Equivalent and Reduced Constraints

We first define the equivalence and the reduction of a linear constraint relative to the domain K.

Definition 3. *The constraint C' of the form $a'x + b'y \leq c'$ is equivalent to the constraint C relative to the domain K iff for all integer point (x, y) in K we have:*

$$ax + by \leq c \Leftrightarrow a'x + b'y \leq c'$$

Definition 4. *The constraint C' corresponds to a reduction of the constraint C relative to the domain K iff C' is equivalent to C relative to K and $\max\{|a|, |b|\}$ does not exceed D.*

To find a reduced constraint C' of C relative to the domain K, we first compute the lower border of the convex hull of the integer points located above the support line of C and in the domain K. Then, we compute the upper border of the convex

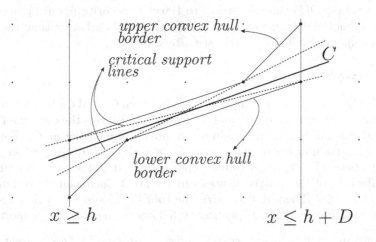

Fig. 2. Example of a reduction of constraint

hull of the integer points located below the support line of C and in the domain K. The support line of the reduced constraint C' corresponds to one of the two *critical support lines*[1] of these borders (see Fig. 2). Each convex hull border has at most $1 + \log(D + 1)$ vertices (see [12]) and we compute one of their critical support lines in $O(\log(D))$ time (see [11]).

3.3   Computing the Convex Hull of Integer Solutions

Let $S = \{C_i\}_{1 \leq i \leq n}$ denote a set of n two-dimensional linear integer constraints. Let A_{max} denote the maximum of the size of each constraint C_i, $i = 1, \cdots, n$. We suppose that the set of solutions of S is bounded. As a result, it is included in a domain $\{(x,y) | h \leq x \leq h + D, \ h' \leq y \leq h' + D\}$. We can transform each constraint C_i of S into a reduced constraint C'_i relative to the domain defined by $\{(x,y) | h \leq x \leq h+D\}$ or to the domain defined by $\{(x,y) | h' \leq y \leq h' + D\}$. Let S' denote the set of reduced constraints, the size of each constraint does not exceed D. We can apply Harvey's algorithm on S' and compute the convex hull of the set of integer solutions in $O(n \log(D))$ time.

4   Integer Convex Hull between Two Constraints

We consider the problem of computing the convex hull of integer points satisfying two linear integer constraints C_1 and C_2 respectively of the form $a_1 x + b_1 y \leq c_1$ and $a_2 x + b_2 y \leq c_2$. Let A_1 (resp. A_2) denote the size of C_1 (resp. of C_2), then A_{max} denotes the maximum value of A_1 and A_2. As we focus on integer points satisfying the constraints, for $j \in \{1, 2\}$, if the gcd of a_j and b_j is not equal to one, we replace w.l.o.g. the inequality of the constraint C_j by $a_j / \gcd(a_j, b_j) \, x + b_j / \gcd(a_j, b_j) \, y \leq \lfloor c_j / \gcd(a_j, b_j) \rfloor$ (the support line of the constraints passes through integer points). We describe Harvey's algorithm [2] and our approach which both run in $O(\log(A_{max}))$ time by considering that arithmetic operations on integers take $O(1)$ time. Contrary to Harvey's algorithm, our approach does not use unimodular transformations and this entails a better time complexity for the problem of reduction as we see afterwards.

4.1   Harvey's Algorithm

Harvey's algorithm assumes that the angle between C_1 and C_2 is less than π (i.e. $a_1 b_2 - b_1 a_2 > 0$), or it swaps C_1 and C_2. It first transforms the constraint C_2 into a vertical one and it sets the origin to an integer point lying on C_1. For this, it applies a unimodular transformation $[\alpha \ \beta | \gamma \ \delta]$ such that $[a_1 \ b_1 | a_2 \ b_2] \cdot [\alpha \ \beta | \gamma \ \delta] = [t \ u | 1 \ 0]$ where $\alpha\delta - \beta\gamma = \pm 1$ (unimodularity), $t \leq 0$ and $u > 0$. In order to find the coefficients of the matrix, it uses the extended Euclidean algorithm on the coefficients of C_2. Then, it translates the coordinate system so that the origin becomes the integer point P defined as follows: P lies on the support line of

---

[1] Critical support lines of two convex polygons are straight lines tangent to both polygons, such that the polygons lie on opposite sides of the lines.

C_1, it satisfies C_2 and it has the greatest abscissa. The constraints C_1 and C_2 become respectively $tx + uy \leq 0$ and $x \leq d$. It then computes in clockwise order the vertices of the convex hull of integer points satisfying C_1 and C_2. The first vertex corresponds to the origin. In order to determine the first edge of the hull, it uses the geometrical interpretation of Prop. 1. If the new edge intersects the support line of C_2 at an integer point, the problem is solved. Otherwise, it iterates with a new constraint C_1 supported by the last edge. Note that it does not have to compute the Klein's polygonal line for the new constraint because it is included in the initial polygonal line. Then, it translates the coordinate system so that the origin becomes the last determined vertex. The determination of the unimodular matrix takes $O(\log(A_2))$ time and the translation of the coordinate system at the first iteration runs in $O(\log(A_1))$ time. Harvey proves that its algorithm runs in $O(\log(A_{max}))$ time.

4.2   A Different Approach

We present our approach. We successively determine the vertices $(K_i)_{1 \leq i \leq m}$ of the convex hull from the support line of C_1 to the support line of C_2. When we find the i-th vertex K_i of the convex hull we determine the vertex K_{i+1} by looking for an integer point satisfying C_1 and C_2 and such that the angle between the two vectors $K_{i-1}K_i$ and K_iK_{i+1} is minimal. We do not succeed in finding such a point in a constant number of operations. Thus, we iteratively compute candidate points P_{ij} located in a search region depending on K_i. When this point is a vertex of the convex hull, we have $K_{i+1} = P_{ij}$. Else, we reduce the search region and we iterate. The (i,j)-th iteration corresponds to the j-th iteration after determining the i-th vertex of the convex hull K_i. Let $\Omega_{u_{ij}v_{ij}}$ denote the search region at the (i,j)-th iteration. This region corresponds to the integer points obtained by positive linear integer combination of the two vectors u_{ij} and v_{ij}, relative to the location of the vertex K_i. We can describe this set as: $\Omega_{u_{ij}v_{ij}} = \{(x,y) \in \mathbb{Z}^2 | (x,y) = K_i + \alpha u_{ij} + \beta v_{ij}, (\alpha, \beta) \in \mathbb{N}^2 \backslash (0,0)\}$. At the (i,j)-th iteration, the vector v_{ij} always corresponds to the Bezout vector of u_{ij} such that $K_i + v_{ij}$ satisfies the two constraints C_1 and C_2 and such that $K_i + v_{ij} + u_{ij}$ does not satisfy the constraint C_2. Such a vector is called a *valid Bezout vector*. We consider two different configurations according to whether the value of the dot product of v_{ij} and (a_2, b_2) is strictly positive or strictly negative. We explain later (see remark 1) that the case where the dot product is equal to zero is trivial. At each iteration, the straight line segment $[K_i, K_i + u_{ij}]$ intersects the support line of the constraint C_2 (this last property is verified afterwards). Let B (resp. B') denote the integer point which lies on the support line of C_1, which satisfies (resp. does not satisfy) the constraint C_2 and which is the closest point to the support line of C_2. The integer point B corresponds to the first vertex of the convex hull, so we set K_1 to B. Moreover, we set the vector u_{11} to the vector BB'.

First Configuration. In this configuration, the dot product of v_{ij} and (a_2, b_2) is strictly positive. We show that in this case the candidate point corresponds to the vertex K_{i+1} of the convex hull. Let us suppose we enter in the (i,j)-th iteration. Let I_{ij} denote the intersection of the support line of C_2 and the straight line of direction vector v_{ij} passing through K_i. The configuration implies that I_{ij} is equal to $K_i + \beta v_{ij}$ where β is a positive value greater than one. We set the candidate point P_{ij} to $K_i + \lfloor \beta \rfloor v_{ij}$. The candidate point P_{ij} corresponds to the $(i+1)$-th vertex of the convex hull, named K_{i+1}. Indeed, let J_{ij} denote the intersection of the support line of C_2 and $[K_i, K_i + u_{ij}]$. By definition of the Bezout vector v_{ij}, no integer point lies in the triangle (K_i, J_{ij}, I_{ij}), except between the points K_i and P_{ij} (see Fig. 3.a). When K_{i+1} lies on the support line of C_2, the algorithm finishes. Otherwise, we iterate the algorithm with a new constraint C_1 supported by $K_i K_{i+1}$. We set the vector $u_{i+1\,1}$ to v_{ij}. Notice that, as claimed in the previous section, the straight line segment $[K_{i+1}, K_{i+1} + u_{i+1\,1}]$ intersects the support line of C_2. Otherwise, the integer point K_{i+1} would not be a vertex of the convex hull.

Second Configuration. In this configuration, the dot product of v_{ij} and (a_2, b_2) is strictly negative. We determine the candidate point P_{ij} of the (i,j)-th iteration. As we cannot decide in a constant number of operations whether this point is a vertex of the convex hull, we iterate the algorithm with a reduced search region. Indeed, this new search region could contain a better candidate point $P_{i\,j+1}$. This means that the angle between u_{ij} and $K_i P_{i\,j+1}$ would be smaller than the angle between u_{ij} and $K_i P_{i\,j}$. Let I_{ij} denote the intersection of the support line of C_2 and the straight line of direction vector v_{ij} passing through $K_i + u_{ij}$. The configuration implies that I_{ij} is equal to $K_i + u_{ij} + \beta v_{ij}$ where β denotes a positive value greater than one. We determine the integer point $P_{ij} = K_i + u_{ij} + \lceil \beta \rceil v_{ij}$ (see Fig. 3.b). We set the vector $u_{i\,j+1}$ to $K_i P_{ij} - v_{ij}$ and we iterate with a new constraint C_1 supported by the straight line segment $[K_i, K_i + u_{i\,j+1}]$. As the support line of C_2 intersects the straight line segments

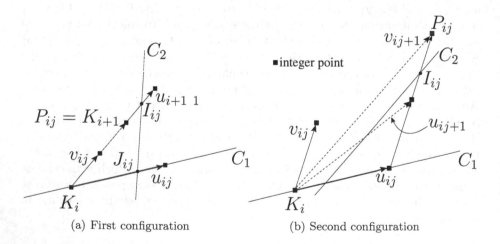

(a) First configuration   (b) Second configuration

Fig. 3. The two configurations of our algorithm

$[K_i + u_{i\ j+1}, K_i + u_{i\ j+1} + v_{ij}]$ and $[K_i, K_i + u_{ij}]$, we can conclude that the straight line segments $[K_i, K_i + u_{i\ j+1}]$ and the support line of C_2 intersect. Notice that, at the $(i, j + 1)$-th iteration, the vector $v_{i\ j+1}$ becomes $u_{ij} + \lceil \beta \rceil v_{ij}$. It means that the cross product $v_{i\ j+1} \wedge v_{ij}$ is equal to one. As a result, at each iteration in the second configuration the angle between (a_2, b_2) and v_{ij} strictly decreases and so in a finite number of iterations in the second configuration we encounter the first configuration.

Remark 1. Notice that the case where the dot product of v_{ij} and (a_2, b_2) is equal to zero is trivial. At the (i, j)-th iteration, let L_1 (resp. L_2) denote the straight line of direction vector v_{ij} passing through K_i (resp. $K_i + u_{ij}$). By definition of Bezout vector, no straight line parallel to L_1 and passing through integer points strictly lies between L_1 and L_2. If the dot product of v_{ij} and (a_2, b_2) is equal to zero, then the support line of C_2 is parallel to v_{ij}. As a result, the support line of C_2 passes through K_i or through $K_i + u_{ij}$ and the algorithm finishes.

Complexity Analysis. Our algorithm can be summed up as follows. The function *Intersection(P,u,C)* computes the floor value and the ceiling value of β such that the point $P + \beta v$ lies on the support line of the constraint C. This function runs in constant time.

```
ALGORITHM FOR THE INTEGER CONVEX HULL BETWEEN TWO CONSTRAINTS:
0   K₁ ← B
1   u₁₁ ← BB'
2   v₁₁ ←BezoutVector(u₁₁) IF (v₁₁ · (a₁,b₁) > c₁) v₁₁ ← −v₁₁
3   (⌊α⌋,⌈α⌉) ← Intersection(K₁ + v₁₁, u₁₁, C₂)
4   v₁₁ ← v₁₁ + ⌊α⌋u₁₁
5   i ← 1  j ← 1
6     WHILE ((a₂,b₂) · Kᵢ ≠ c₂)
7       WHILE(SECOND CONFIGURATION)
8         (⌊β⌋,⌈β⌉) ← Intersection(Kᵢ + uᵢⱼ, vᵢⱼ, C₂)
9         uᵢ ⱼ₊₁ ← uᵢⱼ + (⌈β⌉ − 1)vᵢⱼ
10        vᵢ ⱼ₊₁ ← uᵢⱼ + ⌈β⌉vᵢⱼ
11        j ← j + 1
12      (⌊β⌋,⌈β⌉) ← Intersection(Kᵢ, vᵢⱼ, C₂)
13      Kᵢ₊₁ ← Kᵢ + ⌊β⌋vᵢⱼ
14      uᵢ₊₁ ₁ ← vᵢⱼ
15      (⌊α⌋,⌈α⌉) ← Intersection(Kᵢ − uᵢⱼ, uᵢ₊₁ ₁, C₂)
16      vᵢ₊₁ ₁ ← −uᵢⱼ + ⌊α⌋uᵢ₊₁ ₁
17      i ← i + 1  j ← 1
```

The computation of a valid Bezout vector of u_{11} and the computation of the first vertex runs in $O(\log(A_1))$ time. Then, each iteration in the first or in the second configuration runs in constant time. For the complexity analysis, we have to estimate the number of iterations of our algorithm. We show afterwards that our algorithm describes the convex hull of two linear constraints in $O(\log(A_{max}))$.

Let w denote an integer vector lying on the support line of C_2 such that w is irreducible and such that for all integer point P not satisfying C_1 there exists a positive integer value γ such that $P + \gamma w$ satisfies C_1. At the (i,j)-th iteration, let A_{ij} denote the area defined by the following equalities: $A_{ij} = u_{ij} \wedge w$. We first consider that at the (i,j)-th iteration the points are in the second configuration. According to our algorithm, we determine a candidate point P_{ij} and we begin a new iteration where $u_{i\,j+1}$ is equal to $u_{ij} + \gamma v_{ij}$ and $v_{i\,j+1}$ is equal to $u_{ij} + (\gamma + 1)v_{ij}$, $\gamma \geq 1$. At the $(i\,j+1)$-th iteration, if we are again in the second configuration, it means that the cross product $v_{i\,j+1} \wedge w$ is negative and so $v_{ij} \wedge w$ is less than $-u_{ij} \wedge w/(\gamma + 1)$. The area $A_{i\,j+1}$ corresponds to the cross product $u_{i\,j+1} \wedge w$ and admits an upper bound. Indeed, $A_{i\,j+1}$ is less than $A_{ij}/\gamma + 1$. As $\gamma + 1$ is greater than two, $A_{i\,j+1}$ is less than the half of A_{ij}. In conclusion, in this configuration we have: $A_{i\,j+1} \leq A_{ij}/2$. If we are in the first configuration at the (i,j)-th iteration, we determine the vertex K_{i+1} in a constant number of operations and we begin the $(i+1,1)$-th iteration where $u_{i+1\,1}$ is equal to v_{ij}. It means that the cross product $w \wedge (v_{ij} - u_{ij})$ is strictly positive and so that $v_{ij} \wedge w$ is strictly less than $u_{ij} \wedge w$. As A_{ij} is equal to $u_{ij} \wedge w$ and $A_{i+1\,1}$ is equal to $v_{ij} \wedge w$ we conclude that $A_{i+1\,1}$ is strictly less than A_{ij}. As a result, the number of iterations in second configuration is bounded in $O(\log(A_{max}))$ and the number of iterations in first configuration is bounded by the number of vertices in the hull. As the number of vertices is logarithmic relative to the length of the intersection (see [12]), we conclude that our algorithm computes the full convex hull in $O(\log A_{max})$.

5   Reducing Linear Constraints

We recall that we consider a two-dimensional linear integer constraint C of the form $ax + by \leq c$ where $|a| \leq |b|$. We want to reduce this constraint relative to the domain of size D defined by $\{(x,y)|h \leq x \leq h+D\}$. Let C_L denote the constraint of the form $-x \leq h$ and let C_R denote the constraint of the form $x \leq h + D$. In this section, we study the time complexity of the two previous algorithms when they are used to reduce the constraint C. We show that Harvey's algorithm always runs in $O(\log(A))$ time when our approach runs in $O(\log(D))$ time. Let CH_A (resp. CH_B) denote the lower (resp. upper) border of the convex hull of integer points located above (resp. below) the support line of C.

5.1   Using Harvey's Algorithm

We know that a reduced constraint C' of C passes at least through one vertex of CH_A or of CH_B. Indeed, by definition of equivalent constraints, the support line of the reduced constraint lies between these two convex hulls. Moreover, the size of the reduced constraint does not exceed D and so its support line has to pass at least through one vertex of the convex hulls. As a result, to compute a reduced constraint, Harvey's algorithm has to find at least one vertex of these convex hulls. W.l.o.g., we suppose that it begins by computing CH_A between the constraints C_L and C. Suppose that the angle between C_L and C is less than π.

In this first case, the algorithm applies a unimodular transformation to set (a, b) to $(1, 0)$. We know that the computation of the transformation matrix leads to a $O(\log(A))$ time complexity. Suppose now that the angle between C_L and C is strictly greater than π. In this second case, no transformation is needed but the convex hull is computed from one integer point lying on the support line of C. The computation of the first vertex lying in the domain K takes $O(\log(A))$ time in the worst case because it implies the computation of a Klein polygonal line relative to the coefficients of C. As a result, such a configuration also leads to a $O(\log(A))$ time complexity.

5.2   Improving the Time Complexity

To compute a reduced constraint with our method, we compute the convex hulls CH_A and CH_B and we determine a critical support line of the two convex hulls. We begin by computing the convex hull CH_A. To do that, we compute the convex hull of integer points lying between the support lines of C_L and C until we find a candidate point which does not lie in the domain. Then, we compute the integer convex hull between the last computed candidate point and C_R. For all constraint C, there exists a constraint C' which is equivalent to C and whose size does not exceed D. We show that our algorithm computes the convex hull of integer points which are included in the domain and which satisfy C or C' with the same time complexity. As our algorithm computes this last convex hull in $O(\log(D))$ time, it means that our algorithm computes the convex hull of the constraint C in $O(\log(D))$ time too. Moreover, you can notice that our algorithm does not need the constraint C to have integer coefficients. At the (i, j)-th iteration, let u_{ij} and v_{ij} (resp. u'_{ij} and v'_{ij}) denote the vectors defining the search region during the computation of the hull of C (resp. C'). At the first iteration, u_{11} and v_{11} are equivalent to u'_{11} and v'_{11} by definition of equivalent constraints. Suppose that at the (i, j)-th iteration C is in the first configuration. Thus, C' is also in the first configuration and the next vertex of the convex hull is the same for C and C'. Suppose now that at the (i, j)-th iteration C is in the second configuration. Thus, for all positive integer value α the points $K_i + \alpha v_{ij}$ satisfy C. We determine the candidate point P_{ij} which is equal to $K_i + u_{ij} + \gamma v_{ij}$. For the same reason as above, the points $K_i + \alpha v_{ij}$ also satisfy C'. As a result, either the constraint C' is also in the second configuration and the candidate point P_{ij} is the same as C, or the vector v_{ij} lies on C'. In this particular case, we determine the next vertex of the hull of C' which is equal to $K_i + \delta v_{ij}$ such that this point lies in the domain. Note that in this case the candidate point P_{ij} of C cannot lie both in the domain and between the support line of C and the support line of C'. As a result, we begin a $(i, j + 1)$-th iteration with $u_{i\,j+1} = K_i P_{ij}$, $v_{i\,j+1} = v_{ij}$ and the constraint C_R. The $(i, j + 1)$-th iteration corresponds to a first configuration and so we determine a new vertex K_{i+1} which is equal to $K_i + \delta v_{ij}$. In conclusion, our algorithm computes the convex hull between the constraint C_L and the constraint C or C' in a domain of size D with the same time complexity $O(\log(D))$. When we find a candidate point which does not lie in the domain, we compute the convex hull between the constraint C_R and this point. This second step also runs in $O(\log(D))$ time.

6 Conclusion

Given a two-dimensional linear integer constraint, we describe a method to reduce the constraint into an equivalent one relative to a domain of size D such that the size of the reduced constraint does not exceed the size of the domain. For this, we compute the border of the convex hull of integer points lying in the domain which are located above and below the support line of the constraint. Each critical support line of the two convex hulls corresponds to the support line of a reduced constraint. We prove that our method runs in $O(\log(D))$ time contrary to the other approaches which run in $O(\log(A))$ time. Moreover, we show that such a transformation is useful to compute the integer convex hull of the solution space of a set of linear integer constraints in $O(n\log(D))$ time. Indeed, if the solution space is bounded then it is included in the domain defined by $\{(x,y)|h \leq x \leq h+D, \ h' \leq y \leq h'+D\}$. We can reduce each constraint relative to the domain and then compute the integer convex hull of the solution space. The source code for the computation of the integer convex hull between two linear constraints is available on www.esiee.fr/~charriee/integerConvexHull.

References

1. Chrystal, G.: Algebra- An elementary text-book Prt II, ch. XXXII, pp. 423–452. Adam and Charles Black, Edinburgh (1889)
2. Harvey, W.: Computing two-dimensional integer hulls. SIAM J. Compute. 28(6), 2285–2299 (1999)
3. Hirschberd, D.S., Wong, C.K.: A polynomial algorithm for the knapsack problem in two variables. J. ACM 23(1), 147–154 (1976)
4. Kannan, A.: A polynomial algorithm for the two variable integer programming problem. J. ACM 27, 118–122 (1980)
5. Klein, F.: Ausgewählte Kapitel der Zahlentheorie. Teubner (1907)
6. Lenstra, W.: Integer programming with a fixed number of variables. Math. Oper. Res. 8(4), 538–547 (1983)
7. Mora, T.: Solving Polynomial Equation Systems I: The Kronecker-Duval Philosophy. In: Encyclopedia of Math. and its applications, vol. 88, Cambridge University Press, Cambridge (2002)
8. Scarf, H.E.: Production sets with indivisibilities. Part I: Generalities. Econometrica 49, 1–32 (1981)
9. Scarf, H.E.: Production sets with indivisibilities. Part II: The case of two activities. Econometrica 49, 395–423 (1981)
10. Schrijver, A.: Theory of linear and integer programming. In: Wiley-Interscience Series in Discrete Math., Wiley-Interscience, New York (1986)
11. Toussaint, G.T.: Solving geometric problem with the rotating calipers. In: Proceedings of IEEE MELECON 1983, Greece, pp. A10.02/1-4 (1983)
12. Zolotykh, N. Y.: On the number of vertices in integer linear programming problems. Technical report, University of Nizhni Novgorod (2000)

A Branch & Bound Algorithm for Medical Image Registration

Michael Stiglmayr, Frank Pfeuffer, and Kathrin Klamroth

Department of Mathematics
Friedrich-Alexander University Erlangen-Nuremberg, Germany

Abstract. For a mixed integer programming formulation of the problem of registering two medical images we propose a geometric Branch & Bound algorithm, which applies a geometric branching strategy on the transformation variables. The results show that medium sized problem instances can be solved to global optimality in a reasonable amount of time.

1 Introduction and Literature Review

Medical images have become an indispensable tool for physicians for diagnosis as well as treatment and operation planning. Modern imaging techniques and, in particular, their combination can provide more and more detailed information that allows for a fast advancement of diagnosis and operation techniques. Pre-operative NMR-, CT- and PET-scans, for example, of a patients brain can be combined with intra-operative NMR-scans to show the progress of the operation as well as a possible brain shift, hence allowing for an online adaptation of the operation plan [7].

Registration problems have attracted much attention in the recent literature, see, for example, [9], [16] and [10] for recent surveys. The different solution approaches can be divided into two major classes: On one hand, algorithms using voxel properties work directly on the gray values of the considered images. Maybe the most common voxel based similarity measure is mutual information [15]. On the other hand, feature based registration methods require segmentation as a preprocessing step in order to extract the relevant information for the registration procedure. Feature points, lines, or surfaces are extracted from the images, and their properties are used to define a similarity measure between the images in this case [16].

In this paper, we focus on the second class of methods, namely a mixed integer programming formulation of the point matching problem that combines outlier handling techniques with appropriate registration objectives and constraints. Assuming that the correspondences between feature points in the respective images are not known a priori, the registration process consists of two coupled subproblems: Find corresponding feature points and search for a mapping such that the assigned points are mapped onto each other. In the following, a brief literature review of related approaches is given.

V.E. Brimkov, R.P. Barneva, H.A. Hauptman (Eds.): IWCIA 2008, LNCS 4958, pp. 217–228, 2008.

Mount et al. suggest the application of *partial Hausdorff distances* given by $\delta_H^k(\mathcal{A}, \mathcal{B}) = k_{a \in \mathcal{A}}^{th} \min_{b \in \mathcal{B}} \|a - b\|$, where the operator k^{th} ($1 \le k \le |\mathcal{A}|$) returns the k-th smallest value of the argument set [11]. The authors argue that the use of the kth smallest distance enables the user to specify a percentage of outlier points that should be rejected. However the value k can only be guessed and is not determined by the algorithm. The resulting optimization problem is solved using a geometric branch and bound algorithm to ε-global optimality. B&B methods are frequently used in the field of integer or mixed integer programming. Geometric B&B is an implementation of the B&B idea for continuous optimization. Different from combinatorial B&B algorithms, the computational complexity of geometric B&B algorithms depends on the required solution accuracy. An overview with a comparison of geometric B&B techniques related to matching problems is given in [2,3].

One of the computationally fastest, however, heuristic methods to handle the coupled assignment and transformation problem in point matching is the *iterative closest point algorithm* (ICP) [1]. Instead of trying to solve the two subproblems simultaneously, they are considered separately and iteratively: For each feature point in the reference data set, search for the closest points in the other data set. With the obtained correspondences, find a transformation function that minimizes the distances between assigned points, and iterate. The main disadvantage of this algorithm is that it converges in general only to a stationary point of the problem. Even under some additional regularity assumptions, it can only be guaranteed to yield local optimality [13]. However, due to its computational efficiency the ICP algorithm is used in many applications where computation time is crucial, and whenever a good starting solution is known, see [14] and [8]. The *robust point matching approach* (RPM) [5] can be seen as a further advancement of ICP-based algorithms. A mixed integer programming problem is formulated where binary variables represent the assignments between pairs of points, and continuous variables describe the non-rigid transformation function sought. In order to solve the resulting mixed integer programming problem, the authors suggest to iterate between the solution of assignment subproblems and optimizing the transformation function (in their implementation these functions are modelled by thin plate splines). In addition, a "softassign" approach is used (temporarily and partially relaxing the integrality constraints for the assignment variables) and deterministic annealing is applied to avoid local minima. Nevertheless, this solution method remains heuristic and global convergence can not be guaranteed. In this paper, we suggest to combine the advantages of a global optimization approach with applying a geometric Branch & Bound method to a variation of the RPM [5].

2   Problem Formulation

Since the focus of this paper is on image registration rather than on image segmentation, we assume that feature points $Y = \{y_j \in \mathbb{R}^d : j \in \mathcal{J}\}$, $\mathcal{J} := \{1, \dots, J\}$ in the reference image, and $X = \{x_i \in \mathbb{R}^d : i \in \mathcal{I}\}$, $\mathcal{I} := \{1, \dots, I\}$

in the template image have already been extracted by some available segmentation method. The aim of this paper is the development of global optimization approaches. Therefore, the distance measure used in the objective function of our model should not be based on pre-calculated point correspondences. Instead, the search for an optimized assignment will be performed simultaneously with the calculation of an optimized transformation. The transformation T is constrained to some function space $\mathcal{T}$ which we assume to consist of affine or rigid transformations only, but which could be extended to, for example, radial basis functions. Moreover, the problem formulation should be capable of handling *outliers*, which are points that have (for example, due to occlusion, different display windows or segmentation errors) no correspondence in the other data set. We therefore define assignment variables z_{ij}, $i \in \mathcal{I}$, $j \in \mathcal{J}$, that are used to assign points from X to points in Y and vice versa. If a point is considered as an outlier, it is assigned to the virtual point y_{J+1} or x_{I+1}, respectively: $z_{ij} = 1$ if x_i is assigned to y_j and zero otherwise.

By $Z := (z_{ij})_{i \in \overline{\mathcal{I}}, j \in \overline{\mathcal{J}}}$ ($\overline{\mathcal{I}} := \mathcal{I} \cup \{I+1\}$ and $\overline{\mathcal{J}} := \mathcal{I} \cup \{J+1\}$) we denote the corresponding assignment matrix. Our overall optimization model is a slight modification of the RPM approach [5]:

Problem 1

$$\min_T D_{\mathrm{RPM}}(T) := \min_{T,Z} f_{\mathrm{RPM}}(T, Z) :=$$

$$\min_{T,Z} \sum_{i \in \mathcal{I}} \sum_{j \in \mathcal{J}} z_{ij} \|y_j - T(x_i)\|^2 + \zeta_Y^2 \sum_{j \in \mathcal{J}} z_{I+1,j} + \zeta_X^2 \sum_{i \in \mathcal{I}} z_{i,J+1}$$

$$s.t. \quad \sum_{i \in \overline{\mathcal{I}}} z_{ij} = 1 \qquad \forall j \in \mathcal{J}$$

$$\sum_{j \in \overline{\mathcal{J}}} z_{ij} = 1 \qquad \forall i \in \mathcal{I} \tag{1}$$

$$z_{ij} \in \{0, 1\} \ \forall (i, j) \in \overline{\mathcal{I}} \times \overline{\mathcal{J}}$$

$$T \in \mathcal{T}.$$

By $\mathcal{Z}_\mathbb{B} :- \{Z \in \{0,1\}^{|\overline{\mathcal{I}}| \times |\overline{\mathcal{J}}|} : \sum_{i \in \overline{\mathcal{I}}} z_{ij} - 1, j \in \mathcal{J} \text{ and } \sum_{j \in \overline{\mathcal{J}}} z_{ij} - 1, i \in \mathcal{I}\}$ we denote the set of all feasible assignments in (1).

Note that the dummy points y_{J+1} and x_{I+1} to which outlier points can be assigned without contributing to the distance term in the objective function take care of the outlier handling in this formulation. However, this makes an additional term for outlier control necessary since otherwise, $z_{ij} = 0 \ \forall i \in \mathcal{I}, j \in \mathcal{J}$ (i.e., considering all points y_j and x_i as outliers) would always be an optimal solution of the problem. Instead of adding a gratification for assigned points as in [5], we propose to use a penalty term $+\zeta_Y^2 \sum_{j \in \mathcal{J}} z_{I+1,j} + \zeta_X^2 \sum_{i \in \mathcal{I}} z_{i,J+1}$ with penalty parameters $\zeta_Y^2, \zeta_X^2 > 0$ in the objective function. Note that this does not change the structure of the problem while it facilitates the computation of lower bounds since all terms in the objective function are now nonnegative.

One possibility to simplify the RPM problem is to separate the optimization of the assignment variables and the optimization of the spatial transformation.

In order to analyse the RPM problem for fixed assignment variables or for fixed transformation, respectively, we split the mixed integer programming problem (1) into two subproblems (2) and (3):

For a fixed transformation $\overline{T} \in \mathcal{T}$, (1) is equivalent to a *generalized assignment problem* (2):

Problem 2 (Generalized Assignment Problem)

$$\min_{Z} \sum_{i \in \mathcal{I}} \sum_{j \in \mathcal{J}} z_{ij} d_{ij}(\overline{T}) + \zeta_Y^2 \sum_{j \in \mathcal{J}} z_{I+1,j} + \zeta_X^2 \sum_{i \in \mathcal{I}} z_{i,J+1}$$
$$s.t. \ \ Z \in \mathcal{Z}_{\mathrm{B}} \tag{2}$$

Whereat $\mathcal{Z}_{\mathrm{B}}$ is the polyhedron of feasible assignments and $d_{ij}(\overline{T}) := \left\| y_j - \overline{T}(x_i) \right\|^2$ is the matrix of (constant) point distances. It is easy to see that the constraint matrix of the generalized assignment problem (2) is totally unimodular [12]. Thus the integrality constraints $z_{ij} \in \{0,1\} \ \forall i \in \overline{\mathcal{I}}, j \in \overline{\mathcal{J}}$ can be relaxed to $z_{ij} \in \mathbb{R}_+$, since the polyhedron $\mathcal{Z}_{\mathrm{B}}$ is integral. Therefore, the generalized assignment problem (2) can either be solved directly using linear programming, or it can be reformulated in terms of a transportation problem which can be efficiently solved by means of network flow algorithms.

If, on the other hand, the assignment is fixed to some $\bar{Z} \in \mathcal{Z}_{\mathrm{B}}$, we search for the optimal transformation T solving (constant terms in the objective function are omitted):

Problem 3

$$\min_{T} \sum_{i \in \mathcal{I}, j \in \mathcal{J} : \bar{z}_{ij} = 1} \left\| y_j - T(x_i) \right\|^2$$
$$s.t. \ \ T \in \mathcal{T}. \tag{3}$$

The selection of an appropriate solution method for the least squares problem (3) depends on the considered function space $\mathcal{T}$. However, this optimization problem is convex for all linear parameterized, finite dimensional function spaces $\mathcal{T}$. In [6], different analytical solution methods are discussed for the function space of rigid transformations $\mathcal{T}_{\mathrm{rigid}} = \{T : \mathbb{R}^d \rightarrow \mathbb{R}^d : T(x) = xR + t, R \in \mathrm{SO}(d), t \in \mathbb{R}^d\}$. Modersitzki [10] gives algorithms for general linear parameterized function spaces as well as for thin-plate-splines .

Summarizing the discussion above we can conclude that each subproblem (for fixed transformation variables or fixed assignment variables, respectively) can be efficiently solved. For a fixed transformation we obtain a linear assignment problem, and for a fixed assignment we obtain a continuous, convex optimization problem. The combination of both remains, however, in general an NP-hard problem.

3   Geometric Branch & Bound

Due to the high number of possible assignments it is predictable that the computational expense of a B&B algorithm for the assignment variables is highly dependent on the number of points in the input datasets. For problems where many feature points have to be registered a different approach appears to be more suitable. Whereas the B&B algorithm for the assignment variables examines the logic correspondence between point sets by assigning point pairs, a process called *geometric branching* analyses the geometric correspondence of the sets by hierarchically searching the transformation space.

A geometric B&B algorithm to solve registration problems was proposed by Mount et al. [11] for the minimization of the directed Hausdorff distance. Given an initial subset of the search space in form of a multidimensional interval $[t^{min}, t^{max}]$ in the parameter space, the initial *transformation cell*, the algorithm consecutively subdivides this initial interval into subintervals while calculating lower bounds and feasible solutions (which yield upper bounds) for the subinterval. By comparing the lower bounds with the best solution generated so far, intervals that cannot contain the optimal transformation can be identified and eliminated. During the course of the algorithm the diameter of the intervals considered decreases while they are subdivided. Finally, when the diameter is small enough, the gap between lower bounds and upper bounds will be less than a given error tolerance ε. Then the algorithm stops and returns the best solution generated guaranteeing optimality up to an accuracy of ε.

When applying this procedure to (1), we assume that the search space $\mathcal{T}$ is given in parameterized form by using the parameterization $\pi : \mathcal{P} \to \mathcal{T}, t \mapsto T_t$ with parameter space $\mathcal{P} \subseteq \mathbb{R}^n$. Additionally, let the possible transformation parameters be bounded by a multidimensional interval $\Delta^{init} = [t^{min}, t^{max}] \subseteq \mathcal{P}$. An interval $\Delta^{sub} = [t^{sub,min}, t^{sub,max}] \subseteq \Delta^{init}$ defines the subproblem

$$\min_{t \in \Delta^{sub}} \min_{Z \in \mathcal{Z}_B} \sum_{j \in \mathcal{J}} \sum_{i \in \mathcal{I}} z_{ij} \|y_j - T_t(x_i)\|^2 + \zeta_Y^2 \sum_{j \in \mathcal{J}} z_{I+1,j} + \zeta_X^2 \sum_{i \in \mathcal{I}} z_{i,J+1} \qquad (4)$$

of the registration problem, which we will denote shortly as the subproblem (Δ^{sub}). The transformations that are considered for this subproblem are $\mathcal{T}^{sub} := \pi(\Delta^{sub})$, and the original registration problem corresponds to subproblem (Δ^{init}).

Now the following geometric B&B algorithm can be formulated.

Algorithm 4 (Geometric Branch & Bound). *Let points y_j, $j \in \mathcal{J}$, and x_i, $i \in \mathcal{I}$, a search space $\mathcal{T}$ specified by an interval Δ^{init} in the parameter space, an absolute error tolerance ε_{abs} and a relative error tolerance ε_{rel} be given.*

Step 1: *Initialize the problem list with $S := \{(\Delta^{init})\}$ and the initially best objective function value with $f^\circ := \infty$.*

Step 2: *Repeat while the problem list S is non-empty:*

    a) *Remove some $(\Delta^{sub}) \in S$ from the problem list.*

    b) *Generate a lower bound l for the subproblem (Δ^{sub}). If $l \geq \max \left\{ f^\circ - \varepsilon_{abs}, \frac{f^\circ}{1+\varepsilon_{rel}} \right\}$, the subproblem is pruned; in this case return to Step 2).*

    c) *Generate a feasible solution (T, Z) of subproblem (Δ^{sub}) and denote its objective value by $f = f_{RPM}(T, Z)$. If $f < f^\circ$, the newly generated solution is better than the current best known solution; update $f^\circ := f$ and $(T^\circ, Z^\circ) := (T, Z)$.*

    d)   *If $f \leq l$, the newly generated solution is at least as good as any feasible solution of the subproblem. The subproblem is solved to optimality and can be pruned; return to Step 2).*

    e)   *Split the interval Δ^{sub} into two new subintervals $I^{\mathrm{sub},1}$ and $I^{\mathrm{sub},2}$ and insert the corresponding subproblems into the problem list S.*

Step 3: *Stop with (T°, Z°) as approximately optimal solution.*

To ensure termination of the algorithm, error tolerances $\varepsilon_{\mathrm{abs}}$ and $\varepsilon_{\mathrm{rel}}$ need to be specified. We will discuss the issue of termination later. Thus the geometric B&B algorithm is not an exact procedure, but the returned solution approximates the global optimum in the following sense.

Theorem 5 (see [11]). *Let (T^*, Z^*) be a global optimum with objective value f^*. In case of termination of the algorithm the final solution (T°, Z°) with objective value f° satisfies: $f^\circ - f^* \leq \min\{\varepsilon_{\mathrm{abs}}, f^\circ \varepsilon_{\mathrm{rel}}/(1 + \varepsilon_{\mathrm{rel}})\}$.*

3.1   Computation of Lower Bounds

The efficiency of B&B methods strongly depends on the quality of the applied bounding procedure. We consider for every point $x_i \in X$ the region of possible images of x_i under the transformations in $\pi(\Delta^{\mathrm{sub}}) = \{T_t \in \mathcal{T} : t \in \Delta^{\mathrm{sub}}\}$. Additionally, let the function $\pi : I_0 \times \mathbb{R}^d \to \mathbb{R}^d$ be defined by $\pi(t, x) := T_t(x)$. Similar to [11] we use the following notation. The set $\pi(\Delta^{\mathrm{sub}}, x_i) = \{T_t(x_i) : t \in \Delta^{\mathrm{sub}}\}$ is called the *uncertainty region* of x_i with respect to the subproblem (Δ^{sub}). By estimating lower bounds for the distances from y_j with $j \in \mathcal{J}$ to the uncertainty regions $\pi(\Delta^{\mathrm{sub}}, x_i)$ with $i \in \mathcal{I}$, a lower bound for the subproblem (Δ^{sub}) can be generated.

Theorem 6. *Let a subproblem be given by an interval $\Delta^{\mathrm{sub}} \subseteq \Delta^{\mathrm{init}}$ according to (4). Let lower bounds $D_{ij}^{\mathrm{l}} \leq \|y_j - T_t(x_i)\|^2 \; \forall t \in \Delta^{\mathrm{sub}}$ be known. Then*

$$l := \min_{Z \in \mathcal{Z}_B} \sum_{j \in \mathcal{J}} \sum_{i \in \mathcal{I}} z_{ij} D_{ij}^{\mathrm{l}} + \zeta_Y^2 \sum_{j \in \mathcal{J}} z_{I+1,j}^{\mathrm{sub}} + \zeta_X^2 \sum_{i \in \mathcal{I}} z_{i,J+1}^{\mathrm{sub}} \tag{5}$$

is a lower bound for the optimal objective value of this subproblem.

Proof. Let (T_{t^*}, Z^*) with $t^* \in \Delta^{\mathrm{sub}}$ be an optimal solution of the subproblem and let Z' be an optimal solution of the optimization problem (5). This implies

$$
\begin{aligned}
l &= \sum_{j \in \mathcal{J}} \sum_{i \in \mathcal{I}} z_{ij}' D_{ij}^{\mathrm{l}} + \zeta_Y^2 \sum_{j \in \mathcal{J}} z_{I+1,j}' + \zeta_X^2 \sum_{i \in \mathcal{I}} z_{i,J+1}' \\
&\leq \sum_{j \in \mathcal{J}} \sum_{i \in \mathcal{I}} z_{ij}^* D_{ij}^{\mathrm{l}} + \zeta_Y^2 \sum_{j \in \mathcal{J}} z_{I+1,j}^* + \zeta_X^2 \sum_{i \in \mathcal{I}} z_{i,J+1}^* \\
&\leq \sum_{j \in \mathcal{J}} \sum_{i \in \mathcal{I}} z_{ij}^* \|y_j - T_{t^*}(x_i)\|^2 + \zeta_Y^2 \sum_{j \in \mathcal{J}} z_{I+1,j}^* + \zeta_X^2 \sum_{i \in \mathcal{I}} z_{i,J+1}^*.
\end{aligned}
$$

$\qquad\qquad\qquad\qquad\qquad\qquad\qquad\qquad\qquad\qquad\qquad\qquad\qquad\qquad\qquad\qquad\square$

The best possible lower bound D_{ij}^l is the squared distance between y_j and the uncertainty region $\pi(\Delta^{\mathrm{sub}}, x_i)$ of x_i, i.e., $D_{ij}^l = d\big(y_j, \pi(\Delta^{\mathrm{sub}}, x_i)\big)^2 = \min_{t \in \Delta^{\mathrm{sub}}}$ $\|y_j - T_t(x_i)\|^2$. It is not always possible to calculate this distance efficiently. But if an interval $\Gamma_i = [s_i^{\min}, s_i^{\max}] \subseteq \mathbb{R}^d$ can be found that is a bounding box $\pi(\Delta^{\mathrm{sub}}, x_i) \subseteq \Gamma_i$ for the uncertainty region, a lower bound can easily be calculated by

$$D_{ij}^l = d(y_j, \Gamma_i)^2 = \sum_{k=1}^{d} d_{ijk}^2, \quad \text{with} \quad d_{ijk} = \begin{cases} s_{ik}^{\min} - y_{jk} & \text{if } y_{jk} < s_{ik}^{\min} \\ 0 & \text{if } s_{ik}^{\min} \leq y_{jk} \leq s_{ik}^{\max} \\ y_{jk} - s_{ik}^{\max} & \text{if } s_{ik}^{\max} < y_{jk} \end{cases}$$

The observation that $d(y_j, \Gamma_i) \leq d\big(y_j, \pi(\Delta^{\mathrm{sub}}, x_i)\big)$ is also illustrated in Figure 1. The intervals Γ_i can often be calculated by using interval arithmetic. [4] describes the use of interval arithmetic in the context of geometric B&B algorithms; for details we refer to this publication.

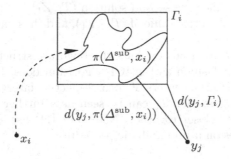

Fig. 1. Lower bounds D_{ij}^l for the squared distances from a point y_j to the uncertainty region of a point x_i: The best possible bound $D_{ij}^l = d(y_j, \pi(\Delta^{\mathrm{sub}}, x_i))^2$ and the bound based on interval arithmetic $D_{ij}^l = d(y_j, \Gamma_i)^2$

As a concrete example, we consider affine transformations $\mathcal{T}_{\mathrm{affine}} = \{T : \mathbb{R}^d \to \mathbb{R}^d : T(x) = xA + t,\ A \in \mathbb{R}^{d \times d},\ t \in \mathbb{R}^d\}$ which can be parameterized linearly by the components of A and t. As a generalization of affine transformations we consider a linear parameterized function space (or rather a subset of it)

$$\mathcal{T}_{\mathrm{lin}} := \Big\{ T_t = \big(\sum_{b \in \mathcal{B}} t_{bk} T^b\big)_{k \in \{1,\dots,d\}} : t \in \Delta^{\mathrm{init}} \Big\}$$

with basis functions T^b, $b \in \mathcal{B}$. The uncertainty region of x_i, $i \in \mathcal{I}$, for the interval $\Delta^{\mathrm{sub}} = [t^{\min}, t^{\max}]$ is itself an interval $[s_i^{\min}, s_i^{\max}]$, where

$$s_{ik}^{\min} = \sum_{\substack{b \in \mathcal{B} \\ T^b(x_i) > 0}} t_{bk}^{\min} T^b(x_i) + \sum_{\substack{b \in \mathcal{B} \\ T^b(x_i) < 0}} t_{bk}^{\max} T^b(x_i) \qquad \forall k \in \{1, \dots, d\}$$

$$s_{ik}^{\max} = \sum_{\substack{b \in \mathcal{B} \\ T^b(x_i) > 0}} t_{bk}^{\max} T^b(x_i) + \sum_{\substack{b \in \mathcal{B} \\ T^b(x_i) < 0}} t_{bk}^{\min} T^b(x_i) \qquad \forall k \in \{1, \dots, d\}.$$

3.2 Computation of Feasible Solutions

Starting with a transformation $T \in \mathcal{T}^{\mathrm{sub}}$, determine the best corresponding assignment, and to this assignment in turn the best transformation. Note that this can again be regarded as one iteration of an ICP algorithm with starting solution T.

A reasonable starting solution is the transformation $T_{t^{\mathrm{m}}}$ corresponding to the center t^{m} of the interval Δ^{sub}. Then compute

$$Z_1 := \operatorname*{arg\,min}_{Z \in \mathcal{Z}_B} \sum_{j \in \mathcal{J}} \sum_{i \in \mathcal{I}} z_{ij} \|y_j - T_{t^{\mathrm{m}}}(x_i)\|^2 + \zeta_Y^2 \sum_{j \in \mathcal{J}} z_{I+1,j} + \zeta_X^2 \sum_{i \in \mathcal{I}} z_{i,J+1}$$

$$T_1 := \operatorname*{arg\,min}_{T \in \mathcal{T}} \sum_{j \in \mathcal{J}} \sum_{i \in \mathcal{I}} z_{ij}^1 \|y_j - T(x_i)\|^2 + \zeta_Y^2 \sum_{j \in \mathcal{J}} z_{I+1,j}^1 + \zeta_X^2 \sum_{i \in \mathcal{I}} z_{i,J+1}^1.$$

The pair $(T_{t^{\mathrm{m}}}, Z_1)$ is a feasible solution of the subproblem (Δ^{sub}), while (T_1, Z_1) is only a feasible solution for the original registration problem (1). This is, however, not a problem since the second solution (T_1, Z_1) has an objective value at least as good as the objective value of $(T_{t^{\mathrm{m}}}, Z_1)$, and thus makes the first solution obsolete.

The generation of solutions exploits the bivariate structure of the objective function in a similar fashion as the ICP algorithm does. Instead of restricting the transformations to the centers of their respective intervals, the computation of the second transformation T_1 can be seen as a further alignment step that brings the point sets closer together. Note that [11] use a different alignment approach, that relies on a probabilistic procedure.

3.3 Control and Acceleration of the Algorithm

The order in which subproblems are considered can be influenced by the order in which subproblems are inserted into the problem list and by the order in which they are removed from this list. We suggest to select the subproblems from the list according to their lower bounds. To generate new subproblems in Step 2e) of Algorithm 4, the current subproblem with $\Delta^{\mathrm{sub}} = [t^{\mathrm{sub,min}}, t^{\mathrm{sub,max}}]$ is subdivided into two subproblems with $\Delta^{\mathrm{sub},1} = [t^{\mathrm{sub},1,\mathrm{min}}, t^{\mathrm{sub},1,\mathrm{max}}]$ and $\Delta^{\mathrm{sub},2} = [t^{\mathrm{sub},2,\mathrm{min}}, t^{\mathrm{sub},2,\mathrm{max}}]$ by choosing a component $\bar{j}$ and splitting the interval into two halves. It is advantageous to split the interval in the component $\bar{j}$ that contributes the most to the size of the uncertainty region and thus to the uncertainty of the lower bound. Let the *size of the subproblem* (Δ^{sub}) be defined as the maximum size of the uncertainty regions $\pi(\Delta^{\mathrm{sub}}, x_i) := \max_k (s_{ik}^{\mathrm{max}} - s_{ik}^{\mathrm{min}})$ $i \in \mathcal{I}$, this strategy reduces the size of the subproblem as much as possible.

3.4 Termination of the Algorithm

In the beginning of Section 3 we have proven that the geometric B&B algorithm returns an optimal solution up to the precision of a given error tolerance, provided that it terminates after finitely many iterations. Because no variables are fixed during the branching process, but the set of feasible transformations is halved instead, it is not obvious that the algorithm terminates in a finite number of steps. In this section we will adapt corresponding results of [11] to prove termination of the algorithm in the setting considered in this paper.

Lemma 7 *Let T be either a linear search space or the space of rigid transformations in $\mathbb{R}^d$. Let (T^*, Z^*) be the global optimal solution of the registration problem (1) with objective value f^*, and let absolute and relative error tolerances $\varepsilon_{\mathrm{abs}}$ and $\varepsilon_{\mathrm{rel}}$ be given that satisfy $\varepsilon_{\mathrm{abs}} > 0$ and $\varepsilon_{\mathrm{rel}} f^* > 0$. Then there exists a constant $R > 0$ such that subproblems of size at most R are not split any further in Algorithm 4.*

Proof. Due to the definition of the size of a rigid subproblem based on the size of a linear subproblem, it is sufficient to consider only linear search spaces. Therefore, let (Δ^{sub}) be a subproblem of size r for a linear search space. Denote its lower bound by l, and denote the assignment that is used to compute this bound by Z' (cf. Theorem 6). Let (T_1, Z_1) be the feasible solution generated for the subproblem starting from the subproblem's center T_{t^m} (cf. Section 3.2). Furthermore, let (T°, Z°) be the best known solution so far (while considering subproblem (Δ^{sub})), and let f° be its objective value.

The diameter of the uncertainty region $\pi(\Delta^{\mathrm{sub}}, x_i)$ is at most $\sqrt{d}r$. This implies

$$\|y_j - T_{t^m}(x_i)\|^2 \le (d(y_j, \pi(\Delta^{\mathrm{sub}}, x_i)) + \sqrt{d}r)^2 \le D_{ij}^l + 2\sqrt{d}Mr + dr^2, \quad (6)$$

where $M > 0$ is an upper bound for $d(y_j, \pi(\Delta^{\mathrm{sub}}, x_i))$. Then we obtain

$$f_{\mathrm{RPM}}(T_1, Z_1) \le f_{\mathrm{RPM}}(T_{t^m}, Z_1) \le f_{\mathrm{RPM}}(T_{t^m}, Z') \overset{(6)}{\le} l + \sum_{j \in \mathcal{J}} \sum_{i \in \mathcal{I}} z'_{ij}(2\sqrt{d}Mr + dr^2)$$

$$\le l + \min\{I, J\}(2\sqrt{d}Mr + dr^2).$$

Since $f^* \le f^\circ$, we have $\varepsilon_{\mathrm{rel}} f^\circ > 0$ by the assumptions of the lemma. Therefore, there exists a constant $R > 0$ such that $\min\{I, J\}(2\sqrt{d}Mr + dr^2) \le \min\left\{\varepsilon_{\mathrm{abs}}, f^\circ \frac{\varepsilon_{\mathrm{rel}}}{1 + \varepsilon_{\mathrm{rel}}}\right\}$ for all $r \le R$. If the subproblem (Δ^{sub}) is subdivided into two new subproblems, their lower bounds l^1 and l^2 satisfy

$$l^1, l^2 \ge l \ge f_{\mathrm{RPM}}(T_1, Z_1) - \min\{I, J\}(2\sqrt{d}Mr + dr^2)$$
$$\ge f_{\mathrm{RPM}}(T^\circ, Z^\circ) - \min\{I, J\}(2\sqrt{d}Mr + dr^2)$$
$$\ge \max\left\{f^\circ - \varepsilon_{\mathrm{abs}}, \frac{f^\circ}{1 + \varepsilon_{\mathrm{rel}}}\right\}.$$

Thus, the new subproblems will be removed in Step 2b) of Algorithm 4.   □

Theorem 8. *Under the assumptions of Lemma 7, Algorithm 4 terminates.*

Proof. It is sufficient to show that the size of every subproblem is at least halved in a finite number m of subdivisions. Then $m \log_2(R^{\text{init}}/R)$ subdivisions are sufficient to decrease the size R^{init} of the initial subproblem (Δ^{init}) to the size R specified in Lemma 7.

Subproblems are split along that side of Δ^{sub} that contributes the most to the size of the subproblem. Due to linearity, this term is decreased by half when the subproblem is split, and in $m = Bd$ consecutive subdivisions the size of the subproblem is at least halved. The proof for rigid transformations can be carried out analogously.   □

The above results allow us to examine the run time of Algorithm 4. In the worst case all subproblems have to be subdivided as long as they are not removed according to Lemma 7. For linear search spaces this may lead to a B&B tree depth of $Bd \log_2(R^{\text{init}}/R)$ and a breadth of $(R_0/R)^{Bd}$. In this case the number of subproblems that have to be considered grows exponentially with the dimension and the number of basis functions. A similar dependence on the dimensionality exists for the search space of rigid transformations.

4   Computational Results

For the solution of the generalized assignment problem (2) a primal-dual algorithm based on a transportation problem formulation was applied. To solve the least squares problem (3) for rigid transformations, an approach based on singular value decompositions as described in [6] was used. To examine how the algorithms handle realistic situations, we applied them to some medical examples. The 2-dimensional image data was provided within the Collaborative Research Center (DFG/SFB 603) at the University of Erlangen-Nuremberg, which we gratefully acknowledge.

Figure 2 shows an MRA (magnetic resonance angiography) and an MRI (magnetic resonance imaging) of the same skull. The MRA displays blood vessels as bright spots. In the MRI, on the other hand, structures of the brain are well visible. By applying a registration algorithm, one image can be generated that combines the advantages of both imaging techniques. To verify that the

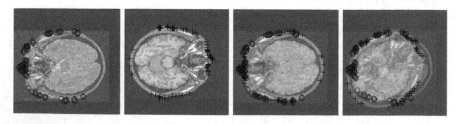

Fig. 2. From left to right: Reference MRA with 100 points, Template MRI with 100 points, Registered image using geometric B&B, Incorrectly registered image using the iterative closest point procedure

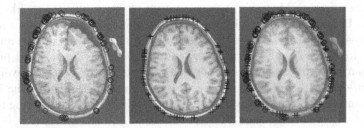

Fig. 3. Left: MRI during the surgery with 100 points. Center: MRI before the surgery with 100 points. Bottom: Registered image using geometric B&B.

geometric B&B algorithm finds a global optimum rather than only a local one, the originally roughly prealigned images have been rotated by 180°.

From both images points lying on the outer contour of the skull were extracted and the 100 points with the highest curvature were chosen. To find the rigid transformation that registers these point sets, the geometric B&B algorithm was applied using a relative error tolerance of $\varepsilon_{rel} = 0.1$ (ε_{abs} sufficiently large, without influence). During the registration, 1485 nodes of the B&B tree were traversed in 139.9 seconds. To compare the resulting registration with the outcome of a local registration algorithm, Figure 2 shows the registration obtained by the B&B method as well as the registered image using the iterative closest point algorithm.

As an example for unimodal, intra-operative registration, Figure 3 shows the skull of a patient before and during brain surgery. In the image that was taken during the surgery the shift of the brain due to the opening of the skull is visible. Using rigid registration, this shift can be measured either directly by comparison of the registered image or by applying methods for non-rigid registration to the now preregistered image to obtain the mapping representing the shift. As described for the example above, 100 points were generated per image. The search space was again the space of rigid transformations. The geometric B&B algorithm solved the problem with error tolerance $\varepsilon_{rel} = 0.1$ (ε_{abs} sufficiently large) in 387.3 seconds traversing 5137 nodes of the B&B tree.

5   Conclusions and Future Research

In this paper we have demonstrated that an exact solution of the strongly non-convex robust point matching problem for medical image registration can be realized using Branch & Bound methods. The numerical results show that medium sized problem instances in 2 dimensions can be solved to global optimality in a reasonable amount of time. Local search methods like the frequently used iterative closest point algorithm may, however, fail to find realistic solutions if the starting solution is too far away from the global optimum.

Future research topics include the advancement of appropriate automatic methods for the selection of (reference or feature) points in the template and the reference image. Moreover, a combination of exact algorithms like the discussed B&B approaches with local search or other heuristic methods seems to be a promising approach to find very good solutions in a reasonable amount of time also for considerably larger data sets or for higher dimensional problems. For this purpose, an exact registration could be computed on a relatively coarse resolution level while on finer resolution levels, local methods like the iterative closest point algorithm are applied using the B&B result as a starting solution.

References

1. Besl, P.J., McKay, N.D.: A method for registration of 3-D shapes. IEEE Trans. Pattern Anal. Mach. Intell. 14(2), 239–256 (1992)
2. Breuel, T.M.: A comparison of search strategies for geometric branch and bound algorithms. In: Heyden, A., Sparr, G., Nielsen, M., Johansen, P. (eds.) ECCV 2002. LNCS, vol. 2352, Springer, Heidelberg (2002)
3. Breuel, T.M.: Implementation techniques for geometric branch-and-bound matching methods. Computer Vision and Image Understanding 90(3), 258–294 (2003)
4. Breuel, T.M.: On the use of interval arithmetic in geometric branch and bound algorithms. Pattern Recognition Letters 24(9-10), 1375–1384 (2003)
5. Chui, H., Rangarajan, A.: A new point matching algorithm for non-rigid registration. Computer Vision and Image Understanding 89(2-3), 114–141 (2003)
6. Eggert, D., Lorusso, A., Fisher, R.: Estimating 3-D rigid body transformations: A comparison of four major algorithms. Machine Vision and Applications 9(5-6), 272–290 (1997)
7. Hastreiter, P., Rezk-Salama, C., Soza, G., Bauer, M., Greiner, G., Fahlbusch, R., Ganslandt, O., Nimsky, C.: Strategies for brain shift evaluation. Medical Image Analysis 8(4), 447–464 (2004)
8. Liu, Y.: Improving ICP with easy implementation for free-form surface matching. Pattern Recognition 37(2), 211–226 (2004)
9. Maintz, J.B.A., Viergever, M.A.: A survey of medical image registration. Medical Image Analysis 2(1), 1–36 (1998)
10. Modersitzki, J.: Numerical Methods for Image Registration. Oxford University Press, New York (2004)
11. Mount, D.M., Netanyahu, N.S., Le Moigne, J.: Efficient algorithms for robust feature matching. Pattern Recognition 32(1), 17–38 (1999)
12. Nemhauser, G.L., Wolsey, L.A.: Integer and combinatorial optimization. Wiley-Interscience, New York, USA (1988)
13. Pfeuffer, F.: Registrierung medizinischer Bilddaten auf Basis verallgemeinerter Zuordnungsprobleme. Master's thesis, Universität Erlangen-Nürnberg, Institut für Angewandte Mathematik (2006)
14. Rusinkiewicz, S., Levoy, M.: Efficient variants of the ICP algorithm. In: Proceedings of the 3. Int. Conf. on 3D Digital Imaging and Modeling (2001)
15. Viola, P.A.: Alignment by Maximization of Mutual Information. PhD thesis, Massachusetts Institute of Technology (1995)
16. Zitová, B., Flusser, J.: Image registration methods: A survey. Image and Vision Computing 21(11), 977–1000 (2003)

Global Optimization for First Order Markov Random Fields with Submodular Priors

Jérôme Darbon

UCLA Mathematics Department, USA
jerome@math.ucla.edu

Abstract. This paper copes with the optimization of Markov Random Fields with pairwise interactions defined on arbitrary graphs. The set of labels is assumed to be linearly ordered and the priors are supposed to be submodular. Under these assumptions we propose an algorithm which computes an *exact* minimizer of the Markovian energy. Our approach relies on mapping the original into a combinatorial one which involves only binary variables. The latter is shown to be exactly solvable via computing a maximum flow. The restatement into a binary combinatorial problem is done by considering the level-sets of the labels instead of the label values themselves. The submodularity of the priors is shown to be a necessary and sufficient condition for the applicability of the proposed approach.

1 Introduction

Many early vision problems can be formulated as an optimization problem. In particular, Markov Random Fields (MRFs) models have been widely used [21] since the seminal work of Geman et al. [9]. These energies are generally a weighted combination of two terms: the fidelity term and the prior. The first one measures the fidelity of the reconstructed solution with the observed data while the second one contains some knowledge on the result. It is generally hard to find a global optimum since these energies are usually non-convex. For some particular cases, computations are tractable using dynamic programming [2]. However for most of problems, considered energies remain difficult to optimize in general and these optimization problems can even be NP-hard [12]. A general practice is to use Simulated Annealing [9,21] although it may be extremely slow in practice. This paper focus on Markovian energies that involves pairwise interactions and any data fidelity. An algorithm that computes a *global* minimizer of a subclass of these energies in more generality that it was previously possible is presented. Compared to non-global optimization algorithms, global minimization algorithms allow to study the practical performance of a model. Besides, the approach proposed in this paper can be seen as a complementary computational point of view to the theoretical work of Nikolova on the property of global minimizers[8,17,18].

Let us define the problem of minimizing a Markovian energy with pairwise interactions in the context of computer vision. Assume that images are defined

V.E. Brimkov, R.P. Barneva, H.A. Hauptman (Eds.): IWCIA 2008, LNCS 4958, pp. 229–237, 2008.
© Springer-Verlag Berlin Heidelberg 2008

on a set of nodes $\mathcal{V}$ with cardinality $|\mathcal{V}|$. The value of the image u at a site $p \in \mathcal{V}$ is denoted by u_p. The lattice is endowed with a neighboring system and the neighborhood relationship between two adjacent sites p and q is denoted by $p \sim q$. Only pairwise interactions are considered, and such a clique is referred to as (p,q) where $p \sim q$. We denote by $\mathcal{E}$ the set of all cliques. Thus we are interested in minimizing the following Markovian energy:

$$E(u|v) = \sum_{p \in \mathcal{V}} f_p(u_p|v_p) + \sum_{(p,q) \in \mathcal{E}} g_{pq}(u_p, u_q) , \qquad (1)$$

where v is the observed image, and the functions $\{f_p\}$ and $\{g_{pq}\}$ are respectively the fidelity terms and the priors.

In the seminal work [19], Picard and Ratliff show how a subclass of this energy can be optimized by computing a maximum-flow/s-t minimum-cut [1] on a graph associated to this energy. Then Greig et. al. use this approach in [10] to study the behavior of the Ising model for binary image restoration. In [4] Boykov et al. applies this technique for computer vision applications along with an excellent approximation result for the non-binary case. In [14], Kolmogorov and Zabih give a sufficient and necessary condition for the optimization of boolean MRF with pairwise and also triplewise interactions via s-t minimum-cut.

Extension of these approaches for exact optimization of MRFs involving more than two labels have been tackled by some authors. Approaches assume that labels can be linearly ordered and there are no assumptions on fidelity terms. In [13], a graph construction is proposed for MRFs where the priors are convex functions of the difference of labels, i.e. $g_{pq}(\cdot - \cdot)$ where g_{pq} are convex functions. The convexity assumption is shown to be sufficient and necessary. In [22], a class of MRFs whose energies can be rewritten as particular boolean MRF associated to each level is studied. In [6], the two above classes of Markovian energies are considered. The above assumptions allow the authors to devise a graph construction scheme for which a s-t minimum-cut yields a global minimizer minimization. Note that the topology of the underlying graph are different for each method but the size, i.e., the number of nodes and arcs, is the same. The optimization approach we propose in this paper can cope with all the above cases.

In this paper it is assumed that u_p takes value in the discrete set discrete $\mathcal{L} \subset \mathbb{R}$ of cardinality $|\mathcal{L}| = L$. This set is assumed to be linearly ordered, i.e., $\mathcal{L} = \{l_0, \ldots l_{L-1}\}$ with $l_i < l_{i+1} \forall i \in [\![0, L-2]\!]$. We also assume that the functions $\{f_p\}$ and $\{g_{pq}\}$ take values in $\mathbb{R}$ and are respectively defined on the discrete sets $\mathcal{L}$ and $\mathcal{L}^2$. Such functions will be referred to as discrete functions. In this paper, the priors $\{g_{pq}\}$ shall be submodular functions. For any positive integer k, a function $g : \mathcal{L}^k \to \mathbb{R}$ is said submodular if and only if it satisfies the following inequality [16]:

$$\forall (x,y) \in (\mathcal{L}^2)^k \quad g(x \vee y) + g(x \wedge y) \leq g(x) + g(y) , \qquad (2)$$

where $(x \vee y)$ and $(x \wedge y)$ respectively corresponds to the component-wise minimum and maximum between x and y, i.e., $\forall p \in \mathcal{V}$ $(x \vee y)_p = \min\{x_p, y_p\}$ and $(x \wedge y)_s = \max\{x_p, y_p\}$. Submodularity can be seen as a general property of *discrete* functions that are analogous to convexity of functions defined on continuous domain [16].

The main theoretical contributions of this paper are the following. First, we propose an algorithm which computes a global minimizer for MRFs with pairwise interactions where priors are submodular functions. No assumption is set on data fidelity terms. Our approach relies on restating this problem into a binary optimization problem that can be exactly solved with a maximum-flow-based approach [4,10,14,19]. Our mapping to the binary formulation makes use of the level sets of the labels. Second, it is shown that submodularity of the priors is a sufficient and necessary conditions for the application is the proposed approach. To our knowledge, these results are new and considerably extend previous available approaches for global MRF optimization. The complexity of our algorithm is pseudo-polynomial [1].

The remainder of this paper is organized as follows. Section 2 describes how one can we rewrite data fidelity and prior terms using the level sets of the variables. These rewritings are the core of our restatement of the original minimization problem to a binary minimization one. In Section 3 we cope with exact optimization of MRFs with submodular priors. Finally we draw some conclusions in Section 4.

2   Development through Level Sets

This Section is devoted to rewrite every single data fidelity term $f_p(\cdot)$ and all prior terms $g_{pq}(\cdot, \cdot)$ appearing in the Markovian energy E defined by equation (1), as a linear combination of binary energies. These restatements will be used for optimizing exactly first order MRF's with submodular priors. This mapping is achieved thanks to the level sets of a label. We first define the notion of level sets and then we give the developments on level sets for functions of one and two variables.

Let us introduce the level set $[x]_\lambda$ of a variable $x \in \mathcal{L}$ at a level $\lambda \in \mathcal{L}$ as follows:

$$[x]_\lambda = \begin{cases} 0 & \text{if } x \leq \lambda, \\ 1 & \text{if } x > \lambda . \end{cases}$$

The level sets of a variable x satisfies an monotone property:

$$\forall \lambda \leq \mu \quad [x]_\lambda \geq [x]_\mu , \tag{3}$$

The original gray-level value x can be reconstructed from its level sets using the following equality as shown in [11,15]:

$$x = \max\{\lambda \in \mathcal{L}, [x]_\lambda = 0\} . \tag{4}$$

Conversely, it is shown in [15] and in [11] that any family of binary variables $\{[x]_\lambda\}_{\lambda=0...L-1}$ which satisfies the monotone properties, given by equation (3), define a label. In other words, knowing the label itself or its binary representation in terms of level sets are equivalent.

The next proposition gives a development for data fidelity term as a summation on the level sets of its variable. It is based on a "discrete" integration of the "discrete" variations of f_p over its level sets.

Proposition 1. *Any data fidelity term* $f_p : \mathcal{L} \mapsto \mathbb{R}$ *rewrites on its level lets as follows:*

$$f_p(x) = \sum_{i=0}^{L-2} D_p(i)[x]_{l_i} + f_p(l_0) , \qquad (5)$$

where $\forall i \in [\![0, L-2]\!] \ D_p(i) = f_p(l_{i+1}) - f_p(l_i)$.

The proof is a straightforward extension of a similar proposition in [5].

Next, we extend the previous result to cope with functions of two variables. A natural way to perform it consists of applying the previous development firstly on the first variable and then on the second one. By rearranging terms it yields the following level sets-based developments.

Proposition 2. *Any prior term* $g_{pq} : \mathcal{L}^2 \mapsto \mathbb{R}$ *rewrites on its level sets as follows:*

$$g_{pq}(x,y) = \sum_{i=0}^{L-2} \sum_{j=0}^{L-2} R_{pq}(i,j)[x]_{l_i}[y]_{l_j} \qquad (6)$$

$$+ \sum_{i=0}^{L-2} \left(D_{pq}^1(i)[x]_{l_i} + D_{pq}^2(i)[y]_{l_i} \right) + C ,$$

where

$$\forall i \in [\![0, L-2]\!] \ D_{pq}^1(l_i) = g(l_{i+1}, l_0) - g(l_i, l_0) ,$$

and

$$\forall i \in [\![0, L-2]\!] \ D_{pq}^2(l_i) = g(l_0, l_{i+1}) - g(l_0, l_i) ,$$

and $C = g_{pq}(l_0, l_0)$ *and more importantly where*

$$\forall (i,j) \in [\![0, L-2]\!]^2 \ R_{pq}(i,j) = g(l_{i+1}, l_{j+1}) - g(l_{i+1}, l_j) - g(l_i, l_{j+1}) + g(l_i, l_j) \quad (7)$$

So far, we have made no assumptions on data fidelity terms and on priors. In other words, results given in Proposition 1 and Proposition 2 hold for any function of one and two variables, respectively. In the next section, we specialize these level sets developments in order to globally optimize MRFs with submodular priors.

3   MRFs with Submodular Priors

In this Section, we assume that all priors $\{g_{st}\}$ are submodular functions and we show that such MRFs can be exactly optimized via computing a maximum flow on an associated graph [4,14,19]. Our approach consist of first applying the

previous proposition to restate the original energy given by Eq. (1) in terms
of binary variables. So we rewrite all data fidelity and prior terms using the
expansions given by Proposition 1 and Proposition 2, respectively. So we get:

$$E(u|v) = \sum_{(p,q) \in \mathcal{E}} \left\{ \sum_{i=0}^{L-2} \sum_{j=0}^{L-2} R_{pq}(i,j)[u_p]_{l_i}[u_q]_{l_j} + \sum_{i=0}^{L-2} D_{pq}^1(i)[u_p]_{l_i} + D_{pq}^2(i)[u_q]_{l_i} \right\}$$
$$+ \sum_{p \in \mathcal{V}} \sum_{i=0}^{L-2} D_p(i)[u_p]_{l_i} + K \ ,$$

where the constant K comes from the constant C in the previous propositions.
Note that the latter rewriting of the energy $E(u|v)$ only involve the level sets of
the image u, i.e., $\{[u]_{l_i}\}_{i=0...L-1}$. So let us define a new energy $\tilde{E}$ whose variables
are now $|L|$ binary images $\{b^i\}_{i=0...L-1}$ as follows:

$$\tilde{E}(\{b^i\}_{i=0...L-1}|v) = \sum_{(p,q) \in \mathcal{E}} \left\{ \sum_{i=0}^{L-2} \sum_{j=0}^{L-2} R_{pq}(i,j)b_p^i b_q^j + \sum_{i=0}^{L-2} D_{pq}^1(i)b_p^i + D_{pq}^2(i)b_q^i \right\}$$
$$+ \sum_{p \in \mathcal{V}} \sum_{i=0}^{L-2} D_p(i)b_p^i + K \ .$$

Now if for all sites $p \in \mathcal{V}$, the families of binary images $\{b^i\}_{i=0...L-1}$ satisfy the
monotone property given by Eq (3), then this family defines an image using the
reconstruction given by Eq (4). However if any of the families $\{[b_s^i]_\lambda\}_{\lambda=0...l_{L-1}}$
violates the monotone property, then a gray level image cannot be defined. Be-
sides note that for any image u we have $E(u|v) = \tilde{E}(\{[u]_\lambda\}_{\lambda=0...l_{L-1}}|v)$. Thus,
if we are able to minimize the energy $E(\{[\cdot]_\lambda\}_{\lambda=l_0...l_{L-1}}|v)$ while preserving the
monotone property, then we get a global minimizer of $E(\cdot|v)$. In order to force
the monotone property to hold we define the following new energy:

$$\tilde{E}_\alpha(\{b_i\}_{i=0...L-1}|v) = \tilde{E}(\{b_i\}_{i=0...L-1}|v) + \sum_{p \in \mathcal{V}} \alpha \sum_{i=0}^{L-2} H(b_p^{i+1} - b_p^i) \ , \quad (8)$$

where $H : \mathbb{R} \mapsto \mathbb{R}$ is the Heaviside function defined as $H(x) = 0$ if $x \leq 0$ and
1 else. It is shown in [6] that if α is set to a sufficiently large finite value, then
we are assured that any global minimizer of $E_\alpha(\{\cdot\}_{i=0...L-1}|v)$ never violates the
monotone property give by Eq. (3).

Now we show that the boolean energy (8) can be optimized via a maximum
flow or by duality a s-t minimum-cut [1]. Following the seminal work of [19] or
equivalently [4,14] it is enough to show that every pairwise interaction terms of
binary variables are submodular. Specializing the definition of submodularity,
Eq. (2) for a binary function f of two variables, i.e., $f : \{0,1\}^2 \rightarrow \mathbb{R}$, we get
that:

$$f(0,0) + f(1,1) \leq f(0,1) + f(1,0) \ . \quad (9)$$

For the case we are considering we shall check the submodularity of the terms $H([u_p]_{l_{i+1}} - [u_p]_{l_i})$ and $R_{pq}(i,j)b_p^i b_q^j$. It is easily seen that the terms $H(b_p^{i+1} - b_p^i)$ satisfy the submodular property; see also [6] for further details. Thus it remains to show the submodularity of the terms $R_{pq}(i,j)b_p^i b_q^j$. Using the inequality (9) it means to show that $[\![0, L-2]\!]^2$ $R(i,j) \leq 0$. This property is assured by the submodularity assumption of the priors, Eq. (2), as shown in the next proposition.

Proposition 3. *Assume* $g : \mathcal{L}^2 \mapsto \mathbb{R}$. *The following two assertions are equivalent:*

1. *g is submodular,*
2. *g writes as*

$$g(x,y) = \sum_{i=0}^{L-2}\sum_{j=0}^{L-2} R(i,j)[x]_{l_i}[y]_{l_j} \qquad (10)$$

$$+ \sum_{i=0}^{L-2} \left(D^+(i)[x]_{l_i} + D^-(i)[y]_{l_j} \right) + C \; ,$$

where $\forall (i,j) \in [\![0, L-2]\!]^2$ $R(i,j) \leq 0$, D^+ *and* D^- *are two functions and* C *is a constant.*

Proof. Case 1) $\Rightarrow$ 2) We apply Proposition 2 to g and we get the form given in 2). It is straightforward to see that any unary function is submodular. The submodularity condition given by Eq. (9) applied for the remaining terms $R(i,j)[x]_{l_i}[y]_{l_j}$, reduces to show that $\forall (i,j) \in [\![0, L-2]\!]^2$ $R(i,j) \leq 0$.
Recall that Eq. (7) of Proposition 2 also states that

$$R(i,j) = g(l_{i+1}, l_{j+1}) - g(l_{i+1}, l_j) - g(l_i, l_{j+1}) + g(l_i, l_j) \; .$$

Now let us introduce the couples $a = (l_i, l_{j+1})$ and $b = (l_{i+1}, l_j)$. Then it is readily seen that $R(i,j)$ rewrites as follows:

$$R(i,j) = g(a \wedge b) - g(a) - g(b) + g(a \vee b) \; .$$

The latter is non-positive due to the submodularity of g. This concludes the proof for the first case.

Case 2) $\Rightarrow$ 1): Let $x \in \mathcal{L}^2$ and $y \in \mathcal{L}^2$. Note that the only interesting case happens when $x \notin \{(x \vee y) \cup (x \wedge y)\}$ (otherwise the submodularity property is obviously satisfied).

Let us denote by $(x_m, y_m) = (x \wedge y)$ and $(x_M, y_M) = (x \vee y)$. We need to show that $g((x_m, y_m)) + g((x_M, y_M)) - g((x_m, y_M)) - g((x_M, y_m)) \leq 0$.

To prove this inequality we write each term in the level-set development form given by Eq. (10). One sees that the constant C and the terms involving the single summation $\left(\sum_{i=0}^{L-2} \cdot \right)$ cancels each other. Thus only the double summation terms remain, i.e., we need to show:

$$\sum_{i=0}^{L-2}\sum_{j=0}^{L-2} R(i,j) \left([x_m]_{l_i}[y_m]_{l_j} + [x_M]_{l_i}[y_M]_{l_j} - [x_M]_{l_i}[y_m]_{l_j} - [x_m]_{l_i}[y_M]_{l_j} \right) \leq 0 \; ,$$

which is equivalent to

$$\sum_{i=0}^{L-2}\sum_{j=0}^{L-2} R(i,j)\left([x_M]_{l_i} - ([x_m]_{l_i})\right)\left([y_M]_{l_j} - ([y_m]_{l_j})\right) \leq 0 .\tag{11}$$

Since $x_M \geq x_m$ and $y_M \geq y_m$ get that

$$\forall i \in \mathcal{L}\ \ ([x_M]_{l_i} \geq ([x_m]_{l_i}) \wedge ([y_M]_{l_i} \geq ([y_m]_{l_i}),$$

and thus every term in the double summation in (11) are non-positive since $R(i,j) \leq 0$. This concludes the proof. □

So far we have shown that the binary energy 8 can be exactly optimized using a maximum flow approach [4,10,14,19]. Minimizing the latter energy is equivalent to minimize a first order MRF with submodular priors. Note that Proposition 3 gives a sufficient and necessary condition for applying the proposed approach. This result highly generalizes the results presented in [6] and [13].

We now consider the case where the priors are a unary function of the difference of the labels. These are widely used in image analysis because it corresponds to regularize the gradient of an image. The most well-known example of such a prior is most probably the Total Variation [20]. Under the above assumption the next proposition shows that only a *convex* regularization of the difference of the labels can be considered using the approach presented in this paper.

Proposition 4. *Assume* $g : \mathcal{L}^2 \to \mathbb{R}$ *is submodular and has the following* $g(x,y) = \tilde{g}(x - y)$ *then* $\tilde{g}$ *is a unary convex function.*

Proof. First we apply Prop 2. Now, due to the form of $\tilde{g}$ we have that $R(i,j) = 2\tilde{g}(i-j) - \tilde{g}(i-j+1) - \tilde{g}(i-j-1)$. We also have $R(i,j) \leq 0$ by the submodularity of g. By letting $k = i - j$ we get that $2\tilde{g}(i-j) \leq \tilde{g}(i-j+1) + \tilde{g}(i-j-1)$ which is exactly the discrete second variation convexity criteria for a unary function [16] applied for $\tilde{g}$. □

Note that although computing a maximum can be performed in polynomial time [1] our approach, like those of [6] and [13], is not. Indeed, an algorithm has a polynomial time if it performs a polynomial number of operations with respect to the number of bits required to describe the optimization problem. The necessary number of bits to describe an integer n is $\lceil log_2\ n \rceil$. The graph we built has for each pixel one node per gray level (i.e., for each pixel we have $(L-1)$ nodes) and is thus exponential with respect to $\lceil log_2\ n \rceil$. This exponential behavior prevents us from applying this approach on very large images (such as 3D volumes) because it requires to much memory. However, The proposed approach is manageable for standard size images and we refer the reader to [6] and [7] for image restoration and time results. Finally note that the maximum flow algorithm described in [3] has been shown to be extremely efficient in practice though its time complexity might not be polynomial. For image processing purposes it outperforms other polynomial algorithms as reported in [3]. This makes the method applicable as reported in [6] and [7].

4   Conclusion

In this paper we have presented a method to globally optimize a Markovian energy with pairwise interactions whose priors are submodular functions. The approach consists of restating the original problem as binary optimization problem that can be efficiently solved using a graph approach. The binarization makes use of the level set of the image. The submodularity of the priors has been shown to be a necessary and sufficient condition for the applicability the proposed approach.

Acknowledgements

The author deeply thanks Marc Sigelle (Télécom Paris, France) for fruitful discussions and careful proofreading, Bülent Sankur (Bogazici Univeristy, Istanbul, Turkey) and TUBITAK for funding his April and May 2006 stay at Bogazici University, during part of this paper was written, and finally the anonymous reviewers for helpful comments. This work was partly done while the author was with the Research and Development Laboratory of EPITA (LRDE), Paris, France. This research was also supported by ONR Grant N000140710810.

References

1. Ahuja, R., Magnanti, T., Orlin, J.: Network Flows: Theory, Algorithms and Applications. Prentice-Hall, Englewood Cliffs (1993)
2. Amini, A., Weymouth, T., Jain, R.: Using Dynamic Programming for Solving Variational Problems in Vision. IEEE Transactions on Pattern Analysis and Machine Interaction 12(9), 855–867 (1990)
3. Boykov, Y., Kolmogorov, V.: An Experimental Comparison of Min-Cut/Max-Flow Algorithms for Energy Minimization in Vision. IEEE Transactions on Pattern Analysis and Machine Interaction 26(9), 1124–1137 (2004)
4. Boykov, Y., Veksler, O., Zabih, R.: Fast Approximate Energy Minimization via Graph Cuts. IEEE Transactions on Pattern Analysis and Machine Interaction 23(11), 1222–1239 (2001)
5. Darbon, J., Sigelle, M.: Image Restoration with Discrete Constrained Total Variation Part I: Fast and Exact Optimization. Journal of Mathematical Imaging and Vision 26(3), 261–276 (2006)
6. Darbon, J., Sigelle, M.: Image Restoration with Discrete Constrained Total Variation Part II: Levelable Functions, Convex Priors and Non-Convex Cases. Journal of Mathematical Imaging and Vision 26(3), 277–291 (2006)
7. Darbon, J., Sigelle, M., Tupin, F.: The use of levelable regularization functions for MRF restoration of SAR images while preserving reflectivity. In: The proceedings of the IS&T/SPIE 19th Annual Symposium Electronic Imaging (oral presentation). Conference on Computational Imaging (E112), San Jose, CA, USA (2007)
8. Durand, S., Nikolova, M.: Stability of the Minimizers of Least Squares with a Non-Convex Regularization. Part II: Global Behavior. Journal of Applied Mathematics and Optimization 53(3), 259–277 (2006)

9. Geman, S., Geman, D.: Stochastic Relaxation, Gibbs Distributions, And The Bayesian Restoration Of Images. IEEE Transactions on Pattern Analysis and Machine Interaction 6(6), 721–741 (1984)
10. Greig, D., Porteous, B., Seheult, A.: Exact maximum a posteriori estimation for binary images. Journal of the Royal Statistics Society 51(2), 271–279 (1989)
11. Guichard, F., Morel, J.M.: Mathematical Morphology "Almost Everywhere". In: Proceedings of International Symposium on Mathematical Morphology, pp. 293–303. Csiro Publishing (2002)
12. Hochbaum, D.S.: An efficient algorithm for image segmentation, Markov random fields and related problems. Journal of the ACM 48(2), 686–701 (2001)
13. Ishikawa, H.: Exact optimization for Markov random fields with convex priors. IEEE Transactions on Pattern Analysis and Machine Interaction 25(10), 1333–1336 (2003)
14. Kolmogorov, V., Zabih, R.: What Energy can be Minimized via Graph Cuts? IEEE Transactions on Pattern Analysis and Machine Interaction 26(2), 147–159 (2004)
15. Maragos, P., Ziff, R.: Threshold superposition in morphological image analysis systems. IEEE Transactions on Pattern Analysis and Machine Intelligence 12(5), 498–504 (1990)
16. Murota, K.: Discrete Convex Optimization. SIAM Society for Industrial and Applied Mathematics (2003)
17. Nikolova, M.: Analysis of the recovery of edges in images and signals by minimizing nonconvex regularized least-squares. SIAM Journal on Multiscale Modeling and Simulation 4(3), 960–991 (2005)
18. Nikolova, M.: Model distortions in Bayesian MAP reconstruction. AIMS Journal on Inverse Problems and Imaging 1(2), 399–422 (2007)
19. Picard, J., Ratlif, H.: Minimum cuts and related problem. Networks 5, 357–370 (1975)
20. Rudin, L., Osher, S., Fatemi, E.: Nonlinear Total Variation Based Noise Removal Algorithms. Physica D. 60, 259–268 (1992)
21. Winkler, G.: Image Analysis, Random Fields and Dynamic Monte Carlo Methods. A Mathematical Introduction. In: Applications of mathematics, 3rd edn., Springer, Heidelberg (2006)
22. Zalesky, B.: Efficient Determination of Gibbs Estimator with Submodular Energy Functions. Technical report, United Institution of Information Problem (2005)

Transformation Polytopes for Line Correspondences in Digital Images

Kristof Teelen and Peter Veelaert

University College Ghent, Engineering Sciences - Ghent University Association,
Schoonmeersstraat 52, B9000 Ghent, Belgium
{Kristof.Teelen,Peter.Veelaert}@hogent.be

Abstract. We present an uncertainty model for geometric transformations, based on polygonal uncertainty regions and transformation polytopes. The main contribution of this paper is a systematic approach for the computation of regions of interest for features by using the uncertainty model. The focus is on the solution of transformation problems for geometric primitives, especially lines, so that regions of interest can be computed for corresponding geometric features in distinct images.

1 Introduction

Quantization, geometric distortion and noise in digital image acquisition invariably lead to uncertainty about the occurrence, location and shape of image features. Coping with feature uncertainty is one of the major challenges in those computer vision applications that want to establish correspondences between features in distinct images, that want to derive geometric relations between features, or that involve geometric reasoning. A good mathematical model for feature uncertainty, however, can greatly improve the performance of a correspondence or geometric reasoning algorithm. For example, small and accurate regions of interest (ROIs) for each feature help to find correspondences more quickly and reliably. Computing ROIs becomes therefore an essential part of real-time image and video processing.

In this work, we examine the problem of computing ROIs from a mathematical viewpoint. Our approach is based on the computation of uncertainty polytopes for the transformations that map features in one image onto features in a second image. In this paper, we give an overview of the uncertainty problems that must be solved to compute a ROI for a certain geometric feature. We extend previous work and show how uncertainty regions and ROIs can be determined for line features.

Several approaches have been used to model geometric uncertainty. Kanatani was one of the first to use statistical inference in a systematic manner to solve uncertainty problems in geometry [3, 4]. Förstner has worked out several simple-to-use tools that are based on statistical inference, and which can be used to represent, analyze and propagate uncertainty through geometric reasoning chains [1]. This work has been further extended to other specific problems [6, 5].

The approach presented in this paper is not based on statistical inference, but on the use of uncertainty regions. That is, instead of estimating a pdf for a feature, we construct

V.E. Brimkov, R.P. Barneva, H.A. Hauptman (Eds.): IWCIA 2008, LNCS 4958, pp. 238–249, 2008.

an uncertainty region in which the feature is likely to occur. Roughly, an uncertainty region is a region which indicates where a pdf is above a certain threshold. Furthermore, instead of using uncertainty ellipses, we use uncertainty polytopes, which provide more flexibility. Using uncertainty polytopes often results in simple computations, and gives the possibility to proceed further in the geometric reasoning chain. Furthermore, in many applications it is more important to have a reliable ROI for a feature, without the exact knowledge of a pdf.

The uncertainty problems described in this work either lead to linear or nonlinear, i.e. quadratic, cases. Although algorithms are known to exist for the solution of quadratic programming problems, they are not useful for most image processing applications, due to their complexity. On the other hand, there exist efficient geometric techniques to solve linear programming problems [2]. In our approach, the focus is therefore on simple methods for the solution of a system of linear equations or inequalities. We will consider solving linear systems as a special case of linear programming, and we also refer to it as such in the remainder of this work.

First, we give an overview of the transformation uncertainty problems in section 2. Next, in section 3, we describe how to compute the ROIs for point features. We go into greater detail for the line transformation problems in section 4, mainly for the cases that can be solved using linear programming. In section 5, we present an application involving the computation of ROIs for line features. Finally, we conclude the paper in section 6.

2 Overview of Transformation Uncertainty Problems

When considering geometric features and their transformations, uncertainty about their exact position must be taken into account. The uncertainty is modeled as a convex uncertainty region in which the primitive must be located. In this section, we will introduce the mathematical notation used in later sections to discuss the transformation problems in greater detail.

In this paper, geometric primitives are used as features, with a focus on points and lines. We will assume from now on that the features, both points and lines, are in general position. We shall not discuss special configurations.

The position or transformation parameters are given either by a single specific point, or by an uncertainty region or polytope in the parameter space. S denotes specific parameters, while U indicates uncertainty regions or polytopes. A convex bounded polygon of a certain size and shape models the uncertainty about the exact location of a feature in the image, which is due to e.g. errors introduced by the digitization process, noise or the feature detector.

In this work, the transformations are limited to affine transformations T, represented as

$$T = \begin{bmatrix} a & b & e \\ c & d & f \\ 0 & 0 & 1 \end{bmatrix}. \tag{1}$$

Affine uncertainty transformations can then be represented as a polytope in 6 dimensions, one for each parameter of the affine transformation T. By sufficiently constraining the parameters, a convex bounded polyhedron or polytope is obtained in the

Table 1. The first column denotes the meaningful situations. The last two columns show whether the problems can be solved for point features using linear programming or not.

Meaningful	Linear	Nonlinear
SSS	$SS?S, SSS?, S?SS$	
USU	$USU?, U?SU$	
SUU	$SUU?, SU?U$	
UUS	$U?US, UU?S$	
UUU		$U?UU, UU?U, UUU?$

parameter space. Some constraints may be known beforehand, e.g. if the expected translation is maximum 100 pixels, then $-100 \le e, f \le 100$.

To denote the transformation problems, we use three letters XYZ to specify a map from X to Z by the transformation Y, i.e. $Y : X \to Z$. For example, SUU denotes a transformation from a feature on a specific location, by a transformation uncertainty polytope into an uncertainty polygon. Table 1 gives an overview of uncertainty problems for point features.

Furthermore, we use a question mark to define a problem in which a feature or region is unknown and should be determined from the other data. For example, $SUU?$ means that a single feature and a transformation uncertainty polytope are given, and that the uncertainty region into which the feature is mapped must be determined.

The cases SSU, USS, SUS are always meaningless. For example, it is impossible to map a single point by a single transformation onto a uncertainty region, which excludes SSU. Some problems may not have a solution in the general. For example, given two arbitrary n-gons in the plane, only when these two polygons are carefully chosen the problem $US?U$ will have a solution. Likewise, we exclude $S?UU$ and $UUS?$.

The situations SUU and UUS are dual to each other. The case $SU?U$ leads to the construction of uncertainty polytopes for transformations (in the case of $SU?U$) or uncertainty polygons for image features (in the case of $SUU?$), while for problems that fall under UUS it is easier to derive such polytopes or polygons for inverse transformations. The situation UUU is the most general, but invariably leads to nonlinear programming problems, as we will illustrate for point features in the following section.

Only a limited number of problems can be solved as linear programming problems. To determine ROIs for features in a second image, several of these problems must be solved. Therefore, we will discuss these cases in greater detail in the following sections for point and line features. We will not consider nonlinear problems into detail in this paper, however, in some cases a simple example is given to illustrate the nonlinearity of a problem.

The solution of the different linear problems requires the use of some properties for affine transformations and affine combinations. From the following observations, we can deduce some useful lemmas.

For two points $p_1 = (x_1, y_1)$ and $p_2 = (x_2, y_2)$, and the affine combination $p = \alpha_1 p_1 + \alpha_2 p_2$, the affine transformation T of the form (1) yields

$$x = \alpha_1(ax_1 + by_1 + e) + \alpha_2(ax_2 + by_2 + e)$$
$$y = \alpha_1(cx_1 + dy_1 + f) + \alpha_2(cx_2 + dy_2 + f).$$

Because $\alpha_1 + \alpha_2 = 1$, this can be rewritten as

$$x = a(\alpha_1 x_1 + \alpha_2 x_2) + b(\alpha_1 y_1 + \alpha_2 y_2) + e$$
$$y = c(\alpha_1 x_1 + \alpha_2 x_2) + d(\alpha_1 y_1 + \alpha_2 y_2) + f.$$

It follows that $\alpha_1 T(p_1) + \alpha_2 T(p_2) = T(\alpha_1 p_1 + \alpha_2 p_2)$. As the convex combination is an affine combination with non-negative coefficients ($0 \leq \alpha_i \leq 1$), the convex combination of points is preserved by an affine transformation. This proves the following lemma.

Lemma 1. *An affine transformation of an affine combination of points is equal to the affine combination of the transformed points, i.e.* $T(\alpha_1 p_1 + \alpha_2 p_2) = \alpha_1 T(p_1) + \alpha_2 T(p_2)$, *provided* $\alpha_1 + \alpha_2 = 1$. *Similarly, an affine transformation of a convex combination of points (with $0 \leq \alpha_i \leq 1$) is equal to the convex combination of the transformed points.*

Lemma 2. *A linear combination of mappings of a point p is equal to the linear combination of transformations applied to the point p, i.e.* $(\alpha_1 T_1 + \alpha_2 T_2)(p) = \alpha_1 T_1(p) + \alpha_2 T_2(p)$.

The proof is obtained along the same lines as for Lemma 1. Since Lemma 2 holds for linear combinations, it also holds for affine and convex combinations.

2.1 Transformation Polytope Duality

Since the situation UUU is unmanageable as a linear problem, when introducing uncertainty polytopes for affine transformations, a choice must be made. Both SUU and UUS are possible situations, where the derived problems can be solved by linear programming. Unfortunately, we cannot convert a transformation polytope that has been found for $SU?U$ into a polytope for the problem $UU?S$.

Suppose we have a transformation polytope of affine transformations, what is the shape of the set of inverse transformations? It is not a polytope, and not even a convex set, as shown by the following example.

The inverse of an affine transformation is also an affine transformation. When the affine transformation T is represented as in (1) then the inverse transformation T^{-1} is given by

$$T^{-1} = \begin{bmatrix} \frac{d}{-bc+ad} & \frac{-b}{-bc+ad} & \frac{-de+bf}{-bc+ad} \\ \frac{-c}{-bc+ad} & \frac{a}{-bc+ad} & \frac{ce-af}{-bc+ad} \\ 0 & 0 & 1 \end{bmatrix} \qquad (2)$$

As an example, we choose a line segment of transformations as transformation polytope. We visualize a projection of the calculated set of inverse transformations upon the parameter plane ab in Fig. 1. The projection is convex as a function, but it is not convex as a curve, which means that also the curve formed by the transformations is not convex. Even if we restrict the transformations to scalings and translations, the set of inverse transformations is not convex. Counterexamples can easily be found for those cases. Only if we restrict the transformations to translations only, the inverse transformations form a polytope.

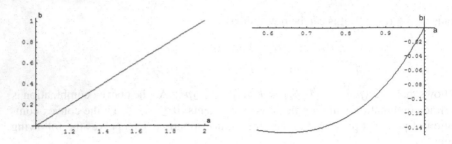

Fig. 1. Example: On the left, we show a projection of the transformation uncertainty polytope onto the ab parameter plane. The projection of the inverse transformation polytope upon the ab parameter plane is shown on the right.

This means that we must either model uncertainty for problems of the form SUU or for problems of the form UUS. The situation SUU is the most natural when considering ROIs, as we want to find out where a specific feature in the first image can be found in a second image.

3   Regions of Interest for Point Features

Suppose we have a set of correspondences between two limited sets of features in two distinct image frames. Assume that the features of one frame can be projected upon the features of the second frame by a transformation map. The exact position of the feature can not be resolved due to errors introduced by the digitization process or by the feature detector. Since the feature position is not exactly known, the transformation map can not be precisely determined. This transformation uncertainty can be described by introducing an uncertainty polytope for the map, as we showed in previous work [7,8, 9,10,11]. With this uncertainty polytope, the ROIs for other features can be determined. The properties of the uncertainty polytope determine the variation of the size and the shape of the ROIs across the image.

The procedure for solving the ROI problem requires solving both the $SU?U$ and $SUU?$ problem. First, we determine the uncertainty of the transformation map by solving $SU?U$. Then the ROIs can be computed by using the uncertainty of the map. In turn, $SU?U$ and $SUU?$ involve $SS?S$ and $SSS?$ as subproblems. Therefore, we first discuss how to solve these problems. Next, a brief overview of the other point transformation problems is given, based on previous work. In particular, we show that the ROI problem has been solved for point features.

SSS?, S?SS and SS?S. Both $SSS?$ and $S?SS$ are trivial problems, which are solved by simply applying the forward and inverse transformation as given by (1) and (2). Also the problem $SS?S$ is easy to solve for points. A closed formula for the transformation mapping three points $(x_1, y_1), \ldots, (x_3, y_3)$ upon three points $(x'_1, y'_1), \ldots, (x'_3, y'_3)$ can easily be derived. Such a transformation exists provided the points are not collinear, i.e. the points must be in general position.

SU?U and SUU? A solution for $SU?U$ and $SUU?$ is required when computing ROIs for the transformed features. Both problems have been solved in previous work [7, 8, 9, 10, 11]. To solve $SU?U$, we must find a transformation polytope mapping point sets into uncertainty regions. It is easy to show that the polytope is the convex hull of the transformations that map source points onto the vertices of the uncertainty polygons in the image plane. Let p_1, p_2, p_3 be three points in the source plane and let q_1^i, q_2^j, q_3^k be the vertices of the corresponding uncertainty polygons in the image plane. Let T_{ijk} be the transformation mapping the points p_1, p_2, p_3 onto the points q_1^i, q_2^j, q_3^k. Then, because of Lemma 2, the transformation uncertainty polytope is the convex hull of the transformations T_{ijk}.

Given a transformation polytope and a set of other feature points, we can determine the ROIs as uncertainty polygons in the second image frame. This follows directly from Lemma 2. When a source point p is projected onto the points q^l in the image plane with the vertices T^v of the transformation polytope, the uncertainty polygons are obtained as the convex hull of the points q^l. We try to find a corresponding feature in the image plane, in a search space limited to only the ROI described by the uncertainty polygon.

USU? When p^i denotes the vertices of the uncertainty polygon in the source plane, the uncertainty region in the image plane is the convex hull of the points $T(p^i)$. This follows from Lemma 1. For the inverse problem $U?SU$, we simply apply the inverse transformation to the vertices q^i of the uncertainty polygon in the image plane. Then the uncertainty region in the source plane is the convex hull of the points $T^{-1}(q^i)$.

UU?U. Let U_1, U_2, U_3 and U_1', U_2', U_3' be convex regions in respectively the source and the image plane. Then the set of affine transformations that map at least one point of U_i into U_i' is not convex, as can be illustrated by a simple example. Fig. 2 shows the points U_2, U_3, U_1', U_3' and the line segments U_1 and U_2' as uncertainty polygons. The projection of several affine transformations in the uncertainty polytope onto the ab parameter plane is shown in Fig. 2. This figure clearly shows that the projected polytope is not convex, which means that also the transformation uncertainty polytope is not convex.

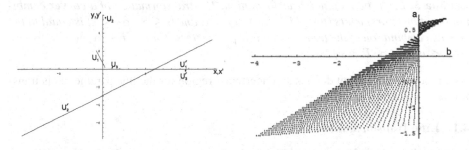

Fig. 2. Example: On the left the uncertainty regions in the source (U_i) and image (U_i') plane are shown. The uncertainty polytope for the transformation from the uncertainty regions U_i into the uncertainty regions U_i' can be computed. The right figure shows the projection of this transformation polytope on the ab parameter plane.

4 Regions of Interest for Lines

Let $px + qy + r = 0$ be the equation of a line, then we represent this line by a parameter point $l = (p, q, r)$ in $\mathbb{R}^3$. Any point of the form $\gamma(p, q, r)$, with γ not zero, represents the same line. An uncertainty region for line parameters is a cone C minus the origin, where C contains rays of the form $(\gamma p, \gamma q, \gamma r)$. To define a convex polyhedral cone C, it is sufficient to define a set D of rays $\gamma(p, q, r)$ such that the minimal cone that contains D and the origin is equal to C.

For a non-singular affine transformation T represented by (1), line parameters are transformed by a linear transformation

$$\begin{bmatrix} p' \\ q' \\ r' \end{bmatrix} = \begin{bmatrix} d & -c & 0 \\ -b & a & 0 \\ bf - de & ce - af & ad - bc \end{bmatrix} \begin{bmatrix} p \\ q \\ r \end{bmatrix} = R \begin{bmatrix} p \\ q \\ r \end{bmatrix} \tag{3}$$

where the matrix R is the transpose of the inverse of T, multiplied with the determinant of T, or the transpose of the adjugate of T.

In principle, since line parameters are transformed by a simple linear transformation R, one could introduce uncertainty polytopes for the entries r_{ij} of the matrix R. In this way, we can proceed by solving problems that are similar to the problems listed in Table 1. The drawback of this approach is that it becomes difficult to combine the ROIs derived for points with the ROIs derived for lines. Therefore, we will try to derive ROIs for lines from the uncertainty polytopes for the affine transformations T.

The transformed line parameters (p', q', r') are given by

$$(dp - cq, -bp + aq, bfp - dep + ceq - afq + adr - bcr). \tag{4}$$

For a given column vector v of line parameters, we shall denote the set of all nonzero scalar multiples of the transformed line parameters (4) as $T < v >$. Let v_1, v_2 be the parameter vectors of two lines. For any linear combination of line parameters we have $R(\alpha_1 v_1 + \alpha_2 v_2) = \alpha_1 R v_1 + \alpha_2 R v_2$. This has an immediate consequence.

Lemma 3. *Let T be an affine transformation. The transformation of a convex combination of line parameters (for which $\alpha_1 + \alpha_2 = 1$, and $0 \leq \alpha_1, \alpha_2 \leq 1$) is equal to the convex combination of the transformed line parameters. That is, $T < \alpha_1 v_1 + \alpha_2 v_2 >= \alpha_1 T < v_1 > + \alpha_2 T < v_2 >$.*

As a result, the cone that defines an uncertainty region for the line parameters is transformed by T into a cone.

4.1 Line Transformations

To solve the region of interest problem for lines, we must solve both $SU?U$ and $SUU?$, where S is a set of lines, and U represents a set of uncertainty regions. $S?SS$, $SS?S$, and $SSS?$ are subproblems of $SU?U$ and $SUU?$ and thus also require a solution.

S?SS and SSS? Both $S?SS$ and $SSS?$ are easy to solve by applying either R or its inverse.

SS?S for n points and m lines with $n + m = 3$. When three lines and their transformed images are given, this problem can easily be reduced to the problem where three points and their images are given, provided the lines are in general position. It suffices to find the affine transformation which maps the three intersection points of the three lines on the image intersection points of the transformed lines.

A more interesting situation occurs when the affine transformation is specified by n points (x_i, y_i) and m lines $y = xp_j + q_j$, with $n + m = 3$, and their images. At first sight, (3) leads to nonlinear equations for the transformation parameters $a, \dots, f$. However, a linear equation can be found by selecting two arbitrary points on the line, writing down the transformed coordinates of the points, and substituting them into the equation of the transformed line.

For example, the line $y = xp_1 + q_1$ contains the points $(0, q_1)$, and $(-q_1/p_1, 0)$. The images of these two points under T as in (1) are the points $(bq_1 + e, dq_1 + f)$ and $(-aq_1/p_1 + e, -cq_1/p_1 + f)$. If we substitute the transformed points into the equation of the transformed line, $y' = x'p_1' + q_1'$, we find

$$dq_1 + f = (bq_1 + e)p_1' + q_1', \ -cq_1/p_1 + f = (-aq_1/p_1 + e)p_1' + q_1' \qquad (5)$$

which are linear equations in the unknowns $a, \dots, f$. Noteworthy in the above derivation is that the exact location of the image points is not specified, only the requirement that the image points must lie on the image line is used.

USU? and U?SU. Suppose we are given a convex polyhedral cone C of line parameters. Then the parameters of the transformed lines also form a cone C', due to Lemma 3. To find the cone C' it is sufficient to transform the vertices of a polytope P that generates C. Likewise, given C' one can find C.

SU?U and SUU? To solve the ROI problem for lines we must solve both $SU?U$ and $SUU?$. The major problem is that an uncertainty polytope for the transformation parameters of T does not correspond to a polytope for the elements r_{ij} of the matrix R as defined in (3). Thus a affine combination of affine transformations $\alpha_1 T_1 + \alpha_2 T_2$ does not correspond to an affine combination $\alpha_1 R_1 + \alpha_2 R_2$ of transformations which transform line parameters into line parameters.

Fortunately, there are some important special cases in which the transformed line parameters do form a polytope.

The line $px + qy + r = 0$ has parameter vector (p, q, r). This vector is transformed into the vector (4). In general this is not a linear expression in the transformation parameters $a, \dots, f$ unless some of the coefficients are either vanishing or fixed. We construct the conflict graph for the transformation parameters in Fig. 3.

The maximal independent sets of this graph are: $\{a, b, e\}$ and $\{c, d, f\}$. Other independent sets are: $\{e, f\}, \{d, f\}, \dots$ Each independent set leads to a special case for which the line parameter vector (4) becomes a linear function of the transformation parameters $a, \dots, f$ of the affine transformation T.

To illustrate the occurrence of linearity, we consider affine transformations of the form

$$T_{abe} = \begin{bmatrix} a & b & e \\ 0 & 1 & 0 \\ 0 & 0 & 1 \end{bmatrix}, T_{aaef} = \begin{bmatrix} a & 0 & e \\ 0 & a & f \\ 0 & 0 & 1 \end{bmatrix}. \qquad (6)$$

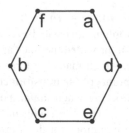

Fig. 3. Conflict graph

The transformation T_{abe} exploits the independent set $\{a, b, e\}$. The image of the line vector (p, q, r) is

$$(p, -bp + aq, -ep + ar), \tag{7}$$

which is linear in a, b, e. The transformation T_{abe} is an affine transformation where the y coordinates are kept constant. In fact, in this case, for lines that are not horizontal we can determine the image of each point of a line, since each points is displaced only horizontally. Therefore, not much is gained by including the uncertainty of lines for transformations of the form T_{abe}.

A more interesting case consists of transformations of the form T_{aaef}, which involve uniform scaling and translation. In this case the image of the line vector (p, q, r) is $(ap, aq, -aep - afq + a^2r)$. Since, any multiple of a line vector represents the same line, we can eliminate a, to obtain the image vector

$$(p, q, -ep - fq + ar), \tag{8}$$

which is a linear function of a, e, f.

Lemma 4. *Let T_1, T_2 be affine transformations that are either of the form T_{abe} or of the form T_{aaef} (both transformations must be of the same form). Let v be a column vector of line parameters. Then the transformed line parameters of an affine combination of transformations is equal to the affine combination of the transformed line parameters. That is, $(\alpha_1 T_1 + \alpha_2 T_2) < v >= \alpha_1(T_1 < v >) + \alpha_2(T_2 < v >)$, with $\alpha_1 + \alpha_2 = 1$.*

Lemma 4 remains valid for convex combinations with $\alpha_1, \alpha_2 \geq 0$. When Lemma 4 holds, we can solve both $SU?U$ and $SUU?$. Suppose we are given a set of line vectors v_i, and a convex polyhedral uncertainty cone C_i' for the image of each line. Furthermore, for each cone C_i', let v_{ij}' be a finite set of line parameters such that C_i' is the smallest cone that contains the lines v_{ij}' and the origin. To solve the problem $SU?U$, for each line v_i, we first solve the problem $SS?S$ for all lines v_{ij}' in the uncertainty cone C_i'. Each pair (v_i, v_{ij}') yields a transformation T_{ij}. The convex hull of the transformations T_{ij} is an uncertainty polytope P_i. The uncertainty polytope for the transformations that map each line into its own uncertainty cone is the intersection of the uncertainty polytopes P_i.

Suppose we are given a uncertainty polytope P for the transformations T and a line vector v. To solve the problem $SUU?$, we compute $T_i < v >$ for each vertex T_i of the polytope P. The uncertainty cone for the transformed image of the line v is the smallest cone that contains all the points $T_i < v >$ and the origin. Note that when the affine transformations are of the form T_{aaef}, then all the lines in the cone have the same slope as the line v.

5 Application Example

In previous work, the use of the cases $SUU?$ and $SU?U$ for point features was demonstrated in practical image processing application as image registration [10] or the comparison of line drawings [8]. Confidence measures can be defined to develop a notice about the likelihood of correspondences between points in two distinct images [11].

In this work, we illustrate the use of uncertain line transformations in image processing. A first advantage is that line features are often more reliable and stable than point features. Lines obtained by detecting edges can be significant features for objects as e.g. buildings or roads. Second, lines can often be positioned more accurately, which leads to less uncertainty and smaller ROIs. A first application of line transformations was presented in [10], which involved only horizontal and vertical lines. In this paper, we present a more general method to obtain ROIs for features in subsequent images.

First, an uncertainty transformation polytope must be computed. This can be done by constraining the parameters of the transformation, based on the expected maximum transformation of the objects in the image. We choose to explicitly compute the uncertainty transformation by solving the problem $SU?U$. The solution requires the information of a limited number of lines. These can be extracted by e.g. looking for remarkable and easy segmentable objects in the image. In the example presented in Fig. 4, we are looking for a transformation T_{aaef} consisting of uniform scaling and translation (6). The information of three lines on the traffic sign in the left image is used to constrain the transformation polytope. In this example, the transformed line parameter $r' = -ep - fq + ar$, as seen in (8), must satisfy $\tau_l \leq r' \leq \tau_u$ for each of the lines so that the polytope is bounded in the parameter space aef. For this type of transformations, the polytope is computed by solving a system of linear inequalities, as explained in section 4.

Fig. 4. Example: the left image shows the lines that are being mapped to the uncertainty regions indicated in the right image

Fig. 5. Example: the left image shows the lines that are being mapped to the uncertainty regions indicated in the right image

Once the uncertainty transformation polytope has been computed, it is used to compute ROIs in the second image. These regions limit the image space in which to look for the features, corresponding to those in the first image. As explained for the case SUU? in the previous section, the vertices of the polytope can be used as transformation parameters to map other lines in the first image to ROIs in the second image. In Fig. 5, we indicated some of the lines for which a correspondence must be found in the left image. When the parameters of these lines are mapped with the vertices of the transformation polytope, we obtain the lines indicated in the right image. Multiple lines are shown, one for each vertex of the transformation polytope. The uncertainty region for each line in the first image is then the convex hull of these mapped lines, i.e. the region between the two outermost lines. This example shows that the ROI for a line is considerably reduced, as line features must only be searched for in the corresponding uncertainty region. Also note that the shape and size of the ROIs will vary across the image, as can be seen for the two lines shown in the example.

6   Concluding Remarks

In this paper, we discuss an uncertainty model for geometric transformations, based on polygonal uncertainty regions and transformation polytopes. The uncertainty model can be used to solve different transformation problems, leading to either linear or nonlinear programming problems. We focus on the problems which can be solved by linear programming.

In previous work, we showed that the concept of uncertainty regions and transformations is indeed useful in several image processing algorithms concerning point features. That work is now extended to transformation problems for line features. Although we cannot represent the uncertainty of an affine transformation by a convex polytope in the general case, there exist some meaningful and important cases in which the transformation problems for line parameters can be solved using linear programming techniques. The uncertainty regions obtained as a solution for the uncertainty transformation problems, indeed prove to be useful as regions of interest in the detection of line features in distinct images.

References

1. Förstner, W.: Uncertainty and Projective Geometry. In: Bayro-Corrochano, E. (ed.) Handbook of Geometric Computing, pp. 493–535. Springer, Heidelberg (2005)
2. Grötschel, M., Lovász, L., Schrijver, A.: Geometric Algorithms and Combinatorial Optimization, 2nd edn. Springer, Heidelberg (1993)
3. Kanatani, K.: Uncertainty Modeling and Model Selection for Geometric Inference. PAMI 26(10), 1307–1319 (2004)
4. Kanatani, K.: Uncertainty Modeling and Geometric Inference. In: Bayro-Corrochano, E. (ed.) Handbook of Geometric Computing, pp. 461–492. Springer, Heidelberg (2005)
5. Perwass, Förstner: Uncertain geometry with circles, spheres and conics. In: Klette, Kozera, Noakes, Weickert (eds.) Geometric Properties for Incomplete Data, pp. 23–41. Springer, Heidelberg (2006)
6. Perwass, Gebken, Sommer: Estimation of Geometric Entities and Operators from Uncertain Data. In: Kropatsch, W.G., Sablatnig, R., Hanbury, A. (eds.) DAGM 2005. LNCS, vol. 3663, pp. 459–467. Springer, Heidelberg (2005)
7. Teelen, K., Veelaert, P.: Computing the Uncertainty of Geometric Primitives and Transformations. In: Proceedings of ProRISC 2004, pp. 317–325 (2004)
8. Teelen, K., Veelaert, P.: Uncertainty of Affine Transformations in Digital Images. In: Proceedings of ACIVS 2004, pp. 23–30 (2004)
9. Teelen, K., Veelaert, P.: Computing the uncertainty of transformations in digital images. In: Proceedings of SPIE Vision Geometry XIII, pp. 1–12
10. Teelen, K., Veelaert, P.: Image Registration Using Uncertainty Transformations. In: Blanc-Talon, J., Philips, W., Popescu, D.C., Scheunders, P. (eds.) ACIVS 2005. LNCS, vol. 3708, pp. 348–355. Springer, Heidelberg (2005)
11. Veelaert, P., Teelen, K.: Consensus sets for affine transformation uncertainty polytopes. Computers & Graphics 30(1), 77–85 (2006)

Linear Boundary and Corner Detection Using Limited Number of Sensor Rows

Bishal Prasad[1], Arijit Bishnu[2], and Tetsuo Asano[3]

[1] NVIDIA graphics Pvt. Ltd., Senapati Bapat Road, Pune-411016, India
[2] Computer Science and Engineering Department, Indian Institute of Technology, Kharagpur, Kharagpur-721302
[3] Japan Advanced Institute of Science and Technology, Japan

Abstract. Linear boundary detection and corner detection are major challenges in computer vision. There exist many solutions for these problems based either on edge detection or interpolation methods but they are inexact in the sense that they do not talk of bounds. The basic objective of the study in this paper is to find out how exactly we can locate or restore linear boundaries or corners in the real plane, rather than pixel domain, by observation at discrete pixels. This paper devises new algorithms for linear boundary detection and corner detection using computational and digital geometric techniques.

1 Introduction

Edges in an image are the set of pixels where the image intensity level undergoes a sharp variation. The edge detectors mostly produce for a pixel, a quantitative value proportional to the chances of that pixel to be an edge pixel and an orientation. Then, a threshold is applied to determine whether a pixel is an edge pixel. The quantitative values and orientations are obtained using some convolution operators, based on derivatives of the image. The edge linking process is usually done by some variants of Hough transform [6]. Note that, for the edge detectors no apriori information is known whereas for the linking process some apriori information as to the nature of the curve of the object is known. In most practical industrial applications, one may assume that the objects to be recognized are known apriori. This gives rise to model based object recognition [4]. Minimizing errors by interpolation techniques and best-fit criteria to an apriori model is the guiding principle behind these recognition techniques. Bern and Goldberg [3] describe one such method where with an apriori knowledge of a rectangular paper they find out the continuous parameters of a best-fit rectangle from discrete observations of certain types of sensors. There are a host of other methods on interpolation techniques and corner detection. Due to shortage of space, we leave out their review here.

The problem we are concerned with in this paper relates to linear boundary detection and corner detection in an image taken by a Charge Coupled Device (CCD) camera.

V.E. Brimkov, R.P. Barneva, H.A. Hauptman (Eds.): IWCIA 2008, LNCS 4958, pp. 250–261, 2008.

1.1 CCD Camera and Image Formation

A CCD camera [1] contains an array of square sensors which emit photo-electrons proportional to the amount of light captured by the sensors. The voltages developed due to emission of photo-electrons from the sensors gets transformed to different levels of intensity in the corresponding pixels of the image. The image formation is performed by projecting the tessellation induced by the grid of sensors (equivalently, pixels) onto the object. To simplify the image formation model, we make the following assumptions:

- The tessellation is assumed to have all the squares of equal size.
- Each square in the tessellation imposed on the object plane has a one-to-one
    correspondence to a sensor pixel i.e., the sensor pixel captures light reflected
    from the corresponding square in the tessellation only.
- There is no loss of light photons in the way from the object to the sensors.

The left part of Fig. 1 shows an object that is a paper whose one side is completely black and the other side is completely white demarcated by a straight line. On the right, is the tessellation induced by the grid sensor pixels on the object. Under the above assumptions, the following observations can be made about the intensity value at any pixel (i, j). Assume that the demarcating line has a slope of $\tan \theta$ and each square grid has a length of a units. For a square grid (pixel) lying in the white region, the intensity is maximum i.e., $a^2 * 1 = a^2$ assuming intensity value of 1 per unit area for the white portion. For a square lying in the black region the intensity is minimum, i.e., $a^2 * 0 = 0$ assuming intensity of 0 for each square. For a square which is intersected by the line, two possible cases are shown in Fig. 2, and the intensity is proportional to the white area.

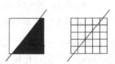

Fig. 1. Left: Paper with one part white and other part black; Right: Grid of sensor pixels

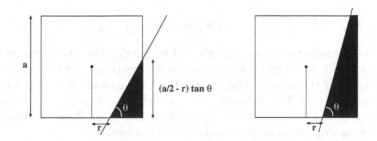

Fig. 2. Two different cases of a line intersecting a sensor

Hence, the intensity at a sensor pixel at position (i,j) is given by the following equation.

$$S_{i,j} = \begin{cases} a^2 - \frac{1}{2}(\frac{a}{2} - r)^2 \tan\theta & \text{if } (\frac{a}{2} - r)\tan\theta < a \\ \frac{1}{2}a^2(1 + \cot\theta) + ar & \text{otherwise .} \end{cases} \tag{1}$$

This is the basic model of image formation that we assume throughout the paper.

1.2 Our Contribution

The basic objective of the study in this paper is to find out how exactly we can locate or restore linear boundaries or corners in $\mathbb{R}^2$ (in the 2D real plane) described by continuous parameters by using observation at discrete pixels. In the first problem dealt in Section 2, we have a rectangular sheet of paper that is partitioned into black and white regions by a straight line on the sheet. With the knowledge that the pixel intensities are formed as mentioned in Section 1.1, we find out the parameters of the equation of the demarcating line. This problem, though very simple, shows the motivation behind our work in this paper. In the second problem in Section 3, we consider the same object but now consider quantization error. So, the pixel intensity values will not be exact. In this case, we can at best hope for a bound on the parameters of the line. In Section 4, we assume the same model of image formation without quantization error but now the input object has a black side and a white side demarcated by two lines meeting at a corner. In this problem, the goal is to locate the real coordinates of the corner from the intensity values.

2   Linear Boundary Detection without Quantization Error

Problem 1. Given a grid of squares each having edge length a with the intensity values at pixels formed as stated in Section 1.1, find the equation of the demarcating line by using as less number of sensors as possible.

Consider two adjacent sensors (shown in Fig. 3) through both of which the unknown line passes. As the line spans through the entire image, such a pair of sensors can be found by scanning an entire row for two neighboring pixels which has intensity value in $(0, a^2)$.

The value of intensity at these squares (proportional to the white area as shown in Fig. 3) in terms of the unknown θ and r are as follows.

$$S_{i,j} = a^2 - \frac{1}{2}(\frac{a}{2} + r)^2 \cot\theta \quad and \quad S_{i+1,j} = \frac{1}{2}(\frac{a}{2} - r)^2 \cot\theta$$

These two intensity values are known and by solving the above two equations we get r and θ. Let the line joining the two centres of neighboring sensors (i, j) and $(i + 1, j)$ cut the demarcating line at **P** which is at a distance of d from the centre of sensor (i, j). The distance d (as shown in Fig. 3) is $d = \frac{a}{2} - r\cot\theta$. This gives the co-ordinates of the point **P** as shown in Fig. 3. Using the coordinates of **P** and θ, the equation of line can be found out exactly.

Result 1. *The number of sensors required to find the demarcating line is two.*

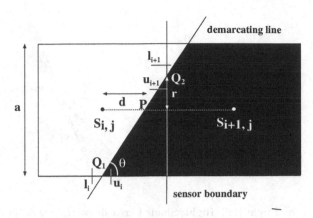

Fig. 3. Boundary sensor pixels. d denotes the distance of the centre of the square grid from P, the point where the demarcating line cuts the line joining the two neighboring centres.

3   Linear Boundary Detection with Quantization Error

Because of quantization, a range of intensity values are mapped to a single value. Owing to this, any point $\mathbf{P}$ as found out in Section 2, does not remain unique, but becomes an interval through which the demarcating line passes.

If the number of bits used for digital value is n and the maximum analog voltage is V, then each discrete digital value can be mapped to a range of real values and that range is $V/(2^n - 1)$. Thus each of the intensity values $S_{i,j}$ is a range

$$[V/(2^n - 1)] * k \le S_{i,j} \le [V/(2^n - 1)] * (k + 1) \qquad where \ 0 \le k \le 2^n - 1$$

Writing d, as defined in the previous section, in terms of $S_{i,j}$ and $S_{i+1,j}$ we get,

$$d = [(S_{i,j} + S_{i+1,j})/a] - (a/2) \tag{2}$$

Hence, d and thus $\mathbf{P}$ lie in a range of $2V/(2^n - 1)a$.

3.1   Lower and Upper Bound on Slope of Line

In a way as mentioned above, we can find intervals around two points Q_1 and Q_2 (as shown in Fig. 3) where the line cuts the boundaries of the sensors. If two such points Q_1 and Q_2 are considered, a candidate line cannot be determined exactly but will pass through the intervals around Q_1 ($[l_i, u_i]$) and Q_2 ($[l_{i+1}, u_{i+1}]$) (as shown in Fig. 3). The candidate lines that can pass through such intervals will stab the shaded region shown in Fig. 4. If we consider an image with N rows, Q_1 and Q_2 can come from any of the $\binom{N}{2}$ rows; and each of them gives rise to a region like the one shown in Fig. 4. The line to be detected has to stab all such regions.

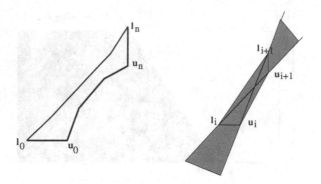

Fig. 4. Left: Left and right hull. Right: Shaded area shows the region pertaining to the candidate lines.

Problem 2. Given a set of ranges for each sensor (pixel) row through which the demarcating line passes, determine the upper and lower bounds on the parameters of the demarcating line.

Let $[l_i, u_i]$ define an interval through which the demarcating line passes. We define two types of convex chains - **Left Hull($\mathcal{LH}$)** and **Right Hull($\mathcal{RH}$)**. $\mathcal{LH}(\mathcal{RH})$ is the lower(upper) part of the convex hull of the points $l_i(r_i)$ where $1 \le i \le N$ and N is the total number of ranges (same as the number of rows). A region $\mathcal{R}$ enclosed by $\mathcal{LH}$ and $\mathcal{RH}$ can be defined as shown in the left part of Fig. 4.

Lemma 1. *Any line passing through the ranges $[l_i, u_i]$ must stab through $\mathcal{R}$.*

Proof. The proof is easy and is omitted. See left part of Fig. 5.   □

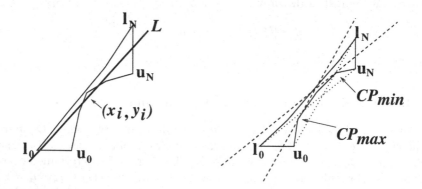

Fig. 5. Left: A line which does not stab the area enclosed by left and right hull Right: Critical points

3.2   Algorithm

In a convex hull, the subsequent line segments in the upper or lower chain have either monotonically increasing or monotonically decreasing slopes. We use this property to find the bound. The line segments in $\mathcal{LH}$ and $\mathcal{RH}$ can be used to formulate an incremental algorithm. The maximum possible slope (L_{max}) of any line passing through $\mathcal{R}$ is the slope of the line connecting l_0 and u_N while the minimum possible slope (L_{min}) is the slope of the line connecting l_N and u_0. To find the upper bound on the slope, scanning starts at the line segment having maximum slope and subsequent line segments are scanned iteratively till a line segment with slope less than L_{max} is found. This slope is an upper bound. Similarly lower bound can be found by iteratively scanning monotonically increasing slopes untill a line segment with slope greater than L_{max} is found. But this bound is quite loose and can be further tightened by considering constant c(intercept) also. We introduce the concept of *critical point* here. Two types of critical points on each of the hulls are defined - CP_{max} and CP_{min}. CP_{max} is a tangential point of contact between $\mathcal{RH}(\mathcal{LH})$ and the line with slope L_{max}. Similarly, CP_{min} is a tangential point of cantact point between $\mathcal{RH}(\mathcal{LH})$ and the line with slope L_{min}. We describe how CP_{max} and CP_{min} can be used to find the upper and lower bounds on the slopes.

While finding a tighter upper bound, we use the loose bound on the slope i.e. m found earlier. To check whether a line $y = mx + c$ is a candidate line, we just need to check whether a line with slope m and passing through CP_{max} passes through the range (l_0, u_0) and the range (l_N, u_N). If this line does not pass through these ranges, then there exists no c corresponding to that value of m and the next slope in the convex hull is considered. Proceeding this way, a value of m can be found for which a legitimate c value exists, and that is the upper bound on the slope. Similarly using point CP_{min}, the lower bound on slope can be found.

The construction of the convex hull chains takes $O(N \log N)$ time [2]. The time complexity of the iterative algorithm described above is $O(N)$. To find out the intervals $[l_i, u_i]$ for each row, the row is to be scanned once taking $O(N)$ time. Under the assumption that there are N columns, the total time taken is $O(N^2)$. Hence, the complete process takes $O(N^2)$ time.

3.3   Proof of Correctness

To prove the correctness of the algorithm, first the choice and use of CP_{max} and CP_{min} is justified. We prove that there exists no point p' (p'') other than CP_{max} (CP_{min}) on $\mathcal{RH}(\mathcal{LH})$ so that if a candidate line l passes through CP_{max} (CP_{min}) and another candidate line l' (l'') passes through p' (p''), then slope of l' (l'') is greater (smaller) than slope of l. The proof for the case of CP_{max} is given. The other can be proved similarly. Let us consider two points on the right hull - one point is CP_{max} and other(called p') is second farthest point from L_{max}. Since p' is second farthest point after CP_{max}, it is easy to see that there is no need to prove it for any other point on the hull. Now considering CP_{max} and p' there can be two cases as shown below in Fig. 6.

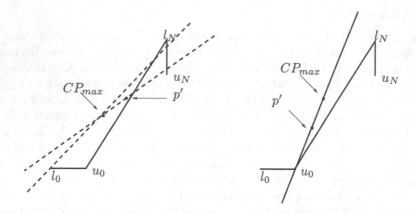

Fig. 6. Two different alignments of points CP_{max} and p'

Case 1: Point p' lies to the right of point CP_{max}. Considering the definition of CP_{max}, it is easy to see that of all the lines which pass through p' and are also candidate lines(i.e. stab through the region $\mathcal{R}$), the line with maximum slope will also have to pass through CP_{max}. If this is not so, it either violates the definition of CP_{max} or does not stab the region $\mathcal{R}$. Thus, there is a candidate line passing through CP_{max} which has atleast as great slope as any candidate line passing through p'. There can be other candidate lines passing through CP_{max} which have greater slope.

Case 2: Point p' lies to the left of point CP_{max}. As can be seen easily from the figure, any candidate line passing through p' must pass through CP_{max}, otherwise it will not stab through $\mathcal{R}$. But considering the definition of CP_{max}, such a line will have a slope greater than L_{max} and hence it can be concluded that no candidate line passes through such a point p'.

From these two cases, it is clear that there exists no point p' other than CP_{max} on the right hull so that if a candidate line l passes through CP_{max} and another candidate line l' passes through p' then slope of l' is greater than slope of l.

From the above proof, we can assert that if we want to find whether there exists a c corresponding to a particular m so that a line $y = mx + c$ is a candidate line or not, then we only need to find whether a line with slope m and passing through CP_{max} (or CP_{min}) passes through ranges (l_0, u_0) and (l_N, u_N). From the above discussions, we have the following result.

Result 2. *The upper and lower bounds on the parameters m and c of the demarcating line $y = mx + c$ can be found in $O(N^2)$ time for an $N \times N$ image by determining the interval $[l_i, u_i]$ for each sensor row through which the demarcating line passes. This interval can be found from the knowledge of the quantization mechanism.* $\qquad\square$

4   Corner Detection without Quantization Error

A *corner* is defined as the intersection of two edges. We propose a method based on manhattan chain and digital straight lines [7].

Problem 3. Given a grid of squares each having edge length a with the intensity values at pixels formed as stated in Section 1.1, find the equation of the two demarcating lines (separating a black and white side) that form a corner.

4.1   Manhattan Chain and Digital Straight Lines

A manhattan chain consists of horizontal and vertical grid edges as shown in Fig. 7. We can have a manhattan chain forming the envelope of a line l as shown in Fig. 7. Using Freeman chain coding [5], a manhattan chain is a string over the alphabets in $\{0, 1, \ldots, 7\}$; the alphabets implying 8 directions. A manhattan chain corresponding to a digital straight line(DSL) has some properties as discussed next. For a detailed discussion on this, refer [7].

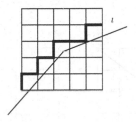

Fig. 7. A manhattan chain(shown in bold) forming an envelope of the two demarcating lines that form a chain denoted as l

A chain code sequence c is a chain code for a DSL if the following conditions are satisfied. The code c_i is the ith reduced form of c.

(A1)   There are at most two different letters a and b in c_n, and if there are two, then $\|a - b\| = 1$ (counting modulo 8 in the case of c_0) where $n \geq 0$.

(A2)   If there are two different letters in c_n, then at least one of them is singular in c_n.

This definition derives a digital straight segment(DSS) property that allows the formulation of a necessary and sufficient condition for such chain code sequences. Let $l(s)$and $r(s)$ denote the run lengths of non-singular letters to the left of the first singular letter, or to the right of the last singular letter, respectively, for a finite word s.

Definition 1. *A finite chain code sequence c satisfies the* DSS *property iff $c_0 = c$ satisfies conditions (A1) and (A2), and any nonempty sequence $c_n = R(c_{n-1})$, for $n \geq 1$, satisfies (A1) and (A2) and the following two conditions:*

(B1) *If c_n contains only one letter a, or two different letters a and $a + 1$, then $l(c_{n-1}) \le a + 1$ and $r(c_n - 1) \le a + 1$.*

(B2) *If c_n contains two different letters a and $a + 1$, and a is non-singular in c_n, then if $l(c_{n-1}) = a + 1$ then c_n starts with a, and if $r(c_n - 1) = a + 1$ then c_n ends with a.*

An online *DSS* recognition algorithm reads the successive chain codes $c_0, c_1, \ldots$ and determines the maximum $k \ge 0$ such that $c_0, c_1, \ldots, c_k$ is a *DSS* but $c_0, c_1, \ldots, c_k, c_{k+1}$ is not. For a review of a host of such algorithms, most of which take linear time, see [7].

4.2   Exact Location of Corner

Based on the image formation model, we will have a manhattan chain that demarcates the region between pixels having intensity in $(0, a^2)$. We analyze this chain for its digital straightness properties to compute the corner. Consider two lines l_1 and l_2 forming a corner point p in Fig. 8.

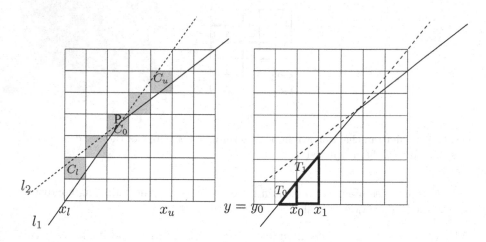

Fig. 8. Left: Corner P with the range $[C_l, C_u]$. Right: Triangles T_0 and T_1 for the corner are shown.

Let l_1 be $y = \alpha_1 x + \beta_1$ and l_2 be $y = \alpha_2 x + \beta_2$. Then it can be clearly seen for this case that for $x \in [x_l, x_u]$, $\|\alpha_1 x + \beta_1 - \alpha_2 x - \beta\| < 1$. In other words, both the lines lie in the same cell (in this section sensor pixels are called as cells) for certain range $[x_l, x_u]$ of x. Now, if the algorithm working on *DSS* property is used to identify the maximal length of straight line l_1, then it will identify the cell C_u as the cell at which the straightness of line l_1 is lost. Similarly, it will return the cell C_l as the cell at which the straightness of line l_2 is lost. But none of these might be the cell in which the corner lies.

4.2.1 Finding a Range for Corner Detection

Lemma 2. *The corner lies between a cell C_l on the lower(left) side and a cell C_u on the upper(right) side.*

Proof. The on-line *DSS* recognition algorithm [7] reads the manhattan chain code one code at a time and decides whether the code read uptil now corresponds to a *DSS*. Since the method is incremental, it is obvious that the corner lies in a cell to the left (assuming that the code is read from left to right) of C_u. Similarly, if the same algorithm is run on the chain code or the line l_2, the corner lies to the right of C_l. Thus, the corner lies in the range of cells C_l and C_u. $\square$

C_l and C_u obtained from Lemma 2 can be used for detecting a range for the corner.

4.2.2 Finding the Exact Cell

Consider the situation depicted in Fig. 8. Each of the cell is a square of side length a. Consider the triangle T_0 formed by the intersection of $l_1, y = y_0$ and $x = x_0$. Let the x-intercept made by l with $y = y_0$ with respect to $x = x_0$ is α. Then, the intensity value for the triangle T_0 is equal to the area of the triangle which is

$$S_0 = \frac{1}{2}\alpha\beta_0 \quad \Longrightarrow \quad \beta_0 = \frac{2S_0}{\alpha} \tag{3}$$

Similarly, for triangle T_1 formed by intersection of $l_1, y = y_0$ and $x = x_1$, its intensity value can be written as

$$S_1 = \frac{1}{2}(\alpha + a)\beta_1 \quad \Longrightarrow \quad \beta_1 = \frac{2S_1}{\alpha + a} \tag{4}$$

If l_1 does not contain a corner between $x = x_{-1}$ and $x = x_1$, then triangles T_1 and T_2 are similar triangles. Using the basic property of similar triangles,

$$\frac{\alpha}{\beta_0} = \frac{\alpha + a}{\beta_1} \tag{5}$$

Using equations 3, 4 and 5 we get

$$\alpha = \frac{a}{\sqrt{\frac{S_1}{S_0}} - 1} \tag{6}$$

Now consider a triangle T_n formed by $l_1, y = y_0$ and $x = x_n$. Using the same logic as above

$$\beta_n = \frac{2S_n}{\alpha + na}$$

Using similarity between triangles T_n and T_0, we get the value of α as

$$\alpha = \frac{na}{\sqrt{\frac{S_n}{S_0}} - 1} \tag{7}$$

If l_1 is a straight line between $x = x_0$ and $x = x_n$, then the two values of α in equations 6 and 7 will be same. Hence, this becomes a test for straightness of line. This test used incrementally, by checking for all possible triangles $T_0, T_1, \ldots, T_n, \ldots$, will lead to find the exact cell which contains the corner. The straightness is tested using each triangle formed by x_i incrementally where $i = 0, 1, \ldots, n, \ldots$. Let x_i be the value for which this test fails. Then the corner lies in the cell lying between $x = x_{i-1}$ and $x = x_i$ through which line l_1 passes. We need to ensure that the initial value of α found by using triangles T_0 and T_1 must be correct i.e. the corner must not lie to the left of $x = x_1$. So, given a range $[C_l, C_u]$, which is the candidate range for existence of a corner, the process of finding α must start from two cells to the left of C_l (or two cells to the right of C_u).

We also need to consider the increment of triangles along the Y-axis and this time α will be the intercept along the Y-axis.

Lemma 3. *Using increment of triangles along both X-axis and Y-axis results in a unique cell which contains the corner.*

Proof. Let line l_1 pass through more than one cell for the same pair of X-axis points (x_i, x_{i+1}). Obviously these cells are vertically on top of each other. Let the Y-axis points for the ceiling of these cells be $y_p, y_{p+1}, y_{p+2}, \ldots, y_{p+r}$. When the triangle increment method described earlier is applied along X-axis, it indicates that one of these vertical cells contains the corner. But, to pinpoint the exact cell, we need to apply this triangle increment method along Y-axis also. At each step, the similar triangle condition is checked. Let the cell whose ceiling is y_{p+q} contains the corner(call it C_q). Then, since the x co-ordinates of these cells does not change, all the cells below C_q will satisfy the similar triangle property but the cell C_q will not. And C_q can be correctly diagnosed as the cell containing the corner. □

This idea of using incremental construction along both the axes is used in the algorithm to find exact cell containing the corner. The algorithm to find the exact cell, given the range $[C_l, C_u]$ as input, is shown below.

Algorithm

1. Move to two cells left(or right) of C_l. Let $x = x_0$ be the ceiling(floor) X-axis of that cell. Let $y = y_0$ be the floor(ceiling) Y-axis of that cell.
2. Let α be the the x-intercept made by the demarcating line $l_1(l_2)$ with $y = y_0$ with respect to $x = x_0$. Calculate the value of α using equation 6. Set $i = 2$.
3. Calculate the new value of α using equation 7 and incrementing i by 1. If this value is same as in step 2, then continue with step 3 else goto 4.
4. If number of cells which lie between x_{i-1} and x_i and though which l_1 passes is more than one then goto 5 else return with the unique cell.
5. Let γ be the intercept made by the damarcating line with $x = x_{-1}$ with respect to $y = y_0$. Repeat steps 3 and 4 for the similar case of γ. Return with unique cell.

time after the manhattan chain has been found since both the on-line *DSS* recognition and increment of triangles can be found in linear time.

4.2.3 Finding the Exact Location

If there is no quantization error, it is possible to find the exact location of the corner in the real plane. Once the cell containing the corner is located, it can be used to efficiently determine the slope of the line l_1 as described in Section 2. Similarly, the slope of line l_2 can be efficiently determined. Once these two lines are determined, their intersection point(i.e. the corner) is easily determined.

Result 3. *The exact location of a corner can be found by analyzing a manhattan chain for* DSS *properties. After that, the minimum number of cells required is four more than the number of cells in the range returned by the DSS algorithm.*   □

5   Conclusion and Discussions

This paper presented ways to use discrete information like intensity values at sensor pixels based on camera models to find out bounds on the continuous parameters of the linear boundary or corner. Algorithms for linear boundary detection both in the absence and presence of quantization error have been proposed. Methods for corner detection in the absence of quantization error has also been proposed. Studies on corner detection in the presence of quantization error is in progress.

References

1. Anatomy of a Camera, OpenDocument, `http://zone.ni.com/devzone/conceptd.nsf/webmain/ba741c90a118ea778625685e00805643?`
2. De Berg, M., Kreveld, M., Overmars, M., Schwarzkopf, O.: Computational Geometry - Algorithms and Applications. Springer, Berlin (2000)
3. Bern, M., Goldberg, D.: Paper positioning System. In: ACM Symposium on Computational Geometry, Barcelona, Spain, pp. 74–81 (June 2002)
4. Chin, R.T., Dyer, C.R.: Model-based recognition in robot vision. ACM Computing Surveys 18(1), 67–108 (1986)
5. Freeman, H.: On the encoding of arbitrary geometric configurations. IRE Trans. Elecron. Comput. EC-10, 260–268 (1961)
6. Illingworth, J., Kittler, J.: A survey of the Hough transform. Computer Vision, Graphics, and Image Processing 44(1), 87–116 (1988)
7. Klette, R., Rosenfeld, A.: Digital Geometry: Geometric Methods for Digital Picture Analysis, New Delhi, India. Morgan Kaufman, Elsevier (2005)

A Convergence Proof for the Horn-Schunck Optical-Flow Computation Scheme Using Neighborhood Decomposition

Yusuke Kameda[1], Atsushi Imiya[2], and Naoya Ohnishi[1]

[1] School of Science and Technology, Chiba University, Japan
[2] Institute of Media and Information Technology, Chiba University, Japan

Abstract. In this paper, we prove the convergence property of the Horn-Schunck optical-flow computation scheme. Horn and Schunck derived a Jacobi-method-based scheme for the computation of optical-flow vectors of each point of an image from a pair of successive digitised images. The basic idea of the Horn-Schunck scheme is to separate the numerical operation into two steps: the computation of the average flow vector in the neighborhood of each point and the refinement of the optical flow vector by the residual of the average flow vectors in the neighborhood. Mitiche and Mansouri proved the convergence property of the Gauss-Seidel- and Jacobi-method-based schemes for the Horn-Schunck-type minimization using algebraic properties of the matrix expression of the scheme and some mathematical assumptions on the system matrix of the problem. In this paper, we derive an alternative proof for the original Horn-Schunck scheme. To prove the convergence property, we develop a method of expressing shift-invariant local operations for digital planar images in the matrix forms. These matrix expressions introduce the norm of the neighborhood operations. The norms of the neighborhood operations allow us to prove the convergence properties of iterative image processing procedures.

1 Introduction

In this paper, we prove the convergence property for the Horn-Schunck optical-flow computation scheme. First, we derive a proof for the original Horn-Schunck scheme. Second, we evaluate the convergence rate. Finally, we introduce a method of selecting the regularization parameter for accurate computation.

The main idea of the Horn-Schunck method for optical-flow computation is the decomposition of the Laplacian to the neighborhood average and the subtraction of the value at each point from the neighborhood average. Then, they derived the Jacobi method for optical-flow computation. Therefore, in this paper, we clarify the mathematical properties and evaluate the operator norm of the neighborhood operations in digital image processing.

In signal processing and analysis, it is well known that a shift-invariant linear operation is expressed as a convolution kernel. Furthermore, a linear transform

V.E. Brimkov, R.P. Barneva, H.A. Hauptman (Eds.): IWCIA 2008, LNCS 4958, pp. 262–273, 2008.

in a finite dimensional space is expressed as a matrix [2,17,6]. It is also possible to express a shift-invariant operation as a band-diagonal matrix [3,2,16]. However, this expression is not usually used in signal processing and analysis. In numerical computation of the partial differential equations, approximations of the partial differentiations in discrete operations are one of the central issues [4,7,17]. The discrete approximations of the partial differentiations are called the neighborhood operations in digital signal and image processing. To analyze and express digital image transformations from the viewpoint of functional analysis, we introduce a method of describing the neighborhood operations in the matrix forms.

Optical flow is an established method of motion analysis in computer vision [8,11,1] and has been introduced to many application areas such as cardiac motion analysis [14,18], robotics [12,15,5], and visualization in chemical sciences [13]. However, there still exist mathematical problems concerning to accurate and stable computation of optical-flow. There are two types of evaluation methods on the schemes for optical-flow computation. The first one is a numerical-based analysis of the accuracy of the solution using normalized phantoms, that is, an evaluation of the differences between the numerical results and the ground truth for the synthetic data images with a predesigned motion field. The second one is a mathematical-theory-based evaluation, that is, clarification of the convergence and stability of the algorithm employing numerical analysis. From the viewpoint of mathematical-theory-based evaluation, we derive the convergence property on a variational optical-flow computation method proposed by Horn and Schunck [8].

Horn and Schunck derived the Jacobi-based-method for the computation of the optical-flow vector of each point [8] as the motion of each point on the image[1]. The basic idea of the Horn-Schunck scheme is to separate the numerical operation into two steps: the computation of the average flow vector in the neighborhood of each point and the refinement of the optical flow vector at each point by the residual of the average flow vectors in the neighborhood. In their paper [8], the mathematical proof for the convergence property of the algorithm was not dealt with. The convergence of the scheme was later examined numerically [1]. Therefore, it might be understood that the convergence of the scheme depends on the input images. The first numerical scheme for computing optical-flow is later

---

[1] In their original paper[8], they said

> We now have a pair of equations for each point in the image. It would be very costly to solve these equations simultaneously by one of the standard methods, such as Gauss-Jordan elimination [11, 13]. The corresponding matrix is sparse and very large since the number of rows and columns equals twice the number of picture cells in the image. Iterative methods, such as the Gauss-Seidel method [11, 13], suggest themselves.

However, the method has the same structure as the iteration form $f_i^{(n+1)} = \frac{1}{2}(f_{i+1}^n + f_{i-1}^n) - g_i$ for solving the equation $g_i = \frac{1}{2}(f_{i+1} - 2f_i + f_{i-1})$, which is derived as the numerical approximation of $g = \frac{d^2}{d^2 x} f$.

extended to the three-dimensional problem for the computation of cardiac optical flow. Mitiche and Mansouri [10] proved the convergence property of the Gauss-Seidel-method-based scheme for the Horn-Schunck-type minimization using the algebraic property that the large system matrix of the problem is symmetry. Furthermore, they proved the convergence property for the Jacobi-type scheme of the Horn-Schunck-type minimization. In this paper, we derive an alternative proof for the original Horn-Schunck scheme and evaluate the convergence rate. Furthermore, we introduce a method of selecting the regularization parameter which guarantees accurate computation.

2   Optical Flow Computation

2.1   Optical Flow and Regularization

For functions in two-dimensional Euclidean space $\mathbf{R}^2$, setting $f(\boldsymbol{x}-\boldsymbol{u},t+1)$ and $f(\boldsymbol{x},t)$ to be the images at times $t+1$ and t, the small displacement $\boldsymbol{u}$ of each point $\boldsymbol{x}$ is called the optical flow of the image f. For a spatio-temporal image $f(\boldsymbol{x},t)$, $\boldsymbol{x} = (x,y)^\top$, the total derivative is given as

$$\frac{d}{dt}f = \nabla f^\top \boldsymbol{u} + \frac{\partial f}{\partial t}\frac{dt}{dt}, \tag{1}$$

where $\boldsymbol{u} = \dot{\boldsymbol{x}}$ is the motion of each point $\boldsymbol{x}$. Optical flow constraint [8,11,1] $\frac{d}{dt}f = 0$ implies that the motion $\boldsymbol{u}$ of the point $\boldsymbol{x}$ is the solution of the singular equation, $\nabla f^\top \boldsymbol{u} + f_t = 0$.

To solve this equation, the regularization method is employed to minimize the criterion

$$J(\boldsymbol{u}) = \int_{\mathbf{R}^2} \left\{ (\nabla f^\top \boldsymbol{u} + f_t)^2 dx + \alpha tr \nabla \boldsymbol{u} \nabla \boldsymbol{u}^\top \right\} d\boldsymbol{x}, \tag{2}$$

where $\boldsymbol{u}$ is the vector gradient of vector $\boldsymbol{u}$, which is given as $\nabla \boldsymbol{u} = (\nabla u, \nabla v)$ for $\boldsymbol{u} = (u,v)^\top$. We call, in this paper, optical-flow computation by the minimization of eq. (2) the Horn-Schunck method. Furthermore, the numerical algorithm to solve eq. (2) is called the Horn-Schunck scheme for optical flow computation.

The Euler-Lagrange equation of the energy function of eq. (2) is

$$\Delta \boldsymbol{u} = \frac{1}{\alpha}(\nabla f^\top \boldsymbol{u} + f_t)\nabla f = \frac{1}{\alpha}(\boldsymbol{S}\boldsymbol{u} + f_t \nabla f), \tag{3}$$

where $\boldsymbol{S} = \nabla f \nabla f^\top$ is called the structure tensor of f at point $\boldsymbol{x}$. We adopt the natural boundary condition $\frac{\partial}{\partial \boldsymbol{n}} f = 0$, where $\boldsymbol{n}$ is the unit outward normal vector on the boundary of the domain.

2.2   The Horn-Schunck Scheme

We assume that the sampled image $f(i,j)$ exists in the $M \times M$ grid region, that is, we express f_{ij} as the value of $f(i,j)$ at the point $(i,j)^\top \in \mathbf{Z}^2$. The natural boundary condition, that is, the Neumann condition, for discrete flow vectors

$$\boldsymbol{u}_{ij} = (u(i,j), v(i,j))^\top = (u_{ij}, v_{ij})^\top, \quad i,j = 1, 2, \cdots, M \tag{4}$$

is

$$u_{11} - u_{22} = 0, \qquad u_{1j} - u_{2j} = 0, \qquad u_{1M} - u_{2M-1} = 0,$$
$$u_{i1} - u_{i2} = 0, \qquad\qquad\qquad u_{Mj} - u_{M-1j} = 0, \tag{5}$$
$$u_{M1} - u_{M-12} = 0, \, u_{Mj} - u_{M-1j} = 0 \, u_{MM} - u_{M-1M-1} = 0.$$

The discrete version of eq. (3) becomes,

$$Lu_{ij} = \frac{1}{\alpha}(S_{ij}u_{ij} + s_{ij}), \quad S_{ij} = \nabla f_{ij}\nabla f_{ij}^{\top}, \quad s_{ij} = (\partial_t f)_{ij}\nabla f_{ij}, \tag{6}$$

where $f_{ij} = f(i, j, t)$ is the sampled function of $f(x, y, t)$ at time t.
Setting $N_4(f_{ij})$ to be the operation to compute

$$N_4(f_{ij}) = \text{av}_4 f_{ij} = \frac{1}{4}\left(f_{i+1\,j} + f_{i-1\,j} + f_{i\,j+1} + f_{i\,j-1}\right), \tag{7}$$

the Laplacian operation L with the four-neighborhood is expressed as

$$Lf_{ij} = \text{av} f_{ij} - f_{ij}. \tag{8}$$

Using N_4, eq. (6) is rewritten as

$$\alpha(N_4(u_{ij}) - u_{ij}) = (S_{ij}u_{ij} + s_{ij}), \; 2 \leq i, j \leq M - 1. \tag{9}$$

Setting $N_4(u_{ij}) = \overline{u}_{ij}$, we have the equation

$$(\alpha I_2 + S)u_{ij} = \alpha\overline{u}_{ij} - s_{ij}. \tag{10}$$

For the matrix $T_{ij} = trS \times I - S_{ij}$ using the relation,

$$(\alpha I + S_{ij})(\alpha I + T_{ij}) = \alpha(\alpha + trS_{ij})I, \tag{11}$$

we have

$$u_{ij} = \overline{u}_{ij} - \frac{1}{\alpha + trS_{ij}}(S_{ij}\overline{u}_{ij} + s_{ij}), \tag{12}$$

and the iteration form

$$u_{ij}^{(m+\frac{1}{2})} = N_4 u_{ij}^{(m)}$$
$$u_{ij}^{(m+1)} = u_{ij}^{(m+\frac{1}{2})} - \frac{1}{\alpha + trS_{ij}}(S_{ij}u_{ij}^{(m+\frac{1}{2})} + s_{ij}). \tag{13}$$

In the original Horn-Schunck scheme, the first equation of the iteration is the weighted summation in the eight-neighborhood of the point $(i, j)^{\top}$. Then, setting N to be an appropriate operation to compute the weighted summation in an appropriate neighborhood, we replace the first equation to

$$u_{ij}^{(m+\frac{1}{2})} = N u_{ij}^{(m)}. \tag{14}$$

Therefore, we have the iteration form

$$
u_{ij}^{(m+1)} = u_{ij}^{(m+\frac{1}{2})} - \frac{1}{\alpha + tr S_{ij}}(S_{ij} u_{ij}^{(m+\frac{1}{2})} + s_{ij}),
$$

$$
u_{ij}^{(m+\frac{1}{2})} = N u_{ij}^{(m)}, \quad \text{if } 2 \leq i, j \leq M - 1,
$$

$$
\begin{aligned}
&u_{11}^{(m+1)} = u_{22}^{(m+1)}, \; u_{1j}^{(m+1)} = u_{2j}^{(m+1)}, \quad u_{1M}^{(m+1)} = u_{2M-1}^{(m+1)}, \\
&u_{i1}^{(m+1)} = u_{i2}^{(m+1)}, \qquad\qquad\qquad\quad u_{Mj} = u_{M-1j}^{(m+1)}, \qquad \text{otherwise.}\,(15) \\
&u_{M1}^{(m+1)} = u_{M-12}^{(m+1)}, \; u_{Mj}^{(m+1)} = u_{M-1j}^{(m+1)}, \; u_{MM}^{(m+1)} = u_{M-1M-1}^{(m+1)},
\end{aligned}
$$

The second term of the right-hand side of eq. (13) is

$$
\frac{1}{\alpha + tr S_{ij}}(S_{ij} u^{(m+\frac{1}{2})} + s_{ij}) = \frac{1}{\alpha + tr S_{ij}}(\nabla f_{ij}^{\top} u^{(m+\frac{1}{2})} + \partial_t f_{ij})\nabla f_{ij}). \tag{16}
$$

Equation (16) implies the next property.

Proposition 1. *If $u_{ij}^{(m+\frac{1}{2})}$ is the solution of the equation*

$$
\nabla f_{ij}^{\top} u + (\partial_t f)_{ij} = 0, \tag{17}
$$

then we have the relation $u_{ij}^{m+1} = u_{ij}^{(m+\frac{1}{2})}$, that is, the iteration does not update the flow vector of the point.

3   Matrix Expression of Problem

3.1   Matrix Expressions of Neighborhood Operations

Since the second-order discrete differentiation is

$$
\partial_2 u = \frac{u(i+1) - 2u(i) + u(i-1)}{2}, \tag{18}
$$

the $M \times M$ second-derivative matrix is tridiagonal [4,7]. For Dirichlet and Neumann boundary conditions, the derivative matrices are

$$
D_1 = \frac{1}{2}\begin{pmatrix} -2 & 1 & 0 & 0 & \cdots & 0 & 0 \\ 1 & -2 & 1 & 0 & \cdots & 0 & 0 \\ 0 & 1 & -2 & 1 & \cdots & 0 & 0 \\ \vdots & \vdots & \vdots & \vdots & \ddots & \vdots & \vdots \\ 0 & 0 & 0 & \cdots & 0 & 1 & -2 \end{pmatrix} \quad D_2 = \frac{1}{2}\begin{pmatrix} -1 & 1 & 0 & 0 & \cdots & 0 & 0 \\ 1 & -2 & 1 & 0 & \cdots & 0 & 0 \\ 0 & 1 & -2 & 1 & \cdots & 0 & 0 \\ \vdots & \vdots & \vdots & \vdots & \ddots & \vdots & \vdots \\ 0 & 0 & 0 & \cdots & 0 & 1 & -1 \end{pmatrix},
$$
$$\tag{19}$$

respectively. Using D_1 and D_2, the discrete Laplacian operations for two-dimensional discrete functions with the Dirichlet and Neumann boundary conditions, are expressed as

$$
L_1 = I_M \otimes D_1 + D_1 \otimes I_M, \quad L_2 = I_M \otimes D_2 + D_2 \otimes I_M, \tag{20}
$$

for $1 \leq i, j \leq M$, respectively, where $\boldsymbol{I}_n$ is the $n \times n$ identity matrix and $\boldsymbol{A} \otimes \boldsymbol{B}$ is the Kronecker product of matrices $\boldsymbol{A}$ and $\boldsymbol{B}$ [4].

Setting

$$
\boldsymbol{B}_1 = \frac{1}{2}
\begin{pmatrix}
0 & 1 & 0 & 0 & \cdots & 0 & 0 \\
1 & 0 & 1 & 0 & \cdots & 0 & 0 \\
0 & 1 & 0 & 1 & \cdots & 0 & 0 \\
\vdots & \vdots & \vdots & \vdots & \ddots & \vdots & \vdots \\
0 & 0 & 0 & \cdots & 0 & 1 & 0
\end{pmatrix}, \quad
\boldsymbol{B}_2 = \frac{1}{2}
\begin{pmatrix}
1 & 1 & 0 & 0 & \cdots & 0 & 0 \\
1 & 0 & 1 & 0 & \cdots & 0 & 0 \\
0 & 1 & 0 & 1 & \cdots & 0 & 0 \\
\vdots & \vdots & \vdots & \vdots & \ddots & \vdots & \vdots \\
0 & 0 & 0 & \cdots & 0 & 1 & 1
\end{pmatrix}, \tag{21}
$$

the matrix $\boldsymbol{N}_\varepsilon$, $\varepsilon \in \{1,2\}$,

$$
\boldsymbol{N}_\varepsilon = (\boldsymbol{B}_\varepsilon \otimes \boldsymbol{I}_M + \boldsymbol{I}_M \otimes \boldsymbol{B}_\varepsilon), \tag{22}
$$

is the averaging operation in the four-neighborhood of each point with Dirichlet and Neumann boundary conditions, respectively.

Let $\rho(\boldsymbol{A})$ be the spectrum of the matrix $\boldsymbol{A}$. Since

$$
\boldsymbol{B}_\varepsilon = \boldsymbol{D}_\varepsilon + \boldsymbol{I}_M, \tag{23}
$$

$\boldsymbol{N}_\varepsilon$ satisfies the property $\rho(\boldsymbol{N}_\varepsilon) < 1$. The discrete Laplacian $\boldsymbol{L}_\varepsilon$ is expressed as

$$
\boldsymbol{L}_\varepsilon \boldsymbol{u} = \boldsymbol{N}_\varepsilon \boldsymbol{u} - \boldsymbol{u}. \tag{24}
$$

3.2   Discrete Model

For the sampled optical flow vector $\boldsymbol{u}_{ij} = (u_{ij}, v_{ij})^\top$, we define two vectorizations of the sampled function as

$$
\boldsymbol{v} =
\begin{pmatrix}
\boldsymbol{u}_{11} \\
\boldsymbol{u}_{12} \\
\vdots \\
\boldsymbol{u}_{MM}
\end{pmatrix}
= \text{vex}\left(\boldsymbol{u}_{11}, \boldsymbol{u}_{12}, \ldots, \boldsymbol{u}_{MM}\right) =
\begin{pmatrix}
u_{11} \\
v_{11} \\
u_{12} \\
v_{12} \\
\vdots \\
u_{MM} \\
v_{MM}
\end{pmatrix} \tag{25}
$$

and

$$
\boldsymbol{u} = \text{vec}(\boldsymbol{u}_{11}, \boldsymbol{u}_{12}, \cdots, \boldsymbol{u}_{MM}) = \text{vec}
\begin{pmatrix}
\boldsymbol{u}_{11}^\top \\
\boldsymbol{u}_{12}^\top \\
\vdots \\
\boldsymbol{u}_{MM}^\top
\end{pmatrix}
= \text{vec}
\begin{pmatrix}
u_{11} & v_{11} \\
u_{12} & v_{12} \\
\vdots & \vdots \\
u_{MM} & v_{MM}
\end{pmatrix}. \tag{26}
$$

For these vectorizations, we define the permutation $\boldsymbol{P}$ as

$$
\boldsymbol{P}\boldsymbol{v} = \boldsymbol{u}. \tag{27}
$$

For the vector function $\boldsymbol{u}_{ij} = (u_{ij}, v_{ij})^\top$ on the discrete plane $\mathbf{Z}^2$, we have the matrix equation for the optical flow computation as

$$L_\varepsilon \boldsymbol{u} = \frac{1}{\alpha}S\boldsymbol{u} + \frac{1}{\alpha}\boldsymbol{s}, \quad \varepsilon \in \{1, 2\} \tag{28}$$

for $1 \le i, j \le M$ and $\varepsilon \in \{1, 2\}$, where

$$\boldsymbol{L} := \boldsymbol{I}_2 \otimes \boldsymbol{L}_\varepsilon \tag{29}$$
$$\boldsymbol{S} = \boldsymbol{P}^\top Diag(\boldsymbol{S}_{11}, \boldsymbol{S}_{12}, \cdots, \boldsymbol{S}_{MM})\boldsymbol{P} \tag{30}$$

$$\boldsymbol{s} = \text{vec}\begin{pmatrix} s_{11}^\top \\ s_{12}^\top \\ \vdots \\ s_{MM}^\top \end{pmatrix} = \boldsymbol{Pt} \tag{31}$$

$$\boldsymbol{t} = \text{vec}\begin{pmatrix} s_{11} & s_{12} & \cdots & s_{MM} \end{pmatrix}. \tag{32}$$

4   The Horn-Schunck Scheme with Four-Neighborhood

Using $\boldsymbol{N}_\varepsilon$, the matrix form of the Horn-Schunck scheme is expressed as

$$\boldsymbol{u}^{(m+1)} = \boldsymbol{N}_4\boldsymbol{u}^{(m)} - \boldsymbol{P}^\top \boldsymbol{F}^{-1}\boldsymbol{P}(\boldsymbol{S}\boldsymbol{N}_4\boldsymbol{u}^{(m)} + \boldsymbol{s}), \tag{33}$$

where

$$\boldsymbol{F} = \alpha\boldsymbol{I} + Diag\,(tr\boldsymbol{S}_{11}\boldsymbol{I}_2, tr\boldsymbol{S}_{12}\boldsymbol{I}_2, \cdots, tr\boldsymbol{S}_{MM}\boldsymbol{I}_2)$$
$$= \alpha\boldsymbol{I} + Diag(\boldsymbol{S}_{ij}) = Diag(\alpha\boldsymbol{I}_2 + \boldsymbol{S}_{ij}) \tag{34}$$
$$\boldsymbol{F}^{-1} = Diag\left(\frac{1}{\alpha + tr\boldsymbol{S}_{11}}\boldsymbol{I}_2, \frac{1}{\alpha + tr\boldsymbol{S}_{12}}\boldsymbol{I}_2, \cdots, \frac{1}{\alpha + tr\boldsymbol{S}_{MM}}\boldsymbol{I}_2\right). \tag{35}$$

Horn and Schunck [8] derived eq. (33) for the pointwise expression. From these expression, we have the relations

$$\begin{aligned} \boldsymbol{u}^{(m+1)} - \boldsymbol{u}^{(m)} &= \boldsymbol{N}_4(\boldsymbol{u}^{(m)} - \boldsymbol{u}^{(m-1)}) - \boldsymbol{P}^\top \boldsymbol{F}^{-1}\boldsymbol{PS}(\boldsymbol{N}_4(\boldsymbol{u}^{(m-1)}) - \boldsymbol{N}_4(\boldsymbol{u}^{(m)})) \\ &= (\boldsymbol{I} - \boldsymbol{P}^\top \boldsymbol{F}^{-1}\boldsymbol{PS})\boldsymbol{N}_4(\boldsymbol{u}^{(m-1)} - \boldsymbol{u}^{(m)}) \\ &= (\boldsymbol{I} - \boldsymbol{P}^\top \boldsymbol{F}^{-1}Diag(\boldsymbol{S}_{ij})\boldsymbol{P})\boldsymbol{N}_4(\boldsymbol{u}^{(m-1)} - \boldsymbol{u}^{(m)}) \\ &= \boldsymbol{P}^\top \left\{\boldsymbol{I} - \boldsymbol{F}^{-1}Diag(\boldsymbol{S}_{ij})\right\}\boldsymbol{PN}_4(\boldsymbol{u}^{(m-1)} - \boldsymbol{u}^{(m)}) \end{aligned} \tag{36}$$

and

$$|\boldsymbol{u}^{(m+1)} - \boldsymbol{u}^{(m)}| \le \rho(\boldsymbol{I} - \boldsymbol{F}^{-1}Diag(\boldsymbol{S}_{ij}))\rho(\boldsymbol{N}_4)|\boldsymbol{u}^{(m)} - \boldsymbol{u}^{(m-1)}| \tag{37}$$

Here, $\rho(\boldsymbol{N}_4) < 1$ and

$$\rho(\boldsymbol{I} - \boldsymbol{F}^{-1}Diag(\boldsymbol{S}_{ij})) = \max_{ij}\left|1 - \frac{tr\boldsymbol{S}_{ij}}{\alpha + tr\boldsymbol{S}_{ij}}\right|. \tag{38}$$

Since $tr\boldsymbol{S} \geq 0$ and $\alpha > 0$, we have

$$0 < \frac{tr\boldsymbol{S}_{ij}}{\alpha + tr\boldsymbol{S}_{ij}} < 1, \tag{39}$$

and

$$\max_{ij} \left| 1 - \frac{tr\boldsymbol{S}_{ij}}{\alpha + tr\boldsymbol{S}_{ij}} \right| < 1 \tag{40}$$

From these analysis, we have the convergence theorem.

Theorem 1. *Let the discrete Laplacian on the plane be*

$$f(i,j) = \frac{1}{4} \left(f(i+1,j) + f(i-1,j) + f(i,j+1) + f(i,j-1) \right) - f(i,j).$$

Then, the classical Horn-Schunck schemes with Dirichlet and Neumann boundary conditions generate sequences of solutions which converge to the solutions of the discrete equation of the Euler-Lagrange equation for $\alpha > 0$.

Equation (36) shows that the convergence rate of the Horn-Schunck scheme depends on the spectral radii of $\boldsymbol{F}^{-1}\boldsymbol{S}$. This expression of the convergence rate implies that the physical dimension of the regularization parameter α is the same as that of $tr\boldsymbol{S}$. Therefore, if $\alpha \ll \max tr\boldsymbol{S}_{ij}$ and $\alpha \gg \min tr\boldsymbol{S}_{ij}$ the vector $\boldsymbol{u}$ becomes the normal vector $\boldsymbol{n}$ and the eigenvector of the operation $\boldsymbol{L}$, respectively. Furthermore, setting $\alpha = k \times tr\boldsymbol{S}$ for $1 \leq k \leq 10$, we have the relation $0.5 \leq \rho \leq 0.9$. From this property of the spectral radii of $\boldsymbol{H}$ and $\boldsymbol{N}_1$, we have the following assertion.

Assertion 1. *For the accurate achievement of the Horn-Schunck scheme for optical flow computation, we are required to select*

$$\alpha = O(tr\boldsymbol{S}_{ij}) = O(|\nabla f|^2) = O(|\nabla f|_{\max}^2). \tag{41}$$

This assertion defines a criterion for the selection of the regularization parameter α.

5   Eight-Neighborhood Scheme

In this section, we derive a matrix expression of the operation corresponding to the original Horn-Schunck scheme and show that the spectrum of this matrix is less than 1, that is, we prove that the original Horn-Schunck scheme converges. Equation (33) is expressed as

$$\boldsymbol{u}^{(m+\frac{1}{2})} = \boldsymbol{N}_4 \boldsymbol{u}^{(m)}$$
$$\boldsymbol{u}^{(m+1)} = \boldsymbol{u}^{(m+\frac{1}{2})} - \boldsymbol{P}^\top \boldsymbol{F}^{-1} \boldsymbol{P}(\boldsymbol{S}\boldsymbol{u}^{(m+\frac{1}{2})} + \boldsymbol{s}). \tag{42}$$

This expression is generalized as

$$\boldsymbol{u}^{(m+\frac{1}{2})} = \boldsymbol{N}\boldsymbol{u}^{(m)}$$
$$\boldsymbol{u}^{(m+1)} = \boldsymbol{u}^{(1+\frac{1}{2})} - \boldsymbol{P}^\top \boldsymbol{F}^{-1} \boldsymbol{P}(\boldsymbol{S}\boldsymbol{u}^{(m+\frac{1}{2})} + \boldsymbol{s}). \tag{43}$$

Using the matrix expression N of the operation to compute the weighted average in the neighborhood of each point, we derive a matrix expression of the operation corresponding to the original Horn-Schunck scheme. Then we prove that the spectrum of this matrix is less than 1, that is, we prove that the original Horn-Schunck scheme converges.

Let M_2 be the matrix operation which computes the average of $f(i,j)$ on four points,

$$(i-1, j+1)^\top,\ (i+1, j+1)^\top$$
$$(i-1, j-1)^\top,\ (i+1, j-1)^\top.$$

These four points are in the four-neighborhood in the coordinate system

$$k = i+j,\quad m = i-j. \tag{44}$$

Furthermore, in the (k, m)-coordinate, the boundary condition is the Dirichlet condition on the edge and the Neumann condition on the vertices. Therefore, M_1 is expressed as

$$M_1 = B_1 \otimes B_1. \tag{45}$$

This geometrical property of the points in the neighborhood of a point on the discrete plane implies that the spectral radius of matrix M_1 is less than 1. Furthermore, the average operation in the eight-neighborhood is expressed as

$$N_8^{a\,b} u = (aN_2 u + bM_1)u,\ \ a+b=1,\ a>b>0. \tag{46}$$

Since

$$\rho(N_8^{a\,b}) \le a\rho(N_1) + b\rho(M_2) < 1, \tag{47}$$

the original Horn-Schunck scheme converges for the Dirichlet condition. In the original Horn-Schunck numerical scheme, the parameters a and b were selected to be $\frac{2}{3}$ and $\frac{1}{3}$. Therefore, we have the local operation on the discrete Laplacian as

$$Lf(i,j) = \mathrm{av}_8 f(i,j) - f(i,j), \tag{48}$$

for

$$\mathrm{av}_8 f(i,j) = \frac{1}{12} \begin{pmatrix} f(i-1,j+1) & +2f(i,j+1) & +f(i+1,j+1) \\ +2f(i-1,j) & & +2f(i+1,j) \\ +f(i+1,j+1) & +2f(i+1,j+1) & +f(i+1,j+1) \end{pmatrix}. \tag{49}$$

If $i,j < 1$ and $M < i,j$, we set $f(i,j) = 0$.

Theorem 2. *The classical Horn-Schunck scheme for the two-dimensional problem generates sequences of solutions which converge to the solutions of the discrete equation of the Euler-Lagrange equation for $\alpha > 0$.*

6   Conclusions

In this paper, we directly proved the convergence property of the optical-flow computation without any assumptions on the system matrices. Furthermore, we introduced an iteration form which does not depend on the images. Moreover, we showed that the selection method of the regularization parameter which guarantees accurate and stable computation.

This research was performed using the support by Grants-in-Aid for Scientific Research from JSPS Japan.

References

1. Barron, J.L., Fleet, D.J., Beauchemin, S.S.: Performance of optical flow techniques. International Journal of Computer Vision 12, 43–77 (1994)
2. Bellman, R.: Introduction to Matrix Analysis, 2nd edn. SIAM, Philadelphia (1987)
3. Bracewell, R.N.: Two Dimensional Imaging. Prentice-Hall, Englewood Cliffs (1995)
4. Demmel, J.W.: Applied Numerical Linear Algebra. SIAM, Philadelphia (1997)
5. Enkelmann, W.: Obstacle detection by evaluation of optical flow fields from image sequences. Image and Vision Computing 9, 160–168 (1991)
6. Grossmann, Ch., Terno, J.: Numerik der Optimierung. Teubner (1997)
7. Grossmann, Ch., Roos, H.-G.: Numerik partieller Differentialgleichungen. Trubner (1994)
8. Horn, B.K.P., Schunck, B.G.: Determining optical flow. Artificial Intelligence 17, 185–204 (1981)
9. Hwang, S.-H., Lee, U.K.: A hierarchical optical flow estimation algorithm based on the interlevel motion smoothness constraint. Pattern Recognition 26, 939–952 (1993)
10. Mitiche, A., Mansouri, A.-R.: On convergence of the Horn and Schunck optical-flow computation. IEEE, Image Processing 13, 848–852 (2004)
11. Nagel, H.-H.: On the estimation of optical flow: Relations between different approaches and some new results. Artificial Intelligence 33, 299–324 (1987)
12. Ohnishi, N., Imiya, A.: Featureless robot navigation using optical flow. Connection Science 17, 23–46 (2005)
13. Ruhnau, P., Knohlberger, T., Schnoerr, C., Nobach, H.: Variational optical flow estimation for particle image velocimetry. Experiments in Fluids 38, 21–32 (2005)
14. Sachse, B.F.: Computational Cardiology, Modeling of Anatomy, Electrophysiology, and Mechanics. In: Sachse, F.B. (ed.) Computational Cardiology. LNCS, vol. 2966, Springer, Heidelberg (2004)
15. Santos-Victor, J., Sandini, G.: Uncalibrated obstacle detection using normal flow. Machine Vision and Applications 9, 130–137 (1996)
16. Strang, G., Nguyen, T.: Wavelets and Filter Banks. Wellesley-Cambridge (1997)
17. Varga, R.S.: Matrix Iteration Analysis, 2nd edn. Springer, Heidelberg (2000)
18. Zhou, Z., Synolakis, C.E., Leahy, R.M., Song, S.M.: Calculation of 3D internal displacement fields from 3D X-ray computer tomographic images. In: Proceedings of Royal Society: Mathematical and Physical Sciences, vol. 449, pp. 537–554 (1995)

Appendix

Original Formulation of Horn and Schunck

In the original paper [8], Horn and Schunck adopted the relation

$$\Delta u_{ij} \cong 3(\text{av}_8 u_{ij} - u_{ij})$$

for

$$\text{av}_8 f(i,j) = \frac{1}{12} \begin{pmatrix} f(i-1,j+1) & +2f(i,j+1) & +f(i+1,j+1) \\ +2f(i-1,j) & & +2f(i+1,j) \\ +f(i+1,j+1) & +2f(i+1,j+1) & +f(i+1,j+1) \end{pmatrix}.$$

If $i,j < 1$ nd $M < i,j$, we set $f(i,j) = 0$. Then, we have the approximate numerical Euler-Lagrange equation

$$3\alpha(\text{av}_8 u - u) = Su + s.$$

Therefore, replacing 3α with β we have the equation

$$\beta(\text{av}_8 u - u) = Su + s.$$

Inverse of a Matrix

For $S = ss^\top$ and $T = trS \times I - S$, we have the relation $S^2 = trS \times S$, $TS = 0$, and $Ts = 0$. Furthermore, for $\mu \geq 0$,

$$(I + \mu S)(I + \mu T) = (1 + \mu trS)I.$$

and

$$(I + \mu S)^{-1} = \frac{1}{1 + trS}(I + \mu T).$$

Since for S_{ij}, we have the orthogonal decomposition

$$S_{ij} = \begin{pmatrix} \frac{\nabla f_{ij}}{|\nabla f_{ij}|} & \frac{\nabla f_{ij}^\perp}{|\nabla f_{ij}^\perp|} \end{pmatrix} \begin{pmatrix} trS_{ij} & 0 \\ 0 & 0 \end{pmatrix} \begin{pmatrix} \frac{\nabla f_{ij}}{|\nabla f_{ij}|} & \frac{\nabla f_{ij}^\perp}{|\nabla f_{ij}^\perp|} \end{pmatrix}^\top$$

and

$$\frac{1}{\alpha + trS_{ij}} I_2 = \begin{pmatrix} \frac{\nabla f_{ij}}{|\nabla f_{ij}|} & \frac{\nabla f_{ij}^\perp}{|\nabla f_{ij}^\perp|} \end{pmatrix} \begin{pmatrix} \frac{1}{\alpha + trS_{ij}} & 0 \\ 0 & \frac{1}{\alpha + trS_{ij}} \end{pmatrix} \begin{pmatrix} \frac{\nabla f_{ij}}{|\nabla f_{ij}|} & \frac{\nabla f_{ij}^\perp}{|\nabla f_{ij}^\perp|} \end{pmatrix}^\top$$

we have the relation

$$F_{ij}^{-1} S_{ij} = \frac{1}{\alpha + trS_{ij}} I_2 S_{ij} = \begin{pmatrix} \frac{\nabla f_{ij}}{|\nabla f_{ij}|} & \frac{\nabla f_{ij}^\perp}{|\nabla f_{ij}^\perp|} \end{pmatrix} \begin{pmatrix} \frac{trS_{ij}}{\alpha + trS_{ij}} & 0 \\ 0 & 0 \end{pmatrix} \begin{pmatrix} \frac{\nabla f_{ij}}{|\nabla f_{ij}|} & \frac{\nabla f_{ij}^\perp}{|\nabla f_{ij}^\perp|} \end{pmatrix}^\top.$$

Therefore,

$$\rho(I_2 - F_{ij}^{-1} S_{ij}) = \left| 1 - \frac{trS_{ij}}{\alpha + trS_{ij}} \right|, \quad \rho(I - P^\top F^{-1} PS) = \max_{ij} \left| 1 - \frac{trS_{ij}}{\alpha + trS_{ij}} \right|.$$

Spectrums of Matrices

For the matrices D_1 and D_2, setting

$$D_1 U = \Lambda_1 U, \quad D_2 V = \Lambda_2 V$$

where U and V are orthogonal matrices, and

$$\Lambda_1 = Diag\left(\lambda_M^1, \lambda_{M-1}^1, \cdots, \lambda_1^1\right), \quad \Lambda_2 = Diag\left(\lambda_M^2, \lambda_{M-1}^2, \cdots, \lambda_1^2\right), \qquad (50)$$

the eigenvalues are

$$\lambda_k^1 = -\left(1 - \cos\frac{\pi}{M+1}k,\right), \quad \lambda_k^2 = -\left(1 - \cos\frac{\pi}{M}k,\right).$$

Since $B_\varepsilon = D_\varepsilon + I$, we have the eigenvalues of B_1 and B_2

$$\mu_k^1 = \cos\frac{\pi}{M+1}k, \quad \mu_k^2 = \cos\frac{\pi}{M}k.$$

Therefore, $\rho(B_1) < 1$ and $\rho(B_2) < 1$. Furthermore, the eigenvalues of N_2 and M_1 are $\mu_{ij} = \mu_i^2 + \mu_j^2$ and $\kappa_{ij} = \mu_i^1 \mu_j^1$. Therefore, we have the relation $\rho(N_2) < 1$ and $\rho(M_1) < 1$.

Topologically Correct 3D Surface Reconstruction and Segmentation from Noisy Samples

Peer Stelldinger

University of Hamburg, 22527 Hamburg, Germany

Abstract. Existing theories on 3D surface reconstruction impose strong constraints on feasible object shapes and often require error-free measurements. Moreover these theories can often only be applied to binary segmentations, i.e. the separation of an object from its background. We use the Delaunay complex and α-shapes to prove that topologically correct segmentations can be obtained under much more realistic conditions. Our key assumption is that sampling points represent object boundaries with a certain maximum error. We use this in the context of digitization, i.e. for the reconstruction based on supercover and m-cell intersection samplings.

1 Introduction

A fundamental question of image analysis is how closely a computed image segmentation corresponds to the underlying real-world partition. Existing geometric sampling theorems are limited to binary partitions, where the space is split into (not necessarily connected) fore- and background components. In this case, the topology of the partition is preserved under various reconstruction schemes when the original regions are sufficiently smooth and the sampling is dense enough, e.g. see [1,2] for the case of 3D surface reconstruction.

However, these results have two important limitations: they do not make any predictions about the consequences of measurement errors, and they are not applicable when there are more regions than just fore- and background. While the second limitation is still valid today, there exist solutions for the first one: recently alternative surface reconstruction methods have been developed, which can deal with measurement errors [8,10].

Digital images consist of a finite number of sampling values in a regular grid. Segmentation means to group the sampling points (i.e. pixels) into meaningful regions. These regions can completely be described by their boundary. Thus segmentation can also be done by reconstructing the segment boundaries based on the subset of sampling points, which lie near the boundary. Our treatment of adaptively placed sampling points on the boundary is inspired by research on laser range scanning. Here, a number of isolated sampling points is scattered over the surface of the object of interest, and the task is to reconstruct the surface from the set of points. A successful solution of this problem is the concept of α-shapes [5,6]. The α-shape is essentially defined as the subset of the Delaunay

V.E. Brimkov, R.P. Barneva, H.A. Hauptman (Eds.): IWCIA 2008, LNCS 4958, pp. 274–285, 2008.

triangulation of the points where the Delaunay cells' radius is below $\alpha \in \mathbb{R}^+$. Under certain conditions, an α-shape is homotopy equivalent or even homeomorphic to the desired object surface. By applying this idea to the problem of image segmentation, a new condition on object shape could recently be derived that ensures homotopy equivalence of the digital segmentation with the original analog partitioning of the space [13]. In this work we prove such properties even when the segmentation is subject to measurement errors.

2   Preliminaries

To segment a geometric image means to partition the image space (i.e. the domain of the image function) into meaningful regions. The image space does not have to be the two-dimensional plane, e.g. for CT or MRT scans it is the three-dimensional space. Each region corresponds to a relevant (part of an) object in the real world and its reconstruction should preserve as much properties as possible. The partition of the image space to be recovered is defined as follows:

Definition 1. *Let the image space I be $\mathbb{R}^n$ with $n \in \mathbb{N}$. A partition of the image space is defined by a finite set of pairwise disjoint regions $R = \{r_i \subset I\}$, such that each region $r_i \in R$ is a connected open set and the union of the closures of the regions covers the whole space, $\bigcup_i \overline{r_i} = I$. The boundary of the partition is $B := \bigcup_i \partial r_i$. Two regions r_i, r_j are called m-*neighbors *if the intersection $\overline{r_i} \cap \overline{r_j}$ contains an m-dimensional manifold with boundary, but no $(m+1)$-dimensional manifold with boundary. Two $(n-1)$-neighbors are also called* direct neighbors.

The simplest case of a partition is a *binary partition*, where the regions can be classified into foreground and background, such that every direct neighbor of a foreground region is a background region and vice versa. Then segmentation means separation of one (not necessarily connected) set from the background. Such a set is called a *shape*.

Most of the known results on topologically correct object or surface reconstruction are restricted to certain subclasses of shapes, having minimal bounds on the surface curvatures, like r-regular sets [9,4,12] or sets with certain local feature size [1,2]. This implies that regions cannot have corners, and junctions of three or more regions are impossible. These restrictions are somewhat relaxed by the notion of *r-halfregular partitions*, where an osculating r-ball must exist at least in the foreground *or* the background, and the topology must not change under either morphological opening or closing with a ball of radius $\leq r$ [11]. Corners are now possible, but the partition is still binary and has no junctions. In this paper, the class of feasible partitions of the space is extended as follows:

Definition 2. *A plane partition of the space is r-stable when its boundary B can be dilated with a closed disc of radius s without changing its homotopy type for any $s \leq r$.*

In other words, we can replace an infinitely thin boundary with a strip of width $2r$ such that the number and enclosure hierarchy of the resulting regions is

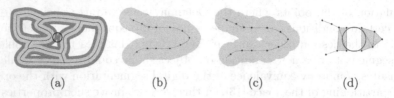

Fig. 1. (a) The homotopy type of an r-stable plane partition does not change when dilated with a disc of radius of at most r (light gray), while dilations with bigger radius (dark gray) may connect different parts as marked by the circle. The α-dilation (b) of the boundary of a two-dimensional α-stable partition may not be homotopy equivalent to the union (c) of the α-discs centered at the boundary sampling points. Thus the α-shape (d), which is always homotopy equivalent to the union of discs (c), may contain unwanted holes consisting of Delaunay triangles of radius greater than α. Thus there exists an α-disc centered in the hole which does not cover any boundary sampling point, as shown in (d).

preserved. In particular, "waists" are forbidden, whereas junctions are allowed, see Fig. 1(a). This includes r-regular and r-halfregular partitions, but also allows non-binary partitions and junctions and models real images much better. Since we want to deal with measurement errors (i.e. noise) when sampling the partition, we define a sampling of the surface as an approximation of the *boundary* of the partition with a finite set of *adaptively placed* sampling points. The sampling points are selected somehow "near" the boundary. We formalize this as follows:

Definition 3. *A finite set of sampling points $S = \{s_i \in \mathbb{R}^2\}$ is called a (p, q)-sampling of the boundary B when the distance of every point $b \in B$ to the nearest point in S is at most p, and the distance of every point $s \in S$ to the nearest point in B is at most q. The elements of S are called* boundary sampling points. *The sampling is said to be* strict *when all boundary sampling points are exactly on the boundary, i.e. $q = 0$.*

Non-zero values of q can be caused by systematic or statistical measurement errors, but also by the sampling method used. Boundary sampling points may be determined in various ways (section 3), but this only matters in so far as it determines the accuracy of the sampling, i.e. the values of p and q. Once computed, we consider boundary sampling points as isolated points that somehow define the digital boundary and connect them by means of the Delaunay complex. Each element of the Delaunay complex is either the convex hull of a finite subset of the sampling points, such that all chosen sampling points lie on the boundary of a common hypersphere and no other sampling point is inside the hypersphere, or the intersection of two other elements of the complex. The hypersphere center of such a *Delaunay cell* is called the *center point* of the cell and the hypershpere radius is also called the *radius* of the cell. An m-dimensional cell of the complex is called m-*cell*. In order to approximate the boundary of the partition, we want to remove those edges and triangles from the Delaunay triangulation that are not related to the boundary. A useful subset of the Delaunay complex is defined by the α-complex introduced in [5]:

Definition 4. *The α-complex $D_\alpha(S)$ of a set of points S is defined as the sub-complex of the Delaunay complex of S which contains all cells C such that*

- *the radius of the smallest sphere containing the sampling points of C is smaller than α, and it contains no other point of S, i.e. $C^0 \cap S = \emptyset$, or*
- *an incident cell C' with higher dimension is in $D_\alpha(S)$.*

The polytope $|D_\alpha(S)|$, i.e. the union of all elements of $D_\alpha(S)$, is called α-shape.

Since cells are removed from the Delaunay complex, the α-complex has holes which hopefully correspond to the regions we are trying to segment. In order to determine when this is the case, the following theorem is of fundamental importance (the proof can be found in [6]):

Theorem 1 (Edelsbrunner). *The union of closed α-balls with centers at the points $s_i \in S$ covers $|D_\alpha|$, and the two sets are homotopy equivalent.*

Consequently, the α-shape $|D_\alpha|$ is homotopy equivalent to the original partition of the space if and only if the dilation of the boundary sampling points with α-balls is homotopy equivalent to the boundary of the partition. This requirement is indeed fulfilled in certain situations: In [4] it is proved that $|D_\alpha|$ is even homeomorphic to B if B is the boundary of a two-dimensional r-regular set with $p < \alpha < r$ and $q = 0$. In three dimensions the authors recently derived an analog result [12]. There the α-shape itself cannot be guaranteed to be homeomorphic, but it can be used to derive a homeomorphic surface approximation in a very simple way: with defining the outer boundary of the α-shape as the union of all triangles of the corresponding α-complex, which can be seen from the outside (i.e. from a point being outside the original object), one gets the following theorem (the proof can be found in [12]):

Theorem 2 (Stelldinger). *Let A be a three-dimensional r-regular set and S be an α-sampling of its boundary ∂A such that $2\alpha < r$. Then the polytope $|D_\alpha|$ is of the same homotopy type as ∂A, and the outer boundary of $|D_\alpha|$ is homeomorphic to ∂A.*

Unfortunately, these approaches no longer apply when the original partition is not r-regular and/or the boundary sampling points are not exactly on the original boundary, i.e. they are noisy. Fig. 1(b)-(d) shows a two-dimensional example where the r-dilation of the boundary is homotopy equivalent to the boundary (i.e. the partition is r-stable), but the dilation of the boundary sampling points is not. This problem has already been solved for the two-dimensional case in [13]. There it is shown that by filling small regions one gets a boundary representation with correct homotopy type such that the separated regions are homeomorphic to the original ones. These small regions can uniquely be identified if the sampling is dense enough. In three or higher dimensional spaces filling such small regions is not enough, since reconstruction artifacts can have a more complicated topological structure, like e.g. tunnels or bridges. The rest of the paper is devoted to the question what can be said under these more general conditions in case of higher dimensional spaces.

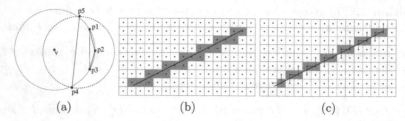

Fig. 2. (a) Any circumcircle around p_4 and p_5 contains p_1, p_2, and p_3 (see text). (b) The *supercover digitization* contains all sampling points whose pixel facets intersect the arc. (c) Where the boundary intersects the dual grid, the nearest sampling points form the *grid intersection digitization*.

As can be seen in Figure 1(b)-(d), if we have an object which is not r-regular, but also if the sampling is not strict, i.e. a (p,q)-sampling with $q > 0$, the complement of the α-shape reconstruction may have new small regions, which lie inside the α-dilation of the original boundary. In order to get a topologically correct boundary reconstruction we must at least fill these small regions. In the following, the components of $|D_\alpha(S)|^c$ will be called α-*holes*. As we will see, the spurious holes are restricted in their size. Thus we define (α, β)-holes in order to distinguish between spurious and wanted α-holes:

Definition 5. *Let $D_\alpha(S)$ be the α-complex of a sampling S and $|D_\alpha(S)|$ be its α-shape. Then the α-holes of $|D_\alpha(S)|$ are the components of $|D_\alpha(S)|^c$. The (α, β)-holes of $|D_\alpha(S)|$ are the α-holes H, where the largest radius of some n-cell in H is at least $\beta \geq \alpha$. The union of the α-shape $|D_\alpha|$ with all α-holes of D_α that are not (α, β)-holes is called the (α, β)-shape reconstruction.*

For simplicity, we also use the term "hole" for the component which contains the infinite region. It is an (α, β)-hole for arbitrary large β. It follows from Theorem 1 that there is a 1-to-1 relation between α-holes and the holes in the union of α-discs around the sampling points. The following lemma establishes that a similar relationship exists for (α, β)-holes (we prove the lemma for the n-dimensional case):

Lemma 1. *An α-hole h is an (α, β)-hole if and only if it contains a point v whose distance from the nearest sampling point is at least β.*

Proof. **I** $(d_H(v \in h, S) \geq \beta \Rightarrow h$ is an (α, β)-hole): when v is in the infinite region, the claim follows immediately. Otherwise, v is contained in some Delaunay n-cell. By assumption, the corners of this triangle must have distance $\geq \beta$ from v. Therefore, the radius of the n-cell must be at least β, and the claim follows. **II** (h is an (α, β)-hole $\Rightarrow \exists v \in h$ with $d_H(v, S) \geq \beta$): by assumption, the closure of h contains a Delaunay n-cell t with radius of at least β. Consider its center point v (i.e. the center of its Delaunay sphere). If it is within the n-cell t, it is also in h and the claim follows. Otherwise, it is at least in some (α, β)-hole, and we must prove that t is in the same hole. Suppose to the contrary that v and t are in different α-holes. Then there exists a Delaunay cell t' (this does not have to be an n-cell)between t and v whose covering radius (i.e. the radius of the smallest covering ball) is smaller than α. The corners of t' cannot be inside the Delaunay sphere of t because otherwise t would not be a Delaunay n-cell. t' cannot contain v because its covering radius would then be at least β. Now consider Figure 2(a). It shows for the two-dimensional case the 2-cell t with corners p_1, p_2, p_3 and its Delaunay circle (gray) with center point v. The points p_4 and p_5 are the end points of one side of t'. Their distance $|p_4p_5|$ must be greater than $|p_1p_3|$. Consequently, any covering circle with radius $\leq \alpha$ (dashed) around p_4 and p_5 contains t, contrary to the condition that it must not contain any other sampling point. This obviously also holds in higher dimensions. The claim follows from the contradiction.   □

Now we can use the notion of (α, β)-holes to "repair" α-complexes that contain too many holes. After filling all α-holes which are not (α, β)-holes we get a one-to-one-mapping of the components of ∂A^c to the components of the complement of the (α, β)-shape reconstruction:

Theorem 3. *Let $\mathcal{P}$ be an r-stable partition of the space $\mathbb{R}^n$, and S be a (p, q)-sampling of $\mathcal{P}$'s boundary B. Then the (α, β)-shape reconstruction $\mathcal{R}$ preserves connectivity and neighborhood relations and defines a one-to one-mapping of the (α, β)-holes of $\mathcal{R}$ to the regions r_i of $\mathcal{P}$, if (1) $p < \alpha \leq r - q$, (2) $\beta = \alpha + p + q$ and (3) every region r_i contains an open γ-disc with $\gamma \geq \beta + q > 2(p + q)$.*

Proof. Let U be the union of open α-balls centered at the points of S. Furthermore, let $B^{\oplus} = B \oplus \mathcal{B}^0_{\alpha+q}$ be the dilation of B with an open $(\alpha + q)$-ball, and $r_i^{\ominus} = r_i \ominus \mathcal{B}_{\alpha+q}$ the erosion of region $r_i \in \mathcal{P}$ with a closed $(\alpha + q)$-ball.

- According to the definition of a (p, q)-sampling, the dilation of B with a closed q-ball covers S. Consequently, $B^{\oplus}$ covers U. Therefore, U cannot have fewer connected components than $B^{\oplus}$. $B^{\oplus}$ has as many components as B due to the r-stability of the partition $\mathcal{P}$. Conversely, since $\alpha > p$, every open α-ball around a point of S intersects B, and the union U of these balls covers the entire boundary B. It follows that U cannot have more components than B. The number of components of B and U is thus equal. Due to the same homotopy types of U and $|D_\alpha|$ (according to Lemma 1), this also holds for the components of $|D_\alpha|$.
- Since $\mathcal{P}$ is r-stable with $r \geq \alpha + q$, each $r_i^{\ominus}$ is a connected set with the same topology as r_i. The intersection $r_i^{\ominus} \cap B^{\oplus}$ is empty, and $r_i^{\ominus}$ cannot intersect $U \subset B^{\oplus}$ and $|D_\alpha| \subset U$. Hence, $r_i^{\ominus}$ is completely contained in a single α-hole of $|D_\alpha|$.
- Due to condition 3, r_i contains a point whose distance from B is at least $\gamma = \beta + q$. Its distance from S is therefore at least $\gamma - q = \beta$. Due to Lemma 1, the α-hole which contains $r_i^{\ominus}$ is therefore also an (α, β)-hole.
- Since $B^{\oplus}$ covers U and U covers B, no (α, β)-hole can intersect both $r_i^{\ominus}$ and $r_j^{\ominus}$ ($i \neq j$). It follows from this and the previous observation, that every region r_i can be mapped to exactly one (α, β)-hole which will be denoted h_i.
- An α-hole that does not intersect any region $r_i^{\ominus}$ must be completely contained within $B^{\oplus}$. Every point $v \in B^{\oplus}$ has a distance $d < \alpha + q$ to the

nearest point of B. In turn, every point in B has a distance of at most p to the nearest point in S. Hence, the distance from v to the nearest point of S is $d' < \alpha + p + q = \beta$. According to Lemma 1, this means that an α-hole contained in $B^{\oplus}$ cannot contain an n-cell with radius β and cannot be an (α, β)-hole.

– The previous observation has two consequences: (i) All holes remaining in $\mathcal{R}$ intersect a region $r_i^{\ominus}$. Therefore, the correspondence between r_i and h_i is 1-to-1, and B and $|\mathcal{R}|$ enclose the same number of regions. (ii) All differences between $\mathcal{R}$ and D_{α} (i.e. all Delaunay cells re-inserted into $\mathcal{R}$) are confined within $B^{\oplus}$. This implies that $|\mathcal{R}|$ cannot have fewer components than $B^{\oplus}$ and B. Since all re-inserted cells are incident to D_{α}, $|\mathcal{R}|$ cannot have more components than $|D_{\alpha}|$, which has as many components as B (see first observation). Hence, B and $|\mathcal{R}|$ have the same number of components.

– Consider the components of the complement $(r_i^{\ominus})^C$ and recall that $r_i^{\ominus}$ is a subset of both r_i and h_i for any i. Since B and $|\mathcal{R}|$ have the same number of components, it is impossible for h_i^C to contain a cell that connects two components of $(r_i^{\ominus})^C$. This means that the sets r_i^C and h_i^C have the same number of components. This finally proves that the constructed one-to-one mapping preserves the neighborhood relations. □

Filling spurious holes in the α-shape reconstruction is a necessary step for getting a topologically correct boundary reconstruction. But for $n \geq 3$ there are also other problems regarding topology: although the (α, β)-shape reconstruction separates the different regions from each other, these regions may have small tunnels and/or other topological changes inside $B^{\oplus}$. In order to identify and remove these cases, we will at first apply a homotopy type preserving thinning:

We will denote an m-cell C in a cell complex D as *simple* if the number of cells of D which contain C is equal to one. Now the containing cell must be an $(m + 1)$-cell and the removal of the two cells does neither change the homotopy type of the complex nor the topology of the background regions. Now the thinning algorithm for the (α, β)-shape reconstruction is as follows:

1. Find all simple m-cells ($n > m \geq 0$) of the given (α, β)-shape reconstruction and put them in a priority queue (the sorting will be discussed below).
2. As long as the queue is not empty:
   (a) Get the m-cell e with the highest priority from the queue.
   (b) If e is not simple anymore, it has lost this property during the removal of other cells. Skip the following and recommence with step 2.
   (c) Otherwise, remove e and the adjacent $(m+1)$-cell $t \in \mathcal{R}$ from the boundary reconstruction.
   (d) Check whether the other cells adjacent to t have now become simple and put them in the queue if this is the case.

Obviously the algorithm terminates for any finite cell complex, and the resulting boundary reconstruction contains no n-cells, i.e. it is thin. Since we want a boundary reconstruction which is as simple as possible in a topological sense (i.e. as few as possible tunnels, etc.), but which still separates the different regions from each other, we want to have a cell complex, where every cell is adjacent to at least two different background components (i.e. regions of the (α, β)-shape reconstruction). Thus we remove every cell, which is not adjacent to two different background regions. Since an $(n-1)$-cell, which is not adjacent to two different background regions, will already be removed by the above thinning algorithm, we only have to check m-cells with $m < n-1$. This can be done locally, since these cells are characterized by having no adjacent $(m+1)$-cell in the complex. Thus the whole algorithm is as follows:

Thinned (α, β)-shape reconstruction algorithm:

1. Given a (p, q)-sampling S of the boundary of some partition of the space, compute the α-complex of S with some $\alpha > p$.
2. Add all cells to the complex, which belong to an α-hole which is no (α, β)-hole for $\beta = \alpha + p + q$.
3. Find all simple m-cells (for any m with $n > m \geq 0$) of the given (α, β)-shape reconstruction and put them in a priority queue (the sorting will be discussed below).
4. As long as the queue is not empty:
   (a) Get the m-cell e with the highest priority from the queue.
   (b) If e is not simple anymore, it has lost this property during the removal of other cells. Skip the following and recommence with step 4.
   (c) Otherwise, remove e and the adjacent $(m+1)$-cell $t \in \mathcal{R}$ from the boundary reconstruction.
   (d) Check whether the other cells adjacent to t have now become simple and put them in the queue if this is the case.
5. For m going from $n-2$ to 0 do:
   (a) Remove all m-cells of the complex, which do not have an adjacent $(m+1)$-cell in the complex.

Theorem 4. *Let $\mathcal{P}$ be an r-stable partition of the space $\mathbb{R}^n$, and S be a (p, q)-sampling of $\mathcal{P}$'s boundary B. Then the thinned (α, β)-shape reconstruction algorithm results in a cell complex D with $|D|$ having the same homotopy type as B, and the components of B^c are topologically equivalent to the components of $|D|^c$, if (1) $p < \alpha \leq r - q$, (2) $\beta = \alpha + p + q$ and (3) every region r_i contains an open γ-disc with $\gamma \geq \beta + q > 2(p+q)$.*

Proof. The resulting reconstruction obviously separates the components of $|D|^c$ from each other, which can be mapped one-to one onto the components of B^c. Since the α-ball reconstruction covers B, the α-shape reconstruction contains a polygonal surface which is of the same homotopy type as B. Thus also the (α, β)-shape reconstruction contains such a polygonal surface. Applying only the thinning algorithm results in a an object B' being a deformation retract of the (α, β)-shape reconstruction. Thus it also contains a polygonal surface B'', which is of the same homotopy type as B. This surface B'' is everywhere thin, i.e. any of its m-cells with $m \leq n-2$ has at least two neighboring $m+1$-cells in B''. Thus B'' remains unchanged during the cell complex simplification.

Any other cell of B' will be removed. Thus $B'' = |D|$. Since the boundaries of any component of B^c and the corresponding component of $|D|^c$ are not only of the same homotopy type but also homeomorphic (because both are composed of components of the same homotopy type, which are all $(n-1)$-dimensional manifolds without boundary), the components are also homeomorphic to each other.   □

The complexity of the algorithm is dominated by the Delaunay tetrahedrization which is known to be $O(n^2)$ in the number of sampling points. As far as region topology is concerned, the ordering of the m-cells in the priority queue is arbitrary. But we think, that orderings should be favored, which lead to visually appealing results, e.g. by emphasizing flat surfaces. This can be done in the following way:

Definition 6. *The minimal (α, β)-shape reconstruction is the result of the thinned (α, β)-shape reconstruction algorithm, when using radii of the simple m-cells as priority, i.e. m-cells with big radius are the first to be removed.*

When only using the thinning algorithm after (α, β)-shape reconstruction, the resulting regions are correctly separated from each other. Moreover by using the cell radii as ordering criterion, the resulting hypersurface is as smooth as possible, since the size of the $(n-1)$-cells is minimized. Thus, since a minimal boundary reconstruction is a shortest possible one with correct topology, the surviving edges connect sampling points closest to each other. Neighboring sampling points therefore align in an optimal way on the thinned boundary.

3   Application to Sampling Schemes

In Theorem 3, p and q are assumed given. We now make their meaning and consequences more intuitive, by computing them for two of the most common sampling schemes. Given a sampling grid S (e.g. a cubic grid), we want to define a subset of sampling points which approxiates the boundaries of a partition of the space. Obviously simple subset sampling of the boundary is not a good choice, since in general hardly no sampling point will lie exactly on this surface. Alternatively one can choose the set of sampling points whose voxels (Voronoi regions) intersect the boundary of the partition. Such a method is called *supercover sampling*, since it is related to supercover digitization [7].

Definition 7. *Let $S \subset \mathbb{R}^n$ be an r'-grid (i.e. the maximal distance from any point to the nearest sampling point is at most r'). The supercover sampling of a set $A \subset \mathbb{R}^n$ based on S is the set S' of all sampling points $s \in S$ whose hypervoxel intersects A, see Figure 2(b).*

Lemma 2. *The supercover sampling S' of a set A based on an r'-grid, is an (r', r')-sampling of A.*

Proof. The distance of any sampling point in S' to the nearest point in A can be at most r', since S' is based on an r'-grid. Since the hypervoxels of the supercover

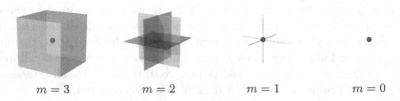

$$m = 3 \qquad m = 2 \qquad m = 1 \qquad m = 0$$

Fig. 3. Given s sampling point (s) in a cubic grid, the shaded region shows the intersection of $\mathcal{V}_S(s)$ with the union of all adjacent m'-cells of the Delaunay complex for $m' \leq m$

sampling cover A, the distance of any point of A to the nearest sampling point in S is also at most r'. □

If the object to be digitized is a curve in $\mathbb{R}^n$, the supercover sampling is $(n-1)$-connected (i.e. connected via cells of dimension of at least $n-1$), since for any point on the curve there exists a neighborhood, which is covered by the hypervoxels. However, one often wants a curve to be represented by a sampling, which is "as thin as possible", i.e. only 0-connected. This is fulfilled for the square grid in two dimensions by the grid intersection sampling [7], see Figure 2(c). It is well-known, that the grid intersection digitization is a subset of the supercover digitization on a square grid. The grid intersection sampling can easily be generalized to arbitrary grids in any dimension:

Definition 8. *Let $S \subset \mathbb{R}^n$ be a sampling grid. Further, for any sampling point $s \in S$ let $G_m(s)$ be the intersection of the hypervoxel $\mathcal{V}_S(s)$ with the the union of all m'-cells, $m' \leq m$, of the Delaunay complex of S being adjacent to s. Then the m-cell intersection sampling of a set $A \subset \mathbb{R}^n$ is defined as the union S' of all sampling points $s \in S$, where $G_m(s) \cap A$ is not empty, see Figure 2(c) for en example of the 1-cell intersection sampling in 2D.*

Thus the 1-cell intersection sampling based on a square grid is equal to the grid intersection sampling. Moreover the 0-cell intersection sampling is the same as the subset sampling and the n-cell intersection sampling equals the supercover sampling in $\mathbb{R}^n$. This directly implies that an m_1-cell intersection sampling is always subset of an m_2-cell intersection sampling of a given object, if they are based on the same sampling grid and if $m_1 < m_2$. Figure 3 shows $G_m(s)$ for different m in case of a cubic grid. While the m-cell intersection sampling of a connected set does not need to be connected in case of $m < n-1$, it is $(n-1)$-connected if $m = n$, since then it is equal to the supercover sampling. Moreover for $m = n-1$ it is 0-connected, if the grid is not degenerated:

Lemma 3. *Let S be the $(n-1)$-cell intersection sampling of a connected set based on a not degenerated grid. Then S is 0-connected.*

Proof. If A is empty or if A consists of only one point, the proof is obvious. Otherwise let x, y be two arbitrary points in A and let P be a path in A from

x to y. If x, y and p lie in the same n-cell of the Delaunay complex, there is nothing to show. Thus let p go through at least two n-cells of the Delaunay complex. Since the union of the $(n-1)$-cells of the complex is equal to the union of all $G_{n-1}(s)$, and since any two sampling points with a common n-cell in the Delaunay complex are always at least 0-connected, the $(n-1)$-cell intersection sampling of p is 0-connected. Thus S must also be 0-connected.   □

Now we will show that the $(n-1)$-cell intersection sampling is of higher sampling accuracy (i.e. lower q) than the supercover sampling, while the sampling density (i.e. the smallest possible value of p) is not as high.

Lemma 4. *Let S' be a not degenerated r'-grid. When each component of a set A is intersected by at least one $(n-1)$-cell of the Delaunay complex of S', the $(n-1)$-cell intersection digitization S of A based on the grid S', is a $(2r', q)$-sampling of A' with $q < r'$. If S' is a Cartesian r'-grid, S is even a $(2r', \frac{\sqrt{n-1}}{\sqrt{n}}r')$-sampling.*

Proof. The set of $(n-1)$-cells the Delaunay complex of a not degenerated grid partitions the space, such that any n-cell has a diameter of at most $2r'$. Thus, since each component of A intersects at least one $(n-1)$-cell, the distance of any point of A to the nearest sampling point of S is at most $2r'$.

For any sampling point $s \in S$ there exists an adjacent $(n-1)$-cell which intersects A inside the hypervoxel of S. Any such intersection point has distance of smaller than r' to s. In case of a Cartesian grid, the sidelength of the $(n-1)$-cells is $\frac{2}{\sqrt{n}}r'$ and the largest distance of a point in the intersection of the hypervoxel and an $(n-1)$-cell of s is $\frac{\sqrt{n-1}}{\sqrt{n}}r'$.   □

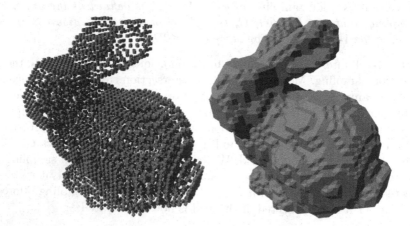

Fig. 4. Artificial boundary samples derived from a CT scan of the Stanford bunny (on courtesy of the Stanford Volume Data Archive graphics.stanford.edu/data/voldata/). Left: Sparse subset of boundary voxels due to 2-cell intersection sampling. Right: Result of thinned (α, β)-shape reconstruction.

We can now simply deal with sampling errors due to noise or blurring. We just have to add the expected positional error caused by these influences to the above computed positional error caused by the sampling.

4   Conclusions

This paper describes how to reconstruct a surface topologically correct from a sufficiently dense set of surface sampling points in the presence of measurement errors due to sampling but also due to noise and other influences. The theorem applies to a much wider class of shapes (r-stable partitions) than previous approaches. The situation in real images is thus modeled much more faithfully because shapes may now have corners and junctions, and standard segmentation algorithms can be used. Moreover an analysis on the amount of the measurement errors is given for the case of some of the mostly used digitization methods.

References

1. Amenta, N., Bern, M., Kamvysselis, M.: A new Voronoi-based Surface Reconstruction Algorithm. In: Proceedings of the 25th annual Conference on Computer Graphics and Interactive Techniques, pp. 415–421 (1998)
2. Amenta, N., Choi, S., Dey, T.K., Leekha, N.: A Simple Algorithm for Homeomorphic Surface Reconstruction. In: Proceedings of the 16th annual Symposium on Computational Geometry, pp. 213–222 (2000)
3. Attali, D.: r-Regular Shape Reconstruction from Unorganized Points. Computational Geometry: Theory and Applications 10(4), 239–247 (1998)
4. Bernardini, F., Bajaj, C.L.: Sampling and Reconstructing Manifolds Using Alpha-Shapes. In: Proc. 9th Canadian Conf. Computational Geometry (1997)
5. Edelsbrunner, H., Mücke, E.P.: Three-dimensional alpha shapes. ACM Trans. Graphics 13, 43–72 (1994)
6. Edelsbrunner, H.: The union of balls and its dual shape. Discrete Comput. Geom. 13, 415–440 (1995)
7. Klette, R., Rosenfeld, A.: Digital Geometry. Morgan Kaufman, San Francisco (2004)
8. Kolluri, R., Shewchuk, J.R., O'Brien, J.F.: Spectral surface reconstruction from noisy point clouds. In: Eurographics Symp. on Geom. Proc. (2004)
9. Latecki, L.J., Conrad, C., Gross, A.: Preserving Topology by a Digitization Process. J. Mathematical Imaging and Vision 8, 131–159 (1998)
10. Mederos, B., Amenta, N., Velho, L., de Figueiredo, L.H.: Surface reconstruction from noisy point clouds. In: Eurographics Symp. on Geom. Proc. (2005)
11. Stelldinger, P.: Digitization of Non-regular Shapes. In: Mathematical Morphology, Proc. of ISMM (2005)
12. Stelldinger, P.: Topologically Correct Surface Reconstruction Using Alpha Shapes and Relations to Ball-Pivoting (submitted, 2007)
13. Stelldinger, P., Koethe, U., Meine, H.: Topologically Correct Image Segmentation Using Alpha Shapes. In: Kuba, A., Nyúl, L.G., Palágyi, K. (eds.) DGCI 2006. LNCS, vol. 4245, Springer, Heidelberg (2006)

Detecting the Most Unusual Part of a Digital Image

Kostadin Koroutchev[1] and Elka Korutcheva[2,*]

[1] EPS, Universidad Autónoma de Madrid,
Cantoblanco, Madrid, 28049, Spain
k.koroutchev@uam.es
[2] Depto. de Física Fundamental,
Universidad Nacional de Educación a Distancia,
c/ Senda del Rey 9, 28080 Madrid, Spain
elka@fisfun.uned.es

Abstract. The purpose of this paper is to introduce an algorithm that can detect the most unusual part of a digital image. The most unusual part of a given shape is defined as a part of the image that has the maximal distance to all non intersecting shapes with the same form.

The method can be used to scan image databases with no clear model of the interesting part or large image databases, as for example medical databases.

1 Introduction

In this paper we are trying to find the most unusual/rare part with predefined size of a given image. If we consider an one-dimensional quasi-periodical image, as for example electrocardiogram (ECG), the most unusual parts with length about one second will be the parts that correspond to rhythm abnormalities [5]. Therefore they are of some interest. Considering two dimensional images, we can suppose that the most unusual part of the image can correspond to something interesting of the image.

Of course, if we have a clear mathematical model of what the interesting part of the image can be, it would be probably better to build a mathematical model that detects those unusual characteristics of the image part that are interesting. However, as in the case of ECG, the part that we are looking for, can not be defined by a clear mathematical model, or just the model can not be available. In such cases the most unusual part can be an interesting instrument for screening images.

To state the problem, we need first of all a definition of the term "most unusual part." Let us chose some shape S within the image A, that could contain that part and let us denote the cut of the figure A with shape S and origin r by $A_S(\rho; r)$, e.g.

$$A_S(\rho; r) \equiv S(\rho)A(\rho + r),$$

---

* E.K. is also with G.Nadjakov Inst. Solid State Physics, Bulgarian Academy of Sciences, 72, Tzarigradsko Schaussee Blvd., 1784 Sofia, Bulgaria.

V.E. Brimkov, R.P. Barneva, H.A. Hauptman (Eds.): IWCIA 2008, LNCS 4958, pp. 286–294, 2008.

where ρ is the in-shape coordinate vector, r is the origin of the cut A_S and we used the characteristic function $S(.)$ of the shape S. Further in this paper we will omit the arguments of A_S. We can suppose that the rarest part is the one that has the largest distance with the rest of the cuts with the same shape.

Speaking mathematically, we can suppose that the most unusual part is located at the point r, defined by:

$$r = \arg\max_{r} \min_{r':|r'-r|>\mathrm{diam}(S)} ||A_S(r) - A_S(r')||. \tag{1}$$

Here we assume that the shifts do not cross the border of the image. The norm $||.||$ is assumed to be L_2 norm[1].

Because the parts of an image that intersect significantly are similar, we do not allow the shapes located at r' and r to intersect, avoiding this by the restriction on $r' : |r' - r| > \mathrm{diam}(S)$.

If we are looking for the part of the image to be rare in a context of an image database, we can assume that further restrictions on r' can be added, for example restricting the search to avoid intersection with several images.

The definition above can be interesting as a mathematical construction, but if we are looking for practical applications, it is too strict and does not correspond exactly to the intuitive notion of the interesting part as there can be several interesting parts. Therefore the correct definition will be to find the outliers of the distribution of the distances between the blocks $||.||$.

If the figure has N^2 points, and $||S|| \ll ||A||$, in order to find deterministically the most interesting part, we need N^4 operations. This is unacceptable even for large images, not concerning image databases. Therefore we are looking for an algorithm that provides an approximate solution of the problem and solves it within some probability limit.

As is defined above in Eq.(1), the problem is very similar to the problem of location of the nearest neighbor between the blocks. This problem has been studied in the literature, concerning Code Book and Fractal Compression [1]. However, the problem of finding r in the above equation, without specifying r', as we show in the present paper, can be solved by using probabilistic methods avoiding slow calculations.

Summarizing the above statements, we are looking for an algorithm that two blocks are similar or different with some probability.

2   The Method

2.1   Projections

The problem in estimating the minima of Eq. (1) is complicated because the block is multidimensional. Therefore we can try to simplify the problem by

---

[1] Similar results are achieved with L_1 norm. The algorithm was not tested with L_{max} norm due to its extreme noise sensitivity. We use L_2 because of its relation with PSNR criteria that closely resembles the human subjective perception.

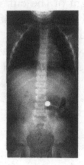

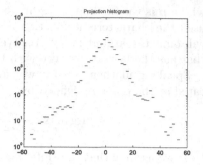

Fig. 1. The original test image. X-ray image of a person with ingested coin.

Fig. 2. The distribution of the projection value for square shape with a size 48x48 pixels

projecting the block $B \equiv A_S(r)$ in one dimension using some projection operator X. For this aim, we consider the following quantity:

$$b = |X.B_1 - X.B| = |X.(B_1 - B)|, \quad |X| = 1. \tag{2}$$

The dot product in the above equation is the sum over all ρ-s:

$$X.B \equiv \sum_\rho X(\rho)B(\rho; r).$$

If X is random, and uniformly distributed on the sphere of corresponding dimension, then the mean value of b is proportional to $|B_1 - B|$; $\langle b \rangle = c|B_1 - B|$ and the coefficient c depends only on the dimensionality of the block. However, when the dimension of the block increases, the two random vectors $(B_1 - B$ and $X)$ are close to orthogonal and the typical projection is small. But if some block is far away from all the other blocks, then with some probability, the projection will be large. The method resembles that of Ref. [4] for finding nearest neighbor.

As mentioned above we ought to look for outliers in the distribution. This would be difficult in the case of many dimensions, but easier in the case of one dimensional projection.

We will regard only projections orthogonal to the vector with components proportional to $X_0(\rho) = 1, \forall \rho$. The projection on the direction of X_0 is proportional to the mean brightness of the area and thus can be considered as not so important characteristics of the image. An alternative interpretation of the above statement is by considering all blocks to differ only by their brightness.

Mathematically the projections orthogonal to X_0 have the property:

$$\sum_\rho X(\rho) = 0. \tag{3}$$

The distribution of the values of the projections satisfying the property (3) is well known and universal [9] for the natural images. The same distribution seems to be valid for a vast majority of the images. The distribution of the projections derived for the X-ray image, shown in Fig. 1, is shown in Fig. 2.

Roughly speaking, the distribution satisfies a power law distribution in log-log scale if the blocks are small enough with exponential drop at the extremes. When the blocks are big enough, the exponential part is predominant.

If A_r and A'_r have similar projections, then they will belong to one and the same or to neighbors bins.

Therefore we can look for blocks that have a minimal number of similar and large projections. But these, due to the universality of the distribution, are exactly the blocks with large projection values.

As a first approximation, we can just consider the projections and score the points according to the bin they belongs to. The distribution can be described by only one parameter that, for convenience, can be chosen to be the standard deviation σ_X of the distribution of $X.B$.

The notion of "large value of the projection" will be different for different projections but will be always proportional to the standard deviation.[2] Therefore we can define a parameter a and score the blocks with $|X.B| > a\sigma_X$.

This procedure consists of the following steps:

0. Construct a figure B with the same shape as A and with all pixels equal to zero.

1. Generate a random projection operator X, with carrier with shape S, zero mean and norm one.

2. Project all blocks (convolute the figure). We denote the resulting figure as C.

3. Calculate the standard derivation σ_X of the result of the convolution.

4. For all points of C with absolute values greater than $a\sigma_X$, increment the corresponding pixel in B.

Repeat steps 1-4 for M number of times.

5. Select the maximal values of B as the most singular part of the image.

The number of iterations M can be fixed empirically or until the changes in B, normalized by that number, become insignificant. Following the algorithm, one can see that the time to perform it is proportional to $MN^2 \log N$. The speed per image of size 1024×2048 on one and the same computer, with S, a square of size 56×56 points, is about 3 seconds compared to about an hour, using the direct search implementing the Eq. (1).[3]

Some results are presented in Figs.3,4, where we used square shapes with different size, 30 projection operators and different values of a.

---

[2] In general, the standard deviation will be larger for projections with larger low-frequency components. That is why we choose the criterion proportional to σ_X and not as absolute value.

[3] If the block is small enough, the convolution can be performed even faster in the space domain and it is possible to improve the execution time.

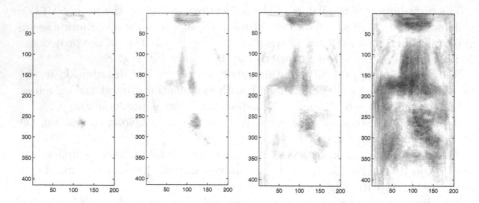

Fig. 3. Score values for different size of the shape (from left to right: 24x24,32x32, 40x40,56x56). The value of the parameter a in all the cases is 22.

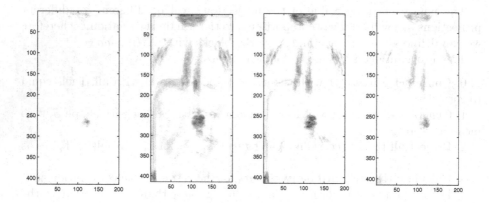

Fig. 4. Score values for different parameter a (from left to right: $a = 8,10,12,16$). The size of the shape is 24x24.

Because the distribution of the projections (Fig. 2) is universal, it is not surprising that the algorithm is operational for different images. We have tested it with some 100 medical Xray images and the results of the visual inspections were good[4].

It can be noted that the number of projection operators is not critical and can be kept relatively low and independent of the size of the block. Note that with significantly large blocks, the results can not be regarded as en edge detector. This empirical observation is not a trivial result at all, indicating that the degrees of freedom are relatively few, even with large enough blocks, something that depends on the statistics of the images and can not be stated in general. With

---

[4] Some of the images require normalization of the projection with the deviation of the block B.

more than 20 projections we achieve satisfactory results, even for areas with more than 3000 pixels. The increment of the number of the projections improves the quality, but with more than 30 projection practically no improvement can be observed.

It is possible to look at that algorithm in a different way. Namely, if we are trying to reconstruct the figure by using some projection operators X_C (for example DCT as in JPEG), then the length of the code, one uses to code a component with distribution like Fig. 2, will be proportional to the logarithm of the probability of some value of the projection $X_C.A$. Therefore, what we are scoring is the block that has some component of the code larger than some length in bits (here we ignore the psychometric aspects of the coding). Effectively we score the blocks with longer coding, e.g. the ones that have lower probability of occurrence.

Using a smoothed version of the above algorithm in step 4, without adding only one or zero, but for example, penalizing the point with the square of the projection difference in respect to the current block divided by σ, and having in mind the universal distribution of the projection, one can compute the penalty function as a function of the value of the projection x, that results to be just $1/2 + x^2/2\sigma^2$. Summing over all projections, we can find that the probability of finding the best block is approximated given by $1/2[1 + \mathrm{erfc}(M(1/2 + x^2/2\sigma^2))]$ as a consequence of the Central Limit Theorem. The above estimation gives an idea why one need few projections to find the rarest block, in sense of the global distribution of the blocks, almost independently of the size of the block. The only dependence of the size of the blocks is given by σ^2 factor, that is proportional to its size. Further, the probability of error will drop better than exponentially with the increment of M.

The non-smoothed version performs somewhat better that the above estimation in the computer experiments.

2.2   Network

The pitfall of the consideration in the previous subsection is that the detected blocks are rare in absolute sense, e.g. in respect to all figures that satisfy the power law or similar distribution of the projections. Actually this is not desirable. If for example in X-ray image appear several spinal segments, although these can be rare in the context of all existing images, they are not rare in the context of thorax or chest X-ray images.

Therefore the parts of the images with many similar projections must "cancel" each other. This gives us the idea to build a network, where its components with similar projection are connected by a negative feedback corresponding to the blocks with similar projections.

As we have seen in the previous section, the small projection values are much more probable and therefore less informative. Using this empirical argument, we can suggest that the connections between the blocks with large projections are more significant.

The network is symmetrical by its nature, because of the reflexivity of the distances. We can try to build it in a way similar to the Hebb network [2] and define Lyapunov o energy function of the network. Thus the network can be described in terms of artificial recursive neural network. Connecting only the elements of the image that produce large projections, the network can be build extremely sparse [10], which makes it feasible in real cases.

Let us try to formalize the above considerations. For each point we define a neuron. The neurons corresponding to some point r and having projection x receive a positive input flux, which is proportional to $-\log p(x)$, where p is the probability of having projection with value x. The same element, if its projection is large, also receives a negative flux from the points r' with nearest projections that satisfy the condition $|r - r'| > \text{diam}(S)$. The flux in general is a function of $p(x)$ and $x' - x$.

As a first approximation we assume that the flux is constant with $p(x)$ and the dependence on $x' - x$ is trivial: the weight is 1 if $|x' - x| < \delta$ and zero otherwise, where δ is some parameter of the model.

In other words, we reformulate our problem in terms of a Hebb-like neural network with external field

$$h = -h_0 \sum_{i=1}^{M} \log p(x_i) \tag{4}$$

and weights

$$w_{rr'} = -\sum_{i=1}^{M} \sum_{\substack{|x_i| > a\sigma_i, \\ |x_i'| > a\sigma_i, \\ |x_i' - x_i| < \delta, \ x_i' x_i > 0}} 1. \tag{5}$$

The extra parameter h_0 balances between the global and the local effects. It can be chosen in a way that the mean fluxes of positive and negative currents are equal in the whole network. The parameter δ, as a proof of concept value, can be assumed to be equal to infinity. So the only parameter, as in the previous case, is a.

The dynamics of the network over time t is given by the following equation [3]:

$$s_r(t+1) = g(\beta[h_r + \sum_{r'} w_{rr'} s_r(t) - T]),$$

where $g(.)$ is a sigmoid function, $s_r(t)$ is the state of the neuron s at position r and time t, β is the inverse temperature and T is the threshold of the system. The result must be insensitive to the particular chose of $g(.)$.

Once the network is constructed, we need to choose its initial state. If the a priori probabilities for all points to be the origin of the rarest block are equal, one can choose $s_r(0) = 1$, $\forall r$. Due to the non-linearity, the analysis of the results is not straightforward. The existence of the attractor is guaranteed by the symmetrical nature of the weights w, which is a necessary condition for the existence of an energy function.

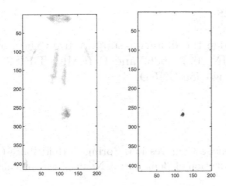

Fig. 5. Comparison between score image (left) and network activity image (right). The size of the area is 24x24 and the parameter $a = 16$.

We can further refine the results of the previous section by fixing the global threshold T in a way to have only some fraction of the excited neurons. Thus we obtain a bump activity of the network, previously considered in [8,6,7]. A sample result is shown in Fig.5.

Regarding the time analysis of the procedure, one can see that the execution times are proportional to the number of the weights w. Having in mind that actually the connectivity is between the blocks, and that we can use a fraction of blocks less than $1/N^2$, the execution time can drop to order inferior to the N^4 limit. Thus, the number of steps to achieve the attractor is of the order $logN$.

3   Discussion and Future Directions

In this paper we present a method to find the most unusual (rare) part in two and higher dimensional images, when its shape is fixed, but in general arbitrary. The method is almost independent on the size of the shape in terms of the execution speed and time. It gives good results on experimental images without predefined model of the interesting event.

One necessary future development of the algorithm is to achieve practical and computable criteria of the "rareness" of the block and comparing the results on large enough database in order to have qualitative measure of the results. The criterion must be different from Eq. (1), because its direct computing tends to be very slow and crispy.

Exact calculus of the probabilistic features of the network in the thermodynamics limit, performed in the sense of probability of finding the outliers, are also of common interest.

Among the future applications of the present method, one could mention the achievement of experiments on different type of images and large image databases and experiments on acceleration of the network due to the special equivalence class construction.

Acknowledgments

The authors acknowledge the financial support from the Spanish Grants TIN 2004-07676-G01-01, TIN 2007-66862 and DGI.M.CyT.FIS2005-1729 - Plan de Promoción de la Investigación UNED.

References

1. Fisher, Y.: Fractal Image Compression. Springer, Heidelberg (1995)
2. Hebb, D.: The Organization of Behavior: A Neurophysiological Theory. Wiley, New York (1949)
3. Hopfield, J.: Neural networks and physical systems with emergent collective computational properties. Proc.Natl.Acad.Sci.USA 79, 2554–2558 (1982)
4. Indyk, P.: Uncertainty Principles, Extractors, and Explicit Embedding of L2 into L1. In: 39th ACM Symposium on Theory of Computing, pp. 615–620 (2007)
5. Keogh, E., Lin, J., Fu, A.: HOT SAX: Efficiently finding the most unusual time series subsequence. In: Fifth IEEE International Conference on Data Mining, pp. 8–14 (2005)
6. Koroutchev, K., Korutcheva, E.: Bump formations in binary attractor neural network. Phys. Rev. E 73, 1–11 (2006)
7. Koroutchev, K., Korutcheva, E.: Bump formations in attractor neural network and their application to image reconstruction. In: The Proceedings of the 9th Granada Seminar on Computational and Statistical Physics, AIP, pp. 242–248 (2006)
8. Roudi, Y., Treves, A.: An associate network with spatially organized connectivity, JSTAT, P07010. pp. 1-20 (2004)
9. Ruderman, D.: The statistics of natural images. Network: Computation in Neural Systems 5, 517–548 (1994)
10. Tsodyks, M., Feigel'man, M.: The enhanced storage capacity in neural networks with low activity level. Europhys. Lett. 6, 101–110 (1988)

Labeling Irregular Graphs with Belief Propagation

Ifeoma Nwogu and Jason J. Corso

State University of New York at Buffalo
Department of Computer Science and Engineering
201 Bell Hall,
Buffalo, NY 14260
{inwogu,jcorso}@cse.buffalo.edu

Abstract. This paper proposes a statistical approach to labeling images using a more natural graphical structure than the pixel grid (or some uniform derivation of it such as square patches of pixels). Typically, low-level vision estimations based on graphical models work on the regular pixel lattice (with a known clique structure and neighborhood). We move away from this regular lattice to more meaningful statistics on which the graphical model, specifically the Markov network is defined. We create the irregular graph based on superpixels, which results in significantly fewer nodes and more natural neighborhood relationships between the nodes of the graph. Superpixels are a local, coherent grouping of pixels which preserves most of the structure necessary for segmentation. Their use reduces the complexity of the inferences made from the graphs with little or no loss of accuracy. Belief propagation (BP) is then used to efficiently find a local maximum of the posterior probability for this Markov network. We apply this statistical inference to finding (labeling) documents in a cluttered room (under moderately different lighting conditions).

1   Introduction

Our goal in this paper is to label (natural) images based on generative models learned from image data in a specific imaging domain, such as labeling an office scene as documents or background (see figure 1). It can be argued that object description and recognition are the key goals in perception. Therefore, the labeling problem of inscribing and affixing tags to objects in images (for identification or description) is at the core of image analysis. But Duncan et al. [3] describe how a discrete model labeling problem (where every point has only a constant number of candidate labels) is NP-complete. The conventional way of solving this discrete labeling in computer vision is by stochastic optimization such as simulated annealing [6]. These are guaranteed to converge to the global optimum under some conditions, but are extremely slow to converge.

However, some efficient approximations based on combinatorial methods have been recently proposed. One such approximation involves viewing the image labeling problem as computing marginalizations in a probability distribution over a Markov random field (MRF). Inspired by the successes of MRF graphs in image analysis, and tractable approximation solutions to inferencing using belief propagation (BP) [10] [9], several other low level vision problems such as denoising, super-resolution, stereo etc., have

V.E. Brimkov, R.P. Barneva, H.A. Hauptman (Eds.): IWCIA 2008, LNCS 4958, pp. 295–305, 2008.
© Springer-Verlag Berlin Heidelberg 2008

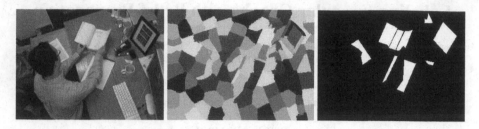

Fig. 1. On the left is a sample of the training data for an office scene with documents, the middle image is its superpixel representation (using normalized cuts, an over-segmentation of the original) and the right image is the manually labeled version of the original showing the documents as the foreground

been tackled by applying BP over Markov networks [5][15]. BP is an iterative sum-product algorithm, for computing marginals of functions on a graphical model. Despite recent advances, inference algorithms based on BP are still often too slow for practical use [4].

In this paper, we present an algorithmic technique that represents our image data with a Markov Random Field (MRF) graphical model defined on a more natural node structure, the superpixels. We infer the labels using belief propagation (BP) but get away from its drawbacks by substantially reducing the node structure of the graph. Thus, we reduce the combinatorial search space and improve the algorithmic running time while preserving the accuracy of the results.

Most stochastic models of images, are defined on the regular pixel grid, which is *not* a natural representation of visual scenes but rather an "artifact" of the image digitization process. We presume that it would be more natural, and more efficient, to work with perceptually meaningful entities called superpixels, obtained from a low-level grouping process [8], [16].

The organization of the paper is as follows: in section 2, we give a brief overview of BP irrespective of the graph structure and describe the process of inferring via message updates; in section 3, we describe our implementation of BP on an irregular graph structure and provide some justification to the use of superpixels; in section 4 we describe our experiments and provide quantitative and qualitative results and finally in section 5 we discuss some of the drawbacks of the technique, prescribe some means of improvement and discuss our plans for future work.

2   Background on Belief Propagation (BP)

A Markov network graphical model is especially useful for low and high level vision problems [5], [4] because the graphical model can explicitly express relationships between the nodes (pixels, patches, superpixels etc). We consider the undirected pairwise MRF graph $G = (V, E)$, V denoting the vertex (or node) set and E, the edge set. Each node $i \in V$ has a set of possible states C and also is affiliated with an observed state r_i. Given an observed state r_i, the goal is to infer some information about the states $c_i \in C$.

The edges in the graph indicate statistical dependencies. In our document-labeling problem, the hidden states C are (1) the documents and (2) the office background.

In general, MRF models of images are defined on the pixel lattice, although this restriction is not imposed by the definition of MRF. Pairwise-MRF models are well suited to our labeling problem because they define (1) the relationship between a node's states and its observed value and (2) the relationship within a clique (a set of pairwise adjacent nodes). We assume in here that the energies due to cliques greater than two are zero.

If these two relationships can be defined statistically as are probabilities, then we can define a joint probability function over the entire graph as:

$$p(c, r) = \frac{1}{Z}p(c_1, c_2 \cdots c_n, r_1, \cdots r_n) \tag{1}$$

$$= \frac{1}{Z}p(c_1)p(r_1|c_1)p(r_2|c_2)p(c_2|c_1) \cdots p(r_n|c_n)p(c_n|c_{n-1}) \tag{2}$$

Z is a normalization constant such that $\sum_{c_1 \in C_1, \cdots, c_n \in C_n} p(c_1, c_2, \cdots, c_n) = 1$. It is important to mention that the derivations in this section are done over a simplified graph (the chain) but the solutions generalize sufficiently to more complex graph structures.

If we let $\psi_{ab}(c_a, c_b) = p(c_a)p(c_b|c_a)$ and $\phi_a(c_a, r_a) = p(r_a|c_a)$ then the marginal probability at any of the nodes is given by:

$$p(c_i) = \tfrac{1}{Z}\sum_{c_1}\sum_{c_2}\cdots\sum_{c_{i-1}}\sum_{c_{i+1}}\cdots\sum_{c_n}\psi_{1,2}(c_1, c_2) \tag{3}$$
$$\psi_{2,3}(c_2, c_3)\cdots\psi_{n-1,n}(c_{n-1}, c_n)$$
$$\phi_1(c_1, r_1)\phi_2(c_2, r_2)\cdots\phi_n(c_n, r_n)$$

$$\text{Let } f(c_2) = \sum_{c_1}\psi_{1,2}(c_1, c_2)$$
$$f(c_3) = \sum_{c_2}\psi_{2,3}(c_2, c_3)f(c_2)$$
$$\vdots \tag{4}$$
$$f(c_n) = \sum_{c_n}\psi_{n-1,n}(c_{n-1}, c_n)f(c_{n-1})$$

The last line in equation (4) shows a recursive definition which we later take advantage of in our implementation. The equation shows how functions of probabilities are propagated from one node to the next. The "probabilities" are now converted to functionals (functions of the initial probabilities).

If we replace the functional $f(c_i)$ with the **message** property m_{ij} where i is the node to which the message are propagated and j is the node from which the message originates, then we can define our marginal probability at a node in terms of message updates. Also, if we replace our probability at a node $p(c_i)$ by the belief at the node $b(c_i)$ (since the computed values are no longer strictly probability distributions), then we can rewrite equation (4) as,

$$b_i(c_i) = \phi_i(c_i, r_i)\prod_{j \in \mathcal{N}(i)} m_{ij}(c_i) \tag{5}$$

where $\mathcal{N}(i)$ is the neighborhood of i. Equation (5) above shows how the derived functions of probabilities (or messages) are propagated along a simplified graph. Under the assumption that our solution so far generalizes to a more complex graph structure, we can now extend our derivation to the joint probability on an MRF graph given as:

$$p(c) = \frac{1}{Z} \prod_{i,j} \psi(c_i, c_j) \prod_k \phi(c_k, r_k) \qquad (6)$$

The joint probability on the MRF graph is described in terms of two **compatibility functions**, (1) between the states and observed nodes and (2) between neighboring nodes.

We illustrate the message propagation process to node 2 in a five-node graph in figure(2). In this simple example, the belief at a node i can now be given as:

$$b_i(c_i) = \sum_{c_j \in C_j, 1 \le j \le 5, j \ne i} p(c_1, \cdots, c_5); \qquad (7)$$

$$= \phi_i(c_i, r_i) \prod_{j \in \mathcal{N}(i)} m_{ji}(c_i)$$

$$m_{ji}(c_i) = \sum_{c_j \in C_j} \phi_j(c_j, r_j) \psi_{ji}(c_j, c_i) \prod_{k \in \mathcal{N}(j) j \ne i} m_{kj}(c_j)$$

Unfortunately, the complexity of general belief propagation is exponential in the size of the largest clique. In many computer vision problems, belief propagation is prohibitively slow. The high-dimensional summation in equation (3) has a complexity of $\mathcal{O}(nM^k)$, where M is the number of possible labels for each variable, k is the maximum clique size in the graph and n is the number of nodes in the graph. By using the message updates, the complexity of the inference (for a non-loopy graph as derived above) is reduced to $\mathcal{O}(nkM^2)$. By extending this derivation to a more complex graph structure, the convergence property of the inference algorithm is removed and it is no longer guaranteed to converge. But in practice the algorithm consistently gives a good solution. Also, by significantly reducing n, we further reduce the algorithmic time.

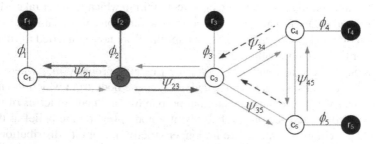

Fig. 2. An example of computing the messages for node 2, involving only $\phi_2, \psi_{2,1}, \psi_{2,3}$ and any messages to node 2 from its neighbors' neighbors

3   Labeling Images Using BP Inference

When doing labeling, the images are first abstracted into a superpixel representation (described in more detail in section (3.1)). The Markov network is defined on this irregular graph, and the compatibility functions are learned from labeled training data (section 3.2). The BP is used to infer the labels on the image.

3.1   The Superpixel Representation

First, we need to chose a representation for the image and scene variables. The image and scenes are arrays of single-valued superpixels. A superpixel is a homogenous segment obtained from a low-level grouping of the underlying pixels. Although irregular when compared to the pixel-grid, we choose the superpixel grid because we believe that it representationally more efficient: i.e. pairwise constraints exist between entire segments, while they only exist for adjacent pixels on the pixel-grid. For a local model such as the MRF model, this property is very appealing in that we can model much longer-range interactions between superpixels segments. The use of superpixels is also computationally more efficient because it reduces the combinatorial complexity of images from hundreds of thousands of pixels to only a few hundred superpixels.

There are many different algorithms that generate superpixels including the segregated weighted algorithm (SWA) [12],[2], normalized cuts [13], constrained Delaunay triangulation [11] etc. It is very important to use near-complete superpixel maps to ensure that the original structures in the images are conserved. Therefore, we use the region segmentation algorithm normalized cuts [13], which was emperically validated and presented in [8]. Figure (3) shows an example of a natural image with its superpixel representation. For building superpixels using normalized cuts, the criterion for partitioning the graph are (1) to minimize the sum of weights of connections across the groups and (2) to maximize the sum of weights of connections within the groups. For completeness, we now give a brief overview of the normalized cuts process.

We begin by defining a similarity matrix as $S = [S_{ij}]$ over an image $I(i, j)$. A similarity matrix is a matrix of scores which express the similarity between any two points in an image. If we define a graph $G(V, E)$ where the node set is defined as the relationship ij between nodes, and all edges $e \in E$ have equal weight, we can define the degree of a node as $d_i = \sum_j S_{ij}$ and the volume of a set in the graph as $vol(A) = \sum_{i \in A} d_i, A \subseteq V$. The cuts in the graph are therefore $cut(A, \bar{A}) = \sum i \in A, j \in \bar{A} S_{ij}$. Given these definitions, normalized cuts are described as the solution to:

$$N_{cut} = cut(A, B) \left(\frac{1}{vol(A)} + \frac{1}{vol(B)} \right) \tag{8}$$

Our implementation of superpixel generation implements an approximate solution using spectral clustering methods. The resulting superpixel representation is an over-segmentation of the original image. We define the Markov network over this representation.

Fig. 3. On the left is an example of a natural image, the middle image is the superpixel representation (using normalized cuts, an over-segmentation of the original) and the right image the superimposition of the two

3.2 Learning the Compatibility Functions

The model we assume for the document labeling images is generated from the training images. The joint probability distribution is modeled using the image data and their corresponding labels. The joint probability is expressed in terms of the two compatibility functions defined in equation(6).

If we let r_i represent a superpixel segment in the training image and c_i, its corresponding label. The first compatibility function $\phi(c_i, r_i)$ can be computed by learning a mixture of Gaussian models. The resulting Gaussian distributions will represent either the document class or the background class. So although we have two real-life classes (document and background), the number of states to be input to the BP problem will have increased based on the output of the Gaussian Mixture Model (GMM), i.e. each component of the GMM represents a distinct label.

The second compatibility function $\psi(c_i, c_j)$ relates the superpixels to each other. We use the simplest interacting Pott's model where the function takes one of 2 values $-1, +1$ and the interactions exists only for amongst neighbors with the same labels. Our compatibility function between superpixels is therefore given as:

$$\psi(c_i, c_j) := \begin{cases} +1 \text{ if } c_i \text{ and } c_j \text{ have the same initial label values,} \\ -1 \text{ otherwise} \end{cases} \tag{9}$$

So given a new image of a cluttered room, we can extract the documents in the image by using the steps given in section (3.3). The distribution of the superpixels $r_i \in R$ given the latent variables $c_i \in C$ can therefore be modeled graphically as:

$$P(R, C) \propto \prod_i \phi(r_i, c_i) \prod_{(c_j, c_k)} \psi(c_j, c_k) \tag{10}$$

Equation (10) can also be viewed as the pairwise-MRF graphical representation of our labeling problem, which can be solved using BP with the two parts of equation (7).

3.3 Putting All Together...

The general strategy for designing the label system can therefore be described as:

1. Use the training data to learn the latent parameters of the system. The number of resulting latent parameter sets will give the number of states required for the inference problem.

2. Using the number of states obtained in the previous steps, design compatibility functions such that eventually, only a single state can be allocated to each superpixel.
3. For the latent variable c_i associated with every superpixel i, use the BP algorithm to choose its best state.
4. If the state values correspond to labeling classes (as in the case of our document labeling system), the selected state variables are converted to their associated class labels

4   Experiments, Results and Discussion

The first round of experiments consisted of testing the BP labeling algorithm on synthetically generated image data, whose values were samples drawn from a known distribution. We first generated synthetic scenes by drawing samples from Gaussian distribution functions, and then added noise to the resulting images. These two datasets (clean and noisy images) represented our observations in a controlled setting. To add X% noise, we randomly selected unique X% of the pixels in the original image and the pixel values were replaced by a random number between 0 and 255;

The scene (or hidden parameters) were represented by the parameters of our generating distributions. We modeled the relationships between the scenes and observations with a pairwise-Markov network and used belief propagation (BP) to find the local maximum of the posterior probability for the original scenes.

Figure (4) shows the results obtained from running the process on the synthetic data. We also present a graph showing the sum-of squared-differences (SSD) between the ground-truth data and varying levels of noise in figure (5).

We then extended this learning based scene-recovery approach to finding documents in a cluttered room (under moderately different lighting conditions). Documents were labeled in images of a cluttered room and used in training to obtain the prior and conditional probability density functions. The labeled images were treated as scenes and the goal was to infer these scenes given a new image from the same imaging domain (pictures of offices) but not from the training set. For inferring scenes from given observations, the computed distributions were used as compatibility functions in the BP message update process.

We learned our distributions from the training data using an EM-GMM algorithm (Expectation Maximization for Gaussian Mixture Models) on 50 office images. The training images consisted of images with different documents in a cluttered background, all taken in one office. The document data was modeled with a mixture of three Gaussian distributions while the background data was modeled with two Gaussian distributions. The resulting parameters (mean μ_i, variance σ_i and prior probability p_i) from training are:

- class 1: $\mu_1 = 6.01$; $\sigma_1 = 9.93$; $p_1 = 0.1689$
- class 2: $\mu_2 = 86.19$; $\sigma_2 = 147.79$; $p_2 = 0.8311$
- class 3: $\mu_3 = 119.44$; $\sigma_3 = 2510.6$; $p_3 = 0.5105$
- class 4: $\mu_4 = 212.05$; $\sigma_4 = 488.21$; $p_4 = 0.2203$
- class 5: $\mu_5 = 190.98$; $\sigma_5 = 2017$; $p_5 = 0.2693$

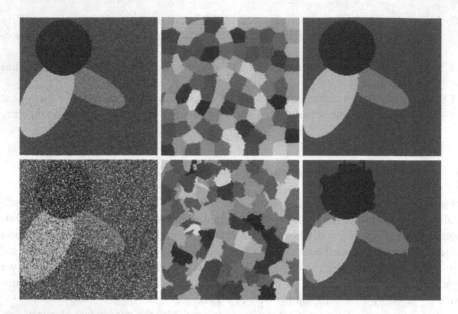

Fig. 4. Top row: the left column shows an original synthetic image created from samples from Gaussian distributions, the middle column is its near-correct superpixel representation and the right column shows the resulting labeled image. Bottom row: the left column shows a noisy version of the synthetic image, the middle column is its superpixel representation and the right column also shows the resulting labeled image.

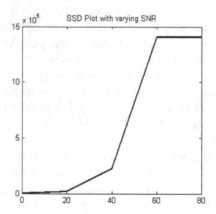

Fig. 5. Quantitative display of how the error increases exponentially with increasing noise

The resulting five classes were then treated as the states that any variable $c_i \in C$ in the MRF graph could take. Classes 1 and 2 correspond to the background while classes 3,4 and 5 are the documents. Unfortunately, because the models were trained separately, the prior probabilities are not reflective of the occurrences in the entire data, only in the background/document data alone.

Fig. 6. The top row shows sample of training data and the corresponding labeled image; the bottom row shows a testing image (taken at a different time, in a different office. The output of the detection is shown.

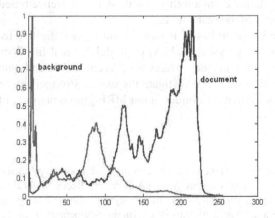

Fig. 7. The distributions of the foreground document and cluttered background classes

For testing, a new image was taken in a completely different office under moderately different lighting conditions and the class labels were extracted. The pictorial results of labeling the real-life rooms with documents are shown in figure (6).

We observed that even with applying relatively weak models (the synthetic images are no longer strongly coupled to the generating distribution), we were able to successfully recover a labeled image for both synthetic and real-life data. A related drawback we faced was the simplicity of our representation. We used grayscale values as the only statistic in our document-finding system and this (as seen in figure (6)b), introduces artifacts into our final labeling solution. Superpixels whose intensity values are close to those of the trained documents can be mis-labeled.

Also, we observed that the use of superpixels reduced the number of nodes significantly, thus reducing the computational time. Also, the segmentation results of our low noise synthetic images and the real-life data were promising with superpixels. A drawback though is the limitation imposed by the superpixel representation. Although we used a well tested and efficient superpixel implementation, we found that as the noise levels increased in the images, the superpixels became more inaccurate and the errors obtained in the representation were propagated into the system. Also, due to the loops in the graph, it does not converge if run long enough, but we can still sufficiently recover the true solution from the graphical structure.

5   Conclusion

In this paper, we have proposed a way of labeling irregular graphs generated by an oversegmentation of an image, using BP inferences on MRF graphs. Because a common limitation of graph models in low level image processing is often due to intractable node size on the graph, we have reduced the computational intensity of the graph model by introducing the use of superpixels, without any loss of generality on the definition of the graph. We reduced the number of node variables from orders of tens of thousands of pixels to about a hundred superpixels.

Furthermore, we define compatibility functions for inference based on learning the statistical distributions of the real-life data.

In the future, we intend to base our statistics on more definitive features of the images (other than simply grayscale values) to model the real-life document and background data.. These could include textures at different scales, and other scale-invariant measurements. We also plan to investigate the use of stronger inference methods by relaxing the assumption that the cliques in our MRF graphs are only of size 2.

References

1. Cao, H., Govindaraju, V.: Handwritten carbon form preprocessing based on markov random field. In: Proc. IEEE Conf. Comput. Vision And Pattern Recogniton (2007)
2. Corso, J.J., Sharon, E., Yuille, A.L.: Multilevel Segmentation and Integrated Bayesian Model Classification with an Application to Brain Tumor Segmentation. In: Larsen, R., Nielsen, M., Sporring, J. (eds.) MICCAI 2006, Part II. LNCS, vol. 4191, pp. 790–798. Springer, Heidelberg (2006)
3. Duncan, R., Qian, J., Zhu, B.: Polynomial time algorithms for three-label point labeling. In: Wang, J. (ed.) COCOON 2001. LNCS, vol. 2108, p. 191. Springer, Heidelberg (2001)
4. Felzenszwalb, P., Huttenlocher, D.: Efficient belief propagation for early vision. In: Proc. IEEE Conf. Comput. Vision And Pattern Recogn. (2004)
5. Freeman, W.T., Pasztor, E.C., Carmichael, O.T.: Learning low-level vision. International Journal of Computer Vision 40(1), 25–47 (2000)

6. Geman, S., Geman, D.: Stochastic relaxation, gibbs distributions, and the bayesian restoration of images. IEEE Trans. on Pattern Analysis and Machine Intelligence 6, 721–741 (1984)
7. Luo, B., Hancock, E.R.: Structural graph matching using the em algorithm and singular value decomposition. IEEE Trans. Pattern Anal. Mach. Intell. 23(10), 1120–1136 (2001)
8. Mori, G., Ren, X., Efros, A.A., Malik, J.: Recovering human body configurations: Combining segmentation and recognition. In: Proc. IEEE Conf. Comput. Vision And Pattern Recogn., vol. 2, pp. 326–333 (2004)
9. Murphy, K.P., Weiss, Y., Jordan, M.I.: Loopy belief propagation for approximate inference: An empirical study. In: Proceedings of Uncertainty in AI, pp. 467–475 (1999)
10. Pearl, J.: Probabilistic Reasoning in Intelligent Systems: Networks of Plausible Inference. Morgan Kaufmann, San Francisco (1988)
11. Ren, X., Fowlkes, C.C., Malik, J.: Scale-invariant contour completion using conditional random fields. In: Proc. 10th Int'l. Conf. Computer Vision, vol. 2, pp. 1214–1221 (2005)
12. Sharon, E., Brandt, A., Basri, R.: Segmentation and boundary detection using multiscale intensity measurements. In: Proc. IEEE Conf. Comput. Vision And Pattern Recogn., vol. 1, pp. 469–476 (2001)
13. Shi, J., Malik, J.: Normalized cuts and image segmentation. IEEE Transactions on Pattern Analysis and Machine Intelligence 22(8), 888–905 (2000)
14. Szeliski, R., Zabih, R., Scharstein, D., Veksler, O., Kolmogorov, V., Agarwala, A., Tappen, M.F., Rother, C.: A comparative study of energy minimization methods for markov random fields. In: Leonardis, A., Bischof, H., Pinz, A. (eds.) ECCV 2006. LNCS, vol. 3952, pp. 16–29. Springer, Heidelberg (2006)
15. Tappen, M.F., Russell, B.C., Freeman, W.T.: Efficient graphical models for processing images. In: Proc. IEEE Conf. Comput. Vision And Pattern Recogniton, pp. 673–680 (2004)
16. Yu, S., Shi, J.: Segmentation with pairwise attraction and repulsion. In: Proceedings of the 8th IEEE IInternational Conference on Computer Vision (ICCV 2001) (July 2001)

Image Registration Using Markov Random Coefficient Fields

Edgar Román Arce-Santana and Alfonso Alba

Facultad de Ciencias
Universidad Autonoma de San Luis Potosi
San Luis Potosi, Mexico
{arce,fac}@fciencias.uaslp.mx

Abstract. Image Registration is central to different applications such as medical analysis, biomedical systems, image guidance, etc. In this paper we propose a new algorithm for multi-modal image registration. A Bayesian formulation is presented in which a likelihood term is defined using an observation model based on linear intensity transformation functions. The coefficients of these transformations are represented as prior information by means of Markov random fields. This probabilistic approach allows one to find optimal estimators by minimizing an energy function in terms of both the parameters that control the affine transformation of one of the images and the coefficient fields of the intensity transformations for each pixel.

Keywords: Image Registration, Markov Random Fields, Bayesian Estimation, Intensity Transformation Function.

1 Introduction

Image registration is the alignment of images that may come from the same or different source. This task is very important to many applications involving image processing or analysis such as medical analysis, biomedical systems, image guidance, depth estimation, and optical flow. A special kind of registrations is called Multimodal Image Registration, in which two o more images coming from different sources are aligned; this process is very useful in computer aided visualization in the medical field.

In the literature, there are basically two classes of methods to register multimodal images: those based on features such as edge locations, landmarks or surfaces [6][7][11], and those based on intensity [1][19][4][16]. Within the intensity methods there are two popular ones. Partitioned Intensity Uniformly (PIU) [19][5], proposed by Woods et al, is one of them. In this method it is assumed that uniform regions in one of the images correspond to regions, also uniform, in the other one. To achieve the registration, a corresponding measure is established based on the statistical characteristics of both images. The goal of this method is to use this measure to minimize the variance of intensity ratios. The other method that has shown good results is the registration based on mutual

V.E. Brimkov, R.P. Barneva, H.A. Hauptman (Eds.): IWCIA 2008, LNCS 4958, pp. 306–317, 2008.

information (MI), proposed by Viola et al [18]. In this method, statistical dependencies between images are compared, establishing a metric based on the entropy of each image and the join entropy. Even though the method is theoretically robust, it is complicated to implement and requires vast computational resources. Another drawback of MI is that it completely ignores spatial information such as edges or homogenous regions.

A method related to the work proposed in this paper is presented in [10]. It focuses only on elastic registration of multimodal images; it uses an iterative scheme that iterates between finding the coefficients of polynomial intensity transformations and registration using the demons method [17]. This method makes the assumption that there are at most two functional dependencies between intensities. This restriction limits its applications since there are cases, as those found in medical imaging, where inhomogeneity and noise are presented in both images to register.

In this work, we present a more general registration method, in which a probabilistic model permits the characterization of the image registration by means of linear intensity transformation functions. Rigourously based on Bayesian estimation, the main goal of this method is to establish the parameters of the affine transformation, and at the same time, determine in a probabilistic framework the coefficient values of these linear functions for each pixel to achieve the image registration. These transformations have the purpose to estimate the adequate intensity changes that match the intensity values between the images. In this approach, the coefficients of the linear intensity transformations (labeled MRCF) are represented as Markov Random Fields (MRF)[2], giving in this way the prior information about the homogeneity of the intensity changes.

The paper is organized as follows: in Section 2, we give an introduction to MRF and present the Bayesian framework of image registration using affine transformation and MRCF; in Section 3, we describe some experiments and results; finally in Section 4, some conclusions are presented.

2 Bayesian Framework for Multimodal Image Registration

2.1 Markov Random Fields

In this subsection, we present the basic definition of Markov Random Fields, for more detail refer to [2][12][8]. Let L be the discrete pixel lattice where 2D images of size $n \times m$ are observed:

$$L = \{(i,j)|1 \leq i \leq n, 1 \leq j \leq m\}. \tag{1}$$

To simplify the notation, the pixels in a $n \times m$ image can be conveniently re-index by a number r taking values in $\{1, 2, ..., n \times m\}$. The sites in L are related to one another via a neighborhood system. A neighborhood system for L is given by

$$N = \{N_r | \forall r \in L\}, \tag{2}$$

where N_r are the sites neighboring r. The neighborhood relationship has the following properties:

- a site is not neighboring to itself: $r \notin N_r$;
- the neighboring relationship is mutual: $r \in N_{r'} \iff r' \in N_r$.

We can define a graph (L, N), where L contains the nodes and N determines the link between the nodes according to a neighborhood system. A clique C for (L, N) is a subset of sites in L such that for all $r, s \in C$ such that $r \neq s$, we have that $r \in N_s$ and $s \in N_r$. In a first order neighborhood system (the four nearest sites to r), cliques may be composed of either single sites $c = \{r\}$, or a pair of neighboring sites $c = \{r, r'\}$, thus the collections of single-cliques C_1 and pair-cliques C_2 are defined as

$$C_1 = \{\{r\} | r \in L\},$$
$$C_2 = \{\{r, s\} | r \in N_s, s \in N_r\}.$$

Let $F = \{F_1, ..., F_{n \times m}\}$ be a family of random variables defined on L, where a realization of F_r can take a value f_r in Ω; we denote a realization of a joint event as $F = f$. F is said to be a Markov random field on L with respect to a neighborhood system N if the following conditions are satisfied:

- $p(f) > 0, \forall f \in F$,
- $p(f_r | f_{L-\{r\}}) = p(f_r | f_{N_r})$.

The Hammersley and Clifford theorem [2] establishes that an MRF has an equivalence with a Gibbs distribution, which has the following form

$$p(f) = \frac{1}{Z_f} \exp\{-\sum_C V_C(f)\}, \tag{3}$$

where Z_f is a normalizing constant, the sum in the exponential ranges over the cliques of the given neighborhood system on L, and $\{V_C\}$ are the potential functions, each one depending on the values of f at the sites that belong to the clique C. These potential functions, together with the neighborhood system, control the appearance of the sample field f.

2.2 Bayesian Estimation

To describe the probabilistic framework for multimodal image registration, we assume first that the observation model in each pixel is given by

$$I_2(T(r)) = g(I_1(r)) + \eta(r), \tag{4}$$

where I_1, I_2 are the images to register; T is the affine transformation that aligns the images I_1, I_2; and $\eta(r) \sim N(0, \sigma^2)$. $g(I_1(r))$ is the intensity transformation function which may be, in general, very complex such as logarithmic, gamma, contrast-stretching, inverse, polynomial, or thresholding transformations, (see more details in [9]). In particular, we model this transformation with a locally linear function given by

$$g(I_1(r)) = K_1(r)I_1(r) + K_2(r), \tag{5}$$

where K_1 and K_2 are Markov random coefficient fields (MRCF) that describe the intensity transformation at each pixel r. Given the observation model (4) and the linear functions (5), one can estimate their parameters using Bayesian estimation theory, following the steps [13]:

1. Find the likelihood of the observation $p(I_1, I_2 | K_1, K_2, T)$.
2. Using the prior distributions $p(K_1, K_2, T)$, find the posterior distribution $p(K_1, K_2, T | I_1, I_2)$.
3. Define an appropriate cost function $C(\widehat{K_1}, \widehat{K_2}, \widehat{T}, K_1, K_2, T)$, that assigns a cost to estimators $\widehat{K_1}, \widehat{K_2}, \widehat{T}$, given that the true values are K_1, K_2, T.
4. Find the optimal estimators K_1^*, K_2^*, T^* by minimizing

$$Q(\widehat{K_1}, \widehat{K_2}, \widehat{T}) = E[C(\widehat{K_1}, \widehat{K_2}, \widehat{T}, K_1, K_2, T) | I_1, I_2]. \tag{6}$$

Now, we proceed to analyze each step in detail.

Assuming that $\eta(r)$ (normal) is known and iid, the likelihood function can be written as

$$p(I_1, I_2 | K_1, K_2, T) = \frac{1}{Z_L} \exp[- \sum_{r \in L} V_T(r)], \tag{7}$$

where

$$V_T(r) = \frac{(I_2(T(r)) - K_1(r)I_1(r) - K_2(r))^2}{2\sigma^2}. \tag{8}$$

In this model, K_1, K_2, and T are assumed independent; hence, one can express $p(K_1, K_2, T)$ as a product of independent probabilities. Now, the probability of T is considered constant, and K_1, K_2 are MRF, resulting in the prior distribution:

$$p(K_1, K_2, T) = p(K_1)p(K_2)p(T)$$
$$= \frac{1}{Z_P} \exp[- \sum_C V_C(K_1) - \sum_C V_C(K_2) + \log p(T)] . \tag{9}$$

Using (8), (9), and the Bayes rule, one finds the posterior distribution as:

$$p(K_1, K_2, T | I_1, I_2) = \frac{1}{Z} \exp[-U(K_1, K_2, T)], \tag{10}$$

where Z is a normalizing constant composed by $1/Z_L$ and $1/Z_P$, and

$$U(K_1, K_2, T) = \sum_{r \in L} V_T(r) + \sum_C V_C(K_1) + \sum_C V_C(K_2) - \kappa, \tag{11}$$

where V_T is given by (8), and κ is a noninformative constant; the potential function V_C considers only cliques of size 2, that is, nearest-pair sites $< r, s >$ which are one unit apart:

$$V_C(K(r), K(s)) = \lambda_{r,s}(K(r) - K(s))^2, \tag{12}$$

where $\lambda_{r,s}$ is a positive regularization parameter that may depend on the sites $< r, s >$; however, we used the same λ for all $< r, s >$ in our implementation.

Let $\theta = [K_1, K_2, T]$ denote the estimator vector, and define the cost function $(1 - \delta(x))$, where

$$\delta(x) = \begin{cases} 1, \text{if } x = 0 \\ 0, \text{otherwise.} \end{cases} \tag{13}$$

To find the optimal estimator θ^*, using this cost function, we see that

$$Q(\widehat{\theta}) = \int_{\theta \in \Theta} (1 - \delta(\widehat{\theta} - \theta))p(\theta|I_1, I_2)d\theta$$

$$= \int_{\theta \in \Theta} p(\theta|I_1, I_2)d\theta - \int_{\theta \in \Theta} \delta(\widehat{\theta} - \theta)p(\theta|I_1, I_2)d\theta$$

$$= 1 - \int_{\theta \in \Theta} \delta(\widehat{\theta} - \theta)p(\theta|I_1, I_2)d\theta. \tag{14}$$

Therefore, to minimize (14), we need to find $\widehat{\theta}$ that maximizes $p(\widehat{\theta}|I_1, I_2)$, which is equivalent to finding

$$K_1^*, K_2^*, T^* = \arg \min_{K_1, K_2, T} U(K_1, K_2, T), \tag{15}$$

which is called maximum a posteriori (MAP) estimator.

2.3 Minimization Algorithm

The minimization of (15) may be achieved using different unconstrained optimization algorithms (see [15]). However, in this paper, we have used an efficient Newtonian gradient descent algorithm (NGD) [14]. This method is based on the idea of moving, in each iteration, in a direction d such that $\nabla U \cdot d < 0$ (i.e., a descent direction). The convergence may be accelerated if one considers each element of $K_1(r), K_2(r)$, and each element of the affine transformation T as the position of a particle of unit mass, subject to a force equal to $-\partial U/\partial K_1(r)$ (respectively, $-\partial U/\partial K_2(r)$, $-\partial U/\partial T$). The equation of motion of these particles may be obtained from Newton's second law. The discretization of these equations gives way to an iterative gradient descent algorithm with inertia:

$$K_1^{(t+h)}(r) = \frac{2}{\alpha h + 1}K_1^{(t)}(r) + \frac{\alpha h - 1}{\alpha h + 1}K_1^{(t-h)}(r)$$
$$- \frac{h^2}{\alpha h + 1}\nabla_{K_1(r)}U^{(t)} \tag{16}$$

$$K_2^{(t+h)}(r) = \frac{2}{\alpha h + 1}K_2^{(t)}(r) + \frac{\alpha h - 1}{\alpha h + 1}K_2^{(t-h)}(r)$$
$$- \frac{h^2}{\alpha h + 1}\nabla_{K_2(r)}U^{(t)} \tag{17}$$

$$T^{(t+h)} = \frac{2}{\alpha h + 1}T^{(t)} + \frac{\alpha h - 1}{\alpha h + 1}T^{(t-h)}$$
$$- \frac{h^2}{\alpha h + 1}\nabla_T U^{(t)}, \tag{18}$$

where α is a friction coefficient, $U^{(t)} = U(K_1^{(t)}(r), K_2^{(t)}(r), T^{(t)})$, and h is the step size. This method differs from the typical gradient descent in that the friction coefficient α allows the algorithm to avoid, in many cases, becoming trapped in local minima. Notice that if $\alpha = 1/h$ the NGD is a typical gradient descent method.

3 Results and Discussion

In the following section, we present some experiments involving different kind of images to test the performance of the algorithm. First, we show the ability of the MRCF to compute the intensity changes required to achieve the image registration between multimodal images and its robustness to noise and inhomogeneities. Second, we compare our algorithm against the method proposed by Viola et al. [18]. All these experiments were performed on a PC-based workstation running at 3.0 GHz.

3.1 Experiments

In order to test the ability of the proposed algorithm to find the coefficients of the linear intensity transformation functions, we built a one-dimensional signal of 126 samples. Fig. 1a shows the signal I_1(thicker line), which is the negative of I_2 (thinner line), and shifted five samples to the left of I_2 . In the plot in Fig. 1b, we can see the thicker line composed by the intensity transformation $K_1(r)I_1(r) + K_2(r)$, and $I_2(r - d)$, where d is the displacement found by the proposed algorithm; to appreciate the matching between the signals, the thicker line is plotted few values below I_2. In fact, the signal I_1 was built by setting $I_1(r) = -I_2(r+d)$; one can observe in Figures 1c and 1d how the MRCF K_1, K_2 approach this transformation.

In order to test the robustness of the algorithm to noise, the following experiment consisted in the registration of the images in Fig. 2a and Fig. 2b; this last one was built artificially. We added normal random values to the image in Fig. 2b. The true relative mean error (TRME) between the true parameter vector $\theta^* = [0.2094, 2.0, -36, -18]$ (corresponding to the angle, scale, and displacements in (x, y)) and the vector values θ_i obtained by the algorithm for different noise standard deviations $\sigma = \{0, 2, 4, 6, 8, 10, 12, 14, 16\}$ is ploted in Fig. 3. This error has the advantage of independently taking into account the unit scales of the quantities to evaluate, and it is computed for each θ_i as follows

$$TRME_i = \frac{\sum_{k=1}^{4} |\frac{\theta_i^*(k) - \theta_i(k)}{\theta_i^*(k)}|}{4}. \tag{19}$$

In all the experiments, we used the same set of values for the parameters of the algorithm; in all tests, the error was less than 3.0%.

We also applied this registration approach to different kinds of brain images coming from different sources or processes. The first experiment consisted in registering a Magnetic Resonance (MR) image in Fig. 4a and the Computed Tomography

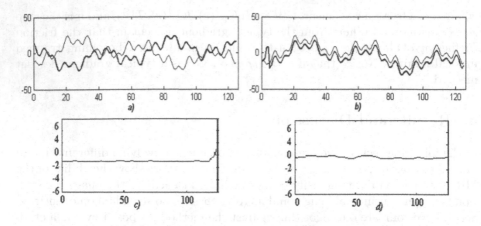

Fig. 1. a) Original signals I_1, I_2; b) aligned signals; c) K_1 field; d) K_2 field

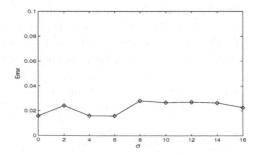

Fig. 2. a), b) images I_1, I_2 to align; c) transformed image I_1 ; d) difference between image transformed I_1 and registered I_2

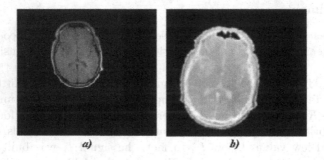

Fig. 3. True relative mean error

(CT) image in Fig. 4b. We can see in Fig. 4c the transformation of the MR-image using the MRCF to match the image in Fig. 4b that together with the estimated affine-parameters produce the superimposed registration shown in Fig. 4d.

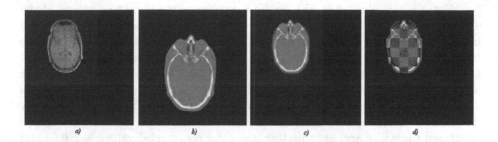

Fig. 4. *a*) MR-image, *b*) CT-image, *c*) transformed MR-image, *d*) superimposed registration

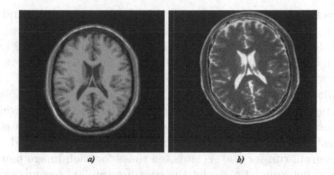

Fig. 5. *a*) T1-image, *b*) T2-image

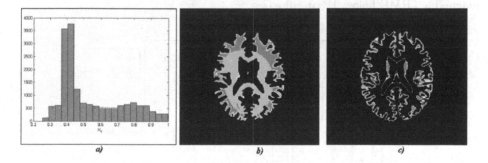

Fig. 6. *a*) Histogram of $0 < K_1 < 1$; quantized values of white and gray natter: *b*) $0.2 \le K_1 < 0.6$, *c*) $0.6 \le K_1 < 1$

In the next experiment, we examine the coefficient values of the fields K_1 and K_2 obtained by registering a synthetic magnetic resonance image spin-lattice relaxation time (T1), and spin-spin relaxation time (T2) obtained from the Brainweb Database [3]; these are shown in Fig. 5. The T1-image was produced with 0% of Gaussian noise and 0% intensity shading (inhomogeneity), while the T2-image with 0% of noise and 40% of inhomogeneity. A histogram of the values of

$0 < K_1(r) < 1$ is plotted in Fig. 6a. These values correspond to a region of the T1-image where it is necessary to reduce the intensity levels in order to match the intensity values of the same regions in the T2-image. We can see that there are two modes localized approximately at 0.4 and 0.8, corresponding respectively to the white and gray matter. These distributions show that it is necessary to have a set of coefficient values (i.e., different intensity transformation functions) to adjust the intensities of T1 to approximate those of T2 in these regions, mainly due to their inhomogeneity. This is more evident in Figures 6b and 6c where we separated the white and gray matter using the K_1-interval values $[0.2, 0.6)$ and $[0.6, 1)$, and thresholded in intervals of 0.1.

3.2   Comparisons

Here we present some comparisons with one of the most popular and referenced algorithms in the literature; this method was presented in [18] and it is based on Mutual Information theory. To do this, we obtained T1 and T2 images from the Brainweb and made several experiments. The first one consists in registering a T1-image with 3% of noise and 20% of inhomogeneity versus a set of T2-images (similar to that shown in Fig. 5) having different level of noise and 40% of inhomogeneity; the set of images were previously transformed using a known affine transformation. In both algorithms, the transformation T was initialized with the identity. Due to the stochastic nature of the MI method, it required to let the program run for 300 seconds ten times for each image pair. However, since MRCF is deterministic, we let the program run 300 seconds once for each image pair. The results are plotted on Figure 7. Notice that MI does not always converge to an acceptable solution in most cases (large variances), while MRCF reached a TRME below 1% in all cases.

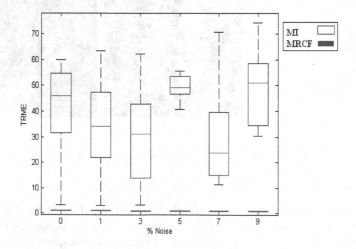

Fig. 7. Boxplot of results obtained by MI and MRCF

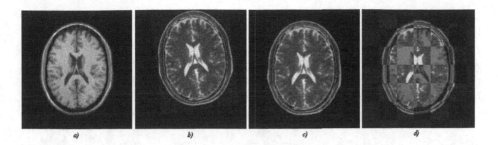

Fig. 8. *a*) T1-image, *b*) T2-image, *c*) transformed T1-image, *d*) superimposed registration

For the final comparison, we choose the hardest case in which a T1-image with 9% of noise and 40% of inhomogeneity was registered with a T2-images with also 9% of noise and 40% of inhomogeneity. The results obtained by MRCF are shown in Fig. 8. The TRME of MRCF was of 1.1865%, while for MI was 71.0916%, both computed in 600 seconds.

4   Conclusions

This work presents an algorithm rigourously based on Bayesian estimation in which two Markov Random Fields (K_1, K_2) represent the coefficients of linear intensity transfer functions applied to each pixel. These functions are included in a very simple observation model (4) that allows one to estimate with high precision the necessary intensity changes and the parameter values of the affine transformation to match the images to register. Another important characteristic of this energy function is that it includes spatial coherence as priori knowledge by means of the MRCF (see equations (11), (12)). Although the resulting posteriori energy function (15) is highly non-linear with respect to the affine transformation parameters, and quadratic with respect to the MRF's K_1, K_2, it was successfully minimized using an efficient, simple, and easy to implement Newtonian gradient descent algorithm.

The paper also presents examples that illustrate the generality of the algorithm to estimate the coefficient values K_1 and K_2 of the local linear functions to approach the intensity transformation needed to achieve the image registration. We showed the performance and stability of the algorithm to get high precision registrations in cases in which radical intensity changes exist, as those shown in Figures 1, 4, 5 and 8. Preliminary results show that the fields K_1 and K_2 may also yield discriminatory information about the different regions in the images, which may be useful for a posterior segmentation process. Finally, we demonstrate the robustness of the proposed algorithm to noise and intensity inhomogeneities, outperforming the MI-algorithm as it was described in [18].

Perspectives for future research include: (1) a generalization of the proposed methodology for the registration of 3D brain images, (2) the addition of a segmentation stage that takes advantage of the MRCF K_1 and K_2, and (3)

the application of MRCF to other problems in computer vision and image processing.

Acknowledgments. For this research work, Edgar Arce-Santana was supported by Grant PROMEP/103.5/04/1387, C06- FAI-11-31.68, and Alfonso Alba by PROMEP/103.5/07/2416. The authors would like to thank Professor Jose Luis Marroquin Zaleta, CIMAT, Mexico, for sharing his insight.

References

1. Banerjee, P.K., Toga, A.W.: Image aligment by integrated rotational and translation transformation matrix. Physics in Medical and Biology 39, 1969–1988 (1994)
2. Besang, J.: Spatial interaction and statistical analysis of lattice sytems. J. Royal Staistical Soc. B 361(2), 192–236 (1974)
3. Cocosco, C.A., Kollokian, V., Kwan, R.K., Evans, A.C.: Brain web: Online interface to a 3DMRI simulated brain database. NeuroImage 5(2), Part 2/4, S425 (1997) (Proceedings of the 3rd International Conference on Functional Mapping of the Human Brain, Copenhagen, May 1997)
4. Ding, E., Kularatna, T., Satter, M.: Volumetric image registration by template matching. In: Medical Imaging 2000, pp. 1235–1246. SPIE, Bellinham, WA (2000)
5. Du, J., Tang, S., Jiang, T., Lu, Z.: Intensity based robust similarity for multimodal image registration. International Journal of Computer Mathematics 83, 49–57 (2006)
6. Fitzpatrick, J., West, J., Maurer, C.: Predicting error in rigid-body, point-based registration. IEEE Tranasctions on Medical Imaging 17, 694–702 (1998)
7. Frantz, S., Rohr, K., Stiehl, H.S., Kim, S.I., Weese, J.: Validation point-based MR/CT registration based on semi-automatic landmark extraction. In: Proceeding of CARS 1999, pp. 233–237. Elsevier, Amsterdam (1999)
8. Geman, S., Geman, D.: Stochastic relaxation, Gibbs distribution, and the Bayesian restoration of images. IEEE Transactions on Pattern Analysis and Machine Intelligence 6(6), 721–741 (1984)
9. Gonzales, R.C., Woods, R.E., Eddins, S.L.: Digital Image Processing Using Matlab. Prentice-Hall, NJ (2004)
10. Guimond, A., Roche, A., Ayache, N., Meunier, J.: Three-Dimensional Multimodal Brain Wraping Using the Demons Algorithm and Adaptive Intensity Corrections. IEEE Transaction on Medical Imaging 20(1), 58–69 (2001)
11. Hsu, L., Loew, M.H., Ostuni, L.J.: Automated registration of CT and MR brain images using hierarchical shape representation. IEEE Engineering in Medicine and Biology Magazine 18, 40–47 (1999)
12. Li, S.Z.: Markov Random Field Modeling in Computer Vision. Springer, Berlin (1995)
13. Marroquin, J.L., Mitter, S., Poggio, T.: Probabilistic Solution of Ill-Posed problems in Computational Vision. J. Am. Statistical Assoc. 82(397), 76–89 (1987)
14. Marroquin, J.L.: Detrministic Interactive Particle Models for Image Processing and Computer Graphics. Graphical Models and Image Processing 55(5), 408–417 (1996)
15. Nocedal, J., Wright, S.J.: Numerical Optimization. Springer, Heidelberg (1999)

16. Shekhar, R., Zagrodsky, V.: Mutual Information-based rigid and non-rigid regis-
    tration of ultrasound volumes. IEEE Transaction on Medical Imaging 21, 9–22
    (2002)
17. Thirion, J.-P.: Image matching as a diffusion process: An analogy with Maxwell's
    demons. Med. Image Anal. 2, 243–260 (1998)
18. Viola, P.A., Wells III, W.M., Atsumi, H., Nakajima, S., Kikinis, R.: Multi-modal
    Volumen Registration by Maximization of Mutual Infromation. Medical Image
    Analysis 1, 5–51 (1996)
19. Woods, R.P., Mazziotta, J.C., Cherry, S.R.: MRI-PET registration with automated
    algorithm. Journal of Computer Asisted Tomography 17, 536–546 (1993)

A Secret Sharing Scheme for Digital Images Based on Two-Dimensional Linear Cellular Automata

Angel Martín del Rey

Department of Applied Mathematics, E.P.S. de Ávila, Universidad de Salamanca
C/Hornos Caleros 50, 05003-Ávila, Spain
delrey@usal.es

Abstract. In this paper a new secret sharing scheme for digital images is introduced. It is based on the use of very simple two-dimensional linear cellular automata and their algebraic properties. It is shown that the scheme presented is ideal and perfect.

1 Introduction

The advent of personal computers and the Internet has made it possible for anyone to distribute worldwide digital information easily and economically. In this new environment, there are several security problems associated with the processing and transmission of digital images over an open network: It is necessary to assure the confidentiality, the integrity and the authenticity of the digital image transmitted. To meet these challenges, a wide variety of cryptographic protocols have appeared in the scientific literature.

Secret sharing schemes are multi-party cryptographic protocols originally related to key establishment. Specifically, they are procedures that allow one to share a secret among a set of participants such that only qualified subsets of these participants (called access structure) can recover the original secret. Moreover, no information about it can be obtained when non-qualified subsets of participants try to recover the secret. The original motivation for secret sharing was to safeguard cryptographic keys from loss. Currently, they have many applications in different areas such as access control, opening a safety deposit box, etc. The basic example of secret sharing schemes are the (m, n)-threshold schemes, where m and n are integer numbers such that $1 \leq m \leq n$. Those are methods by which a third trusted party (called the dealer) computes n secret shares (or shadows) from an initial secret S and securely distributes them among the n participants. Only subsets of m or more of these participants who pool their shares may easily recover the original secret, but any group knowing only $m - 1$ or fewer shares are unable to recover the secret. Of special interest are $(2, n)$-threshold schemes.

A secret sharing scheme is said to be perfect if the shares corresponding to each unauthorized subset of participants provide absolutely no information about the secret S. Moreover, when the sizes of the secret and the shares are equal, the secret sharing scheme is called ideal.

V.E. Brimkov, R.P. Barneva, H.A. Hauptman (Eds.): IWCIA 2008, LNCS 4958, pp. 318–329, 2008.

These schemes were introduced independently by Shamir (see Ref. [14]) and Bakley (see Ref. [2]) in 1979 and they are based on the use of Lagrange interpolation polynomial and the intersection of affine hyperplanes, respectively. Since then several proposals have been appeared in the literature based on different mathematical primitives: matrix theory, prime numbers, etc. (see, for example [12]) Those protocols are specially designed for digital data instead of digital images. Due to the main characteristic of digital images (they have a large amount of datum and the difference between two neighboring datum is very small), it is very difficult to apply directly traditional secret sharing schemes to digital images.

The first proposal to share digital images was due to Naor and Shamir (see [10]) and it is called Visual Cryptography. It is based on visual threshold schemes k of n, i.e. the secret image is divided in n shares such that each of them is photocopied in a transparency and then, the original image is recovered by superimposing any k transparencies but no less. In the last years several construction methods based on Visual Cryptography have been proposed (see, for example, [5,6,9,15,18,23]). Besides the Visual Cryptography, another protocols to share digital images have been designed based on vector quantization ([4]), Shamir's ideas ([16,20,22]), Sharing circle ([7,13]) etc.

The main goal of this work is to propose a novel secret sharing scheme based on the use of a particular type of two-dimensional discrete dynamical system called linear cellular automata. Specifically, our scheme is a $(2,n)$ secret sharing scheme whose main advantages are: (1) There is no data expansion; (2) There is no loss of resolution in the recovered image; (3) It is secure against the most important cryptanalytic attacks; (4) The computational complexity is low since the rules governing the cellular automata used involves only XOR operations. The first proposal based on cellular automata to share a secret was introduced in [1]. This scheme is a degenerate secret sharing scheme whose computational complexity is higher than the algorithm proposed in this paper. Moreover, the access structure given by proposal in [1] is more restricted.

The rest of the paper is organized as follows: In section 2 the basic theory about two-dimensional linear cellular automata is introduced; the secret sharing scheme is presented in section 3; in section 4, the security analysis of the protocol is shown; the generalization of the protocol to obtain a different access structure is presented in section 5 and finally, the conclusions and further work are shown in section 6.

2   Two-Dimensional Linear Cellular Automata

Two-dimensional linear cellular automata (LCA) are finite state machines formed by a collection of $r \times c$ memory units called cells which are uniformly arranged into a rectangular lattice. At each time step, they are endowed with a state from the state set given by the finite field $\mathbb{F}_2 = \{0, 1\}$ (see, for example, [17,19,21]). The state of a particular cell is updated synchronously according to a specified linear deterministic function, whose variables are the states of the neighbor cells at the previous time step. The set of all neighbors is called the neighborhood and usually Moore neighborhoods (formed by the main cell and its eight nearest cells around it) are considered. As a consequence, the local transition function is as follows:

$$
\begin{aligned}
s_{i,j}^{t+1} &= f\left(s_{i-1,j-1}^{t}, s_{i-1,j}^{t}, s_{i-1,j+1}^{t}, s_{i,j-1}^{t}, s_{i,j}^{t}, s_{i,j+1}^{t}, s_{i+1,j-1}^{t}, s_{i+1,j}^{t}, s_{i+1,j+1}^{t}\right) \\
&= a_0 \cdot s_{i-1,j-1}^{t} + a_1 \cdot s_{i-1,j}^{t} + a_2 \cdot s_{i-1,j+1}^{t} + a_3 \cdot s_{i,j-1}^{t} + a_4 \cdot s_{i,j}^{t} + a_5 \cdot s_{i,j+1}^{t} \\
&\quad + a_6 \cdot s_{i+1,j-1}^{t} + a_7 \cdot s_{i+1,j}^{t} + a_8 \cdot s_{i+1,j+1}^{t} \,(\mathrm{mod}\,2)\,,
\end{aligned}
\tag{1}
$$

where $s_{i,j}^{t}$ stands for the state of the cell (i,j) at time t, and $a_k \in \mathbb{F}_2, 0 \le k \le 8$. Note that there are $2^9 = 512$ possible LCA.

To assure a well-defined evolution of the LCA, it is necessary to establish some type of boundary conditions. In this work we will consider null boundary conditions: $s_{i,j}^{t} = 0$ if and only if $i < 1$ or $i > r$ or $j < 1$ or $j > c$.

The configuration of the LCA at a time step t is the matrix $C^t = \left(s_{ij}^{t}\right)$, where $1 \le i \le r, 1 \le j \le c$, which is formed by all the states of the cells of the LCA at time t. The transformation Φ which yields the configuration at the next time step during the evolution of the LCA is called the global transition function, that is: $C^{t+1} = \Phi\left(C^t\right)$.

Our work deals with those LCA with the following local transition functions:

$$
s_{i,j}^{t+1} = s_{i,j-1}^{t} + s_{i,j+1}^{t} \,(\mathrm{mod}\,2)
\tag{2}
$$

$$
s_{i,j}^{t+1} = s_{i,j-1}^{t} + s_{i,j}^{t} + s_{i,j+1}^{t} \,(\mathrm{mod}\,2)\,,
\tag{3}
$$

We can interpret them in terms of Linear Algebra (see Ref. [3]): Their evolutions are given by the matrix expression: $F_i^{t+1} = M \cdot F_i^{t} \,(\mathrm{mod}\,2)$, where F_i^{t} is the i-th row of C^t and M is called the transition matrix. Its explicit expression is:

$$
M = \begin{pmatrix}
\alpha & 1 & 0 & \cdots & 0 & 0 \\
1 & \alpha & 1 & \cdots & 0 & 0 \\
0 & 1 & \alpha & \cdots & 0 & 0 \\
\vdots & \vdots & \vdots & \ddots & \vdots & \vdots \\
0 & 0 & 0 & \cdots & \alpha & 1 \\
0 & 0 & 0 & \cdots & 1 & \alpha
\end{pmatrix},
\tag{4}
$$

with $\alpha = 0$ for the local transition function (2) and $\alpha = 1$ for (3). Moreover, the LCA with local rule (2) is non-reversible iff k is even, whereas the LCA given by (3) is non-reversible iff $k \equiv 2 \,(\mathrm{mod}\,3)$ (see Ref. [11]).

A simple computation shows that $\Phi_0 + \Phi_1 = Id\,(\mathrm{mod}\,2)$, where Φ_0 and Φ_1 are the global transition functions of LCA with local transition rules (2) and (3) respectively. Moreover, the following result holds:

Theorem 1. $\Phi_0^t + \Phi_1^t = Id\,(\mathrm{mod}\,2)$ *if and only if* $t = 2^m$, *with* $m \in \mathbb{Z}^+$.

Proof. Let M_0 and M_1 be the characteristic matrices of LCA given by (2) and (3), respectively. Then:

$$Id \equiv (M_0 + M_1)^t \equiv M_0^t + \sum_{j=1}^{t-1} \left(\binom{t}{j} M_0^j \cdot M_1^{t-j} \right) + M_0^t \pmod 2$$

$$\equiv M_1^{2^m} + M_0^{2^m} + \sum_{j=1}^{2^m-1} \left(\binom{2^m}{j} M_0^j \cdot M_1^{2^m-j} \right) \pmod 2$$

$$\equiv M_0^{2^m} + M_1^{2^m} \pmod 2, \tag{5}$$

and we conclude.

Note that, in general, the evolution of a LCA considers that the configuration at time $t+1$ of the LCA depends only on its configuration at the previous time step t, that is, $C^{t+1} = \Phi(C^t)$. This is the standard paradigm for the evolution of cellular automata; nevertheless, one can also assume that C^{t+1} not only depends on C^t, but also on the configurations at the previous time step: $C^{t+1} = \Psi(C^t, C^{t-1})$. This new kind of LCA is called second order memory LCA (MLCA for short). Specifically, this work deals with MLCA whose local transition function is of the following form:

$$s_{i,j}^{t+1} = f\left(s_{i-1,j-1}^t, s_{i-1,j}^t, s_{i-1,j+1}^t, s_{i,j-1}^t, s_{i,j}^t, s_{i,j+1}^t, s_{i+1,j-1}^t, s_{i+1,j}^t, s_{i+1,j+1}^t\right)$$
$$+ s_{i,j}^{t-1} \pmod 2, \tag{6}$$

where f is the local linear transition function defined in (1). This cellular automata was introduced by Fredkin (see [8]) and it is reversible, that is, the evolution backwards is possible by means of the inverse cellular automata with local transition function:

$$s_{i,j}^{t+1} = -f\left(s_{i-1,j-1}^t, s_{i-1,j}^t, s_{i-1,j+1}^t, s_{i,j-1}^t, s_{i,j}^t, s_{i,j+1}^t, s_{i+1,j-1}^t, s_{i+1,j}^t, s_{i+1,j+1}^t\right)$$
$$+ s_{i,j}^{t-1} \pmod 2. \tag{7}$$

3 The Protocol to Share Digital Images

In this section we will describe the protocol to share a secret digital image I among a set of n participants (n even): $P_1^1, P_1^2, \ldots, P_{n/2}^1, P_{n/2}^2$. Specifically, it is a $(2, n)$-threshold scheme, and it consists of three phases: The setup and sharing phases, which are carried out by the dealer, and the recovery phase, which is carried out by the participants.

3.1 The Setup Phase

In this phase the dealer chooses the cellular automata and the parameters that will be used during the protocol. In this case three cellular automata are considered: two LCA and one MLCA. They are defined by the following local transition functions:

$$s_{ij}^{t+1} = s_{i-1,j-1}^t + s_{i-1,j}^t + s_{i-1,j+1}^t + s_{i,j-1}^t + s_{i,j}^t$$
$$+ s_{i,j+1}^t + s_{i+1,j-1}^t + s_{i+1,j}^t + s_{i+1,j+1}^t + s_{i,j}^{t-1} \pmod 2, \tag{8}$$

$$s_{i,j}^{t+1} = s_{i,j-1}^t + s_{i,j+1}^t \pmod 2, \tag{9}$$

$$s_{i,j}^{t+1} = s_{i,j-1}^t + s_{i,j}^t + s_{i,j+1}^t \pmod 2. \tag{10}$$

Note that these functions only involve XOR operations. Also, the dealer randomly chooses $n/2$ positive integer numbers $1 < m_1 < m_2 < \ldots < m_{n/2}$.

Moreover, let I be the secret digital image which is defined by $r \times c$ pixels and a palette of 2^u colors ($u = 1$ for black and white images, $u = 8$ for gray-level images, and $u = 24$ for full color images). It can be easily converted into a binary matrix J_I of order $r \times u \cdot c$ as follows: Set K_I the matrix of order $r \times c$ whose (i,j)-th coefficient stands for the color code of the (i,j) pixel of the image; then, J_I is obtained from K_I by simply considering the binary representation of each coefficient. In this way a digital image can be represented as a configuration of a cellular automata with state set $\mathbb{F}_2$.

3.2   The Sharing Phase

There are three steps in this phase:

- The dealer divides the image I into two subimages of the same size: I_1 and I_2, defined by $r/2 \times c$ pixels. If r is odd, then a new row of random pixels can be added to the end of the image I to obtain a new image, which can be also called I, with an even number of rows. Let J_{I_1} and J_{I_2} be the binary matrices associated to these subimages, then the dealer computes the following configurations using the MLCA defined by (8):

$$C^0 = J_{I_1}, C^1 = J_{I_2}, C^t = \Psi\left(C^{t-1}, C^{t-2}\right), 2 \leq t \leq N, \tag{11}$$

  where Ψ stands for the global function of the MLCA and $N > 2$ is a secret parameter. Thus a confused image $\tilde{I}$ is obtained from the union of the subimages defined by C^{N-1} and C^N. Set $J_{\tilde{I}}$ the associated binary matrix to $\tilde{I}$.
- The dealer computes the following configurations using the non-reversible LCA defined by (9) and (10): $\Phi_0^{2^{m_i}}\left(J_{\tilde{I}}\right), \Phi_1^{2^{m_i}}\left(J_{\tilde{I}}\right)$, where $1 \leq i \leq n/2$. Note that the order of the binary matrix $J_{\tilde{I}}$ is $r \times (u \cdot c)$ and $u \cdot c$ must be an even integer number such that $u \cdot c \equiv 2 \pmod 3$. If $u \cdot c$ does not satisfy these conditions, a suitable number of bits must be padded to obtain the desirable bitlength.
- The dealer securely distributes the parameter N and the images obtained from the shares between the participants: $P_i^1 \leftrightarrow \Phi_0^{2^{m_i}}\left(J_{\tilde{I}}\right), P_i^2 \leftrightarrow \Phi_1^{2^{m_i}}\left(J_{\tilde{I}}\right)$, with $1 \leq i \leq n/2$.

Once the shares are computed, the dealer can destroy the numbers $m_1, \ldots, m_{n/2}$.

3.3   The Recovery Phase

In this phase the authorized pairs of users, P_i^1, P_i^2, can recover the secret digital image I as follows:

- First of all, they compute the confused image $\tilde{I}$ by the following XOR operation: $J_{\tilde{I}} = \Phi_0^{2^{m_i}} (J_{\tilde{I}}) + \Phi_1^{2^{m_i}} (J_{\tilde{I}}) \,(\mathrm{mod}\, 2)$. Note that the algebraic property stated in Theorem 1 is used.
- Subsequently, the confused image $\tilde{I}$ is divided into two subimages: $\tilde{I}_1$ and $\tilde{I}_2$ of order $r/2 \times c$ and the secret image I is obtained by computing the inverse evolution of the MLCA as follows: $C^0 = J_{\tilde{I}_2}, C^1 = J_{\tilde{I}_1}, C^t = \Psi^{-1} \left(C^{t-1}, C^{t-2} \right)$, with $2 \le t \le N$.

4   The Security Analysis

4.1   Main Properties of the Protocol

The secret sharing scheme proposed in this work is perfect since it involves non-reversible LCA. Consequently, from one configuration of the form $\Phi_0^{2^m} (J_{\tilde{I}})$ or $\Phi_1^{2^m} (J_{\tilde{I}})$ it is impossible to determine the initial configuration $J_{\tilde{I}}$. Moreover, this scheme can be also considered as ideal since in "good" cases (when the bitlengh of the rows of I satisfies the conditions stated in the sharing phase), the size of the shares is exactly the same than the size of the secret to be shared. Otherwise, the bitlength between the secret and the shares differs in a few bits. Consequently, there is not a significant data expansion. Moreover, since the images are considered as configurations of cellular automata, no loss of resolution is presented in the recovered image.

4.2   Statistical Analysis

We have performed a statistical analysis in order to prove the confusion and diffusion properties of the proposed protocol, which allows it to strongly resist statistical attacks. Specifically the histograms of original image and the shares are checked and the correlation coefficients are computed.

Let us consider the 256 gray-scale image of size 128×128 given in Figure 1-(a). Its histogram is shown in Figure 1-(a'). The shared images computed by means of LCA Φ_0 and the following artificially chosen parameters: $N = 5, m_1 = 1, m_2 = 2, m_3 = 3$ are shown in Figure 1-(b),(c) and (d), whereas their corresponding histograms are shown in Figure 1-(b'),(c') and (d'). The shares obtained from LCA Φ_1 and their histograms are shown in Figure 2. From the figures one can see that the histograms of the shares are fairly uniform and they are significantly different from that of the original image. It demonstrates that the secret sharing algorithm has covered up all the characters of the original image and shows good performance of balanced 0-1 ratio and zero correlation.

The following procedure will be carryied out to test the correlation between two adjacent pixels in the original and the shared images: First of all randomly

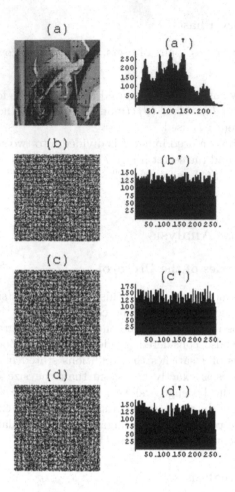

Fig. 1. (a) Lena's picture defined by 128×128 pixels and 256 gray levels; (b) Share 1.1 computed for $m_1 = 1$ and Φ_0; (c) Share 2.1 computed for $m_2 = 2$ and Φ_0; (d) Share 3.1 computed for $m_3 = 3$ and Φ_0; (a') Histogram of Lena image; (b') Histogram of Share 1.1; (c') Histogram of Share 2.1; (d') Histogram of Share 3.1

select 1.000 pairs of two adjacent pixels from the image, and then, calculate the correlation coefficient of each pair by using the following formula:

$$r_{xy} = \frac{cov\,(x,y)}{\sqrt{D\,(x)}\sqrt{D\,(y)}},\tag{12}$$

where x and y are the grey-scale values of the two adjacent pixels in the image and:

$$cov\,(x,y) = \frac{1}{N}\sum_{i=1}^{N}(x_i - E\,(x))\,(y_i - E\,(y)),\tag{13}$$

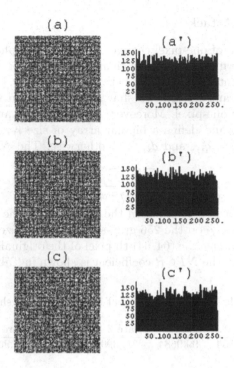

Fig. 2. (a) Share 1.2 computed for $m_1 = 1$ and Φ_1; (b) Share 2.2 computed for $m_2 = 2$ and Φ_1; (c) Share 3.2 computed for $m_3 = 3$ and Φ_1; (a') Histogram of Share 1.2; (b') Histogram of Share 2.2; (c') Histogram of Share 3.2.

$$E\left(x\right) = \frac{1}{N}\sum_{i=1}^{N} x_i, \quad D\left(x\right) = \frac{1}{N}\sum_{i=1}^{N}\left(x_i - E\left(x\right)\right)^2. \qquad (14)$$

As a consequence, the results obtained are shown in Table 1.

Table 1. Correlation coefficients of two adjacent pixels

	Horizontal	Vertical	Diagonal
Lena image	0.9418	0.8510	0.8441
Share 1.1	0.0239	0.0448	0.0018
Share 2.1	−0.0107	0.0108	−0.0587
Share 3.1	0.0728	0.0288	0.0259
Share 1.2	0.0087	0.0039	−0.0145
Share 2.2	0.0099	−0.0214	0.0123
Share 3.2	0.0674	−0.0495	−0.0238

The correlations coefficients of the original image and shared images are far apart (note that the correlation coefficients of Lena's picture are close to 1, whereas the corresponding coefficients of the shared images are very close to 0). Consequently, the secret sharing algorithm satisfies zero co-correlation.

4.3   Differential Attack

To test the influence of one-pixel change on the whole ciphered image, a usual measure is used: the number of pixels change rate, $NPCR$, which measures the percentage of different pixel numbers between two images.

Suppose that S_1 and S_2 are two shares obtained from two original images which differ in only one-pixel. Moreover, set $K_1 = (p_{ij})$ and $K_2 = (q_{ij})$ their associated matrices, and define a bipolar array of size $r \times c$, $D = (d_{ij})$, such that $d_{ij} = 0$ if $p_{ij} = q_{ij}$, and $d_{ij} = 1$ otherwise. The $NPCR$ is defined as follows:

$$NPCR = \frac{\sum_{i=1}^{r} \sum_{j=1}^{c} d_{ij}}{r \cdot c} \times 100\%. \tag{15}$$

Some tests have been performed on the proposed scheme about the influence of only one-pixel change on the 256 gray-scale image of size 128×128 given in Figure 1-(a). If we change the $(64, 64)$-th pixel of the original image and its value passes from 29 to 30, the $NPCR$ coefficient is shown in Table 2:

Table 2. $NPCR$ coefficient of the computed shares

	Share 1.1	Share 2.1	Share 3.1	Share 1.2	Share 2.2	Share 3.2
$NPCR$	99.9634	99.9634	99.9390	99.9634	99.9634	99.9329

4.4   Computational Complexity

The number of bit operations involving the sharing phase and the recovery phase are different. Specifically, the following results hold:

- In the sharing phase, the number of bit operations necessary to compute the confused image $\tilde{I}$ is $9 \cdot u \cdot N \cdot r \cdot c$. Moreover, to compute all shared images it is necessary to obtain the $2^{m_n/2}$-th configuration of both LCA. As a simple calculus shows the number of bit operations necessary to compute the shares are: $3 \cdot 2^{m_n/2} \cdot u \cdot r \cdot c - 2^{m_n/2} \cdot r$. Consequently, the whole process takes

$$3 \cdot u \cdot (3 \cdot N + 2^{m_n/2}) \cdot r \cdot c - 2^{2+m_n/2} \cdot r. \tag{16}$$

  bit operations.
- In the recovery phase, to get the secret image I, a qualified pair of users has to perform a XOR operation between their shares ($u \cdot r \cdot c$ bit operations) and to compute inverse iteration by means of the inverse MLCA ($9 \cdot u \cdot r \cdot c$ bit operations). Consequently, to recover the original image $u \cdot (9 \cdot N + 1) \cdot r \cdot c$ bit operations are necessary.

Note that the recovery phase is faster than the sharing phase, and it is an important and desirable feature of such protocols.

5   Generalization of the Protocol

As is stated in the protocol introduced above, only specific pairs of participants: those of the form

$$\{P_i^1, P_i^2\}, \quad 1 \le i \le n/2, \tag{17}$$

can recover the secret image by pooling their shares. As a consequence there exist only $n/2$ qualified subsets of participants. Nevertheless, this protocol can be easily generalized to obtain an access structure with $(n/2)^2$ qualified subsets:

$$\{P_i^1, P_j^2\}, \quad 1 \le i \le n/2, 1 \le j \le n/2, \tag{18}$$

as follows: The setup phase is exactly the same than in the first proposal. In the sharing phase, the dealer distributes the following data among the participants:

$$P_i^1 \leftrightarrow \{\Phi_0^{2^{m_i}}(J_{\bar{I}}), m_i\}, P_i^2 \leftrightarrow \{\Phi_1^{2^{m_i}}(J_{\bar{I}}), m_i\}, \quad 1 \le i \le n/2. \tag{19}$$

Finally, in the recovery phase the participants P_i^1, P_j^2 recover the secret image I as follows:

- They compare the integer numbers m_i and m_j.
- If $m_i < m_j$, then P_i^1 computes $\Phi_0^{2^{m_i} + m_j - m_i}(J_{\bar{I}}) = \Phi_0^{2^{m_j}}(J_{\bar{I}})$. On the other hand, if $m_j < m_i$, then P_j^2 computes $\Phi_1^{2^{m_j} + m_i - m_j}(J_{\bar{I}}) = \Phi_1^{2^{m_i}}(J_{\bar{I}})$.
- The secret is obtained by simply applying the inverse MLCA and computing: $\Phi_0^{2^{m_j}}(J_{\bar{I}}) + \Phi_1^{2^{m_j}}(J_{\bar{I}}) \, (\text{mod } 2)$ in the first case, and $S = \Phi_0^{2^{m_i}}(J_{\bar{I}}) + \Phi_1^{2^{m_i}}(J_{\bar{I}}) \, (\text{mod } 2)$ in the second case.

Note that this modification of the protocol allows one to obtain an access structure given by two disjoint subsets,

$$A_1 = \{P_i^1, 1 \le i \le n/2\}, A_2 = \{P_j^2, 1 \le j \le n/2\}, \tag{20}$$

such that to recover the secret it is necessary that two participants (one from A_1 and other from A_2) pool their shares.

6   Conclusions

In this work a novel $(2, n)$-secret sharing scheme has been presented. It is based on the use of three linear two-dimensional cellular automata: one of them is a reversible memory linear cellular automata and the another are non-reversible linear cellular automata. The protocol consists of three phases: The setup phase where the local transition functions are determined; The sharing phase where the evolutions of the cellular automata are computed to obtain the shares; and the recovery phase where the original image is computed by means of a XOR operation and the evolution of the inverse memory cellular automata.

Its main features are the following:

- There is neither data expansion nor loss of resolution since the image is considered as a configuration of a cellular automata.
- It is a perfect scheme since it involves non-reversible cellular automata. Moreover, the shares exhibits good statistical properties.
- The computational complexity is low since the rules governing the cellular automata used involves only XOR operations.

Further work is aimed at designing an $(2, n)$-threshold scheme with a more general access structure and a secret sharing (m, n)-scheme, with $m > 2$.

Acknowledgements

This work has been partially supported by the Consejería de Educación (Junta de Castilla y León, Spain) under grant SA110A06.

References

1. Álvarez, G., Hernández, L., Martín, A.: A New Secret Sharing Scheme for Images based on Additive 2-Dimensional Cellular Automata. In: Marques, J.S., Pérez de la Blanca, N., Pina, P. (eds.) IbPRIA 2005. LNCS, vol. 3522, pp. 411–418. Springer, Heidelberg (2005)
2. Blakley, G.R.: Safeguarding cryptographic keys. AFIPS Conf. Proc. 48, 313–317 (1979)
3. Chaudhuri, P., Chowdhury, D., Nandi, S., Chattopadhyay, S.: Additive cellular automata. In: Theory and Applications, vol. 1, IEEE Computer Society Press, Los Alamitos (1997)
4. Chang, C., Hwang, R.: Sharing secret images using shadow codebooks, Inform. Sciences 111, 335–345 (1998)
5. Chen, Y., Chan, Y., Huang, Ch., Tsai, M., Chu, Y.: A multiple-level visual secret-sharing scheme without image size expansion. Inform. Sciences 177, 4696–4710 (2007)
6. Cimato, S., De Prisco, R., De Santis, A.: Colored visual cryptography without color darkening. Theor. Comput. Sci. 374, 261–276 (2007)
7. Feng, J., Wu, H., Tsai, C., Chu, Y.: A new multi-secret image sharing scheme using Lagrange's interpolation. J. Syst. Software 76, 327–329 (2005)
8. Fredkin, E.: Digital mechanics. An informal process based on reversible universal cellular automata, Physica D 45, 254–270 (1990)
9. Lin, S., Lin, J.: VCPSS: A two-in-one two-decoding-options image sharing method combining visual cryptography (VC) and polynomial-style sharing (PSS) approaches. Pattern Recogn. 40, 3652–3666 (2007)
10. Naor, M., Shamir, A.: Visual cryptography. In: De Santis, A. (ed.) EUROCRYPT 1994. LNCS, vol. 950, pp. 1–12. Springer, Heidelberg (1995)
11. Martín del Rey, A., Rodríguez Sánchez, G.: On the reversibility of 150 Wolfram cellular automata. Int. J. Mod. Phys. C 17, 975–984 (2006)
12. Menezes, A., van Oorschot, P., Vanstone, S.: Handbook of Applied Cryptography. CRC Press, Boca Raton, FL (1997)

13. De Santis, A., Masucci, B.: New results on non-perfect sharing of multiple secrets. J. Syst. Software 80, 216–223 (2007)
14. Shamir, A.: How to share a secret, Commun. ACM 22, 612–613 (1979)
15. Shyu, S.: Efficient visual secret sharing scheme for color images. Pattern Recogn. 39, 866–880 (2006)
16. Thien, C., Lin, J.: Secret image sharing. Compup. Graph. 26, 765–770 (2002)
17. Toffoli, T., Margolus, N.: Cellular automata machines. The MIT Press, Cambridge, MA (1987)
18. Tsai, D., Chen, T., Horng, G.: A cheating prevention scheme for binary visual cryptography with homogeneous secret images. Pattern Recogn. 40, 2356–2366 (2007)
19. von Neumann, J.: Theory of self-reproducing automata (edited and completed by A.W. Burks). University of Illinois Press, Illinois (1966)
20. Wang, R., Su, C.: Secret image sharing with smaller shadow images. Pattern Recogn. Lett. 27, 551–555 (2006)
21. Wolfram, S.: A new kind of science, Wolfram Media. Champaign, Illinois (2002)
22. Wu, Y., Thien, L., Lin, J.: Sharing and hiding secret images with size constrain. Pattern Recogn. 37, 1377–1385 (2004)
23. Yang, C., Chen, T.: Reduce shadow size in aspect ratio invariant visual secret sharing schemes using a square block-wise operation. Pattern Recogn. 39, 1300–1314 (2006)

Pure 2D Picture Grammars (P2DPG) and P2DPG with Regular Control

K.G. Subramanian[1,2], Atulya K. Nagar[2], and M. Geethalakshmi[3]

[1] School of Mathematical Sciences,
Universiti Sains Malaysia, 11800 Penang, Malaysia
kgs@usm.my
[2] Deanery of Business and Computer Sciences,
Liverpool Hope University, Hope Park, Liverpool L16 9JD UK
nagara@hope.ac.uk
[3] Department of Mathematics,
Dr. MGR Janaki College, Chennai 28, India

Abstract. In this note a new model of grammatical picture generation is introduced. The model is based on the notion of pure context-free grammars of formal string language theory. The resulting model, called Pure 2D context-free grammar (CFG), generates rectangular picture arrays of symbols. The generative power of this model in comparison to certain other related models is examined. Also we associate a regular control language with a Pure 2D CFG and notice that the generative power increases. Certain closure properties are obtained.

1 Introduction

Theoretical studies on digital pictures and picture analysis include syntactic techniques as one of the main areas of study. In the problem of generation and description of picture patterns considered as connected, digitized, finite arrays of symbols, syntactic methods have played a significant role on account of their structure-handling ability. Several picture language generating devices have been introduced in the literature based on generalizing to two dimensions different kinds of grammars like the Chomskian string grammars, the Lindenmayer systems (L systems) and so on and adapting the techniques and results of formal string language theory. See for example [5,6,14,1].

One of the earliest picture models was proposed by Siromoney et al [9], motivated by certain floor designs called "kolam" patterns. In this two-dimensional model, which we call as Siromoney matrix grammar, generation of rectangular arrays takes place in two phases with a sequential mode of rewriting in the first phase generating strings of intermediate symbols and a parallel mode of rewriting these strings in the second phase to yield rectangular picture patterns. Recently there has been a renewed interest in the study of Siromoney matrix grammars [12,13].

Another very general rectangular array generating model, called extended controlled tabled L array system (ECTLAS) was proposed by Siromoney and

V.E. Brimkov, R.P. Barneva, H.A. Hauptman (Eds.): IWCIA 2008, LNCS 4958, pp. 330–341, 2008.

Siromoney [10], incorporating into arrays the developmental type of generation used in the well-known biologically motivated L-systems. Here the symbols either on the left, right, up or down borders of a rectangular array are rewritten simultaneously by equal length strings to generate rectangular picture arrays.

Pure context-free grammars [4] which make use of only one kind of symbols, called terminal symbols, unlike the Chomskian grammars, have been investigated in formal string language theory for their language generating power and other properties. In this note we introduce a new two-dimensional grammar, called Pure 2D Context-free grammar (CFG), for picture array generation based on pure context-free rules. Unlike the models in [9,10], we allow rewriting any column or any row of the rectangular array rewritten and do not prescribe any priority of rewriting columns and rows as in [9] in which the second phase of generation can take place only after the first phase is over. We compare the generative power of the new model with those in [9,10,11,2]. Certain closure properties are obtained. Also we associate a regular control language with a Pure 2D CFG and notice that the generative power increases. Interpretation of the letter symbols in picture arrays by primitive patterns is a well-known technique to obtain interesting classes of "kolam" [9] pictures or "chain code" [3,12] pictures and so on. We indicate here chain code interpretation of the picture arrays generated by Pure 2D CF grammars.

2   Preliminaries

Let Σ be a finite alphabet. A word or string $w = a_1 a_2 \ldots a_n$ $(n \geq 1)$ over Σ is a sequence of symbols from Σ. The length of a word w is denoted by $|w|$. The set of all words over Σ, including the empty word λ with no symbols, is denoted by Σ^*. We call words of Σ^* as horizontal words. For any word $w = a_1 a_2 \ldots a_n$, we denote by w^T the vertical word

$$
\begin{array}{c}
a_1 \\
a_2 \\
\vdots \\
a_n
\end{array}
$$

We also define $(w^T)^T = w$. We set λ^T as λ itself. A rectangular $m \times n$ array M over Σ is of the form

$$
M = \begin{array}{ccc}
a_{11} & \cdots & a_{1n} \\
\vdots & \ddots & \vdots \\
a_{m1} & \cdots & a_{mn}
\end{array}
$$

where each $a_{ij} \in \Sigma, 1 \leq i \leq m, 1 \leq j \leq n$. The set of all rectangular arrays over Σ is denoted by Σ^{**}, which includes the empty array λ. $\Sigma^{**} - \{\lambda\} = \Sigma^{++}$. We denote respectively by $\circ$ and $\diamond$ the *column concatenation* and *row concatenation* of arrays in V^{**}. In contrast to the case of strings, these operations are partially defined, namely, for any $X, Y \in V^{**}$, $X \circ Y$ is defined if and only if X and Y

have the same number of rows. Similarly $X \diamond Y$ is defined if and only if X and Y have the same number of columns.

We refer to [5,6] for array grammars. For notions of formal language theory we refer to [8]. We briefly recall pure context-free grammars [4] and the rectangular picture generating models in [9,10,11,1,2].

Definition 1 ([4]). *A pure context-free grammar is $G = (\Sigma, P, \Omega)$ where Σ is a finite alphabet, Ω is a set of axiom words and P is a finite set of context-free rules of the form $a \to \alpha, a \in \Sigma, \alpha \in \Sigma^*$. Derivations are done as in a context-free grammar except that unlike a context-free grammar, there is only one kind of symbol, namely the terminal symbol. The language generated consists of all words generated from each axiom word.*

Example 1. The pure context-free grammar $G = (\{a, b, c\}, \{c \to acb\}, \{acb\})$ generates the language $\{a^n cb^n / n \geq 1\}$.

We restrict ourselves to recalling Tabled $0L$ array systems (T0LAS) introduced in [10] for generating rectangular picture arrays.

Definition 2. *A tabled $0L$ array system (T0LAS) is $G = (T, \mathcal{P}, M_0)$ where*
- *T is a finite nonempty set (the alphabet of G);*
- *$\mathcal{P}$ is a finite set of tables, $\{t_1, t_2, \ldots, t_k\}$, and each t_i, $i = 1, \ldots, k$, is a left, right, up, or down table consisting respectively, of a finite set of left, right, up, or down rules only. The rules within a table are context-free in nature but all right hand sides of rules within the same table are of the same length;*
- *$M_0 \in \Sigma^{++}$ is an axiom array of G.*

A derivation in G takes place as follows: Starting with a rectangular array $M_1 \in \Sigma^{++}$, all the symbols of either the rightmost or leftmost column or the uppermost or lowermost row of M_1 are rewritten in parallel respectively by the rules of a left or a right table or an up or a down table to yield a rectangular array M_2. A set $\mathcal{M}(G)$ of rectangular arrays is called a Tabled $0L$ array language (T0LAL) if and only if there exists a tabled $0L$ array system G such that $\mathcal{M}(G) = \{M | M_0 \Rightarrow^* M, M \in T^{**}\}$. The family of Tabled $0L$ array languages is denoted by $L(T0LAL)$.

In the 2D grammar model introduced in [9], which we call as Siromoney Matrix grammar, a horizontal word $S_{i1} \ldots S_{in}$ over intermediate symbols is generated by a Chomskian grammar. Then from each intermediate symbol S_{ij} a vertical word of the same length over terminal symbols is derived to constitute the jth column of the rectangular array generated. We recall this model restricting to regular and context-free cases.

Definition 3. *A Siromoney matrix grammar is a $2-tuple$ (G_1, G_2) where*

$G_1 = (H_1, I_1, P_1, S)$ is a regular or context-free grammar,
H_1 is a finite set of horizontal nonterminals,
$I_1 = \{S_1, S_2, \cdots, S_k\}$, a finite set of intermediates, $H_1 \cap I_1 = \emptyset$,
P_1 is a finite set of production rules called horizontal production rules,
S is the start symbol, $S \in H_1$,

$G_2 = (G_{21}, G_{22}, \cdots, G_{2k})$ where
$G_{2i} = (V_{2i}, T, P_{2i}, S_i), 1 \leq i \leq k$ are regular grammars,
V_{2i} is a finite set of vertical nonterminals, $V_{2i} \cap V_{2j} = \emptyset$, $i \neq j$,
T is a finite set of terminals,
P_{2i} is a finite set of right linear production rules of the form
$X \longrightarrow aY$ or $X \longrightarrow a$ where $X, Y \in V_{2i}, a \in T$
$S_i \in V_{2i}$ is the start symbol of G_{2i}.

The type of G_1 gives the type of G , so we speak about regular, context-free Siromoney matrix grammars if G_1 is regular, context-free respectively. Derivations are defined as follows: First a string $S_{i1}S_{i2} \cdots S_{in} \in I_1^*$ is generated horizontally using the horizontal production rules of P_1 in G_1. That is, $S \Rightarrow S_{i1}S_{i2} \cdots S_{in} \in I_1^*$. Vertical derivations proceed as follows: We write

$$A_{i1} \cdots A_{in}$$

$$\Downarrow$$

$$a_{i1} \cdots a_{in}$$

$$B_{i1} \cdots B_{in}$$

if $A_{ij} \rightarrow a_{ij}B_{ij}$ are rules in $P_{2j}, 1 \leq j \leq n$. The derivation terminates if $A_j \rightarrow a_{mj}$ are all terminal rules in G_2.

The set $L(G)$ of picture arrays generated by G consists of all $m \times n$ arrays $[a_{ij}]$ such that $1 \leq i \leq m$, $1 \leq j \leq n$ and $S \Rightarrow_{G_1}^* S_{i1}S_{i2} \cdots S_{in} \Rightarrow_{G_2}^* [a_{ij}]$. We denote the picture language classes of regular, CF Siromoney Matrix grammars by $RML, CFML$ respectively.

The regular/context-free Siromoney Matrix grammars were extended in [11] by specifying a finite set of tables of rules in the second phase of generation with each table having either right-linear nonterminal rules or right-linear terminal rules. The resulting families of picture array languages are denoted by TRML and TCFML and are known to properly include RML and CFML respectively.

Based on a well known characterization of recognizable string languages in terms of local languages and projections, an interesting model of Tiling Recognizable languages describing rectangular picture arrays was introduced in [1,2]. We now recall briefly these notions.

Given a rectangular picture array p of size $m \times n$ over an alphabet Σ, $\hat{p}$ is an $(m+2) \times (n+2)$ picture array obtained by surrounding p by the special symbol $\# \notin \Sigma$ in its border. A square picture array of size 2×2 is called a tile. The set of all tiles which are sub-pictures of p is denoted by $B_{2 \times 2}(p)$.

Definition 4. Let Γ be a finite alphabet. A two-dimensional language or picture array language $L \subseteq \Gamma^{**}$ is local if there exists a finite set Θ of tiles over the alphabet $\Gamma \cup \{\#\}$ such that $L = \{p \in \Gamma^{**}/B_{2 \times 2}(\hat{p})\} \subseteq \Gamma^{**}$. The family of local picture array languages will be denoted by LOC.

Definition 5. *A tiling system (TS) is a 4-tuple $T = (\Sigma, \Gamma, \Theta, \pi)$ where Σ and Γ are two finite alphabets, Θ is a finite set of tiles over the alphabet $\Gamma \cup \{\#\}$ and $\pi : \Gamma \rightarrow \Sigma$ is a projection.*

The tiling system T recognizes a picture array language L over the alphabet Σ as follows: $L = \pi(L')$ where $L' = L(\Theta)$ is the local two-dimensional language over Γ corresponding to the set of tiles Θ. We write $L = L(T)$ and we say that L is the language recognized by T. A picture array language $L \subseteq \Sigma^{**}$ is tiling recognizable if there exists a tiling system T such that $L = L(T)$. The family of tiling recognizable picture array languages is denoted by REC.

3   Pure 2D Picture Grammars

We now introduce a new two-dimensional grammar for picture generation. The salient feature of this model is that the shearing effect in replacing a subarray of a given rectangular array is taken care of by rewriting a row or column of symbols in parallel by equal length strings and by using only terminal symbols as in a pure string grammar. This new model is related to the model T0LAS in [10] in the sense that a column or row of symbols of a rectangular array is rewritten in parallel. This feature incorporates into arrays the parallel rewriting feature of the well-known and widely investigated Lindenmayer systems [7]. But the difference between this new model and the T0LAS in [10] is that the rewriting is done only at the "edges" of a rectangular array in a T0LAS whereas here we allow rewriting in parallel of any column or row of symbols. We now define the new grammar model.

Definition 6. *A Pure 2D Context-free grammar (P2DCFG) is a 4-tuple $G = (\Sigma, P_c, P_r, \mathcal{M}_I)$ where*

- *Σ is a finite set of symbols ;*
- *$P_c = \{t_{c_i}/1 \leq i \leq m\}, P_r = \{t_{r_j}/1 \leq j \leq n\};$*

Each $t_{c_i}, (1 \leq i \leq m)$, called a column table, is a set of context-free rules of the form $a \rightarrow \alpha, a \in \Sigma, \alpha \in \Sigma^*$ such that for any two rules $a \rightarrow \alpha, b \rightarrow \beta$ in t_{c_i}, we have $|\alpha| = |\beta|$ where $|\alpha|$ denotes the length of $|\alpha|$;

Each $t_{r_j}, (1 \leq j \leq n)$, called a row table, is a set of context-free rules of the form $c \rightarrow \gamma^T, c \in \Sigma$ and $\gamma \in \Sigma^*$ such that for any two rules $c \rightarrow \gamma^T, d \rightarrow \delta^T$ in t_{r_j}, we have $|\gamma| = |\delta|$;

- *$\mathcal{M}_I \subseteq \Sigma^{**} - \{\lambda\}$ is a finite set of axiom arrays.*

Derivations are defined as follows: For any two arrays M_1, M_2, we write $M_1 \Rightarrow M_2$ if M_2 is obtained from M_1 by either rewriting a column of M_1 by rules of some column table t_{c_i} in P_c or a row of M_1 by rules of some row table t_{r_j} in P_r. $\Rightarrow^*$ is the reflexive transitive closure of $\Rightarrow$.

The picture array language $L(G)$ generated by G is the set of rectangular picture arrays $\{M/M_0 \Rightarrow^* M \in \Sigma^{**}, \text{ for some } M_0 \in \mathcal{M}_I\}$. The family of picture array languages generated by Pure 2D Context-free grammars is denoted by P2DCFL.

$$
M_0 \Rightarrow
\begin{array}{c}
 \\
 \\
x\ b\ b\ x \\
z\ y\ y\ z \\
x\ b\ b\ x \\
 \\

\end{array}
\Rightarrow
\begin{array}{c}
 \\
x\ b\ b\ x \\
x\ b\ b\ x \\
z\ y\ y\ z \\
x\ b\ b\ x \\
x\ b\ b\ x \\

\end{array}
\Rightarrow
\begin{array}{c}
x\ b\ b\ x \\
x\ b\ b\ x \\
x\ b\ b\ x \\
z\ y\ y\ z \\
x\ b\ b\ x \\
x\ b\ b\ x \\
x\ b\ b\ x
\end{array}
\Rightarrow
\begin{array}{c}
x\ b\ b\ b\ x \\
x\ b\ b\ b\ x \\
x\ b\ b\ b\ x \\
z\ y\ y\ y\ z \\
x\ b\ b\ b\ x \\
x\ b\ b\ b\ x \\
x\ b\ b\ b\ x
\end{array}
= M_1
$$

Fig. 1. Derivation $M_{01} \Rightarrow^* M_1$

$$
\begin{array}{c}
x\ x\ x\ y\ x\ x\ x \\
b\ b\ b\ z\ b\ b\ b \\
b\ b\ b\ z\ b\ b\ b \\
b\ b\ b\ z\ b\ b\ b \\
b\ b\ b\ z\ b\ b\ b \\
b\ b\ b\ z\ b\ b\ b
\end{array}
$$

Fig. 2. A picture array M_2

Example 2. Consider the Pure 2D Context-free grammar $G = (\Sigma_1, P_{c_1}, P_{r_1}, \{M_{01}\})$ where $\Sigma_1 = \{x, y, z, b\}$, $P_{c_1} = \{t_{c_1}\}$, $P_{r_1} = \{t_{r_1}\}$

$$
t_{c_1} = \{b \to bb, y \to yy\}, \; t_{r_1} = \left\{ y \to \begin{array}{c} b \\ y \\ b \end{array}, z \to \begin{array}{c} x \\ z \\ x \end{array} \right\}, \; M_{01} = \begin{array}{c} x\ b\ x \\ z\ y\ z \\ x\ b\ x \end{array}
$$

A sample derivation $M_{01} \Rightarrow M_1$, on using $t_{c_1}, t_{r_1}, t_{r_1}, t_{c_1}$ in this order, is given in Figure 1:

Each of the arrays occurring in the derivation given belongs to the picture language generated by G_1.

Example 3. Consider the Pure 2D Context-free grammar $G = (\Sigma_2, P_{c_2}, P_{r_2}, \{M_{02}\})$ where $\Sigma_2 = \{x, y, z, b\}$, $P_{c_2} = \{t_{c_2}\}$, $P_{r_2} = \{t_{r_2}\}$

$$
t_{c_2} = \{y \to xyx, z \to bzb\} \; t_{r_2} = \left\{ x \to \begin{array}{c} x \\ b \end{array}, y \to \begin{array}{c} y \\ z \end{array} \right\} \; M_{02} = \begin{array}{c} x\ y\ x \\ b\ z\ b \end{array}
$$

G_2 generates picture arrays M_2 of the form shown in Figure 2.

Here again we note that the number of rows in the generated picture array need not have any proportion to the number of columns but will have an equal number of columns to the left and right of the middle column $(yz \dots z)^T$.

4   Comparisons and Closure Results

We now compare the new 2D grammar model introduced here with those in [9,10,1,2].

Theorem 1. *The family of P2DCFL is incomparable with the families of RML and CFML but not disjoint with these families.*

Proof. The picture language consisting of rectangular arrays over a single symbol a of all sizes $m \times n(m, n \geq 1)$ is generated by a regular Siromoney matrix grammar G. In fact the language of horizontal words generated in the first phase of G_1 is $\{S_1^n/n \geq 1\}$ where S_1 is an intermediate symbol and the language of vertical words generated by S_1 in the second phase is $\{(a^n)^T/n \geq 1\}$. A corresponding Pure 2D CF grammar consists of a column table with the rule $a \to aa$ and a row table with the rule $a \to \dfrac{a}{a}$ and axiom array a. The incomparability with $CFML$ is due to the fact that it is known [9] that the picture languages in examples 2 and 3 cannot be generated by any context-free Siromoney matrix grammar and hence by any regular Siromoney matrix grammar since each of the generated pictures of example 2, has an equal number of rows above and below the middle row $zy \ldots yz$ and in example 3, each of the generated pictures has an equal number of columns to the left and right of the middle column $(yz \ldots z)^T$. On the other hand a picture language consisting of rectangular arrays of the form $M_1 \circ M_2$ where M_1 and M_2 are rectangular arrays over the symbols a, b respectively with equal number of columns can be generated by a context-free Siromoney matrix grammar with the language of horizontal words $S_1^n S_2^n$ (S_1, S_2 are intermediate symbols) in the first phase and S_1, S_2 generating vertical words over a, b respectively. This picture language, cannot be generated by any Pure 2D context-free grammar since the string language $\{a^n b^n/n \geq 1\}$ is not a pure CFL [4] and an argument similar to this can be done in the two-dimensional case also. The incomparability with RML can be seen by considering a picture language with rectangular arrays each row of which is a word in $a^3 b^3 (ab)^*$, known [4] to be not a Pure CFL.

Theorem 2. *The family of P2DCFL is incomparable with the families of TRML and TCFML but not disjoint with these families.*

Proof. In view of the proper inclusions $RML \subset TRML, CFML \subset TCFML$ and incomparability (Theorem 1) of P2DCFL with RML and CFML , it is enough to note that the picture array language of example 2 generating picture arrays as shown in Figure 1 can neither belong to TRML nor to TCFML, in view of the fact that in the picture arrays in Figure 1 each has an equal number of rows above and below the middle row $zy \ldots yz$.

Theorem 3. *Every language in the family $L(T0LAL)$ is a coding of a Pure 2D CFL.*

Proof. Let L be a picture array language generated by a $T0LAS$ [10] $G = (T, \mathcal{P}, M_0)$. We construct a Pure 2D CFG G' as follows: For each symbol a in the alphabet T of G, we introduce a new distinct symbol A. Let $T' = \{A/a \in T\}$. Each rule of the form $a \to a_1 a_2 \cdots a_m b, A, B \in T', a_i (1 \leq i \leq m), b \in T$ in a right table t, is replaced by a rule $A \to a_1 a_2 \cdots a_m B, A, B \in T', a_i (1 \leq i \leq m), b \in T$. Each rule of the form $a \to a_1 a_2 \cdots a_m b, A, B \in T', a_i (1 \leq i \leq m), b \in T$ in a down table t, is replaced by a rule $A \to (a_1 a_2 \cdots a_m B)^T, A, B \in T', a_i (1 \leq i \leq m), b \in T$. Likewise the rules in left and up tables are replaced by rules constructed with a similar idea. Then $G' = (T \cup T', \mathcal{P}', \{M_0'\})$ where $\mathcal{P}'$ consists of the tables of G with each table having the rules replaced as mentioned above. The modified left and right tables of G become the column tables of G' and the modified up and down tables of G the row tables of G'. The axiom array M_0' is M_0 with its border symbols replaced by the new symbols. Define a coding c (a letter to letter mapping) by $c(A) = a$ where A is the new symbol introduced corresponding to a. It can be seen that $c(L(G')) = L$.

Theorem 4. *The family of Pure 2D Context-free languages is incomparable with LOC and REC.*

Proof. The language of square picture arrays with $1s$ in the main diagonal and $0s$ in other positions is known [1] to be in LOC and the language of square picture arrays over $0s$ is known [1] to be in REC but both these languages cannot be generated by any P2DCFG for the simple reason that the language of square arrays cannot be generated by a P2DCFG as the rewriting of a column and of a row are independent. On the other hand a picture array language L_1 consisting of arrays $M = M_1 \circ c \circ M_1$ where M_1 is a string over a (M is a picture array with only one row) is generated by a P2DCFG with a column rule $c \to aca$ but L_1 is known [1] to be not in REC and hence not in LOC.

It is a well-known tool in formal language theory [8] to control the sequence of application of rules of a grammar by requiring the control words to belong to a language. Generally, if the control words constitute a regular language, the generative power of a grammar might not increase. Here we associate a regular control language with a Pure 2D CFG and notice that the generative power increases.

Definition 7. *A Pure 2D Context-free grammar with a regular control is $G_c = (G, Lab(G), \mathcal{C})$ where G is a Pure 2D Context-free grammar, $Lab(G)$ is a set of labels of the tables of G and $\mathcal{C} \subseteq Lab(G)^*$ is a regular (string) language. The words in $Lab(G)^*$ are called control words of G. Derivations $M_1 \Rightarrow_w M_2$ in G_c are done as in G except that if $w \in Lab(G)^*$ and $w = l_1 l_2 \ldots l_m$ then the tables of rules with labels $l_1, l_2, \ldots l_m$ are successively applied starting with M_1 to yield M_2. The picture array language generated by G_c consists of all picture arrays obtained from the axiom array of G with the derivations controlled as described above. We denote the family of picture array languages generated by Pure 2D Context-free grammars with a regular control by $(R)P2DCFL$.*

Lemma 1. *The Pure 2D Context-free grammar G in example 2 with a regular control language $\{(l_1 l_2)^n / n \geq 1\}$ on the labels l_1, l_2 of the tables t_{c_1}, t_{r_1} respectively, generates picture arrays as shown in Figure 1 but with sizes $(2n+1) \times (n+2), n \geq 1$, and thus having a proportion between the height (the number of rows in a picture array) and width (the number of columns in a picture array). In fact the number of rows above and below the middle row $zy \ldots yz$ equals the number of columns between the leftmost and rightmost columns, namely, $(x \ldots xzx \ldots x)^T$.*

Proof. The tables of rules generating the picture array language in example 2 are $t_{c_1} = \{b \to bb, y \to yy\}$, $t_{r_1} = \left\{ \begin{matrix} b & x \\ y \to y, z \to z \\ b & x \end{matrix} \right\}$. Since the control language on the labels of the tables consists of words $\{(l_1 l_2)^n / n \geq 1\}$, an application of the rules of the table t_{c_1} is immediately followed by an application of the rules of the table t_{r_1} so that the array rewritten grows one column followed by one row above and one row below the middle row $zy \ldots yz$. The resulting array is then collected in the language generated. This process is repeated so that the arrays generated have a proportion between the width and height as mentioned in the statement of the theorem.

Theorem 5. *The family of P2DCFL is properly contained in $(R)P2DCFL$.*

Proof. The containment follows since every P2DCFL is generated by a P2DCFG G and the regular control language is $Lab(\mathcal{C})^*$ itself. The proper containment is a consequence of the Lemma 1.

Generating "square arrays" over one symbol a is of interest in picture array generation. Such square arrays can be generated by a 'simple' P2DCFG with a regular control.

Theorem 6. *The picture array language consisting of square arrays over one symbol a is generated by a P2DCFG with a regular control.*

Proof. The P2DCFG $(\{a\}, \{t_{c_1}\}, \{t_{r_1}\}, a)$ where $t_{c_1} = \{a \to aa\}$, $t_{r_1} = \{a \to \begin{matrix} a \\ a \end{matrix}\}$ with the regular control language $\{(l_1 l_2)^n / n \geq 1\}$ where l_1, l_2 are respectively the labels of t_{c_1}, t_{r_1} can be seen to generate the picture array language consisting of square arrays over one symbol a.

We now examine some of the closure properties of P2DCFL. We also consider operations of transposition, reflection about base, reflection about leg. The operation of transposition of a rectangular array interchanges the rows and columns. The operation of reflection about the base reflects the rectangular array about the bottommost row and of reflection about the leg reflects the rectangular array about the leftmost column.

Theorem 7. *The family of P2DCFL is not closed under union, column catenation, row catenation but is closed under projection, transposition, reflection about the base and reflection about the leg.*

Proof. Let the alphabet be $\{a, b, c, x, y\}$. Non-closure under union follows by the fact that $L_1 = \{X_1 \circ (c^n)^T \circ Y_1 / X_1 \in \{a\}^{++}, Y_1 \in \{b\}^{++}, |X_1|_c = |Y_1|_c\}$ where $|X|_c$ stands for the number of columns of X, is generated by a P2DCFG with a column table consisting of a rule $c \to acb$ and a row table with rules $a \to \begin{matrix} a \\ a \end{matrix}, b \to \begin{matrix} b \\ b \end{matrix}, c \to \begin{matrix} c \\ c \end{matrix}$. Likewise $L_2 = \{X_2 \circ (c^n)^T \circ Y_2 / X_2 \in \{x\}^{++}, Y_2 \in \{y\}^{++}, |X_2|_c = |Y_2|_c\}$ is also generated by a similar P2DCFG. It can be seen that $L_1 \cup L_2$ cannot be generated by any P2DCFG, since such a grammar will require a column table with rules of the forms $c \to acb$ and $c \to xcy$. But then this will yield arrays not in the union.

Non-closure under column catenation of arrays can be seen by considering $L_1 \circ L_2$ and noting that any P2DCFG generating $L_1 \circ L_2$ will again require a column table with rules $c \to acb$ and $c \to xcy$ but then this will lead to generating arrays not in the column catenation $L_1 \circ L_2$. Non-closure under row catenation can be seen in a similar manner.

If L is a picture array language generated by a P2DCFG G and L^T is the transposition of L, then the P2DCFG G' to generate L^T is formed by taking the column tables of G as row tables and row tables as column tables but for a rule $a \to \alpha$ in a column table of G, the rule $a \to \alpha^T$ ($\alpha \in \Sigma^{**}$) is added in the corresponding row table of G' and likewise for a rule $b \to \beta^T$ ($\beta \in \Sigma^{**}$) in a row table of G, the rule $b \to \beta$ is added in the corresponding column table of G'. Closure under the operations of reflection about base, reflection about leg can be seen in a similar manner.

5   Interpretations of Picture Arrays

The idea of interpreting letter symbols in a picture array by primitive patterns is a well-known technique to obtain interesting classes of "kolam" [9] pictures or chain code [3] pictures and so on. We can employ here this technique to generate such picture patterns as an application of the Pure 2D CF grammars. Each symbol of a rectangular array is considered to occupy a unit square in the rectangular grid so that each row or column of symbols in the array respectively occupies a horizontal or vertical sequence of adjacent unit squares. A mapping i, called an interpretation, from the alphabet $\Sigma = \{a_1, a_2, \ldots a_n\}$ of a Pure2DCFG G to a set of primitive picture patterns $\{p_1, p_2, \ldots p_m\}$ is defined such that for $1 \leq i \leq n$, $i(a_i) = p_j$, for some $1 \leq j \leq m$. A primitive picture pattern could be a blank. Given a picture array M over Σ, $i(M)$ is obtained by replacing every symbol $a \in M$ by the corresponding picture pattern $i(a)$. For instance, in Example 2, if we define, using two chain code primitives, namely, $|$, $-$ the

Fig. 3. The alphabetic letter H

interpretation mapping i by $i(x) = i(z) = \;\mid\;$, $i(y) = -$ and $i(b) = blank$ then the interpretation $i(M_1)$ of M_1 in Figure 1 will give a picture of the alphabetic letter H (Figure 3).

Likewise if the primitive picture patterns are those used in "kolam" pictures, we can obtain "kolam" patterns from Pure 2D CFL via suitable interpretation.

6   Conclusion

The picture array generating model based on pure context-free grammars introduced here does not prescribe a priority of rewriting column or row unlike [9,11] and does not allow rewriting only the borders of an array as in [10]. But it requires a "control" to maintain a "proportion" between the number of columns and the number of rows. In the case of string grammars, the class of pure CFLs [4], is included in the class of CFLs. Here we have seen that the family of Pure 2D CFLs becomes incomparable with the family of CFMLs introduced in [9]. But we can extend the model of Pure 2D CFG by allowing nonterminal symbols as well and this might increase the power of this model. It remains to be seen in future whether this kind of an extension will be more powerful than the 2D model in [10]. Also it remains to examine whether other properties [4] of pure string languages carry over to the Pure 2D Context-free grammars.

Acknowledgements

The authors are grateful to the referees for their very useful comments which improved the presentation of the paper.

References

1. Giammarresi, D., Restivo, A.: Two-dimensional languages. In: Rozenberg, G., Salomaa, A. (eds.) Handbook of Formal Languages, vol. 3, pp. 215–267. Springer, Heidelberg (1997)
2. Giammarresi, D., Restivo, A.: Recognizable Picture Languages. International Journal of Pattern Recognition and artificial Intelligence (Special issue on Parallel Image Processing), Nivat, M., Saoudi, A., Wang, P.S.P. (eds.), 31–46 (1992)
3. Maurer, H.A., Rozenberg, G., Welzl, E.: Chain-code picture languages. Lecture notes in Computer science, vol. 153, pp. 232–244. Springer, Heidelberg (1983)
4. Maurer, H.A., Salomaa, A., Wood, D.: Pure Grammars. Information and Control 44, 47–72 (1980)
5. Rosenfeld, A.: Picture Languages - Formal Models for Picture Recognition. Academic Press, New York (1979)
6. Rosenfeld, A., Siromoney, R.: Picture languages - a survey. Languages of design 1, 229–245 (1993)
7. Rozenberg, G., Salomaa, A.: The Mathematical Theory of L systems. Academic Press, New York (1980)
8. Salomaa, A.: Formal languages. Academic Press, London (1973)

9. Siromoney, G., Siromoney, R., Krithivasan, K.: Abstract families of matrices and picture languages. Computer Graphics and Image Processing 1, 234–307 (1972)
10. Siromoney, R., Siromoney, G.: Extended Controlled Tabled L- arrays. Information and Control 35(2), 119–138 (1977)
11. Siromoney, R., Subramanian, K.G., Rangarajan, K.: Parallel/Sequential rectangular arrays with tables. International Journal of Computer Mathematics, 143–158 (1977)
12. Stiebe, R.: Picture generation using matrix systems. Journal of Information Processing and Cybernetics 28, 311–327 (1992)
13. Stiebe, R.: Slender Siromoney Matrix Languages. In: Proceedings of the 1st International Conference on Language and Automata: Theory and Applications, Tarragona, Spain (2007)
14. Wang, P.S.P.: Array grammars, Patterns and recognizers. World Scientific, Singapore (1989)

A Deterministic Turing Machine for Context Sensitive Translation of Braille Codes to Urdu Text

Muhammad Abuzar Fahiem

Lahore College for Women University, Lahore, Pakistan
University of Engineering and Technology, Lahore, Pakistan
abuzar@uet.edu.pk

Abstract. In this paper we have developed a context sensitive translator for optically recognizing the Braille codes for Urdu language. Urdu is a context sensitive language and have different glyphs of an alphabet depending upon the position of use. Our research is aimed at bridging the gap between blind and sighted people. We developed Braille codes for Urdu, scanned the Braille, recognized it optically, devised a deterministic Turing machine for context sensitive translation and generated the output in Urdu.

Keywords: optical recognition, context sensitive translation, Braille, deterministic Turing machine, Urdu glyphs.

1   Introduction

Braille, the language of blind, was developed by Louis Braille in 1821. Each Braille cell consists of six raised dots in a two column format, each column containing three dots. There is an extended version of Braille too, having eight dots with four dots in each column. A typical Braille page is 280x292 mm with approximately 40 Braille cells per line and 25 lines. A Braille dot is raised by approximately 0.5 mm and the other dimensions are as shown in Fig. 1.

Braille is understandable by visually impaired people however sighted people need not be able to understand these codes. So there is a communication gap between blind and sighted people. A lot of effort has been made by different researches to bridge this gap. These efforts may generally be categorized as:

1)   Text to Braille translation
2)   Braille to text translation

Our technique is concerned with Braille to text translation and the novelty of the technique is that it is designed for Urdu language in which characters have different glyphs depending upon the context. So a context sensitive translation is required which is not the case for English.

In the following sections we will discuss different issues regarding our technique in detail. Section 2 deals with different techniques employed for text to Braille translation while section 3 is dedicated to Braille to text translation techniques. Section 4 depicts the general structure of Urdu glyphs and their Braille codes. Section 5 explains our technique for optical recognition of Braille cells while section 6 demonstrates the

V.E. Brimkov, R.P. Barneva, H.A. Hauptman (Eds.): IWCIA 2008, LNCS 4958, pp. 342–351, 2008.
© Springer-Verlag Berlin Heidelberg 2008

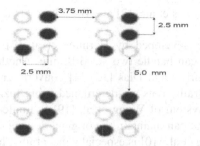

Fig. 1. Braille cell dimensions

context sensitive translation of the Braille codes into Urdu. The discussion is concluded with a summary in section 7 followed by future recommendations in section 8.

2 Text to Braille Translation

The systems developed for text to Braille translation deal with the computerized production of Braille codes from text. A Braille printer or press is attached with such a system and text is automatically printed in Braille format. One of such system was developed by Das et al. [4] and the major feature of the system is that it can handle grade 1 as well as grade 2 Braille. The system also supports different Indian languages. Another system developed by Blenkhorn and Evans [2] for text to Braille translation is unique in a sense that it preserves the formatting (italic, bold, etc.) of the text into Braille by inserting different Braille markers as formatting letters. The systems proposed by Watanabe et al. [18] and Otsuka et al. [16] convert Japanese text into Braille using neural networks. Yet another system developed by Hara et al. [8] is dedicated to the conversion of mathematical expressions in LATEX to Braille.

Some other researches have worked on the material to be used for Braille cells and the way it is used. One of such efforts is by Lee and Lucyszyn [11] who developed a microchemical refreshable Braille cell. Their main emphasis is on selection and heat treatment of the material for Braille cell. Another Braille display developed by Nobels et al. [15] used electromagnetic actuators. This is a low cost, small size refreshable display.

3 Braille to Text Translations

Recognizing and translating Braille code into text have always been of great interest to researchers. Different techniques for Braille OCR have been proposed and implemented in the past. One of the very first systems to convert Braille into text was developed by Dubus et al. [5] and the technique adopted deals with basic operations of image processing to recognize Braille characters. Yet some other very early systems were developed by Mennens et al. [12] and Blenkhorn [1] to handle Standard English Braille.

An OCR for Braille developed by Hermida et al. [7] uses scanned images of Braille as an input and can handle two sided Braille. Another optical recognizer for Braille designed by Murray and Dias [13, 14] converts grade 1 and grade 2 types Braille into text. The Braille dots are determined using fuzzy logic to accommodate positional errors. The system of Wong et al. [19] is limited to single sided Braille, uses neural networks and can maintain the original document layout. A Braille translator developed by Jiang et al. [10] is especially meant for Chinese Braille.

An interesting Braille translator of Germagnoli and Magenes [6] converts the Italian Braille codes into sound using artificial tactile sensors.

4   Urdu Glyphs and Braille Codes

Urdu language is different from English and the other similar languages in two ways: It is right to left language and each character may have one out of four shapes depending upon the context in which it is used. These shapes are known as glyphs. For example 'Beh' second alphabet in Urdu may have four shapes or glyphs. It may have ﺐ or ﺑ or ﺒ or ﺏ shape if is used as isolated or as final character of a word or as initial character of a word or in between other characters within a word, respectively. Ishida [9] and Bhurgri [3] gave a very good study on Urdu glyphs, their shapes and their implementation in Microsoft word. We developed Braille codes for each of the characters in Urdu and a standard for the use of Unicode for Urdu [17] in our Braille translator. The Braille codes were developed in consultation with The Directorate of Special Education and The Training College for Teachers of Blind, Pakistan. We performed an extensive analysis of the experiences encountered by relevant people, sighted as well as blind, for this standardization.

Table 1 Represents the Braille codes for Urdu alphabets and the Unicode for each glyph. Some of the alphabets have only one or two shapes independent of the position. We have treated these alphabets like others and used single Unicode representation for each glyph. For example ﻉ has only one shape but we have assumed four glyphs and used the same Unicode for all the glyphs for the sake of uniformity.

5   Optical Recognition of Braille Cells

We have used scanned images of Braille as input in our technique. The image is grayscaled, binarized and the position of the dots is determined by the standard distances for Braille. A threshold value of 84 for binarization of grascale images is chosen after a series of experiments on scanned images of the Braille. The main emphasis of our technique is not on the issues of optical recognition; instead we have concentrated on the context sensitivity of Urdu Braille. Interested readers may consult the papers mentioned in section 3, typically [12] fro error detection in the Braille. Sample execution of our approach is shown in Fig. 2.

Table 1. Braille and Unicode Representations of Urdu Glyphs

Index (i)	Braille Code	Urdu Alphabet	Glyphs (Unicode for Urdu Glyphs)			
			Isolated (j=0)	Final (j=1)	Initial (j=2)	Medial (j=3)
0		آ	FE81	FE82	FE81	FE81
1		ا	FE8D	FE8E	FE8D	FE8D
2		ب	FE8F	FE90	FE91	FE92
3		پ	FB56	FB57	FB58	FB59
4		ت	FE95	FE96	FE97	FE98
5		ٹ	FB66	FB67	FB68	FB69
6		ث	FE99	FE9A	FE9B	FE9C
7		ج	FE9D	FE9E	FE9F	FEA0
8		چ	FB7A	FB7B	FB7C	FB7D
9		ح	FEA1	FEA2	FEA3	FEA4
10		خ	FEA5	FEA6	FEA7	FEA8
11		د	FEA9	FEAA	FEA9	FEA9
12		ڈ	FB88	FB89	FB88	FB88
13		ذ	FEAB	FEAC	FEAB	FEAB
14		ر	FEAD	FEAE	FEAD	FEAD
15		ڑ	FB8C	FB8D	FB8C	FB8C

Table 1. (*continued*)

Index (i)	Braille Code	Urdu Alphabet	Glyphs (Unicode for Urdu Glyphs)			
			Isolated (j=0)	Final (j=1)	Initial (j=2)	Medial (j=3)
16	⠿	ز	FEAF	FEB0	FEAF	FEAF
17	⠿	ژ	FB8A	FB8B	FB8A	FB8A
18	⠿	س	FEB1	FEB2	FEB3	FEB4
19	⠿	ش	FEB5	FEB6	FEB7	FEB8
20	⠿	ص	FEB9	FEBA	FEBB	FEBC
21	⠿	ض	FEBD	FEBE	FEBF	FEC0
22	⠿	ط	FEC1	FEC2	FEC3	FEC4
23	⠿	ظ	FEC5	FEC6	FEC7	FEC8
24	⠿	ع	FEC9	FECA	FECB	FECC
25	⠿	غ	FECD	FECE	FECF	FED0
26	⠿	ف	FED1	FED2	FED3	FED4
27	⠿	ق	FED5	FED6	FED7	FED8
28	⠿	ک	FB8E	FB8F	FB90	FB91
29	⠿	گ	FB92	FB93	FB94	FB95
30	⠿	ل	FEDD	FEDE	FEDF	FEE0
31	⠿	م	FEE1	FEE2	FEE3	FEE4

Table 1. (*continued*)

Index (i)	Braille Code	Urdu Alphabet	Glyphs (Unicode for Urdu Glyphs)			
			Isolated (j=0)	Final (j=1)	Initial (j=2)	Medial (j=3)
32		ن	FEE5	FEE6	FEE7	FEE8
33		و	FEED	FEEE	FEED	FEED
34		ه	FBA6	FBA7	FBA8	FBA9
35		ء	FE80	FE80	FE80	FE80
36		ى	FBFC	FBFD	FBFE	FBFF
37		ﮮ	FBAE	FBAF	FBAE	FBAE

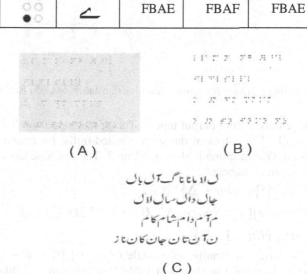

(A)   (B)

ل الا باتا تا گ۔ آل بال
جال دال سال الل
م آم دام شام کام
ن آن تان جان کان نا ز

(C)

Fig. 2. Sample execution (A) Scanned Braille image in grayscale (B) Binarized image (C) Urdu translation

6 Context Sensitive Translation

We have developed a deterministic Turing machine for context sensitive translation of Braille into Urdu. An interesting point during translation is that Urdu Braille is written from left to right while Urdu language itself is from right to left. To handle this

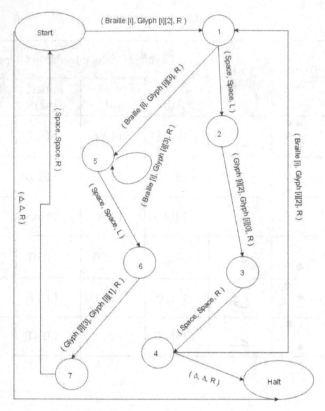

Fig. 3. Turing machine for context sensitive translation of Urdu Braille

issue, we read the Unicode from output tape of Turing machine from left to right and inserted it into a stack. The output of the stack resulted in the formation of text from right to left direction. Our machine is shown in Fig. 3. The machine has the following states and input / output alphabets.

Input: $\Sigma = \{Braille[i]\,,\,Space,\,\Delta\};\,0 \le i \le 37$

Output: $\Gamma = \{Glyph[i][j]\,,\,Space,\,\Delta\};\,0 \le i \le 37\,,\,0 \le j \le 3$

Sates: $q = \{Start\,,\,Halt\,,\,1\cdots7\}$

Braille [i] is the index of Braille code while Glyph [i] [j] is the corresponding glyph for a particular value of i as shown in Table 1. Δ represents the blank or end of input.

Following is the algorithm that we used for context sensitive translation:

```
integer: previous, current, next

previous = 38

current = getInput()

next = getInput()
```

```
while !endOfInput
    if previous = = 38 && next = = 38
        writeOutput (glyph[current][0])
    if previous != 38 && next != 38
        writeOutput (glyph[current][3])
    if previous != 38 && next = = 38
        writeOutput (glyph[current][1])
    if previous = = 38 && next != 38
        writeOutput (glyph[current][2])
    previous = current
    current = next
    next = getInput()
/*
38 is the index of Space
writeOutput writes unicode of appropriate glyph shown
in Table 1
getInput returns the index of corresponding Braille
code
*/
```

7 Summary

There is a gap between the sighted and the blind people and we have developed a system to bridge this gap. First of all we have developed Braille code for Urdu and then standardized the Unicode representation for Urdu glyphs. Our system translates Braille for Urdu into Urdu language. The translation is context sensitive as Urdu has different glyphs of a character depending upon the position of its appearance.

8 Future Recommendations

In this paper we have dealt with only single sided Braille for Urdu. Our work can be extended to handle double sided Braille. Moreover we have not incorporated Urdu numeric yet, and this can be accommodated in the future.

Acknowledgements

We are very thankful to Mr. Khalid Mehmood and Mr. Khursheed Aslam for their help in this work. Moreover we acknowledge The Directorate of Special Education and The Training College for Teachers of Blind, Pakistan for their support in developing Braille codes and standardizing the Unicode for Urdu glyphs.

References

1. Blenkhorn, P.: A System for Converting Braille into Print. IEEE Transactions on Rehabilitation Engineering 3(2), 215–221 (1995)
2. Blenkhorn, P., Evans, G.: Automated Braille Production from Word-Processed Documents. IEEE Transactions on Neural Systems and Rehabilitation Engineering 9(1), 81–85 (2001)
3. Bhurgri, A. M.: Enabling Pakistani Languages through Unicode.
   http://download.microsoft.com/download/1/4/2/142aef9f-1a74-4a24-b1f4-782d48d41a6d/PakLang
4. Das, P.K., Das, R., Chaudhuri, A.: A Computerised Braille Transcriptor for the Visually Handicapped. In: Engineering in Medicine and Biology Society, 1995 and 14th Conference of the Biomedical Engineering Society of India. An International Meeting, Proceedings of the First Regional Conference, IEEE, Los Alamitos (1995)
5. Dubus, J.P., Benjelloun, M., Devlamink, V., Wauquier, F., Altmayer, P.: Image Processing Techniques to Perform an Autonomous System to Translate Relief Braille into Black-Ink Called Lectobraille. In: IEEE Engineering in Medicine & Biology Society 10th Annual International Conference (1988)
6. Germagnoli, F., Magenes, G.: A Computerized System for Helping Blind People to Learn Braille Code. In: IEEE Engineering in Medicine & Biology Society 15th Annual International Conference (1993)
7. Hermida, X.F., Rodriguez, A.C., Rodriguez, F.M.: A Braille O.C.R. for Blind Peo-ple. In: Proceedings of the International Conference on Signal Processing Application and Technology (ICSPAT) (1996)
8. Hara, S., Ohtake, N., Higuchi, M., Miyazaki, N., Watanabe, A., Kusunoki, K., Sato, H.: MathBraille; a System to Transform LATEX Documents into Braille. ACM SIGCAPH Computers and the Physically Handicapped (2000)
9. Ishida, R.: Implementing Urdu in Unicode (2004),
   http://people.w3.org/rishida/scripts/urdu/urdu-in-unicode.pdf
10. Jiang, M., Zhu, X., Gielen, G., Drabek, E., Xia, Y., Tan, G., Bao, T.: Braille to Print Translations for Chinese. Information and Software Technology 44, 91–100 (2002)
11. Lee, J.S., Lucyszyn, S.: A Micromachined Refreshable Braille Cell. Journal of Microelectromechanical Systems 14(4), 673–682 (2005)
12. Mennens, J., Tichelen, L.V., Francois, G., Engelen, J.J.: Optical Recognition of Braille Writing Using Standard Equipment. IEEE Transactions on Rehabilitation Engineering 2(4), 207–212 (1994)
13. Murray, I., Dias, T.: A Portable Device for Optically Recognizing Braille-Part I: Hardware Development. In: 7th Australian and New Zealand Intelligent Information Systems Conference (2001)
14. Murray, I., Dias, T.: A Portable Device for Optically Recognizing Braille-Part II: Software Development. In: 7th Australian and New Zealand Intelligent Information Systems Conference (2001)
15. Nobels, T., Allemeersch, F., Hameyer, K.: Design of a High Power Density Electromagnetic Actuator for a Portable Braille Display. In: 10th International Power Electronics and Motion Control Conference EPE - PEMC (2002)

16. Otsuka, K., Kishida, S., Watanabe, A.: Application Of' Neural Networks to Braille Transcription. In: Proceedings of the 9th International Conference on Neural information Processing (ICONIP 2002) (2002)
17. The Unicode Character Code Charts by Script,
    http://www.unicode.org/charts/
18. Watanabe, T., Kisa, K., Nishimura, K., Kishida, S.: Application of Neural Networks into Japanese-to-Braille Translation. In: The 47th IEEE International Midwest Symposium on Circuits and Systems (2004)
19. Wong, L., Abdulla, W., Hussmann, S.: A Software Algorithm Prototype for Optical Recognition of Embossed Braille. In: Proceedings of the 17th International Conference on Pattern Recognition (ICPR 2004) (2004)

Rewriting P Systems Generating Iso-picture Languages

S. Annadurai[1], D.G. Thomas[2], V.R. Dare[2], and T. Kalyani[1]

[1] Department of Mathematics
St. Joseph's College of Engineering, Chennai - 119
[2] Department of Mathematics
Madras Christian College, Chennai - 59
dgthomasmcc@yahoo.com

Abstract. Membrane Computing is a branch of natural computing aiming to abstract computing ideas for the structure and the functioning of living cells as well as from the way the cells are organized in tissues or higher order structures. We consider iso-picture languages introduced in [2,3] and the possibility to handle them with P systems. In this paper we introduce regular iso-array rewriting P system, context-free iso-array rewriting P system and Basic puzzle iso- array rewriting P system and they are compared for generative power.

Keywords: Iso-array languages, membrane computing, P system.

1 Introduction

The present paper brings together two areas of theoretical computer science which were not very much linked before, membrane computing and picture grammars - the latter are considered here in the form of 2D iso-array grammars.

Membrane computing deals with distributed computing models inspired from the structure and the functioning of the living cell [6]. Very briefly, in the compartments (also called regions) defined by a hierarchical arrangement of membranes, one processes multisets of objects by evolution rules associated with the membranes. One of the branches of membrane computing is concerned with objects described by strings, and then one considers usual sets of strings (languages) instead of multisets of objects. These strings are processed by rewriting or other string handling operations [1].

The research and development of multi-dimensional pattern recognition, scene analysis, computer vision and image processing have progressed very rapidly in recent years. Among various models employed for pattern representation and analysis, the array grammar has attracted more and more attention because it has several advantages over others [7,8]. Motivated by problems in tiling, Nivat et al. [5] proposed a class of grammars called puzzle grammars for generating connected arrays of unit cells. It has been shown that Basic Puzzle Grammars have higher generative power than regular array grammars [9].

Iso-arrays are made up of isosceles right angled triangles and an iso-picture is a picture formed by catenating iso-arrays of same size. We introduced the notion of iso-arrays, iso-pictures and iso-picture languages in [2]. A motivation for

V.E. Brimkov, R.P. Barneva, H.A. Hauptman (Eds.): IWCIA 2008, LNCS 4958, pp. 352–362, 2008.
© Springer-Verlag Berlin Heidelberg 2008

this study is that one can generate some interesting iso-picture languages which cannot be generated by earlier models available in the literature. In particular iso-picture languages include more picture languages like hexagonal picture languages, rectangular picture languages, languages of rhombuses and triangles. One application of the study of iso-picture languages is its use in the generation of interesting kolam patterns. Another application of this study lies in the area of tiling rectangular plane.

In this paper we introduce rewriting P systems to generate iso-picture languages. We show that Basic Puzzle Iso-Array rewriting P systems have higher generative power than Regular Iso-Array rewriting P systems.

2   Preliminaries

In this section we recall the notions of iso-pictures, iso-picture languages and iso-triangular tiling systems proposed in [2,3].

Let $\Sigma = \{ {}^{a_1}\!\!\triangle^{a_3}_{a_2}, {}^{b_2}\!\!\nabla^{b_3}_{b_1}, {}^{c_3}\!\!\triangleleft_{c_1}^{c_2}, {}^{d_1}\!\!\triangleright^{d_2}_{d_3} \}$ be a finite set of labeled isosceles right angled triangular tiles of dimensions $\frac{1}{\sqrt{2}}, \frac{1}{\sqrt{2}}$ and 1 unit, obtained by intersecting a unit square by its diagonals. Gluable rules of tile A are as follows: Tiles which can be glued with A are B, C and D by the rules $\{(a_1, b_1), (a_2, b_2), (a_3, b_3), (a_3, c_1), (a_1, d_3)\}$. In a similar way the gluable rules can be defined for the remaining tiles.

Definition 1. *An iso-array is an isosceles right-angled triangular arrangement of elements of Σ, whose equal sides are denoted as S_1 and S_3 and the unequal side as S_2. An iso-array of size m consists of m tiles along the side S_2 and it contains m^2 gluable elements of Σ. Iso-arrays can be classified as U-iso-array, D-iso-array, R-iso-array and L-iso-array, if tiles A, B, D and C are used in side S_2 respectively.*

For example the U-iso-array of size 3 (U_3), D-iso-array of size 3 (D_3), L-iso-array of size 3 (L_3) and R-iso-array of size 3 (R_3) are shown in Fig. 1.

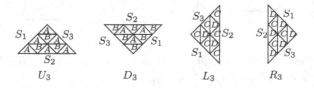

U_3 D_3 L_3 R_3

Fig. 1.

Iso-arrays of same size can be catenated using the following catenation operations. Horizontal catenation $\ominus$ is defined between U and D iso-arrays of same size. Right catenation $\oslash$ is defined between any two gluable iso-arrays of same size. This catenation includes the following:
(a) $D \oslash U$ (b) $U \oslash R$ (c) $D \oslash L$ (d) $R \oslash L$
In a similar way vertical $\odot$ and left $\obslash$ catenations can be defined.

Definition 2. *An iso-picture is a picture formed by catenating iso-arrays of same size. It is said to be of size (n, m) if there are n iso-arrays of size m catenated to form the iso-picture. The number of tiles in any iso-picture of size (n, m) is nm^2.*

An element of an iso-picture p of size (n, m) is represented as $p(i, j, k)$, where i is the ith iso-array of the picture and j is the jth row of the ith iso-array and k is the kth element of jth row of the ith iso-array, where $i = 1, 2, \ldots, n$, $j = 1, 2, \ldots, m$ and $k = 1, 2, \ldots, 2j - 1$. The set of all iso-pictures over the alphabet Σ is denoted by Σ_I^{**}. An iso-picture language L over Σ is a subset of Σ_I^{**}.

Definition 3. *Let p be an iso-picture of size (n, m). We denote by $B_{n', m'}(p)$, the set of all sub iso-pictures of p of size (n', m'), where $n' \leq n$, $m' \leq m$. $\hat{p}$ is an iso-picture obtained by surrounding p with a special boundary symbols*

 $\notin \Sigma$.

Definition 4. *A Regular Iso-Array Grammar (RIAG) is a structure $G = (N, T, P, S)$ where $N = \{\triangle, \ldots\}$ and $T = \{\triangle, \ldots\}$ are finite sets of symbols (isosceles right angled triangular tiles*

$\triangle, \triangledown, \triangleleft, \triangleright$, $\triangle, \triangledown, \triangleleft, \triangleright$ *); $N \cap T = \phi$.*

Elements of N and T are called non terminals and terminals respectively, $S \in N$ is the start symbol or the axiom. P consists of rules of the following forms

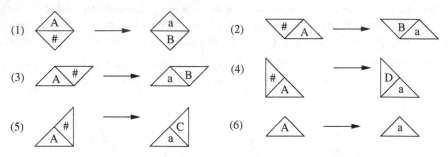

where A, B, C and D are non terminals and a is a terminal. Similar rules can be given for the other tiles $\triangledown, \triangleleft$ and $\triangleright$.

Definition 5. *A Context-Free Iso-Array grammar (CFIAG) is a structure $G = (N, T, P, S)$ where $N = \{\triangle, \ldots\}$ and $T = \{\triangle, \ldots\}$ are finite nonempty set of symbols (isosceles right angled triangular tiles*

$\triangle, \triangledown, \triangleleft, \triangleright$, $\triangle, \triangledown, \triangleleft, \triangleright$ *); $N \cap T = \phi$. Elements of N and T are called non terminals and terminals, respectively, $S \in N$ is the start symbol or the axiom. P consists of rules of the form $\alpha \to \beta$, where α and β are finite connected* *array of one or more triangular tiles over* $V \cup T \cup \{$ $\}$ *and satisfy the following conditions:*

1. *The shapes of α and β are identical.*
2. *α contains exactly one nonterminal and possibly one or more #s.*
3. *Terminals in α are not rewritten.*
4. *The application of the rule $\alpha \to \beta$ preserves the connectedness of the host array (that is, the application of the rule to a connected array results in a connected array).*

The rule $\alpha \to \beta$ is applicable to a finite connected array γ over $V \cup T \cup$ $\{$ $\}$ if α is a subarray of γ and in a direct derivation step, one of the occurrences of α is replaced by β, yielding a finite connected array δ. We write $\gamma \Rightarrow_G \delta$.

Definition 6. *A Basic Puzzle Iso-Array Grammar (BPIAG), is a structure* $G = (N, T, R, S)$ *where* $N = \{$ $, \dots \}$ *and* $T = \{$ $, \dots \}$ *are finite sets of symbols (isosceles right angled triangular tiles* $); N \cap T = \phi.$
Elements of N and T are called non terminals and terminals, respectively, $S \in N$ is the start symbol or the axiom. P consists of rules of the form

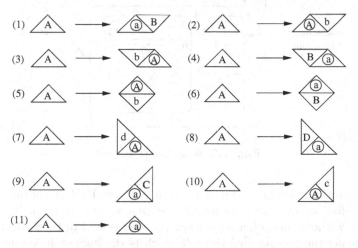

Similar rules can be given for the other tiles *and* .

Derivations begin with S written in a unit cell in the two-dimensional plane, with all other cells containing the blank symbol $\{$ $\}$ not in $N \cup T$. In a derivation step, denoted by $\Rightarrow$, a nonterminal in a cell is replaced by the right hand member of a rule, whose left hand side is . In this replacement, the circled symbol of the right side of the rule used occupies the cell of the replaced symbol and the non-circled symbol of the right side occupies the cell to the right or the left or above or below the cell of the replaced symbol, depending on the type of the rule used. The replacement is possible and defined only if the cell to be filled in by the non-circled symbol contains a blank symbol.

3   Iso-array Rewriting P Systems

In this section, we recall the notion of rewriting P system [4,6] and introduce the notion of iso-array rewriting P system. We give some examples of iso-array rewriting P systems.

P system [6] is a new compatibility model of a distributed parallel type based on the notion of a membrane structure. Such a structure consists of computing cells which are organized hierarchically by the inclusion relation. Each cell is enclosed by its membrane. Each cell is an independent computing agent with its own computing program, which produces objects. The interaction between cells consists of the exchange of objects through membranes.

A membrane structure can be represented in a natural way as a Venn diagram. (Fig. 2).

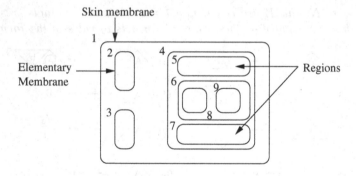

Fig. 2. A membrane structure

The membranes are labeled in one-to-one manner. Each membrane identifies a region delimited by it and the membranes placed directly inside it (if any). A membrane without any other membrane inside it is said to be elementary.

The membrane surrounding the cell which is the highest in the hierarchy is called the skin membrane.

In the regions delimited by the membranes we place multisets of objects from a specified finite set V together with evolution rules for these objects.

In this paper, we concentrate on rewriting P systems.

Definition 7. *A rewriting P system of degree n, $n \geq 1$, is a construct*

$$\pi = (V, T, \mu, L_1, \ldots, L_n, R_1, \ldots, R_n)$$

where

(i) V *is the total alphabet of the system;*

(ii) $T \subseteq V$ *is the terminal alphabet;*

(iii) μ *is a membrane structure;*

(iv) L_i, $1 \leq i \leq n$, *are finite languages over V representing the strings initially present in the regions $1, \ldots, n$ of μ;*

(v) R_i, $1 \leq i \leq n$, *are finite sets of rewriting rules of the form $X \to v(tar)$, where $X \in V, v \in V^*$, and $tar \in \{here, out, in\}$.*

We process string objects in rewriting P systems with rules of the form $X \to v(tar)$, where $X \to v$ is a usual context-free rule and $tar \in \{here, in, out\}$ is a target indication specifying the region where the result of rewriting should go. All strings are processed in parallel, but each single string is rewritten by only one rule. In other words, the parallelism is maximal at the level of strings and rules, but the rewriting is sequential at the level of the symbols from each string.

The application of a rule $u \to v$ in a region containing a multiset μ results in subtracting from μ the multiset identified by u, and then in following the prescriptions of v. If an object appears in v in the form $(a, here)$, then it remains in the same region; if it appears in the form (a, out), then a copy of object a will be introduced in the region of the membrane place directly outside the region of the rule $u \to v$; if it appears in the form (a, in_i), then a copy of a is introduced in the region of the membrane with label i.

Definition 8. *An iso-array rewriting P system is a construct*

$$\pi = (V, T, \triangle_{\#A}, \triangledown^{\#B}, \triangleleft_{\#C}, \triangleright_{\#D}, \mu, F_1, \ldots, F_m, R_1, \ldots, R_m, i_0)$$

where V is the total alphabet consisting of isosceles right angled triangles

$\triangle_A, \triangledown_B, \triangleleft_C, \triangleright_D$, $T \subseteq V$ *is the terminal alphabet,* $\triangle_{\#A}, \triangledown^{\#B}, \triangleleft_{\#C}, \triangleright_{\#D}$ *are the blank symbols, μ is a membrane structure with m membranes labeled in a one-to-one way with $1, 2, \ldots, m$, $F_1, \ldots, F_m$ are finite sets of iso-arrays over V associated with the m regions of μ, $R_1, \ldots, R_m$ are finite sets of iso-array rewriting rules over V associated with the m regions of μ; the rules have attached targets here, out, in, hence they are of the form $A_{i,1} \to B_{i,1}$ (tar); finally, i_0 is the label of an elementary membrane of μ (the output membrane).*

According to the form of its rules, an iso-array rewriting P system can be regular (REG) or context-free (CF).

(i) *A rule is called regular if it is one of the forms given in Definition 4.*

(ii) *A rule is called context-free if it is one of the forms given in Definition 5.*

A computation in an iso-array rewriting P system is defined in the same way as in a string rewriting P system with the successful computations being the halting ones: each iso-array, from each region of the system, which can be rewritten by a rule associated with that region (membrane), should be rewritten; this means that one rule is applied (the rewriting is sequential at the level of iso-arrays); the iso-array obtained by rewriting is placed in the region indicated by the target associated with the used rule (here means that the iso-array remains in the same region, out means that the iso-array exits the current membrane - thus, if the rewriting was done in the skin membrane, then it can exit the system; iso-arrays leaving the system are "lost" in the environment), and in means that the iso-array is immediately sent to one of the directly lower membranes, nondeterministically chosen if several exist (if no internal membrane exists, then a rule with the target indication in cannot be used).

A computation is successful only if it stops, a configuration is reached where no rule can be applied to the existing iso-arrays. The result of a halting computation consists of the iso-arrays composed only of symbols from T placed in the membrane with label i_0 in the halting configuration.

The set of all such iso-arrays computed (we also say generated) by a system Π is denoted by IAL(Π). The family of all iso-array languages IAL(Π) generated by system Π as above, with at most m membranes, with rules of type $\alpha \in \{REG, CF\}$ is denoted by $IARP_m(\alpha)$.

We briefly discuss several examples, both in order to illustrate the previous definitions and to shed some light on the power of iso-array rewriting P systems.

Example 1. Consider a regular iso-array rewriting P system

$$\pi_1 = (\{\ \triangle_A,\ \triangledown_B,\ \triangleleft_C,\ \triangleright_D,\ \triangle_a,\ \triangledown_b,\ \triangleleft_c,\ \triangleright_d\ \}, \{\ \triangle_a,\ \triangledown_b,\ \triangleleft_c,\ \triangleright_d\ \},$$

$$\triangle_{\#A},\ \triangledown_{\#B},\ \triangleleft_{\#C},\ \triangleright_{\#D}\ [_1[_2[_3]_3]_2]_1, \{\ \triangle_{D/A}\ \}, \phi, \phi, R_1, R_2, R_3, 3) \text{ where}$$

$$R_1 = \left\{\ \boxed{\substack{\#C \\ D}} \longrightarrow \boxed{\substack{C \\ d}}\ (\text{in}),\ \boxed{\substack{\#D \\ C}} \longrightarrow \boxed{\substack{D \\ c}}\ (\text{in})\ \right\}$$

$$R_2 = \left\{\ \substack{B\ \#A} \longrightarrow \substack{b\ A}\ (\text{out}),\ \substack{B\ \#A} \longrightarrow \substack{b\ a}\ (\text{in}),\ \substack{A\ \#B} \longrightarrow \substack{a\ B}\ (\text{out}),\ \substack{A\ \#B} \longrightarrow \substack{a\ b}\ (\text{in})\ \right\}$$

$$R_3 = \left\{\ \boxed{D} \rightarrow \boxed{d},\ \boxed{C} \rightarrow \boxed{c}\ \right\}$$

A member of the system is shown above. The nonterminals $\triangle_A$ and $\triangledown_B$ take care of growing the horizontal arm (in membrane 2) and the nonterminals $\triangleleft_C$ and take care of growing the vertical arm (in the skin membrane) step by step; at any time, the iso-array may be sent from membrane 2 to membrane 3 by using any one of the rules $\boxed{\begin{smallmatrix}B\\\#A\end{smallmatrix}} \longrightarrow \boxed{\begin{smallmatrix}b\\a\end{smallmatrix}}$ (in), $\boxed{\begin{smallmatrix}\#B\\A\end{smallmatrix}} \longrightarrow \boxed{\begin{smallmatrix}B\\a\end{smallmatrix}}$ (in) and the computation stops after using one of the rules $\boxed{D} \longrightarrow \boxed{d}$, $\boxed{C} \longrightarrow \boxed{c}$ from membrane 3. $IAL(\pi_1)$ consists of all L-shaped angles with equal arms, each arm being of length at least one.

Example 2. $IAL(\pi_1)$ can be generated by a context-free iso-array rewriting P system as follows:

$$\pi_2 = (\{ \boxed{A}, \boxed{B}, \boxed{C}, \boxed{D}, \boxed{a}, \boxed{b}, \boxed{c}, \boxed{d} \}, \{ \boxed{a}, \boxed{b}, \boxed{c}, \boxed{d} \},$$

$$\boxed{\#A}, \boxed{\#B}, \boxed{\#C}, \boxed{\#D} \left\{ \boxed{\begin{smallmatrix}D\\C\\d\\A\end{smallmatrix}} \right\}, \phi, \phi, R_1, R_2, R_3, 3)$$

Starting from the unique iso-array present initially in region 1, one grows step by step the two arms of an L-shaped angle, with one pixel up in the skin membrane and with one pixel to the right in membrane 2, at any moment, from membrane 2 we can send the array to membrane 3 (at that step three pixels are added to the horizontal arm) and the computation stops. Thus $IAL(\pi_2)$ consists of all L-shaped angles with equal arms, each arm being of length at least two.

4   Basic Puzzle Iso-array Rewriting P System

In this section, we introduce the notion of basic puzzle iso-array rewriting P system and compare it with regular iso-array rewriting P system for generative power. We obtain some interesting results.

Definition 9. *A basic puzzle iso-array rewriting P system is an iso-array rewriting P system, in which the rules are of the forms as in Definition 6.*

Derivations begin with S written in a unit cell in the two-dimensional plane, with all other cells containing the blank symbol { $\boxed{\#A}, \boxed{\#B}, \boxed{\#C}, \boxed{\#D}$ } not in $N \cup T$. In a derivation step, denoted by $\Rightarrow$, a nonterminal $\boxed{A}$ in a cell is replaced by the right hand member of a rule, whose left hand side is $\boxed{A}$. In this replacement, the circled symbol of the right side of the rule used occupies the cell of the replaced symbol and the non-circled symbol of the right side occupies the cell to the right or the left or above or below the cell of the replaced symbol, depending on the type of the rule used. The replacement is possible and defined only if the cell to be filled in by the non-circled symbol contains a blank symbol.

Example 3. A basic puzzle iso-array rewriting P system generating a sequence of overlapping right angled triangles is given below:

$\pi_3 = (\{$ ▲A, ▽B, ◿a, ◺b $\}, \{$ ◿a, ◺b $\}, \{$ ▲A $\},$
$\phi, \phi, R_1, R_2, R_3, 3)$ where

$$R_1 = \left\{ \text{▲A} \to \text{◿ⓐB} \text{ (here)}, \quad \text{▽B} \to \text{▽Ⓑa} \text{ (in)} \right\}$$

$$R_2 = \left\{ \text{▽B} \to \text{◆Aⓑ} \text{ (in)}, \quad \text{▽B} \to \text{◆Aⓑ} \text{ (out)} \right\}$$

$$R_3 = \left\{ \text{▲A} \to \text{◿ⓐ} \right\}$$

A sample derivation is shown below:

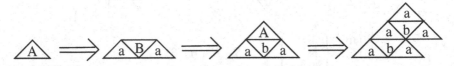

Theorem 1. *The class of picture languages generated by regular iso-array rewriting P systems is strictly included in the class of picture languages generated by basic puzzle iso-array rewriting P systems.*
i.e., $\mathcal{L}(IARP_m(REG)) \subset \mathcal{L}(IARP_m(BPG))$.

Proof. We first note that every regular iso-array rewriting P system is a basic puzzle iso-array rewriting P system. Rules of regular iso-array rewriting P system in the form given in Definition 4 can be respectively written in the form of rules of a basic puzzle iso-array rewriting P system as

(1) ◆A# → ◆ⓐB   (2) ▽#A → ▽Bⓐ

(3) ◿A# → ◿ⓐB   (4) ◺#A → ◺Dⓐ

(5) ◹#A → ◹Cⓐ   (6) ▲A → ◿ⓐ

Similar rules can be given for the other tiles. This proves the inclusion part.

   Proper inclusion can be seen from the fact that the class of picture languages given in Example 3, cannot generated by any regular iso-array rewriting P system as the junctions in ◿aⓑa cannot be handled by regular iso-array rewriting P system.

   It is known that context-free puzzle grammars coincide with context-free array grammars [9] in generating power. The same is true for iso-array grammars also. Hence we call the iso-array rewriting P system of degree m with context-free puzzle iso-array grammar (CFPIAG) rules or context-free iso-array grammar

(CFIAG) rules as context-free (CF) iso-array rewriting P system of degree m, denoted by $IARP_m(CF)$.

Theorem 2

(i) $\mathcal{L}(IARP_m(REG)) \subset \mathcal{L}(IARP_m(BPG)) \subset \mathcal{L}(IARP_m(CF))$
(ii) $\mathcal{L}(IARP_1(X)) = \mathcal{L}(X),\ X \in \{RIAG, BPIAG, CFIAG\}$

where $\mathcal{L}(RIAG)$ *and* $\mathcal{L}(CFIAG)$ *are the iso-array language classes of regular array grammars and CF array grammars and* $\mathcal{L}(BPIAG)$ *is the iso-array language class of basic puzzle grammars.*
   The inclusions in (i) and the equality in (ii) are clear from the definitions.

Theorem 3

 (i) $\mathcal{L}(IARP_3(REG)) - \mathcal{L}(RIAG) \neq \phi$
 (ii) $\mathcal{L}(IARP_3(BPG)) - \mathcal{L}(BPIAG) \neq \phi$
 (iii) $\mathcal{L}(IARP_3(BPG)) \supset \mathcal{L}(IARP_3(REG))$

Proof. (i) The iso-array language $IAL(\pi_1)$ consisting of all L-shaped angles with equal arms, each arm being of length at least one can be generated by the regular iso-array rewriting P system given in Example 1. But this picture language $IAL(\pi_1)$ cannot be generated by any regular iso-array grammar, as the rules of regular iso-array grammar cannot maintain proportion.
(ii) The iso-array language consisting of iso-picture arrays describing token L with equal arms and single protrusions is generated by the following iso-array rewriting P system π with basic puzzle iso-array grammar rules.

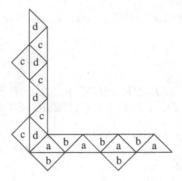

Token L with equal arms and protrusions. But the language class cannot be generated by BPIAG as maintaining equal arms is not possible with BPIAG rules. (iii) It follows from Theorem 1.

5   Conclusion

We have introduced iso-array rewriting P systems namely regular, context-free and basic puzzle iso-array rewriting P systems to generate iso-picture languages. It is proved that basic puzzle iso-array rewriting P systems have more generative power than regular iso-array rewriting P systems.

References

1. Ceterchi, R., Mutyam, M., Păun, Gh., Subramanian, K.G.: Array-rewriting P systems. Natural Computing 2, 229–249 (2003)
2. Kalyani, T., Dare, V.R., Thomas, D.G.: Local and Recognizable Iso Picture Languages. In: Pal, N.R., Kasabov, N., Mudi, R.K., Pal, S., Parui, S.K. (eds.) ICONIP 2004. LNCS, vol. 3316, pp. 738–743. Springer, Heidelberg (2004)
3. Kalyani, T., Dare, V.R., Thomas, D.G.: Iso-array Acceptors and Learning. In: Sakakibara, Y., Kobayashi, S., Sato, K., Nishino, T., Tomita, E. (eds.) ICGI 2006. LNCS (LNAI), vol. 4201, pp. 327–339. Springer, Heidelberg (2006)
4. Martin-Vide, C., Păun, Gh.: String objects in P-systems. In: Proc. Algebraic Systems, Formal Languages and Computations Workshop, RIMS Kokyuroku. Kyoto University. Kyoto, pp. 161–169 (2000)
5. Nivat, M., Saoudi, A., Subramanian, K.G., Siromoney, R., Dare, V.R.: Puzzle grammars and context free array grammars. International Journal of Pattern Recognition and Artificial Intelligence 5, 663–676 (1995)
6. Păun, Gh.: Computing with membranes. Journal of Computer and System Sciences 61(1), 108–143 (2000)
7. Siromoney, R.: Advances in array languages. In: Ehrig, H., Nagl, M., Rosenfeld, A., Rozenberg, G. (eds.) Graph Grammars 1986. LNCS, vol. 291, pp. 549–563. Springer, Heidelberg (1987)
8. Siromoney, R., Subramanian, K.G., Dare, V.R., Thomas, D.G.: Some results on picture languages. International Journal of Pattern Recognition 32, 295–304 (1999)
9. Subramanian, K.G., Siromoney, R., Dare, V.R., Saoudi, A.: Basic Puzzle Languages. International Journal of Pattern Recognition and Artificial Intelligence 95, 763–775 (1995)

Reconstructing a Matrix with a Given List of Coefficients and Prescribed Row and Column Sums Is NP-Hard

Yan Gerard

LAIC, Auvergne University, Ensemble universitaire des Cézeaux BP86,
63172 Aubière, France
gerard@laic.u-clermont1.fr

Abstract. We investigate a natural generalization of the problem of reconstruction of a binary matrix A with prescribed row and column sums: we consider an integer matrix whose list of coefficients is given in the input. The question is to organize the coefficients in the matrix in order to obtain prescribed row and column sums. We prove that this problem is NP-complete by reducing it to a 2D problem of Discrete Tomography with 3 directions of projections.

Keywords: discrete tomography, combinatorial matrix theory, NP-completeness.

1   Introduction

Reconstruction of binary matrices with prescribed line sums is a subject which has drawn attention of many mathematicians and computer scientists since the fifties. The initial problem was the reconstruction of a binary matrix with prescribed row and column sums. We call it for short *Gale-Ryser problem* from the name of the two pioneers which have both and independently discovered in 1957 that the question can be solved in polynomial time [5,12]. This result is the starting point of many generalizations which have been considered in combinatorial frameworks (Time-Table problems [4,10], Combinatorial Matrix Theory[1]...) or under pression of Electronic Microscopy, Medical Imaging (Discrete Tomography [9,6]...). Most of these generalizations are known today as NP-hard. Some of these complexities remain open problems such as for instance the multi-atomic problem of reconstruction with 3 kinds of atoms [7,3].

We consider in this paper a new generalization in relation with Combinatorial Matrix Theory or Tomography. The input is made of a row sum vector R, a column sum vector S and an initial integer matrix X_0. The question is to permute the coefficients of X_0 in order to obtain a matrix X with the prescribed line sums R and S. If we consider an initial matrix X_0 with only binary coefficients, we fall in the well-known framework of initial Gale-Ryser problem. With coefficients which are just assumed to be integers, the problem becomes more complicated. As the coefficients are not restricted to $\{0,1\}$, we can consider matrices X_0

V.E. Brimkov, R.P. Barneva, H.A. Hauptman (Eds.): IWCIA 2008, LNCS 4958, pp. 363–371, 2008.
© Springer-Verlag Berlin Heidelberg 2008

and X as gray-levels images. In this framework, the problem is to permute the pixels of an image in order to obtain prescribed row and column sums. It places the problem on the boundary of Discrete Tomography (the set of the coefficients remains discrete since it is finite) and Computerized Tomography since we can work with any gray-level image. We are also clearly in the framework of Combinatorial Image Theory or again Combinatorial Matrix Theory. This last field provides an interpretation of our problem in terms of flows in a bipartite graph: imagine that r_i passengers are waiting in m airports ($1 \leq i \leq m$) and that a flight supervisor has the mission to send them in n other airports. s_j (with $1 \leq j \leq n$) passengers exactly are waited in the arrival airport j (without taking account of their origin). The aircraft is made of exactly mn planes with several given capacities that sum is equal to $\sum_{i=1}^{m} r_i = \sum_{j=1}^{n} s_j$. How to organize the aerial traffic between the airports in order to transport the $\sum_{i=1}^{m} r_i = \sum_{j=1}^{n} s_j$ passengers from their starting place to an arrival (their should be exactly one plane for deserving each pair of airports)?

The task of this paper is to prove that this generalization of Gale-Ryser problem is NP-hard. We obtain this result by reducing it to a 2-dimensional problem of Discrete Tomography with an horizontal, a vertical and a diagonal direction of projection.

2   NewP and HardP

Let $X \in M_{m,n}(\mathbb{N})$ be an integer matrix with m rows and n columns. The coefficient in row i and column j is denoted $x_{i,j}$. The row sum vector of X is denoted $R(X)$: by definition its ith coordinate is $r_i = \sum_{j=1}^{n} x_{i,j}$. In the same way, the column sum vector is denoted $S(X)$ and its jth coordinate is $s_j = \sum_{i=1}^{m} x_{i,j}$.

We introduce also the set of matrices $P(X)$ of all matrices obtained by permutation of the coefficients of X. All these matrices have exactly the same coefficients but not in the same places.

Example. Matrix $X' = \begin{pmatrix} 4 & 2 & 4 \\ 6 & 0 & 5 \\ 0 & 4 & 1 \end{pmatrix}$ is in $P(X)$ where $X = \begin{pmatrix} 0 & 4 & 4 \\ 4 & 6 & 2 \\ 0 & 5 & 1 \end{pmatrix}$.

We focus our attention on the following generalization of Gale-Ryser problem (we call it NewP).

Problem 1 (NewP). Input: Two integer vectors $R \in \mathbb{N}^m$, $S \in \mathbb{N}^n$ and an integer matrix $X_0 \in M_{m,n}(\mathbb{N})$.
Output: Does there exist a matrix $X \in P(X_0)$ verifying $R(X) = R$ and $S(X) = S$?

NewP is a problem of consistency. The question is to know if there exists a permutation of coefficients of X_0 allowing to obtain the prescribed row and column sums R and S. If we restrict the coefficients of X_0 from $\mathbb{N}$ to $\{0, 1\}$, NewP is reduced to the Gale-Ryser problem.

The input X_0 is given here under the form of a matrix but it could be as well a list of mn coefficients since their initial order in X_0 is of no importance for the problem. This choice is however more convenient than a list because it allows to limit the objects used in the paper to matrices.

$NewP$ is clearly in NP because we can check in polynomial time if a given matrix X is a solution of an instance. Our goal is to prove that $NewP$ is NP-complete. Such a proof requires to provide a polynomial encoding in an instance of $NewP$ of a second problem already known as NP-hard . The second problem that we are going to use is a problem of Discrete Tomography with three directions (horizontal, vertical and diagonal). It requires to introduce a notation for the third direction of projection. Given matrix X, we denote $T(X)$ the antidiagonal sum vector that k^{th} coordinate is $t_k = \sum_{i+j=k+1} x_{i,j}$ (the +1 in condition $i + j = k + 1$ is necessary to have an index k starting from 1).

Example. With matrix $X = \begin{pmatrix} 2 & 0 & 2 \\ 1 & 3 & 2 \\ 0 & 1 & 1 \end{pmatrix}$ we have $R(X) = (4, 6, 2)$, $S(X) = (3, 4, 5)$ and $T(X) = (2, 1, 5, 3, 1)$.

With these notations, we have the following consistency problem:

Problem 2 (HardP). Input: Three integer vectors $R \in \mathbb{N}^m$, $S \in \mathbb{N}^n$ and $T \in \mathbb{N}^{m+n-1}$.
Output: Does there exist a binary matrix $Y \in M_{m,n}(\{0, 1\})$ verifying $R(Y) = R$, $S(Y) = S$ and $T(Y) = T$?

HardP is NP-hard even by considering square matrices $(m = n)$ [8]. We can also notice that it can be understood as a problem of reconstruction on the hexagonal grid using projections along the three principal axes [11].

3   Tools

3.1   Barycenter of a Matrix

We introduce the notion of barycenter of a matrix. This notion is related to classical notion of barycenter by considering that the coefficient $x_{i,j}$ is the weight of point (i, j). Thus the barycenter $G(A)$ of matrix A is

$$(i_G(A), j_G(A)) = \sum_{1 \leq i \leq m, 1 \leq j \leq n} a_{i,j}(i, j)/W$$

where W is the sum of all coefficients ($W = \sum_{1 \leq i \leq m, 1 \leq j \leq n} a_{i,j}$). We can easily assume that W is different from 0 - we just have to avoid the trivial case of null matrix- since we work with matrices having only positive coefficients.

We notice that the two coordinates of the barycenter of matrix A can be deduced from the row and column sums vectors: The first coordinate of the barycenter of A is $i_G(A) = \sum_{1 \leq i \leq m, 1 \leq j \leq n} i a_{i,j}/W = \sum_{i=1}^{m} i r_i/W$ with $W =$

$\sum_{i=1}^{m} r_i = \sum_{i=1}^{n} s_i$. The same computation on second coordinate provides j_G $(A) = \sum_{j=1}^{n} j s_j / W$. It means that if A and A' are two solutions of problem *NewP*, they have the same barycenter.

We introduce now the coordinate $k_G(A)$ of barycenter of matrix A according to diagonal direction: it is just the sum the two coordinates according to index i and j. Hence, we have $k_G(A) = i_G(A) + j_G(A)$ namely $k_G(A) = \sum_{1 \le i \le m, 1 \le j \le n} (i + j) a_{i,j} / W$. By adding the coefficients $a_{i,j}$ with a sum of indices $i + j = k$, we can rewrite it $k_G(A) = \sum_{k=2}^{m+n} k t_{k-1} / W$ where t_k is the kth coordinate of the antidiagonal line sums $(T(A) = (t_k)_{1 \le k \le m+n-1})$. It proves the next lemma:

Lemma 1. *By denoting r_i, s_j and t_k the coordinates of the line sums vectors $R(A)$, $S(A)$, $T(A)$ of any matrix A, we have $\sum_{i=1}^{m} i r_i + \sum_{j=1}^{n} j s_j = \sum_{k=2}^{m+n} k t_{k-1}$.*

Lemma 1 puts in relation the diagonal coordinate of the barycenter of matrix A with its horizontal and vertical coordinates. It will be more practical in the following to avoid the denominator $/W$ from the coordinates $i_G(A)$, $j_G(A)$ and $k_G(A)$. Thus we introduce notation $I_G(A) = W i_G(A)$, $J_G(A) = W j_G(A)$ and $K_G(A) = W k_G(A)$ with the advantage that they depend linearly on A.

3.2   Technical Lemma

We introduce a technical lemma. The proposition just says that with two sequences of positive integers $(f_i)_{1 \le i \le d} \in \mathbb{N}^d$ and $(g_i)_{1 \le i \le d} \in \mathbb{N}^d$, the sum $\sum_{i=1}^{d} f_i g_{\sigma_i}$ where σ is a permutation of the indices is maximal over the set of all permutations if and only if the order $(\le)$ of f_i fits with the one of g_{σ_i}. In other words, the sum is maximal by multiplying the greatest f_i by the greatest g_i, the second greatest f_i by the second greatest g_i...

Lemma 2 (Technical). *Let $(f_i)_{1 \le i \le d} \in \mathbb{N}^d$, $(g_i)_{1 \le i \le d} \in \mathbb{N}^d$ and σ be a permutation of $\{1...d\}$. The sequence g_{σ_i} is increasing relatively to the order of $(f_i)_{1 \le i \le d}$ ($f_i < f_{i'}$ implies $g_{\sigma_i} \le g_{\sigma_{i'}}$) if and only if $\sum_{i=1}^{d} f_i g_{\sigma_i}$ is maximal over the set of permutations (for any permutation σ', $\sum_{i=1}^{d} f_i g_{\sigma'_i} \le \sum_{i=1}^{d} f_i g_{\sigma_i}$).*

3.3   Increasing and Maximal Matrices

A matrix A is called *maximal* if for any matrix A' in $P(A)$, we have $k_G(A') \le k_G(A)$ (or $K_G(A') \le K_G(A)$ since the sum of all coefficients is the same).

We say that a matrix A is *increasing* if the sequence $a_{i,j}$ of its coefficients is increasing from antidiagonal to antidiagonal namely if $i + j < i' + j'$ implies $a_{i,j} \le a_{i',j'}$.

Example. Matrix $A = \begin{pmatrix} 0 & 2 & 2 & 5 \\ 1 & 3 & 3 & 8 \\ 2 & 4 & 6 & 8 \end{pmatrix}$ is increasing while matrix $A' = \begin{pmatrix} 0 & 0 & 1 & 1 \\ 0 & 1 & 2 & 3 \\ 2 & 3 & 3 & 4 \end{pmatrix}$ is

not increasing since there is 2 in antidiagonal $i + j = 4$ ($i = 3$, $j = 1$) and a 1 in antidiagonal $i + j = 5$ ($i = 1$, $j = 4$).

Lemma 2 with $d = mn$, $f_{i,j} = i + j$ and $g_{i,j} = x_{i,j}$ implies that increasing matrices are maximal and conversely:

Lemma 3. *A matrix is increasing if and only if it is maximal.*

We introduce now the matrices I, J and especially $I + J$: the coefficients of ith row of matrix I are equal to i. The coefficients of jth column of matrix J are equal to j so that the coefficient of row i and column j of matrix $I + J$ is $i + j$. The matrix $2(I + J)$ is clearly increasing and it remains true by adding some 1 in some places:

Lemma 4. *For any binary matrix $B \in M_{m,n}(\{0,1\})$, the matrix $2(I + J) + B$ is increasing.*

Example. $I = \begin{pmatrix} 1\,1\,1\,1 \\ 2\,2\,2\,2 \\ 3\,3\,3\,3 \end{pmatrix}$, $J = \begin{pmatrix} 1\,2\,3\,4 \\ 1\,2\,3\,4 \\ 1\,2\,3\,4 \end{pmatrix}$, $I + J = \begin{pmatrix} 2\,3\,4\,5 \\ 3\,4\,5\,6 \\ 4\,5\,6\,7 \end{pmatrix}$. If we choose

$B = \begin{pmatrix} 1\,0\,1\,1 \\ 0\,1\,0\,1 \\ 0\,1\,0\,0 \end{pmatrix}$ then $2(I + J) + B = \begin{pmatrix} 5\ \ 6\ \ 9\ \ 11 \\ 6\ \ 9\ \ 10\ 13 \\ 8\ 11\ 12\ 14 \end{pmatrix}$ which is increasing.

It follows from lemma 3 and 4 that any matrix $A = 2(I+J)+B$ with a binary B is maximal. We can notice that the coefficients belonging to the antidiagonal $i + j = k$ are either equal to $2k$ either equal to $2k + 1$. Among all matrices of $P(A)$ (we recall that they are obtained by permutation of the coefficients), only a few of them are increasing and it follows from the disjoined ranges $[2k, 2k + 1]$ of antidiagonal k in A that they differ from A only by moving the coefficients along antidiagonals. Thus they can all be written as $2(I + J) + B'$ where B' is a binary matrix with exactly the same number of 1s than B in any antidiagonal. This last condition can be rewritten as $T(B) = T(B')$ and it proves the next lemma:

Lemma 5. *Let $A = 2(I+J)+B$ where B is a binary matrix $B \in M_{m,n}(\{0,1\})$. A matrix A' of $P(A)$ is increasing if and only if $A' = 2(I + J) + B'$ where B' is a binary matrix verifying $T(B) = T(B')$.*

It means that if we look at the matrices A' of $P(A)$ (A is the sum of a binary matrix B and $2(I+J)$), only a few of them can have the same barycenter on the same antidiagonal as A. Let us consider these matrices A' of $P(A)$ that barycenter is on the same antidiagonal as A. As A is increasing and thus maximal, A' should be also maximal and thus increasing. It constraints the choice of A' to a very restrictive set of matrices. A' is necessarily related to A by permutations of coefficients in each antidiagonal. In other words, a coefficient of A in antidiagonal k can not belong to a different antidiagonal $k' \neq k$ in A'. It follows that A' is of the same form than A: it can be written $B' + 2(I + J)$ with a binary matrix B' verifying $T(B') = T(B)$.

4   Reduction

The reduction that we provide is directly inspired from [2] where a construction using barycentric optimality allows to reduce *HardP*. The result is the NP-hardness of the problem of reconstruction of a 3D lattice set with a prescribed number of points in the slices parallel to the planes of coordinates. Nevertheless the NP-completeness of $NewP$ is not a consequence of this result but, with some work, an undirect consequence of the proof given in [2].

Let us consider an instance of $HardP$: we have three integer vectors $R_H \in \mathbb{N}^m$, $S_H \in \mathbb{N}^n$ and $T_H \in \mathbb{N}^{m+n-1}$ (index H is used in reference to $HardP$ while letter N is used for $NewP$). The first step of the reduction is the construction of an instance of $NewP$: $R_N \in \mathbb{N}^m$, $S_N \in \mathbb{N}^n$ and an integer matrix $X_0 \in M_{m,n}(\mathbb{N})$.

4.1   Construction of an Instance of $NewP$

We decompose the construction in three steps which provide two different cases.

- The first step of the construction is to check that the condition $\sum_{i=1}^{m} i r_i + \sum_{j=1}^{n} j s_j = \sum_{k=2}^{m+n} k t_{k-1}$ is verified by the coordinates of R_H, S_H and T_H. This condition can be checked in polynomial time. If it is not satisfied, lemma 1 guarantees that the instance R_H, S_H, T_H of $HardP$ has no solution. In this case, we construct an instance of $NewP$ which is clearly inconsistent by taking for instance the zero matrix as X_0 and two vectors R_N and S_N with at least one of their coordinates different from 0. Otherwise, we go to the second step...
- Second step, we check the existence of a binary matrix B verifying $T(B) = T_H$: the binary matrix B exists if and only if the coordinates t_k of the antidiagonal sum vector do not exceed the cardinality of the antidiagonal $i + j = k + 1$ namely $t_k \leq k$, $t_k \leq m$, $t_k \leq n$ and $t_k \leq m + n - k$ for any k from 1 to $2n - 1$. This condition can again be checked in polynomial time. If B does not exist, the instance of $HardP$ is not consistent. In this case, we construct again an inconsistent instance of $NewP$.
- We assume now that the condition $\sum_{i=1}^{m} i r_i + \sum_{j=1}^{n} j s_j = \sum_{k=2}^{m+n} k t_{k-1}$ is satisfied and that there exists a binary matrix B with $T(B) = T_H$. A possible choice for B among others is to take the matrix defined by $b_{i,j} = 1$ for $i \leq t_{i+j-1} + max\{i+j-n, 0\}$ and 0 otherwise but it is of minor importance. With this matrix B, we introduce $A = 2(I + J) + B$. We take this matrix A as matrix of coefficients X_0 (we construct it in polynomial time). For the row and column sum vectors, we take $R_N = R_H + R(2(I + J))$, $S_N = S_H + S(2(I + J))$.

The row vector sum $R_N \in \mathbb{N}^n$, the column vector sum $S_N \in \mathbb{N}^n$ and the integer matrix $X_0 \in M_{m,n}(\mathbb{N})$ is our instance of $NewP$ encoding the initial instance of $HardP$.

Its computation can be done in polynomial time (which is a necessary condition to prove NP-hardness). It remains now to prove that the consistency of the $HardP$ instance R_H, S_H, T_H is equivalent to the consistency of the $NewP$ instance R_N, S_N, X_0 namely that there exist a solution for one instance if and only if there exist a solution for the other.

4.2 Consistency of HardP Instance Implies Consistency of NewP Instance

We assume that the *HardP* instance R_H, S_H, T_H is consistent: there exists a binary matrix Y verifying $R(Y) = R_H$, $S(Y) = S_H$ and $T(Y) = T_H$. It follows that we are in the last case of construction of *NewP* instance: $R_N = R_H + R(2(I + J))$, $S_N = S_H + S(2(I + J))$ and a matrix of coefficients $X_0 = A = B + 2(I + J)$ where the binary matrix B verifies $T(B) = T_H$.

We show now that the matrix $X = Y + 2(I + J)$ is solution of the *NewP* instance R_N, S_N and X_0: we have first $R(X) = R(Y) + R(2(I + J)) = R_H + R(2(I+J)) = R_N$, $S(X) = S(Y) + S(2(I+J)) = S_H + S(2(I+J)) = S_N$. At last, we have $T(Y) = T(B)$. Hence matrices $2(I+J) + Y = X$ and $2(I+J) + B = A = X_0$ have exactly the same coefficients, which can be formulated as $X \in P(X_0)$. We conclude that X is a solution of instance R_N, S_N and X_0.

4.3 Consistency of NewP Instance Implies Consistency of HardP Instance

Let us assume that there exists a matrix X solution of our instance R_N, S_N, X_0 of *NewP*. Since the first and second cases of construction provide unconsistent instances, we are necessarily in the third case of construction: a binary matrix B with $T(B) = T_H$ exists and we have $\sum_{i=1}^{m} ir_i + \sum_{j=1}^{n} js_j = \sum_{k=2}^{m+n} kt_{k-1}$ where r_i, s_j and t_k are the coordinates of R_H, S_H and T_H. As in this case, we have chosen $X_0 = A$, the solution X is in $P(A)$ and verifies $R(X) = R_N$, $S(X) = S_N$.

Let us consider the diagonal position of the barycenter of X or more precisely $K_G(X)$. We are going to prove that it is equal to $K_G(A)$. As starting point, we have $R(X) = R_N = R(2(I + J)) + R_H$ and $S(X) = S_N = S(2(I + J)) + S_H$ which lead to $R(X - 2(I + J)) = R_H$ and $S(X - 2(I + J)) = S_H$. As the coordinates of R_H and S_H are respectively r_i and s_j, we have $I_G(X - 2(I+J)) = \sum_{i=1}^{m} ir_i$ and $J_G(X - 2(I + J)) = \sum_{j=1}^{n} js_j$. It leads to $K_G(X - 2(I + J)) = I_G(X - 2(I + J)) + J_G(X - 2(I + J)) = \sum_{i=1}^{m} ir_i + \sum_{j=1}^{n} js_j$. Thus we have (i) $K_G(X) = K_G(2(I + J)) + \sum_{i=1}^{m} ir_i + \sum_{j=1}^{n} js_j$

We can consider now the diagonal position of the barycenter of matrix A given by $K_G(A)$. We have $K_G(A) = K_G(2(I + J)) + K_G(B)$ since $A = 2(I + J) + B$. As $T(B) = T_H$, we have $K_G(B) = \sum_{k=2}^{m+n} kt_{k-1}$. Hence we have (ii) $K_G(A) = K_G(2(I + J)) + \sum_{k=2}^{m+n} kt_{k-1}$.

We can now associate the equalities (i) and (ii) with our initial condition $\sum_{i=1}^{m} ir_i + \sum_{j=1}^{n} js_j = \sum_{k=2}^{m+n} kt_{k-1}$: we obtain $K_G(A) = K_G(X)$. Lemma 4 says that matrix A is increasing. Lemma 3 implies that A is maximal. With equality $K_G(X) = K_G(A)$ and $X \in P(A)$, we have that matrix X is also maximal. It follows again from Lemma 3 but in the converse sense that X is increasing. We have the conditions to apply Lemma 5 with matrices X and A: matrix X is in $P(A)$ and X is increasing. It follows that X is the sum of $2(I + J)$ and a binary matrix Y verifying $T(Y) = T(B) = T_H$.

As conclusion, matrix $Y = X - 2(I + J)$ is binary and verifies $R(Y) = R(X - 2(I+J)) = R_N - R(2(I+J)) = R_H, S(Y) = S(Y-2(I+J)) = S_N - S(2(I+J)) = S_H, T(Y) = T_H$. Thus matrix Y is a solution of instance R_H, S_H, T_H of HardP.

4.4 Result

Based on the NP-completeness of *HardP* class of problems, we have proved that reconstructing a matrix with given coefficients and with prescribed row and column sums is NP-hard (and it is in NP):

Theorem 1. NewP *class of problems is NP-complete.*

5   Next Challenge

We have proved that given an integer matrix X_0, a row sum and a column sum vector, it is NP-hard to know whether there exist a permutation of the coefficients of X_0 providing the given line sums. If the matrix X_0 is binary, there are only 0s and 1s to put in the right place: we are in the framework of classical Gale-Ryser problem and this subclass of problems can be solved in polynomial time. It leads of course to the question: what happens if we consider a matrix with only 0s, 1s and 2s ? More generally, what is the complexity of the subclass S_k of problems where the coefficients of matrix X_0 are integers between 0 and k. We know hat S_1 is polynomial and a challenge would be to determine the complexity of classes S_k for $k > 1$.

References

1. Brualdi, R.: Combinatorial Matrix Theory. Cambridge University Press, Cambridge (1991)
2. Brunetti, S., Lungo, A.D., Gerard, Y.: On the computational complexity of reconstructing three-dimensional lattice sets from their two dimensional x-rays. Journal of Linear Algebra and Application 339(1-3), 59–73 (2001)
3. Chrobak, M., Durr, C.: Reconstructing polyatomic structures from discrete x-rays: Np-completeness proof for three atoms. Theoretical Computer Science 259, 81–98 (2001)
4. Even, S., Itai, A., Shamir, A.: On the complexity of timetable and multicommodity flow problems. SIAM Journal of Computing 5, 691–703 (1976)
5. Gale, D.: A theorem on flows in networks. Pacific Journal of Mathematics 7, 1073–1082 (1957)
6. Gardner, R.J.: Geometric Tomography. Encyclopedia of Mathematics and Its Applications, vol. 58. Cambridge University Press, Cambridge (1995)
7. Gardner, R.J., Gritzmann, P., Prangenberg, D.: On the computational complexity of determining polyatomic structures by x-rays. Theoretical Computer Science 233(1-2), 91–106 (2000)
8. Gritzmann, P.: On the reconstruction of finite lattice sets from their x-rays. In: Ahronovitz, E. (ed.) DGCI 1997. LNCS, vol. 1347, pp. 19–32. Springer, Heidelberg (1997)

9. Herman, G., Kuba, A.: Discrete Tomography: Foundations, Algorithms and Applications. Applied and Numerical Harmonic Analysis. Birkhaüser (1999)
10. Irving, M., Jerrum, R.: Three-dimensional statistical data security problems. SIAM Journal on Computing 23, 166–184 (1993)
11. Kong, T., Herman, G.: On which grids can tomographic equivalence of binary pictures be characterized in terms of elementary switching operations? International Journal of Imaging Systems and Technology 9, 118–125 (1998)
12. Ryser, H.: Combinatorial properties of matrices of zeros and ones. Canadian Journal of Mathematics 9, 371–377 (1957)

A Reasoning Framework
for Solving Nonograms

K. Joost Batenburg[1] and Walter A. Kosters[2]

[1] Vision Lab, Department of Physics
University of Antwerp, Belgium
joost.batenburg@ua.ac.be
[2] Leiden Institute of Advanced Computer Science
Leiden University, The Netherlands
kosters@liacs.nl

Abstract. Nonograms, also known as Japanese puzzles, are logic puzzles that are sold by many news paper vendors. The challenge is to fill a grid with black and white pixels in such a way that a given description for each row and column, indicating the lengths of consecutive segments of black pixels, is adhered to. Although the Nonograms in puzzle books can usually be solved by hand, the general problem of solving Nonograms is NP-hard. In this paper, we propose a local reasoning framework that can be used to deduce the value of certain pixels in the puzzle, given a partial filling. By iterating this procedure, starting from an empty grid, it is often possible to solve the puzzle completely. Our approach is based on ideas from dynamic programming, 2-satisfiability problems, and network flows. Our experimental results demonstrate that the approach is capable of solving a variety of Nonograms that cannot be solved by simple logic reasoning within individual rows and columns, without resorting to branching operations. Moreover, all the computations involved in the solution process can be performed in polynomial time.

1   Introduction

A *Nonogram*, also known as a *Japanese puzzle* in some countries, is a kind of logic puzzle, where the goal is to draw a rectangular image that adheres to certain row and column constraints. Usually, the image is black-and-white, although Nonograms with more than two grey values exist as well. Fig. 1 shows an example of a Nonogram. The puzzle has a rectangular shape, which is subdivided in unit cells. We will also refer to these cells as *pixels*. For each row and each column, a *description* is given. The description indicates the length of the consecutive segments of black pixels along the corresponding line. For example, the description "1, 1" in the first row indicates that when traversing the pixels in that row from left to right, there should first be zero or more white pixels, followed by one black pixel. Then, at least one white pixel must occur, followed by exactly one black pixel. There may be additional white pixels at the end of the line. The symbol ϵ denotes the empty description, leading to an all white line. The goal

V.E. Brimkov, R.P. Barneva, H.A. Hauptman (Eds.): IWCIA 2008, LNCS 4958, pp. 372–383, 2008.

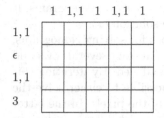

	0	1	0	1	0
	0	0	0	0	0
	x	x	0	x	x
	x	x	1	x	x

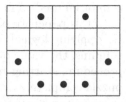

Fig. 1. A simple 4×5 Nonogram: a) original puzzle; b) partial solution (1 = black, 0 = white, x = yet unknown); c) final solution (dots denote black pixels)

of the puzzle is to colour all pixels with either black or white, in such a way that each horizontal and vertical line is consistent with the given description. As we shall see later, when using only information concerning single rows and columns, puzzles can often be solved partially (see the picture in the middle). For instance, one can infer that the middle pixel in the bottom row must be black. Using 2-satisfiability (2-SAT) rules we can completely solve this simple puzzle. More complicated puzzles require more sophisticated techniques, as we will also demonstrate.

Nonograms can be considered as a generalization of a well-known problem in Discrete Tomography: reconstructing hv-convex sets (where the black pixels in each row and column must be consecutive). For this Discrete Tomography problem, the description for each line consists of a single number, indicating the length of the segment of black pixels along that line. The problem of reconstructing hv-convex polyominoes can be solved in polynomial time [6,2], whereas the reconstruction problem for general hv-convex sets is NP-hard [9]. Therefore, the reconstruction problem for Nonograms is also NP-hard (and, clearly, NP-complete). In [8] this is shown through the more general concept of parsimonious reductions.

The Nonogram problem can also be related to several *job scheduling problems*, where each row corresponds to a single processor and the jobs for the processors are indicated by the row descriptions. In such scheduling problems, the type of constraints that occur in Nonograms only apply to the rows, or the columns, but not both.

There can be considerable differences in the difficulty level of Nonograms. On the one hand, the Nonograms that appear in newspapers can typically be solved by applying a series of simple logical rules, each of which considers only a single horizontal or vertical line. Later on we will refer to them as being *simple*. These puzzles will always have a unique solution. On the other hand, large random puzzles can be very difficult to solve, even using a computer, and may have many different solutions. Clearly, the fact that solving Nonograms is NP-hard indicates that not all puzzles can be solved using simple logic reasoning.

Although several implementations of Nonogram solvers can be found on the Internet (see [7] for a list of solvers), we have not found studies of this problem in the scientific literature. In [4], an evolutionary algorithm is described for solving

Nonograms. Although this algorithm is quite effective at solving Nonograms, it cannot be used to find *all* solutions, if more than one solution exists.

In this paper we propose a local reasoning framework for solving Nonograms. By applying logical rules, which may involve information from several rows and columns, the value of certain unknown pixels can be deduced. By iterating this procedure, starting from an empty grid, it is often possible to either solve the puzzle completely or to determine a substantial part of the pixels. In the latter case one can distinguish between situations where there exist different solutions (that can sometimes be enumerated), and situations where one cannot infer anything anymore.

The paper is organized as follows: Section 2 introduces Nonograms in a formal, somewhat more general context; in Section 3 we show solutions to some relaxed versions (i.e., single lines, and the Discrete Tomography version); we combine these techniques into a general framework in Section 4 and Section 5, also incorporating 2-SAT rules; experiments are shown in Section 6; Section 7 concludes.

2   Notation and Concepts

We first define notation for a single line (i.e., row or column) of a Nonogram. Put $\Sigma = \{0, 1\}$. The symbols "0" and "1" represent the white (0) and black (1) pixels in the puzzle. In addition, we introduce a special symbol, "x", indicating that a pixel is not decided yet. Put $\Gamma = \{0, 1, x\}$. For $\ell \geq 0$, let Σ^ℓ (resp. Γ^ℓ) denote the set of all strings over Σ (resp. Γ) of length ℓ.

For describing a Nonogram, we introduce more general concepts of row and column descriptions, such that Nonograms are in fact a special case. Most of the concepts in this paper can be applied to all logic problems that follow the more general definitions.

A *description* d of length $k > 0$ is an ordered series $(d_1, d_2, \ldots, d_k)$ with $d_j = \sigma_j\{a_j, b_j\}$, where $\sigma_j \in \Sigma$ and $a_j, b_j \in \{0, 1, 2, \ldots\}$ with $a_j \leq b_j$ $(j = 1, 2, \ldots, k)$. Let D_k denote the (infinite) set of all descriptions of length k, and put $D = \cup_{k=0}^\infty D_k$, where D_0 consists of the empty description ϵ. A single $d_j = \sigma_j\{a_j, b_j\}$ is called a *segment description*. The perhaps somewhat confusing curly braces are used here in order to stick to the conventions from regular expressions; so, in $\sigma_j\{a_j, b_j\}$ they do not refer to a set, but to an ordered pair. We will sometimes write σ^* as a shortcut for $\sigma\{0, \infty\}$ (for $\sigma \in \Sigma$) and σ^+ as a shortcut for $\sigma\{1, \infty\}$, where ∞ is suitably large number. And finally, we put σ^a as a shortcut for $\sigma\{a, a\}$ $(a \in \{0, 1, 2, \ldots\})$, and we sometimes omit parentheses and commas; also σ^0 is omitted.

A finite string s over Σ *adheres* to a description d (as defined above) if s first has between a_1 and b_1 σ_1s (boundaries included), then between a_2 and b_2 σ_2s, $\ldots$, and ends with between a_k and b_k σ_ks. Example: again take $\Sigma = \{0, 1\}$, and assume that the description is

$$(0\{0, \infty\}, 1\{a_1, a_1\}, 0\{1, \infty\}, 1\{a_2, a_2\}, 0\{1, \infty\}, \ldots, 1\{a_r, a_r\}, 0\{0, \infty\}).$$

This is precisely the Nonogram-type description $a_1, a_2, \ldots, a_r$ for a line (row or column). Note that it has length $2r + 1$ and can also be written as

$$(0^*, 1^{a_1}, 0^+, 1^{a_2}, 0^+, \ldots, 1^{a_r}, 0^*) = 0^* 1^{a_1} 0^+ 1^{a_2} 0^+ \ldots 1^{a_r} 0^*.$$

We denote the set of all Nonogram-type descriptions by $D_{\text{nonogram}} \subseteq D$. In the sequel we will concentrate on this type of description.

Suppose we have a string s over Γ. If zero or more xs are replaced with elements from Σ, the resulting string is called a *specification* of s. A specification to a string over Σ (i.e., no longer containing any "x" symbols) is called a *complete specification* or *fix*. If a string s has a fix that adheres to a given description d, s is called *fixable with respect to d*. The boolean function $Fix(s, d)$ is **true** if and only if s is fixable with respect to d.

A *Nonogram description* N consists of $m > 0$ row descriptions $r_1, r_2, \ldots, r_m \in D_{\text{nonogram}}$ and $n > 0$ column descriptions $c_1, c_2, \ldots, c_n \in D_{\text{nonogram}}$. A *partial filling* is a $m \times n$ matrix over Γ. The set of all these partial fillings is denoted by $\Gamma^{m \times n}$; its elements can also be considered as strings of length $m \times n$. If such a filling contains no xs, it is called a *complete filling* or *full fix*. A complete filling F *adheres* to the Nonogram description N if the ith row of F adheres to r_i (for all $i = 1, 2, \ldots, m$) and the jth column of F adheres to c_j (for all $j = 1, 2, \ldots, n$). We generalize the concepts of specification, fix and fixable in the natural way.

3 Partial Solution Methods

In this section we study two relaxations of the original problem. In Section 3.1 we confine the puzzle to a single line. In Section 3.2 we only require that the total number of black pixels in each line (i.e., row or column) adheres to its description. Clearly, any pixel that can only have a single value in all solutions of the relaxation, must also have this same value in any solution of the complete Nonogram. For both relaxations we show that such pixels can be found efficiently.

3.1 Solving a Single Line

We will now describe a recursive algorithm to decide fixability for a single line. This algorithm can be implemented by dynamic programming. First we introduce some notations. For a string $s = s_1 s_2 \ldots s_\ell$ of length ℓ over Γ we define its prefix of length i by $s^{(i)} = s_1 s_2 \ldots s_i$ ($1 \le i \le \ell$), so $s = s^{(\ell)}$; $s^{(0)}$ is the empty string. Similarly, for a description $d = (d_1, d_2, \ldots, d_k)$, we put $d^{(j)} = (d_1, d_2, \ldots, d_j)$ for $1 \le j \le k$, so $d = d^{(k)}$; $d^{(0)} = \epsilon$ is the empty description. Furthermore, let $A_j = \sum_{p=1}^{j} a_p$ and $B_j = \sum_{p=1}^{j} b_p$; put $A_0 = B_0 = 0$. We note that a string of length $\ell < A_k$ is certainly not fixable with respect to d, simply because it has too few elements; similarly, a string of length $\ell > B_k$ is not fixable with respect to d. Finally, for a given string s of length ℓ, let $L_i^\sigma(s)$ denote the largest index $h \le i$ such that $s_h \ne \sigma$ and $s_h \ne$ x, if such an index exists, and 0 otherwise ($\sigma \in \Sigma$, $1 \le i \le \ell$). We will put $Fix(i, j) = Fix(s^{(i)}, d^{(j)})$, and are interested in $Fix(\ell, k)$. As boundary values we note that $Fix(0, j) = $ **true** if and only if $A_j = 0$ ($j = 0, 1, 2, \ldots, k$); and $Fix(i, 0) = $ **false** for $i = 1, 2, \ldots, \ell$ (by the way, these last values are never used). We clearly have $Fix(i, j) = $ **false** if $i < A_j$ or $i > B_j$ ($0 \le i \le \ell$, $0 \le j \le k$), as indicated above.

Our main recursion is:

$$Fix(i,j) = \bigvee_{p \,=\, \max(i - b_j,\, A_{j-1},\, L_i^{\sigma_j}(s))}^{\min(i - a_j,\, B_{j-1})} Fix(p, j-1) \qquad (1)$$

This holds for i and j with $1 \le i \le \ell$, $1 \le j \le k$ and $A_j \le i \le B_j$. Note that an empty disjunction is `false`; this happens for example if $L_i^{\sigma_j}(s) \ge i - a_j + 1$. For $j = 1$ we have $Fix(i, 1) = $ `true` if and only if $L_i^{\sigma_1}(s) = 0$.

The validity of the recursion can be shown as follows. The last part of $s^{(i)}$ must consist of between a_j and b_j σ_js, say we want σ_js at positions $p+1, p+2, \ldots, i$. We then must have $a_j \le i - p \le b_j$. Also note that all elements $s_{p+1}, s_{p+2}, \ldots, s_i$ must be either x or σ_j; this holds exactly if $L_i^{\sigma_j}(s) \le p$. Finally, the first part of $s^{(i)}$, i.e., $s^{(p)}$, must adhere to $d^{(j-1)}$. Clearly, p must be between A_{j-1} and B_{j-1}, otherwise this would not be possible. Note that the A_j and B_j represent general tomographic restrictions, in some sense.

It is natural to implement this recursive formula by means of dynamic programming, using lazy evaluation: once a `true` $Fix(p, j-1)$ is found, the others need not be computed.

Now given a string s over Γ that is fixable with respect to a description d, it is easy to find those string elements x that have the same value from Σ in every fix: these elements are then set at that value. Indeed, during the computation of $Fix(s, d)$ (which of course yields `true`), one can keep track of all possible specifications that lead to a fix. In Equation (1) those $Fix(p, j-1)$ that are `true` correspond with a fix, where the string elements $s_{p+1}, s_{p+2}, \ldots, s_i$ are all equal to σ_j. Now one only has to verify, for each string element of s, whether precisely one element from Σ is allowed. In practice this can be realized by using a separate string, whose elements are filled when specifying s, and where those elements that are filled only once are tagged. Note that for this purpose lazy evaluation is not an option, since we need to examine all fixes. As an example, if the description for a five character string $s = s_1 s_2 s_3 s_4 s_5$ over $\{0, 1, x\}$ is $0^* 1^3 0^*$ (cf. the bottom row from the example in Section 1), one can derive that s_3 must be equal to 1. The algorithm that performs this operation is called *Settle*, and the resulting string s' is denoted by $s' = Settle(s, d)$.

The complexity of the computation of $Fix(\ell, k)$ is bounded by $k \cdot \ell^2$, and is in practice, especially when using lazy evaluation, much lower.

3.2 Discrete Tomography Problem

The Nonogram problem can be considered as a special case of a well-known problem from Discrete Tomography (DT), which deals with the reconstruction of a binary image from its horizontal and vertical linesums. These horizontal and vertical linesums can be easily computed from the Nonogram descriptions, by adding the segment descriptions for each line. (In the more general setting from Section 2 we get lower and upper bounds for the linesums.) Suppose that we have a partially filled Nonogram $X \in \Gamma^{m \times n}$, which we would like to extend

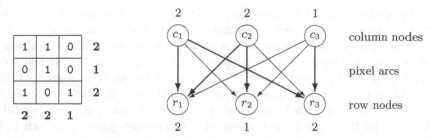

Fig. 2. a) DT problem and one of its solutions, where bold figures denote the linesums; b) associated network of the DT problem

further. Clearly, any solution of the Nonogram must also be a solution of the corresponding DT problem. The DT problem can be solved in polynomial time, even if an arbitrary subset of the image is kept fixed. It is also possible to compute the set of all pixels that must have the same value in all solutions of the DT problem in polynomial time. These pixels can be fixed immediately in the partial Nonogram solution. The paper [1] gives a constructive procedure for finding all such pixels.

Extendibility to a solution of the DT problem can easily be checked using network flow methods. We refer to [3] for the details of this model. Fig. 2a shows a simple 3×3 DT problem. We put linesums to the right of the rows and below the columns, to distinguish them from our earlier descriptions. This problem can be modelled as the transportation problem in Fig. 2b, which can be solved efficiently by network flow methods; thick arcs denote the solution. If none of the pixels are fixed, each pixel arc has a capacity of one. To fix a pixel at value $v \in \{0, 1\}$, we simply set the capacity of the corresponding pixel arc to 0 and subtract/add v to the surplus/demand at the corresponding column and row nodes. The resulting transportation problem has a solution if and only if the partial filling can be extended to a complete filling satisfying the DT constraints.

4   Combining the Partial Methods Using 2-SAT

The method from Section 3.1 can only take into account the description of a single line. On the other hand, the discrete tomography approach from Section 3.2 can deal with all lines simultaneously, but only incorporates partial knowledge from the descriptions. We will now describe how the information from different lines, and from different relaxations of the Nonogram problem, can be combined.

Consider the example in Fig. 3a (which is the same as that from Fig. 1). Using only the information from single lines, or from the discrete tomography problem, the values of the remaining undecided pixels cannot be derived. Four of the undecided pixels are denoted by the variables a, b, c and d respectively, which can take the values 0 (**false**) or 1 (**true**).

Using the partial solution methods, dependencies can be derived between pairs of undecided pixels. For example, on the bottom row, the description dictates that $c \Rightarrow d$ (or, equivalently, $\neg c \vee d$). Similarly, one can deduce that $c \Rightarrow \neg a$

(first column), $\neg a \Rightarrow b$ (third row) and $b \Rightarrow \neg d$. This provides us with both implications $c \Rightarrow d$ and $c \Rightarrow \neg d$, resulting in the conclusion that c must be 0.

Note that any such implication relation between two variables can be written in one of the forms $x \vee y$, $x \vee \neg y$, $\neg x \vee y$ or $\neg x \vee \neg y$. This is the standard form of a *2-SAT clause*, see [5]. The 2-SAT problem is to decide whether or not there exists an assignment of truth values to all the variables, such that a given conjunction of such clauses is simultaneously satisfied. It can be solved in polynomial time, using the concept of a *dependency graph*, as shown in Fig. 3 for our simple example.

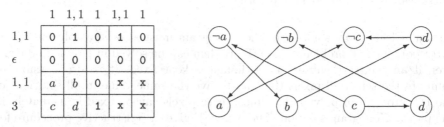

Fig. 3. a) Partially solved Nonogram; b) (part of) its corresponding dependency graph

However, when solving a Nonogram, the goal is not to find an assignment of all variables that satisfies the 2-SAT constraints. Rather, we search for variables that must have the same truth value in *all* satisfying assignments. Assume that at least one such assignment exists. Then a variable x is **false** in all satisfying assignments if and only if there is a path from x to $\neg x$ in the dependency graph. Alternatively, x must be **true** in all satisfying assignments if and only if there is such a path from $\neg x$ to x.

This provides a polynomial-time algorithm for finding all variables that must have the same value in all satisfying assignments of the 2-SAT problem. In the example from Fig. 3a, many more 2-SAT clauses can be found from the single rows and columns, or from the discrete tomography problem.

Our procedure for combining the information from the subproblems (one for each line, and a complete DT problem) is as follows: for each pair of undecided pixels (x, y) involved in the subproblem, all four assignments are tested. For each assignment, a fixability test is performed. Each such test that returns **false** provides an additional 2-SAT clause (e.g., $x \vee \neg y$). The resulting dependency graph captures information from all subproblems simultaneously. If one considers this process as "guessing", it can also be performed in a way similar to the *Settle* operation. Indeed, when computing the *Fix* value for a line, one can keep track of all pairs of pixels, and determine those values of pairs that cannot occur.

Although the 2-SAT approach is a powerful way to combine the knowledge from different partial problems, it generally does not capture all information that is present. For example, the three character string s over $\{0, 1, \mathtt{x}\}$ with description $d = 0^*1^10^*$ yields rules that do not forbid the fix 000, which is not a good fix. If one introduces clauses that can involve three variables, this leads to a clause $s_1 \neq 0 \vee s_2 \neq 0 \vee s_3 \neq 0$, which is in 3-SAT format. The general

3-SAT problem is NP-hard. Therefore, we chose the 2-SAT model, for which polynomial-time algorithms can be used.

5   Iterative Solving of Nonograms

Each relaxation of the Nonogram problem, such as the single line and discrete tomography relaxations from Section 3, can be used to deduce the value of certain pixels. By using such methods iteratively, filling in the new known pixels in each iteration, it is often possible to deduce even more pixel values. For clearness' sake, we now focus on the iterative application of the *Settle* operation from Section 3.1. It can be combined with alternative relaxations to form a more complete iterative algorithm. The *Settle* operation produces, given a string s over Γ and a description d, the string where all string elements that have the same value in every fix are set: $s \leftarrow Settle(s, d)$. Given $X \in \Gamma^{m \times n}$ and a Nonogram description N, we can repeat the *Settle* operation for all rows and columns of X (using the appropriate descriptions from N) until no new, previously unknown pixels are set. Note that we can use several heuristics to determine the order in which lines are examined. This operation is called *FullSettle*: $X \leftarrow FullSettle(X, N) \in \Gamma^{m \times n}$. If X now happens to be in $\Sigma^{m \times n}$, the puzzle is solved. Such a puzzle is called *simple*.

Note that the *Settle* operation, or rather the induced *Fix* operations, can also be used to detect certain contradictions, i.e., unsolvable puzzles. Indeed, if some line s of a proposed solution satisfies $Fix(s, d) = \texttt{false}$, this solution cannot be completed.

Now, given $X \in \Gamma^{m \times n}$ and a Nonogram description N such that $X = FullSettle(X, N)$, we can *harvest* 2-SAT expressions, leading to a 2-SAT problem Π, and set all elements from X that have the same value in all solutions to Π as in Section 4. This operation is called *2SATSolve*: $X \leftarrow 2SATSolve(X, N)$. The operations *FullSettle* and *2SATSolve* can be intertwined, until no further progress is made; this combination is called *Solver0*: $X \leftarrow Solver0(X, N)$. Again, if the resulting X happens to be in $\Sigma^{m \times n}$, the puzzle is solved. Such a puzzle is called *0–Solvable*. Note that this whole process takes polynomial time (expressed in height and width of the puzzle).

Now suppose that $X = Solver0(X, N)$, but the puzzle is not solved yet. We now consider one unknown element X_{ij} from X. In a copy Y of X we *try* both $Y_{ij} = 0$ and $Y_{ij} = 1$. If for one of these $Solver0(Y, N)$ gives a contradiction, we know that X_{ij} must have the other value. We can, again in some heuristic order, examine all unknown pixels. Note that only those pixels that occur in a 2-SAT clause need to be examined. This procedure can be repeated, until no further progress is made, again intertwined with the use of *Solver0*. This procedure is called *Solver1*. If a Nonogram can be solved in this way, it is called *1–Solvable*.

This process can be repeated with respect to the depth of the "tries", and it can then solve any Nonogram. In this way, we could define *k–Solvable*. Indeed, we then basically implement full backtracking. However, for our current purposes we only allow for (in depth!) one try, thereby inferring polynomial time complexity.

	2	1	1,1	1	1
1		0		0	
2					
1		0		0	x
2					
1		0		0	

Fig. 4. Partially solved 5×5 Nonogram, where the fact that pixel x must be white (0) is hard to infer

So several tries are possible, but at each moment at most one pixel is currently "tried".

To summarize these efforts, Nonograms can be simple (if *FullSettle* solves them), 0–*Solvable* (if *Solver0* solves them), 1–*Solvable* (if *Solver1* solves them), or more complicated. Many puzzles from newspapers are simple; the example from Fig. 1 is 0–*Solvable*. Also note that *FullSettle*, *Solver0* and *Solver1* are capable of providing partial solutions. It is possible to define the difficulty level of a Nonogram as the minimum number of tries necessary to solve it. The puzzle from Fig. 1 has level 1.

There are relatively small Nonograms for which *Solver1* cannot make any progress, even though it is still possible to infer the value of certain pixels. In Fig. 4 we show an example of such a Nonogram, where one can prove that the rightmost pixel in the third row must be white, yet *Solver1* fails to infer the value of any more pixels.

6   Experimental Results

We will describe several experiments with the techniques from the previous sections. All considered puzzles will have at least one solution: the image that was used to construct the puzzle. We mention the observation that in our experience most puzzles from newspapers are of simple type. Although one could attribute this to the relatively small size of these puzzles, this is contradicted by the example from Fig. 1, which shows that small puzzles do not have to be of simple type. Nevertheless, the larger the puzzle, the more complicated it can be. All puzzles of simple type can be solved very fast using our proposed framework, as the operation described in Section 3.1 effectively captures all information contained in the description of each single horizontal or vertical line.

As an illustrative example for a more difficult puzzle, we mention the 30×30 Nonogram from Fig. 5. It was randomly generated with 50 % black pixels. Using only *FullSettle* just 11 pixels are found. Using *Solver0* (which only takes approximately one second on a modern PC), the puzzle is solved but for 15 pixels. One can verify that there are 6 different solutions, where it turns out that for all 15 unknown pixels both black and white can occur. This Nonogram was also

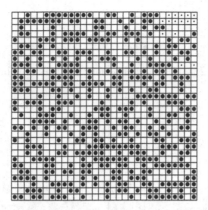

Fig. 5. Randomly generated partially solved 30×30 Nonogram, with 50 % black pixels; the 15 small dots denote the unknown pixels; 6 solutions

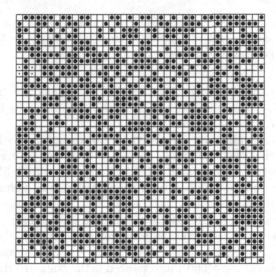

Fig. 6. Randomly generated partially solved 40×40 Nonogram, with 881 black pixels; the 4 small dots denote a pure switching component, leading to 2 solutions

included in [4], where an evolutionary algorithm was used to find one of the six solutions. A clear advantage of our reasoning framework over the algorithm presented in the former paper is that our approach finds the set of all solutions, along with a proof that there are no others. In addition, our method is much faster: seconds versus hours. On the other hand, both approaches can be considered as complementary, as the evolutionary algorithm can sometimes find a solution that cannot be deduced using our reasoning approach.

Most descriptions can be deduced from the figure. The description of row 1 is: 1, 8, 2, 1, 3, 2, of row 2: 7, 2, 2, 1, 1, 2, 2, of row 4: 2, 1, 4, 1, 2, 3, 1, 1, 1 and of row

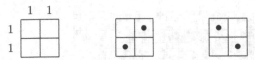

Fig. 7. Pure 2×2 switching component, with its 2 solutions

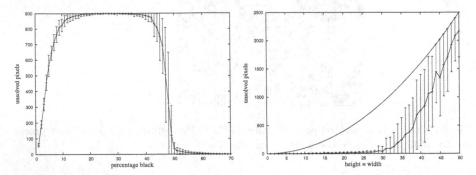

Fig. 8. a) Average number of unsolved pixels with standard deviation, for randomly generated 30×30 puzzles, with different percentages of black pixels; b) idem, for different sizes, with a fixed percentage of 50 %

9: 3, 1, 4, 1, 1, 4, 1; the descriptions of the last 6 columns are: 2, 4, 1, 1, 3, 1, 1, 1; 1, 3, 1, 2, 2, 1, 1, 2, 3; 1, 1, 1, 2, 2, 1, 1, 1, 1, 1; 1, 1, 1, 3, 2, 1, 1, 3, 3; 1, 1, 2, 1, 1, 4, 3, 1, 2; and 1, 1, 1, 1, 1, 3, 5, respectively.

In Fig. 6 we see a randomly generated 40×40 Nonogram, with 881, i.e., 55 %, black pixels. In this case the puzzle has a nearly unique solution: there is only one pure 2×2 *switching component* (cf. Fig. 7), so there are 2 different solutions. The descriptions can be deduced from the figure. Again, *Solver0* is necessary: *FullSettle* finds 101 pixels.

In Fig. 8a we see the results of 7,000 runs. For each p in $\{1, 2, 3, \ldots, 70\}$ the algorithm has been run 100 times on a randomly generated 30×30 puzzle, with p % black pixels. The piecewise linear curve connects the averages of the number of unsolved pixels (at most 900); also plotted is the standard deviation per percentage, truncated at 0 and 900. For small and large percentages the puzzles are solvable, in some cases leaving small switching components (cf. Fig. 5). Finally in Fig. 8b we plot, for each size s in $\{1, 2, 3, \ldots, 50\}$ these same quantities for randomly generated square $s \times s$ puzzles, all with 50 % black pixels. The smooth curve depicts the total number of pixels, i.e., s^2.

7   Conclusions

The general Nonogram problem is known to be NP-hard. However, it appears that in practice many instances can be solved quickly. In this paper we presented a general framework for solving Nonograms. By combining several relaxations, that can each be solved in polynomial time, a solution of the Nonogram is computed iteratively. The different solution methods are combined using a 2-SAT formulation. We demonstrated that this approach can solve a variety of interesting Nonograms. More importantly, the algorithm generates a logical proof for all pixels that are decided. Even if the puzzle cannot be solved completely, it may still be still be possible to decide the value of a substantial part of the pixels. The class of Nonograms that can be solved effectively using our approach includes the simple puzzles that can be found in puzzle books, but also includes random puzzles, which can often not be solved by simple logic reasoning, considering one line at a time.

Our framework is quite general. For example, as indicated in Section 2, the concept of a *description* can be generalized in a straightforward manner. In future work, we intend to study several such generalizations.

References

1. Aharoni, R., Herman, G., Kuba, A.: Binary vectors partially determined by linear equation systems. Discrete Math. 171, 1–16 (1997)
2. Barcucci, E., Del Lungo, A., Nivat, M., Pinzani, R.: Reconstructing convex polyominoes from horizontal and vertical projections. Theoret. Comp. Sci. 155, 321–347 (1996)
3. Batenburg, K.J.: A network flow algorithm for reconstructing binary images from discrete X-rays. J. Math. Imaging Vision 27(2), 175–191 (2007)
4. Batenburg, K.J., Kosters, W.A.: A discrete tomography approach to Japanese puzzles. In: Proceedings of the 16th Belgium-Netherlands Conference on Artificial Intelligence, BNAIC, pp. 243–250 (2004)
5. Garey, M., Johnson, D.: Computers and Intractability, A Guide to the Theory of NP-Completeness. W.H. Freeman, New York (1979)
6. Kuba, A., Balogh, E.: Reconstruction of convex 2D discrete sets in polynomial time. Theoret. Comp. Sci. 283(1), 223–242 (2002)
7. Simpson, S.: Nonogram solver (2007),
   www.comp.lancs.ac.uk/~ss/nonogram/links.html
8. Ueda, N., Nagao, T.: NP-completeness results for nonogram via parsimonious reductions (preprint, 1996),
   citeseer.ist.psu.edu/ueda96npcompleteness.html
9. Woeginger, G.: The reconstruction of polyominoes from their orthogonal projections. Inform. Process. Lett. 77, 225–229 (2001)

A Memetic Algorithm for
Binary Image Reconstruction

Vito Di Gesù[1,2], Giosuè Lo Bosco[1], Filippo Millonzi[1], and Cesare Valenti[1]

[1] Dipartimento di Matematica e Applicazioni
[2] Centro Interdipartimentale Tecnologie della Conoscenza
Università di Palermo, via Archirafi 34, 90123, Italy
{digesu,lobosco,millonzi,cvalenti}@math.unipa.it

Abstract. This paper deals with a memetic algorithm for the reconstruction of binary images, by using their projections along four directions. The algorithm generates by network flows a set of initial images according to two of the input projections and lets them evolve toward a solution that can be optimal or close to the optimum. Switch and compactness operators improve the quality of the reconstructed images which belong to a given generation, while the selection of the best image addresses the evolution to an optimal output.[1]

1 Introduction

Discrete tomography (DT) is a particular case of computerized tomography that deals with structures which have a few internal density values (usually a couple), and sometimes it is possible to reduce the number of projections. Fundamental works on discrete tomography have been reported in [12]. In this contribution we will consider only the reconstruction of binary images, which correspond to just a couple of density values.

It has been proved that it is possible to state in polynomial time whether there exists any object compatible with a given pair of projections [10,20]; vice versa, if we consider a bigger set of projections, the complexity increases because the image must satisfy a greater number of constraints. For example, the complexity of the reconstruction process becomes NP-hard in the case of at least three projections along non parallel directions [11]. Moreover, the number of images compatible with a couple of projections is normally huge because they can be very different from each other [21]. Therefore, the exact reconstruction requires additional information: not only a big enough number of projections, but also the geometric and topological properties of the object [15].

The algorithm presented here tries to reconstruct binary images that satisfy four projections, by creating an initial *population* through network flows. This

---

[1] This work makes use of results produced by the PI2S2 Project managed by the Consorzio COMETA, a project co-funded by the Italian Ministry of University and Research (MIUR) within the Piano Operativo Nazionale "Ricerca Scientifica, Sviluppo Tecnologico, Alta Formazione" (PON 2000-2006). More information is available at http://www.pi2s2.it and http://www.consorzio-cometa.it.

V.E. Brimkov, R.P. Barneva, H.A. Hauptman (Eds.): IWCIA 2008, LNCS 4958, pp. 384–395, 2008.

population reaches the desired solution through particular operators. Sect. 2 introduces the reconstruction methods present in literature and our new approach; basic notations are described, too. Experimental results are presented in Sect. 3, while Sect. 4 reports conclusions, future progresses and possible applications.

2 A Memetic Reconstruction Method

Usually, DT reconstruction algorithms are developed to process particular images such as *hv-convex polyominoes*, which are connected sets with 4-connected rows and columns, and *periodic images* [9], which have repetitions of pixels along some directions.

A well known reconstruction algorithm of hv-convex polyominoes is based on the filling of empty parts of the images and is taken back to the satisfiability of Boolean formulas; its time complexity is $O(mn \min\{m^2, n^2\})$ though it does not guarantee the accuracy of the final result [2,5,14]. In the case of noisy projections affected by quantization and instrumental errors, a quantitative estimate of the stability of these methods has been proposed in [7]. A memetic approach uses two projections and proper operators on noiseless images [3]. This last work differs from our method both on the number of projections and on the operators used. For example, the individuals obtained by crossover and mutation have to be further processed in order to verify the input projections. These evolutionary operators are performed with probabilities that are automatically tuned for each generation according to the fitness values of the previous generation. Moreover, a hash table is used to apply all possible switches on the current individual.

Memetic algorithms [18] (MAs) were introduced in 1989 and belong to the class of evolutionary algorithms which explore the solutions space of the problem by the generation of proper *agents* and the application of cooperative and competitive operations [6]. MAs use a fitness function to evaluate the quality of the agents, optimization operators to improve each single agent and a selection process to direct the whole population toward an optimal solution. Therefore, MAs induce a *vertical* evolution (between consecutive generations) and an *horizontal* evolution (within the same generation). These methods have been proved to be efficient in solving a variety of tasks such as the traveling salesman problem [22], the quadratic assignment problem [16] and the graph bi-partitioning problem [17].

2.1 Basic Notations

A binary image I of size $n \times m$ can be stored as a matrix $A = \{a_{ij}\}$ with elements equal to 0, if the corresponding pixel in I is black (i.e. it belongs to the background), or equal to 1, if the pixel is white (i.e. it belongs to the foreground).

Given a direction $\mathbf{v} \equiv (r, s)$ with $r, s \in \mathbb{Z}$ and $|r| + |s| \neq 0$, we define the *projection line* of A through a_{ij} along $\mathbf{v}$ the subset of A:

$$\ell_{\mathbf{v}}(i, j) = \{a_{i'j'} \in A : i' = i + zs, \ j' = j - zr \text{ with } z \in \mathbb{Z}\}.$$

By varying $a_{ij} \in A$, we obtain $t(\mathbf{v}) < \infty$ distinct projection lines parallel to $\mathbf{v}$, each one indicated with $\mathcal{L}_k^{\mathbf{v}}$, where $k = 1, \ldots, t(\mathbf{v})$.

Let us define the k-th projection of A along $\mathbf{v}$ as:

$$p_k^{\mathbf{v}} = \sum_{a_{ij} \in \mathcal{L}_k^{\mathbf{v}}} a_{ij}.$$

Informally, $p_k^{\mathbf{v}}$ coincides with the number of 1's on $\mathcal{L}_k^{\mathbf{v}}$. The projection of A along $\mathbf{v}$ is therefore the vector:

$$P_{\mathbf{v}} = \left(p_1^{\mathbf{v}}, p_2^{\mathbf{v}}, \ldots, p_{t(\mathbf{v})}^{\mathbf{v}} \right).$$

In order to simplify the method, we chose the following four directions in our projections:

$$\mathbf{v}_1 \equiv (1, 0), \quad \mathbf{v}_2 \equiv (0, 1), \quad \mathbf{v}_3 \equiv (1, 1), \quad \mathbf{v}_4 \equiv (1, -1).$$

An example of image together with its projections along the horizontal and vertical directions ($\mathbf{v}_1$ and $\mathbf{v}_2$ respectively) can be found in Fig. 1a-b.

It could be that no object exists compatible with a given set of projections. For example, there is no image that simultaneously satisfies $P_{\mathbf{v}_1} = (4, 4, 3, 1)$ and $P_{\mathbf{v}_2} = (2, 2, 4, 4)$. We assume, in the following, that the projections are satisfied by at least one image.

2.2   The Proposed Method

The memetic algorithm presented here takes the projections $Q_{\mathbf{v}_i}$ of the image to be reconstructed as input. It creates an initial population of *agents* by using their corresponding network flows (see next Sect. 2.4). Should the solution be already present in this population, then the algorithm ends, otherwise it continues with a new generation. Each one of them starts with the application of vertical and horizontal *crossovers* between random pairs of agents; just a couple of agents are selected among the parents and their four descendants and subsequently *mutation* and *compactness* are applied if they enhance the agents. A *switch* operation further improves the quality of the agents soon after these three operators (i.e. crossover, mutation and compactness). At the end of this process, the algorithm checks if it has already found a solution or if it has to continue with a successive generation. Anyway, a halt condition is established by the total number ng of generations.

We will see that each agent always maintains the same number of white pixels as the image to be reconstructed; moreover, the continuous selection of the best agents makes the size of the population constant. The choice of the operators and their order of application has been obtained on an experimental basis, by considering a variety of combinations on a training set containing only 20% of the whole database. We want to stress that, unlike the genetic algorithms and the method proposed in [3], our algorithm always applies the crossover, mutation, compactness and switch techniques.

2.3   Fitness Function

Given two projections $P_{\mathbf{v}}$ and $Q_{\mathbf{v}}$, let us consider their l_1 distance:

$$l_1(P_{\mathbf{v}}, Q_{\mathbf{v}}) = \sum_{k=1}^{t(\mathbf{v})} |p_k^{\mathbf{v}} - q_k^{\mathbf{v}}|.$$

By representing an agent with its corresponding binary matrix A, we define its fitness function as:

$$\mathcal{F}(A) = \sum_{i=1}^{4} l_1(P_{\mathbf{v}_i}, Q_{\mathbf{v}_i}),$$

where $P_{\mathbf{v}_i}$ is the projection of A and $Q_{\mathbf{v}_i}$ is the input projection, both taken along the same direction $\mathbf{v}_i$. The goal of the algorithm is the minimization of $\mathcal{F}$.

2.4   Initial Population

A convenient representation of I is given by the network flow G with one source S, one sink T and two layers of arcs between S and T: the first layer of *row-nodes* $\{R_1, R_2, \ldots, R_m\}$ and the second one of *column-nodes* $\{C_1, C_2, \ldots, C_n\}$. Each arc $\widehat{XY}$ of G has a flow:

$$f_{\widehat{SR_i}} = p_i^{\mathbf{v}_1}, \quad f_{\widehat{R_iC_j}} = a_{ij}, \quad f_{\widehat{C_jT}} = p_j^{\mathbf{v}_2}$$

and capacity:

$$c_{\widehat{SR_i}} = p_i^{\mathbf{v}_1}, \quad c_{\widehat{R_iC_j}} = 1, \quad c_{\widehat{C_jT}} = p_j^{\mathbf{v}_2}$$

where $i = 1, \ldots, m$ and $j = 1, \ldots, n$ (see Fig. 1c). It is noteworthy that in order to construct the network flow we consider only the directions $\mathbf{v}_1$ and $\mathbf{v}_2$. Moreover, the maximal flow through G corresponds to an image that satisfies $P_{\mathbf{v}_1}$ e $P_{\mathbf{v}_2}$ [1].

The Ford-Fulkerson algorithm [8] can be applied to compute the maximal flow of the network. Anyway, we do not act a breadth first search to find the augmenting paths, but we randomly select those arcs which have the biggest residual capacity; this does not guarantee a maximal flow, but assures a variability in the initial population and speeds up the whole method. Should the flow not be maximal, then white pixels are randomly added to the image so as to reach the correct number w of white pixels that must be present in the input image (this amount of pixels is equal to the integral of any projection).

2.5   Crossover

Our crossover can be applied vertically or horizontally: we define here the former, while the latter just operates on the transpose of the matrices. Given two parents A_1 and A_2, their vertical offsprings B_1 and B_2 are obtained by swapping the columns $A_1^{(j)}$ and $A_2^{(j)}$, where j corresponds to the positions of 1 in a random binary *mask* $M = (M_1, M_2, \ldots, M_n)$ (see Fig. 2):

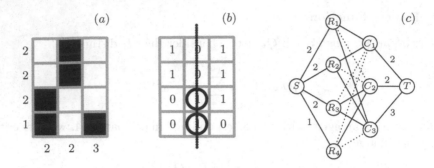

Fig. 1. A binary image (a) and its matrix representation (b). The projection lines $\ell_{v_2}(3,2)$ and $\ell_{v_2}(4,2)$ are equivalent because they intercept the same pixels (marked by the circles). The dashed arcs $\widehat{R_i C_j}$ in the network flow indicate black pixels; the remaining arcs correspond to white pixels (c).

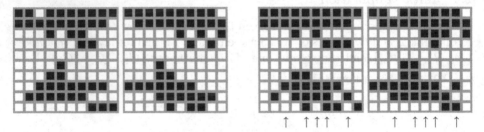

Fig. 2. Two parents (*left*) and their vertical offsprings, obtained through the mask $M = (0, 0, 1, 0, 1, 1, 1, 0, 1, 0)$ represented by the arrows (*right*)

$$B_1^{(j)} = \begin{cases} A_1^{(j)} \text{ if } M_j = 0 \\ A_2^{(j)} \text{ if } M_j = 1 \end{cases} \quad \text{and} \quad B_2^{(j)} = \begin{cases} A_2^{(j)} \text{ if } M_j = 0 \\ A_1^{(j)} \text{ if } M_j = 1 \end{cases}$$

It must be noted that the vertical (horizontal) crossover maintains the vertical (horizontal) projection P_{v_2} (P_{v_1}).

2.6 Mutation

Mutation modifies no more than $\rho = min\{\lfloor \frac{m \times n}{20} \rfloor, m \times n - w, w\}$ pixels, which correspond at most to 5% of the image. This threshold has been experimentally determined since it returns better fitness values. The operator locates $\kappa \leq \rho$ white and κ black pixels in a random fashion and inverts their color (see Fig. 3).

2.7 Switch

The switch operator maintains both the horizontal and vertical projections, while swapping the values of two pairs of pixels. The *elementary switch* operator on A swaps a_{ij} with a_{ik} and a_{hj} with a_{hk}, where the following constraint must hold:

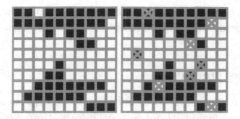

Fig. 3. The mutated image (*left*) has been obtained by swapping the pixels depicted by the crosses (*right*)

$$\begin{cases} a_{ij} = a_{hk} = 1 \\ a_{ik} = a_{hj} = 0 \end{cases} \quad \text{or} \quad \begin{cases} a_{ij} = a_{hk} = 0 \\ a_{ik} = a_{hj} = 1 \end{cases}$$

As in Fig. 4, this operation swaps two *compatible arcs* $\widehat{R_iC_j}$ and $\widehat{R_hC_k}$ in the network flow G of A so that $R_i \neq R_h$, $C_j \neq C_k$ and both $\widehat{R_iC_k}$ and $\widehat{R_hC_j}$ do not already belong to G.

An exhaustive search of compatible arcs $\widehat{R_iC_j}$ and $\widehat{R_hC_k}$ has a time complexity equal to $O(\min\{m \times n^2, n \times m^2\})$, but usually we do not have to explore the whole space because, by permuting the rows or the columns, we find their proper combination after just a few iterations. The projections P_{V_1} and P_{V_2} are maintained, while generally P_{V_3} and P_{V_4} change their values.

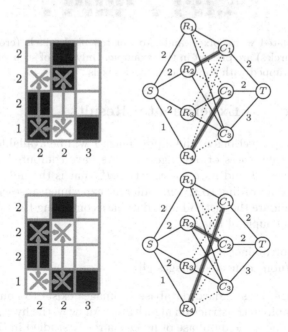

Fig. 4. Net representation of an elementary switch. The swapped arcs have been marked in gray.

Notice that any couple of images that satisfy the same set of two projections can be transformed one into each other by a finite sequence of elementary switches [13,19].

2.8 Compactness Operator

The algorithm till now described is suitable to reconstruct objects with more components and holes, even without a priori information. Nevertheless, the image to be reconstructed generally does not have *isolated* pixels, where a pixel is called isolated if it is surrounded by 8 pixels with opposite color.

The compactness operator usually speeds up the convergence of the memetic algorithm and improves the result by locating and eliminating as many isolated pixels as possible. Let $\sigma = \min\{w', b'\}$, where w' and b' are the number of isolated white and black pixels, respectively. The operator randomly locates σ isolated white pixels and σ isolated black pixels and sets their values to the value of their neighbors (see Fig. 5).

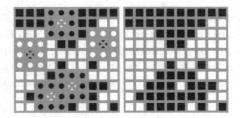

Fig. 5. Three isolated white pixels and two isolated black pixels (crosses) and their neighborhoods (circles) are present in this example. Only two of such white pixels have been randomly swapped with the isolated black pixels.

3 Database and Experimental Results

Though only the projections of the input image I will be available in real cases, to evaluate the robustness of our algorithm we have computed the symmetric difference ε between I and its reconstruction I', that is the number of unequal pixels normalized according to $n \times m$. Different experimental sessions have been carried out to validate the method on a database containing the following classes (Fig. 6 depicts a couple of images):

1. hv-convex polyominoes;
2. non-convex/non-connected images [4].

To our knowledge, no standard database of images exists to compare different discrete tomography reconstruction algorithms. Anyway, the hv-convex polyominoes come from the same database of images already studied in [7], where some of the methods reported in Sect. 2 have been considered. We have used 10 subsets of 50 images, ranging from 10×10 to 100×100 pixels, with a linear step of 10 pixels. In order to calculate the best parameters that achieve a "satisfactory" final error the algorithm was performed 9 times on each class of images, by varying at each execution the number ng of generations and the size na of the population. For each session, we computed the average error $\bar{\varepsilon}$ and the number $N(\bar{\varepsilon})$ of solutions with a reconstruction error less than $\bar{\varepsilon}$. Better results have been obtained for images that satisfy strong topological and geometric constraints, as in the case of the hv-convex polyominoes. Tables 1-2 refer to the last plots of Figs. 7-8 and summarize the results for both classes of images, with 100×100 pixels. The bars represent the reconstruction error within the first and third quartiles, the horizontal line inside each bar indicates the median reconstruction error $m(\varepsilon)$, while the horizontal lines outside each bar indicate the inter-quartile range (small circles are outliers beyond that range). For example, 86% of the hv-convex solutions obtained through 1000 individuals and 1500 generations have an error ε less than the average error $\bar{\varepsilon}=0.480\%$.

One last note regards the execution time of the algorithm: it has been implemented in (interpreted) MatLab language and requires about 10 seconds on a standard personal computer per generation to elaborate an image with 100×100 pixels and 1000 individuals.

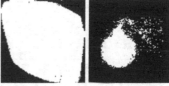

Fig. 6. A 100×100 hv-convex polyomino and a 100×100 non-convex/non-connected image (*left*), extracted from our test database. Their reconstructed versions (*right*) have errors equal to $\varepsilon=0.36\%$ and $\varepsilon=4.87\%$ respectively.

Table 1. HV-convex polyominoes with 100×100 pixels

na	500	500	500	750	750	750	1000	1000	1000
ng	500	1000	1500	500	1000	1500	500	1000	1500
$m(\varepsilon)$	0.015%	0.020%	0.005%	0.010%	0.005%	0.000%	0.005%	0.005%	0.000%
$\bar{\varepsilon}$	0.532%	0.530%	0.502%	0.536%	0.493%	0.524%	0.529%	0.521%	0.480%
$N(\bar{\varepsilon})$	84%	82%	84%	84%	84%	84%	84%	84%	86%

Table 2. NC images with 100×100 pixels

na	500	500	500	750	750	750	1000	1000	1000
ng	500	1000	1500	500	1000	1500	500	1000	1500
$m(\varepsilon)$	4.635%	3.840%	3.595%	3.920%	3.395%	3.700%	3.145%	3.305%	4.275%
$\bar{\varepsilon}$	8.097%	7.523%	7.425%	7.615%	7.388%	7.228%	7.569%	7.349%	7.449%
$N(\bar{\varepsilon})$	58%	58%	58%	58%	58%	58%	58%	58%	56%

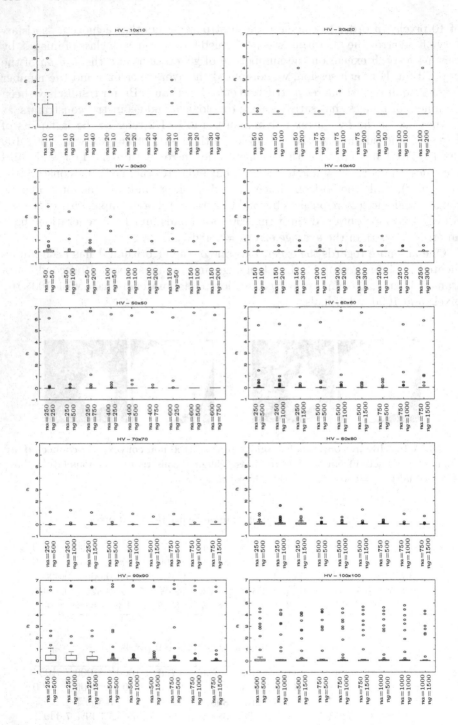

Fig. 7. Complete results for hv-convex polyominoes. The size of the images is indicated on the title of the plots.

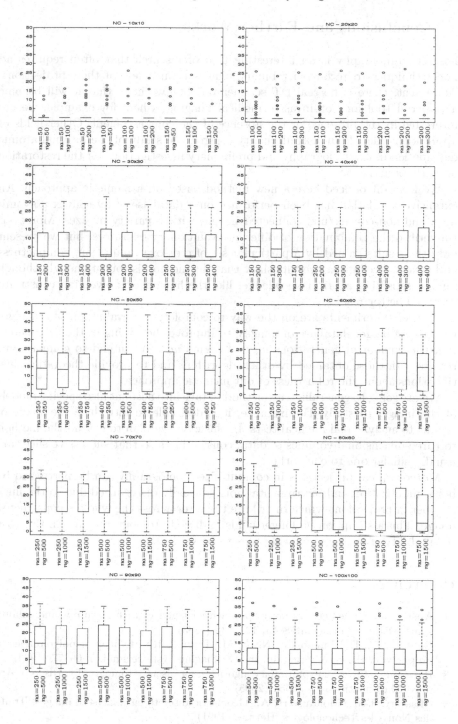

Fig. 8. Complete results for non-convex/non-connected images. The size of the images is indicated on the title of the plots.

4   Conclusions and Further Works

Discrete tomography is an interesting field of research that often requires ad hoc techniques to reconstruct binary images from a few of their projections. At present, there is no real DT scanner yet, but we hope that it will become a powerful tool for non-invasive medical imaging, when, for example, we are interested in the presence or the absence of the contrast agent. Other fields of applications include industrial quality control, the study of the internal atomic distribution in crystallography and preliminary investigations for the restoration of works of art.

We have introduced here a new method based on a memetic approach. An extensive simulation has been performed on a database of hv-convex polyominoes and non-convex/non-connected images with a variety of sizes. Moreover, we verified the combination of different evolutionary operators, but we present only the results gained by a particular set of crossover, mutation, compactness and switch. Preliminary results show that the method is robust enough, though images satisfying a priori knowledge, like the hv-convex polyomino constraint returned better results.

Further experiments are on the way to complete the evaluation of our algorithm on a bigger database of images, to improve the tuning of the parameters (i.e. the number of generations ng and agents na). We intend to compare our results with those obtained by other algorithms and to provide theoretical justifications about the robustness of our memetic approach.

We are going to generalize the method to take into account also specific models of the images to study. For instance, it is desirable to introduce the shape of organs to analyze real medical tests. In this case, we plan to extend the method to three-dimensional volumes of data, considering also that successive slices are usually similar one to each other.

Moreover, real projections in computerized tomography can be perturbed by both quantization and instrumental noises, usually disregarded due to the huge number of projections which distribute the error across the whole reconstructed image. This assumption no longer holds in discrete tomography and in this case the stability of our method still needs to be verified.

References

1. Anstee, R.P.: The network flows approach for matrices with given row and column sums. Discrete Mathematics 44, 125–138 (1983)
2. Barcucci, E., Del Lungo, A., Nivat, M., Pinzani, R.: Medians of polyominoes: a property for the reconstruction. Int. J. Imaging Syst. Technol. 9, 69–77 (1998)
3. Batenburg, J.K.: An evolutionary algorithm for discrete tomography. Discrete Applied Mathematics 151, 36–54 (2005)
4. Chassery, J.-M., Montanvert, A.: Géométrie discrète en analyse d'images. Traité des Nouvelles Techonologies, Hermes (1991)
5. Chrobak, M., Dürr, C.: Reconstructing hv-convex polyominoes from orthogonal projections. Information Processing Letters 69, 283–291 (1999)

6. Corne, D., Dorigo, M., Glover, F.: New ideas in optimization. McGraw-Hill, New York (1999)
7. Di Gesù, V., Valenti, C.: The stability problem and noise projections in discrete tomography. Journal of Visual Languages and Computing 15(5), 361–371 (2004)
8. Edmonds, J., Karp, R.M.: Theoretical improvements in algorithmic efficiency for network flow problems. Journal of the ACM 19(2), 248–264 (1972)
9. Frosini, A., Nivat, M., Vuillon, L.: An introductive analysis of periodical discrete sets from a tomographical point of view. Theoretical Computer Science 347(1–2), 370–392 (2005)
10. Gale, D.: A theorem on flows in networks. Pacific J. Math. 7, 1073–1082 (1957)
11. Gardner, R.J., Gritzmann, P., Prangenberg, D.: On the computational complexity of reconstructing lattice sets from their X-rays. Discrete Mathematics 202, 45–71 (1999)
12. Herman, G.T., Kuba, A. (eds.): Discrete Tomography: Foundations, Algorithms, and Applications. Birkhäuser, Basel (1999)
13. Kong, T.Y., Herman, G.T.: Tomographic equivalence and switching operations. In: Discrete Tomography: Foundations, Algorithms, and Applications, pp. 59–84. Birkhäuser, Basel (1999)
14. Kuba, A.: The reconstruction of two-directional connected binary patterns from their two orthogonal projections. Computer Vision, Graphics, and Image Processing 27, 249–265 (1984)
15. Matej, S., Vardi, A., Herman, G.T., Vardi, E.: Binary Tomography Using Gibbs Priors. In: Discrete Tomography: Foundations, Algorithms, and Applications, pp. 191–212. Birkhäuser, Basel (1999)
16. Merz, P., Freisleben, B.: Fitness landscape analysis and memetic algorithms for the quadratic assignment problem. IEEE Transactions on Evolutionary Computation 4(4), 337–352 (2000)
17. Merz, P., Freisleben, B.: Memetic algorithms and the fitness landscape of the graph bi-partitioning problem. In: Eiben, A.E., Bäck, T., Schoenauer, M., Schwefel, H.-P. (eds.) PPSN 1998. LNCS, vol. 1498, pp. 765–774. Springer, Heidelberg (1998)
18. Moscato, P.: On evolution, search, optimization, genetic algorithms and martial arts: Towards memetic algorithms. Caltech Concurrent Computation Program, C3P Report 826 (1989)
19. Ryser, H.J.: Combinatorial mathematics. The carus mathematical monographs, vol. 14, ch. 6, MAA (1963)
20. Ryser, H.J.: Combinatorial properties of matrices of zeros and ones. Canadian J. Math. 9, 371–377 (1957)
21. Wang, B., Zhang, F.: On the precise number of (0,1)-matrices in A(R,S). Discrete Mathematics 187, 211–220 (1998)
22. Zou, P., Zhou, Z., Chen, G., Yao, X.: A novel memetic algorithm with random multi-local-search: A case study of TSP. In: Proc. of Congress on Evolutionary Computation, vol. 2, pp. 2335–2340. IEEE Press, Los Alamitos (2004)

Personal Identification Based on Weighting Key Point Scheme for Hand Image

Dongbing Pu[1,2], Shuang Qi[2,3], Chunguang Zhou[1], and Yinghua Lu[2,*]

[1] College of Computer Science & Technology, Jilin University, Changchun,
Jilin Province, China
[2] College of Computer, Northeast Normal University, Changchun, Jilin Province, China
[3] Key Laboratory for Applied Statistics of MOE, China
{pudb,qis180,luyh}@nenu.edu.cn

Abstract. Biometrics-based personal identification is regarded as an effective method for automatically recognizing a person's identity with a high confidence. This paper presents a novel approach for personal identification using weighting relative distance of key point scheme on hand images. In contrast with the existing approaches, this system extracts multimodal features, including hand shape and palmprint to facilitate the task of coarse-to-fine dynamic identification. Five hand geometrical features are used to guide the selection of a small set of similar candidate samples at the coarse level matching stage. In the fine level matching stage, the weighting relative distance of key point approach is proposed to extract palmprint texture.

Keywords: personal identification, weighting key point, relative distance, coarse-to-fine.

1 Introduction

With the wide spread utilization of biometric identification systems, establishing the authenticity of biometric data itself has emerged as an important research issue. Numerous distinguishing traits have been used for personal identification including fingerprint, face, voice, iris, hand geometry and so on. Due to its stability and uniqueness, palmprint can be considered as one of the reliable means distinguishing a man from others, and it also can be easily integrated with the existing identification system to provide enhanced level of confidence in personal identification [5].

Two kinds of biometric features can be extracted from the hand images; (i) palmprint features, which are composed of principal lines, wrinkles, minutiae, delta points, etc.. (ii) hand geometry features which include area/size of palm and length and width of fingers. How to extract these features is a key step for identification. The available approaches of personal recognition based on a palmprint image can be divided into three categories on the basis of the type of extracted features. These categories are as follows: the line-based approaches [14] and [16], [20], the texture-based approaches [4], [17] and [19] and appearance-based approaches [2], [6] and [15].

---

* Corresponding author.

V.E. Brimkov, R.P. Barneva, H.A. Hauptman (Eds.): IWCIA 2008, LNCS 4958, pp. 396–407, 2008.

A biometric system based on a single biometric characteristic is regarded as a unimodal system. However, a single physiological or behavioral characteristic of a person may fail to be sufficient for recognition. For this reason, multimodal biometric systems, which integrate two or more different biometric characteristics, are being developed to increase the accuracy of decisions [13].

In this paper, we propose a multimodal biometric personal identification system based on weighting relative distance of key point scheme for hand image, in which, both hand geometrical features and palmprint region of interest (ROI) features are employed and a coarse-to-fine dynamic identification strategy is adopted to implement a reliable and real time personal identification system. The block-diagram of the proposed system is shown in Fig. 1, where hand geometrical features and texture features are stored in Database1 and Database2, respectively. Firstly, hand image is captured by a flatbed scanner. Then, a series of preprocessing operations are employed for the segmentation of ROI and geometry features are also obtained in this process. The hand shape geometry features are first used for coarse-level identification. And the weighting relative distance of key point approach is applied to extract the features of ROI features for fine level identification. At decision stage, the Mahalanobis distance matching mechanism is employed to output the identification result.

The paper is organized as follows. Section 2 introduces the image acquisition and the pre-processing. Section 3 describes the feature extraction based on weighting relative distance of key point method briefly. The process of identification is depicted in Section 4. The experimental results are reported in Section 5. Finally, the conclusions are summarized in Section 6.

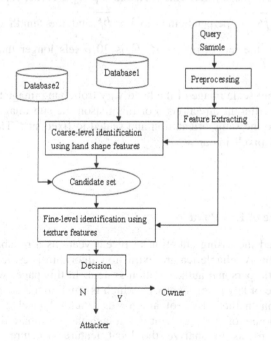

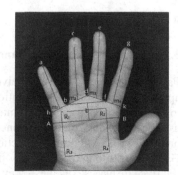

Fig. 1. Block diagram of the proposed identification system

Fig. 2. Shows the process of handprint segmentation

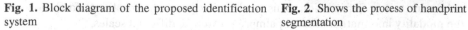

2 Images Acquisition and Pre-processing

The hand image is acquired with a desktop optical scanner at a resolution of 300 dpi. The size of the gray-scale image is 500×500 pixels. The right hands of users are placed on the scanner with the fingers spreading naturally. There are no pegs or other hand-position constrainers on the scanner. Our image acquisition setup is inherently simple and does not employ any special illumination.

The extraction of ROI which contains palm-lines and creases is necessary for extracting palmprint features. The finger-webs location algorithm proposed by Lin [9] and [11] is used to obtain the seven key points (a-g), which is shown in Fig. 2.And the regions of interest $R_1 R_2 R_3 R_4$ are segmented and five hand shape features are also extracted in following process:

1. Find the point h and k which are the intersection of lines $\overline{db}, \overline{df}$ with hand contour. Then compute the midpoints m_1, m_2, m_3 and m_4 of lines $\overline{hb}, \overline{db}, \overline{df}$ and $\overline{fk}$.

2. Find line $\overline{AB}$ which is parallel to line $\overline{bf}$, and the distance L between line $\overline{AB}$ and line $\overline{bf}$ is 50 pixels.

3. Form five length features by computing length of lines of $\overline{am_1}, \overline{cm_2}, \overline{em_3}, \overline{gm_4}$ and $\overline{AB}$ which will be considered as geometry features at the coarse level matching stage. Locate the top left corner R_1 and top right corner R_2 of ROI. As shown in Fig. 2, line $\overline{fR_2}$ is perpendicular to line $\overline{bf}$ and the length of line $\overline{fR_2}$ is 20 pixels. In addition, the length of line $R_1 \ R_2$ is 20 pixels longer than line $\overline{bf}$.

The sizes of the ROIs on the gray-scale image of the hand vary from hand image to hand image, and the ROIs lie in different directions. For this reason the sub-images defined by the ROIs are rotated to the same orientation and scaled to fixed size. The ROIs are normalized to 128×128 pixels in our work.

3 Feature Extraction

3.1 Weighting Relative Distance of Key Point

Palmprint recognition has received increasing attention in recent years as a reliable approach to personal identification. A reliable feature extraction algorithm is critical to the performance of an automatic personal authentication system. In this paper we use the weighting relative distance of key points approach which is ameliorated based on Ref. [18] as feature extraction method. First of all, we use multi-channel 2-D Gabor filters to extract the key points of the palmprint image and then compute the relative distances of these key points to analyze the local feature structure of palmprint texture information. This method integrates the location information and the modality information of local palmprint texture at different scales.

In our method, we use multi-channel 2-D Gabor filters [1], [3], [7], [8], [10] and [12] to extract the texture information in various directions and different scales in ROI of palmprint. Each channel corresponds to a direction in a different scale.

In (x, y) coordinate system, a 2-D Gabor filter in a certain channel can be given by

$$G(x, y, T, \theta) = e^{-(x'^2/2\alpha^2)-(y'^2/2\beta^2)} e^{-(2\pi i/T)x'} \tag{1}$$

where α and β are the width and height (standard deviation) of a 2-D Gaussian function, respectively. $i = \sqrt{-1}$ is the imaginary unit, T is the periods (or wavelength) in spatial domain, x' and y' are given by

$$\begin{bmatrix} x' \\ y' \end{bmatrix} = \begin{bmatrix} \cos\theta & \sin\theta \\ -\sin\theta & \cos\theta \end{bmatrix} \begin{bmatrix} x \\ y \end{bmatrix} \tag{2}$$

where θ is the angle between is (x, y) coordinates system and (x', y') coordinates system. Fig. 3 (a-d) show the filtered images in some channels.

As the symmetry of 2-D Gabor filters, we set four different values for $\theta : 0°, 45°, 90°$ and $135°$. In order to extract texture information in different scales, the wavelength is set to four discrete values: $8, 16, 32$ and 64, so that there are $4 \times 4 = 16$ channels. After performing the filtering, we obtain 32 filtered images.

In order to capture the local texture information of palmprint, we segment the filtered images into 16 sub-images. In each 32×32 sub-image, it is proper to regard the points with the largest coefficient as the feature points of the palmprint image. Here, we choose 64 feature points in every sub-image from each channel. Then the barycenter of these feature points is taken as the key point.

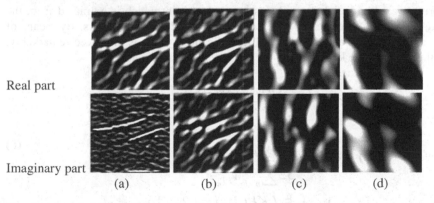

Real part

Imaginary part

(a)   (b)   (c)   (d)

Fig. 3. (a), (b), (c), and (d) are the filtered images respectively in $T = 8$, $\theta = 90°$; $T = 16$, $\theta = 45°$; $T = 32$, $\theta = 0°$; $T = 64$, $\theta = 135°$ from the same person

Let $F(x, y)$ be the absolute value of a filtered image, and

$$LFP = \{(x_1, y_1), (x_2, y_2), \ldots, (x_{64}, y_{64})\} \tag{3}$$

is defined as the locations of feature point, then the location of key point (x_{kp}, y_{kp}) in one of 16 channels can be given by

$$
\begin{aligned}
x_{KP} &= \frac{\sum_{i=1}^{64} x_i F(x_i, y_i)}{\sum_{i=1}^{64} F(x_i, y_i)}, \\
y_{KP} &= \frac{\sum_{i=1}^{64} y_i F(x_i, y_i)}{\sum_{i=1}^{64} F(x_i, y_i)},
\end{aligned}
\qquad (x_i, y_i) \in LFP. \tag{4}
$$

Because there are 32 filtered images (16 channels), we can get 32 key points in each sub-images. The location of these key points (LKP) in the jth sub-image is defined as

$$LKP = \{(x_{KP_1}, y_{KP_1})_j, (x_{KP_2}, y_{KP_2})_j, \ldots, (x_{KP_{32}}, y_{KP_{32}})_j\}, \; j = 1,2,\ldots16 \tag{5}$$

As above, we extract 512 ($32 \times 16 = 512$) key points in the whole palmprint image.

In our method, the location of key points may be easily affected by ROI localization, such as the variation of pressure and noise, but these factors have little influence on the relative distances of key points, so we make use of these relative distances as the palmprint feature vectors which are ready to be used in subsequent process.

Before computing the relative distances of the key points, we should find the center of the key points. Because the channels of multi-channel 2-D Gabor filters are used to describe different scales and various directions texture information in ROI of palmprint, we add a weighting value on each key point which is extracted from the filtered image in different channels of multi-channel 2-D Gabor filters. By means of weighting, the performance of personal identification system is improved remarkably. Here, the center of the key points in the jth sub-image is defined as

$$
\begin{aligned}
O_j(x) &= \frac{\sum_{n=1}^{32} (w_{KP_n} \times x_{KP_n})_j}{32 \sum_{n=1}^{32} w_{KP_n}}, \\
O_j(y) &= \frac{\sum_{n=1}^{32} (w_{KP_n} \times y_{KP_n})_j}{32 \sum_{n=1}^{32} w_{KP_n}},
\end{aligned}
\tag{6}
$$

$$(x_{KP_n}, y_{KP_n})_j \in LKP(j), \; j = 1,2,\ldots16.$$

Each channel of multi-channel 2-D Gabor filters contributes different effect-degree to the capability of describing local texture of palmprint. Based on this theory, we attempt to use weighting strategy on each key point. Because in this paper we regard the weighting relative distance of key points as feature vector, and a single key point can't realize this algorithm, the weighting strategy is firstly assign weighted values to every four channels with man-made factor. Then we use Eq. (7) to compute the weighted value of each key point in the jth sub-image before we find out the center of key points in different channels.

$$w_{KP_n} = \frac{10 \times \sqrt{a_k \times l_q}}{4}, \tag{7}$$

$$k = 1,2,3,4, q = 1,2,3,4.$$

Where w_{KP_n} is the weighted value of each key point. When $T = 8,16,32,64$, a_k is the weighted value in $\theta = 0°,45°,90°,135°$ channels separately. When $\theta = 0°,45°,90°,135°$, l_q is the weighted value in $T = 8,16,32,64$ channels separately.

The distance between the center of the key points $O_j(x, y)$ and each key point is defined as the relative distance.

$$D_j(n) = \sqrt{((x_{KP_n})_j - O_j(x))^2 + ((y_{KP_n})_j - O(y))^2}, n = 1,2,\ldots,32. \tag{8}$$

Therefore, every sub-image will obtain 32 relative distances, and the total number of relative distance is 512.

3.2 Extraction of Hand Geometry Features

The hand geometry features such as length and width of fingers, thickness and relative location of these features, form the other set to discriminate features from the hand images. Five hand shape length values are obtained in the pre-processing block.

4 Identification

4.1 Coarse-Level Identification

Though the geometrical length features are not so discriminative but it can be used in coarse level matching to facilitate the system to work on a small candidates.

If the distance is smaller than pre-defined threshold value, record the index number of the template into an index vector R for fine-level identification.

4.2 Fine-Level Identification

The index vector R has been recorded in coarse-level identification stage. In this section, the testing image will be further matched with the templates whose index numbers are in R. The Mahalanobis distances, which are considered the most robust techniques defined in Eq. (9), are used to measure the similarity between query sample and the template at fine-level identification stage.

$$d_{ij}^2(M) = (X_i - X_j)'\Sigma^{-1}(X_i - X_j)$$

(9)

Where Σ is a covariance matrix whose size is $p \times p$, $\Sigma = (\sigma_{ij})_{p \times p}$.

5   Experimental Results

In this section, our proposed approach is performed to evaluate the effectiveness and accuracy. The hand images database contains 1000 hand images collected from 100 individuals' right hand using our flatted scanner. The size of all images is 500×500 and the resolution is 300 dip. Each image was processed by the procedures involving pre-processing, segmentation and feature extraction.

5.1   The Experiment of Weighting Strategy

In the system-design phase we performed several verification tests to find the optimum values for the system parameters. In this phase, the database includes 10 different persons with 10 samples, and three samples of per person are considered as genuine samples and remaining seven images are used for impostor test.

Multi-channel 2-D Gabor filters describes the texture information in various directions and different scales in ROI of palmprint. Each channel corresponds to a direction in a different scale. We attempt to choose the channels which can improve the performance of our system better. As the symmetry of 2-D Gabor filters, we set four different values for θ : $0°$, $45°$, $90°$ and $135°$. In order to extract texture information in different scales, the wavelength is set to six discrete values: $2, 4, 8, 16, 32$ and 64, respectively. Because in this paper we regard the weighting relative distance of key points as feature vector, but a single key point in one channel can't realize this algorithm, we do the experimentation, aimed at selecting better channels and assigning the weight, in every four channels. The experiment results are shown in Fig. 4 which shows distributions of intra-class and inter-class matching distance in different four channels. From the results, we can find that the distance between the intra-class and the inter-class distribution and the portion that overlaps between the intra-class and the inter-class varies following the wavelength transforms, at the same time, θ sets to four discrete values: $0°, 45°, 90°$ and $135°$ immovably. The distance between the intra-class and the inter-class distribution is larger, and the portion that overlaps between the intra-class and the inter-class is smaller, the discrimination of the channel is better. Thus, we decide the wavelength is set to four values: $8, 16, 32$ and 64, and θ set four different values for θ : $0°, 45°, 90°$ and $135°$ based on experimental results.

In selection of channel phase, we confirm the wavelengths and angles of Multi-channel 2-D Gabor filters which can improve the accuracy of our proposed system

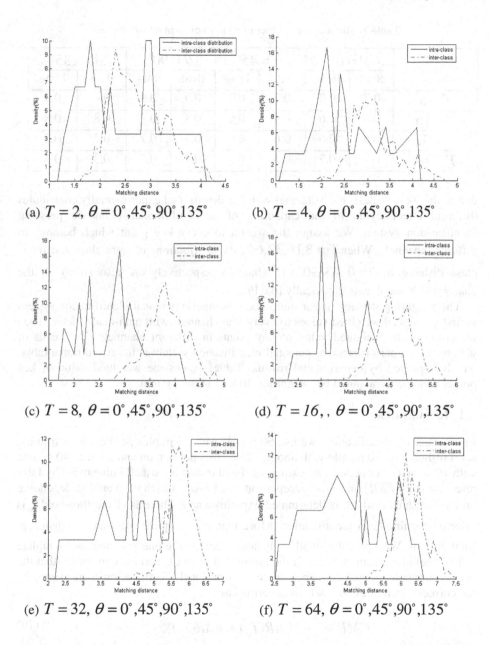

(a) $T = 2$, $\theta = 0°,45°,90°,135°$

(b) $T = 4$, $\theta = 0°,45°,90°,135°$

(c) $T = 8$, $\theta = 0°,45°,90°,135°$

(d) $T = 16$, , $\theta = 0°,45°,90°,135°$

(e) $T = 32$, $\theta = 0°,45°,90°,135°$

(f) $T = 64$, $\theta = 0°,45°,90°,135°$

Fig. 4. Distributions of intra-class and inter-class distance, (a) is the distribution when $T = 2$, $\theta = 0°,45°,90°,135°$, (b)(c)(d)(e)(f) is the distribution when $T = 4,8,16,32,64$ respectively

Table 1. The weighted values of all key points in jth sub-image

	$\theta = 0°$		$\theta = 45°$		$\theta = 90°$		$\theta = 135°$	
	Real	Imag	Real	Imag	Real	Imag	Real	Imag
$T = 8$	0.7	0.7	0.7	0.7	0.9	0.9	0.9	0.9
$T = 16$	0.6	0.6	0.6	0.6	0.8	0.8	0.8	0.8
$T = 32$	0.6	0.6	0.6	0.6	0.7	0.7	0.7	0.7
$T = 64$	0.3	0.3	0.3	0.3	0.3	0.3	0.3	0.3

distinctly. Nevertheless, each channel which is determined experimentally contributes different effect-degree to the capability of our multimodal biometric personal identification system. We assign the weight to every key point which belongs to different channels. When $T = 8,16,32,64$, the distributions of intra-class and inter-class distance in $\theta = 0°,45°,90°,135°$ channels respectively are also shown in the charts which are plotted statistically like Fig. 4.

The weighting strategy in our multimodal biometric personal identification system is firstly assigned weighted values to every four channels with man-made factor. Then we compute the weighted value of key points in different channels. By means of weighting, the performance of personal identification system is improved remarkably, which is proved by experimental results. Table1 shows the weighted value of key points which are computed by weighting strategy in our system.

5.2 Identification

In the stage of identification, we use 60 people with 3 samples per person as training set. Furthermore, 60 people with another 7 samples per person and another 40 people with 10 samples per persons as testing set. Final results are usually quantified by false rejection rate (FRR) and false acceptation rate (FAR) which are variable depending on the threshold T which is determined experimentally. There is also a threshold T_0 is selected for fine-level identification. More than one template may smaller than T_0 at final outputs. We select the smallest distance between the query sample and template as the final identification result. If the number of output is zero, it illuminates that the query sample is an attacker. The accuracy of personal identification is measured by the correct match rate CMR which is defined as:

$$CMR = 1 - (FRR(T_0) + FAR(T_0)), \tag{10}$$

The identification result based on different T_0 in fine-level stage is list in Table 2 and the corresponding ROC of FAR and FRR is depicted in Fig. 7(c).We use relative distance of key point, weighting relative distance of key point, coarse-to-fine dynamic identification method to test the validity of our system. Table 3 shows the comparison identification results.

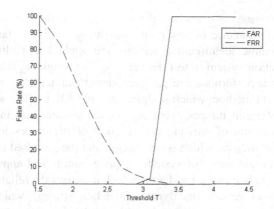

Fig. 5. The distributions of *FRR* and *FAR* used the weighting relative distance of key point and geometrical feature method

Table 2. *FAR* and *FRR* using different T_0 in Coarse-to-fine identification stage

T_0	$FRR(\%)$	$FAR(\%)$
1.5	100	0
1.8	83.24	0
2.1	54.33	0
2.4	29.47	0
2.7	9.47	0
2.9	5.26	0
3.1	1.73	1.86
3.4	0	100

Table 3. Comparison of identification results

method	$FRR(\%)$	$FAR(\%)$	Recognition (%)
Relative distance of key point	8.17	8.17	83.66
Weighting key point	5.26	6.06	88.68
Coarse-to-fine dynamic identification	1.73	1.86	96.41

When we use the proposed method to verify identity, the *CMR* can reach the value 96.41%, when $T_0 = 3.1$, $FRR = 1.73\%$, $FAR = 1.86\%$.

6 Conclusions

This paper proposes a novel feature extraction method for palmprint recognition based on matching the weighting relative distances of key points. The proposed method has been tested and evaluated on our private databases. In the stage of experiment, relative distance of key point, weighting relative distance of key point, coarse-to-fine dynamic identification method are applied to multimodal biometric personal identification system to test the validity. The weighting strategy in proposed system improves the performance of personal identification system more than relative distance of key point method, which is shown in table 3.But failure identification may occur in some palmprint images, there are pseudo texture lines in the side of ROI because of the variation of pressure and tensility or sometimes hand moves while acquiring palmprint images, which is one reason that the proposed system can't reach very high CMR. In addition, this system adopts a multimodal approach, rather than concentrating just on one of the hand area, which increases the reliability of decisions. Our system adopts a coarse-to-fine dynamic matching strategy, which implements the real-time of system. The experimental results show that the proposed multimodal personal identification approach based on weighting relative distance of key point is feasible and reliable.

References

1. Ahmadian, M.A.: An Efficient Texture Classification Algorithm Using Gabor Wavelet. In: Proceedings of the 25 Annual International Conference of the IEEE EMBS Cancun, Mexico, pp. 17–21 (2003)
2. Connie, T., Teoh, A., Goh, M., Ngo, D.: Palmhashing: a novel approach for cancelable biometrics. Inf. Process. Lett. 93(1), 1–5 (2005)
3. Daugman, J.: Complete Discrete 2D Gabor Transforms by Neural Networks for image Analysis and Compressin. IEEE Transactions on Acoustic, Speed and Signal Processing 7(36), 1169–1179 (1988)
4. Han, C.C., Cheng, H.L., Lin, C.L., Fan, K.C.: Personal Authentication Using Palm-print Features. Patt. Recog. 36, 371–381 (2003)
5. Jain, A.K., Bolle, R., Pankanti, S. (eds.): Biometrics: Personal Identification in Networked Society. Kluwer Academic Publishers, Dordrecht (1999)
6. Jing, X.Y., Zhang, D.: A Face and Palmprint Recognition Approach Based on Discriminant DCT Feature Extraction. IEEE Transaction on systems, Man and Cybernetics 34, 2405–2415 (2004)
7. Kong, W.K., Zhang, D., Li, W.: Palmprint Feature Extraction Using 2-D Gabor filters. Pattern Recognition 36, 2339–2347 (2003)
8. Lee, T.S.: Image Representation Using 2-D Gabor Wavelets. IEEE Trans. PAMI 18, 959–971 (1996)
9. Lin, C.-L., Chuang, T.C., Fan, K.-C.: Palmprint Verification Using Hierarchical Decomposition. Pattern Recognition 38, 2639–2652 (2005)
10. Ma, L., Wang, Y., Tan, T.: Iris recognition based on multichannel Gabor filtering. In: Proceedings of ACCV 2002, vol. I, pp. 279–283 (2002)
11. Ribaric, S., Fratric, I.: A Biometric Identification System Based on Eigenpalm and Eigenfinger Features. IEEE Trans. PAMI 27, 1698–1709 (2005)
12. Sanchez-Avila, C., Sanchez-Reillo, R.: Two Different Approaches for Iris Recognition using Gaobr Filters and Multiscale Zero-crossing Representation. Pattern Recognition 38, 231–240 (2005)

13. Savic, T., Pavesic, N.: Personal recognition based on an image of the palmar surface of the hand. Pattern Recognition (2007), doi: 10.1016/j.patcog.2007.03.005
14. Wu, X., Wang, K.: A Novel Approach of Palm-line Extraction. In: Proceedings of International Conference on Image Processing, New York (2004)
15. Wu, X., Zhang, D., Wang, K.: Fisherpalms Based Palmprint Recognition. Pattern Recognition Letters 24, 2829–2838 (2003)
16. Wu, X., Zhang, D., Wang, K., Huang, B.: Palmprint Classification Using Principal Lines. Pattern Recognition 37, 1987–1998 (2004)
17. You, J., Li, W., Zhang, D.: Hierarchical Palmprint Identification Via Multiple Feature Extraction. Pattern Recognition 35, 847–859 (2003)
18. Yu, L., Zhang, D., Wang, K.Q.: The relative distance of key point based iris recognition. Pattern Recognition 40, 423–430 (2007)
19. Zhang, D., Kong, W.K., You, J., Wong, M.: On-line Palmprint Identification. IEEE Trans. PAMI 25, 1041–1050 (2003)
20. Zhang, D., Shu, W.: Two novel characteristics in palmprint verification: datum point invariance line feature matching. Pattern Recognition 32(4), 691–702 (1999)

A Min-Cost-Max-Flow Based Algorithm for Reconstructing Binary Image from Two Projections Using Similar Images

Vedhanayagam Masilamani and Kamala Krithivasan

Dept. of Computer Science and Engineering
Indian Institute of Technology Madras
Chennai India 600036
kamala@iitm.ac.in

Abstract. The aim of this paper is to study the reconstruction of binary images from two projections using a priori images that are similar to the unknown image. Reconstruction of images from a few projections is preferred to reduce radiation hazards. It is well known that the problem of reconstructing images from a few projections is ill-posed. To handle the ill-posedness of the problem, a priori information such as convexity, connectivity and periodicity are used to limit the number of possible solutions. We use a priori images that are similar to the unknown image, to reduce the class of images having the same two projections. The a priori similar images may be obtained in many ways such as by considering images of neighboring slices or images of the same slice, taken in previous time instances. In this paper, we give a polynomial time algorithm to reconstruct binary image from two projections such that the reconstructed image is optimally close to the a priori similar images. We obtain a solution to our problem by reducing our problem to min cost integral max flow problem.

Keywords: binary matrix reconstruction, computed tomography, discrete tomography, min cost integral max flow problem.

1 Introduction

Discrete Tomography (DT) is an emerging reconstruction technique that reconstructs discrete images from a few projections of the images. As the Computed Tomography requires hundreds of projections to reconstruct images of interior of objects, the object is exposed to more X-ray energy, which causes some side effects such as cancer in medical imaging and destruction of atomic structure in crystalline structure reconstruction. As more projections require more X-ray energy to be transmitted into the object, one of the ways to reduce radiation hazards is to reconstruct images from a few projections. The area of *discrete tomography* is concerned about reconstruction of a discrete object or its geometrical properties from its projections or some other information. This has application in fields such as: image processing [12], statistical data security [6], biplane angiography [10], graph theory, crystallography, medical imaging [4], neutron imaging [8] etc. [5] gives the fundamentals related to this topic.

V.E. Brimkov, R.P. Barneva, H.A. Hauptman (Eds.): IWCIA 2008, LNCS 4958, pp. 408–419, 2008.

Here we consider the problem of reconstructing bi-level image from its projections along row and column, and a priori information namely a set of images that are similar to the unknown image.

An important area where binary image reconstruction obtained is medical imaging, in particular, Digital subtraction angiography [4]. In Digital subtraction angiography, the reconstructed image is the difference between images acquired before and after intra-arterial injection of radio-opaque contrast medium and hence if the difference of a few projections of those two images are given, binary image can be reconstructed. Another area where binary image reconstruction obtained is crystallography. Peter Schwander and Larry Shepp proposed a model that identifies each possible atom location with a cell of integer lattice Z^3 and the electron beams with lines parallel to given direction. The value 1 in a cell of Z^3 denotes the presence of atom in the corresponding location of crystal and the value 0 in a cell of Z^3 denotes the absence of atom in the corresponding location of the crystal. The number of atoms that are present in the line passing through the crystal defines the projection of the structure along the line [7].

The problem of reconstructing 3D-binary matrix is reduced to reconstructing 2D-binary matrix. Reconstructing 2D-binary matrix was studied much before the emergence of its practical application. In 1957 Ryser [11] and Gale [2] gave a necessary and sufficient condition for a pair of vectors being the projections of binary matrices along horizontal and vertical directions. The projections in horizontal and vertical directions are equal to row and column sums of the matrix. They have also given necessary and sufficient conditions for existence of unique 2D-binary matrix which has a given pair of row sum and column sum. In general, the class of binary matrices having same row and column sums is very large. Though the reconstructed matrix and the original matrix have same projections, they may be very different. One of the main issues in *Discrete Tomography* is to reconstruct the object which is more close to the original object with few projections only. One approach to reduce the class of possible solutions is to use some a priori information about the objects. For instance, convex binary matrices have been reconstructed uniquely from projections taken in some prescribed set of four directions in [3]. An another approach is given in [9], where the class of binary matrices having same projections is assumed to have some Gibs distribution. By using this information, object which is close to the original unknown object is reconstructed.

In this paper, We consider the first approach, namely, a set of a priori images that are similar to the unknown image, to limit the possible solutions of 2D-Binary images having given projections. In practice, It is possible to obtain images that are similar to unknown image. One such situation is that the images of the same slice taken in previous time instances may be considered as similar images, and the another situation is that the images of adjacent slices may be considered as similar images. As the patients who have undergone diagnosis may need to undergo diagnosis periodically (more in case of CT-Angiography), images taken in the previous time instances can be used to take images at current instance.

In the next section, we give notations and definitions. In section 3, we give the algorithm. In section 4, we give some results obtained by simulation studies. In section 5, we briefly discuss the correctness and complexity of the proposed algorithm. . The paper concludes with a brief remark in section 6.

2 Notations and Definitions

Let $A = (a_{i,j})$ and $B = (b_{i,j})$ be two binary image of order $m \times n$ where $1 \leq i \leq m$ and $1 \leq j \leq n$. . The images A and B are said to be similar if $\sum_{i=1}^{m} \sum_{j=1}^{n} |a_{i,j} - b_{i,j}|$ is small.

The row and column projections of $A = (a_{i,j})$ are $R = (r_1, ..., r_m)$ and $C = (c_1, ..., c_n)$ respectively, where $r_i = \sum_{j=1}^{n} a_{i,j}$ and $c_j = \sum_{i=1}^{m} a_{i,j}$

Two integral vectors $R = (r_1, ..., r_m)$ and $C = (c_1, ..., c_n)$ are said to be consistent if $\sum_{i=1}^{m} r_i = \sum_{j=1}^{n} c_j$. For binary matrices T and S of size $m \times n$, we define $|T - S| = \sum_{i=1}^{m} \sum_{j=1}^{n} |T(i,j) - S(i,j)|$

For two integral vectors $R = (r_1, ..., r_m)$ and $C = (c_1, ..., c_n)$, $\Gamma_{(R,C)}$ denotes the class of all binary matrices having row sum R and column sum C.

3 Reconstruction Problem

Given row projection $R = (r_i)$, column projection $C = (c_j)$ and a set of images $S = (S_k)$, where $1 \leq k \leq l$, that are similar to unknown bi-level image $A = (a_{i,j})$, the goal is to obtain a bi-level image $B = (b_{i,j})$ such that R and C are row and column projections of $B = (b_{i,j})$ respectively, and

$$B = arg[\min_{T \in \Gamma_{(R,C)}} \sum_{k=1}^{l} (|T - S_k|)]$$

We construct directed network G for the given a priori image-set S and the projections $R = (r_1, r_2, \ldots, r_m)$ and $C = (c_1, c_2, \ldots, c_n)$ as follows:
$G = (V, E, C', C'')$ be a weighted directed graph where
$V = U \cup W \cup \{s, t\}$
$U = \{ u_i \mid 1 \leq i \leq m \}$
$W = \{ w_j \mid 1 \leq j \leq n \}$
$E = \{(u_i, w_j) \mid 1 \leq i \leq m, 1 \leq j \leq n\}$
$\quad \cup \{(s, u_i) \mid 1 \leq i \leq m \} \cup \{(w_j, t) \mid 1 \leq j \leq n \}$

We define the cost associated with each edge as follows: For each $1 \leq i \leq m$ and $1 \leq j \leq n$,
$C'(u_i, w_j) = - \sum_{k=1}^{m} S_k(i,j)$, $C'(s, u_i) = -1$ and $C'(w_j, t) = -1$
where s is the source and t is the sink.

We define the capacity associated with each edge as follows: For each $1 \leq i \leq m$ and $1 \leq j \leq n$,
$C''(u_i, w_j) = 1$, $C''(s, u_i) = r_i$ and $C''(w_j, t) = c_j$

For a binary matrix $A = (a_{i,j})$ of size $m \times n$, the set of all locations with pixel value 0 is denoted by $A_0 = \{(i,j) | a_{i,j} = 0, 1 \leq i \leq m, 1 \leq j \leq n\}$, and the set of all locations with pixel value 1 is denoted by
$A_1 = \{(i,j) | a_{i,j} = 1, 1 \leq i \leq m, 1 \leq j \leq n\}$.

Lemma 1. *Let B and B' be two binary matrices, having same row projection R and column projection C. Let $S = (S_k)$, where $1 \leq k \leq l$, be the class of images that are*

*similar to the unknown image whose row and column projections are also R and C.
Then*

$$\sum_{(i,j)\in B_0'}\sum_{k=1}^{l}S_k(i,j)-\sum_{(i,j)\in B_0}\sum_{k=1}^{l}S_k(i,j)=\sum_{(i,j)\in B_1}\sum_{k=1}^{l}S_k(i,j)-\sum_{(i,j)\in B_1'}\sum_{k=1}^{l}S_k(i,j)$$

Proof

From the definition of B_0, B_1, B_0' and B_1', we get

$$\sum_{(i,j)\in B_0}\sum_{k=1}^{l}S_k(i,j)+\sum_{(i,j)\in B_1}\sum_{k=1}^{l}S_k(i,j)=\sum_{i=1}^{m}\sum_{j=1}^{n}\sum_{k=1}^{l}S_k(i,j) \qquad (1)$$

$$\sum_{(i,j)\in B_0'}\sum_{k=1}^{l}S_k(i,j)+\sum_{(i,j)\in B_1'}\sum_{k=1}^{l}S_k(i,j)=\sum_{i=1}^{m}\sum_{j=1}^{n}\sum_{k=1}^{l}S_k(i,j) \qquad (2)$$

From (1) and (2),

$$\sum_{(i,j)\in B_0'}\sum_{k=1}^{l}S_k(i,j)-\sum_{(i,j)\in B_0}\sum_{k=1}^{l}S_k(i,j)=\sum_{(i,j)\in B_1}\sum_{k=1}^{l}S_k(i,j)-\sum_{(i,j)\in B_1'}\sum_{k=1}^{l}S_k(i,j)$$

Hence the Lemma.

Lemma 2. *Let R and C be two integral vectors, $S = (S_k)$, where $1 \leq k \leq l$, be the class of images. Let G be the network associated with R, C and S. Maximum-Flow f of G has minimum cost iff $\sum_{i=1}^{m}\sum_{j=1}^{n}f(u_i,w_j)C(u_i,w_j)$ is minimum.*

Proof

Cost of flow is

$$\sum_{e\in E}f(e)C(e)=\sum_{i=1}^{m}f(s,u_i)C(s,u_i)+\sum_{i=1}^{m}\sum_{j=1}^{n}f(u_i,w_j)C(u_i,w_j)+\sum_{j=1}^{n}f(w_j,t)C(w_j,t)$$

As the index set for minimization is the set of all binary matrices having R and C as projections, the first and second terms of right hand side of above equation are constants. Hence

$$\sum_{e\in E}f(e)C(e) \text{ is min iff } \sum_{i=1}^{m}\sum_{j=1}^{n}f(u_i,w_j)C(u_i,w_j) \text{ is minimum}$$

Hence the proof.

Theorem 1. *: Let $S = (S_k)$, where $1 \leq k \leq l$, be a set of binary images, and $R = (r_1, r_2, \ldots, r_m)$ and $C = (c_1, c_2, \ldots, c_n)$ be two integral vectors. There* *exists a binary image $B = (b_{i,j})$ such that row and column projections of B are R and C respectively, and*

$$B = arg[\min_{T \in \Gamma_{(R,C)}} \sum_{k=1}^{l}(|T - S_k|)]$$

iff *R and C are consistent and max flow value for the network G corresponds to R, C and S is $|f| = \sum_{i=1}^{m} r_i$, and cost of the flow is minimum.*

Proof

$\Longrightarrow$:

Let $R = (r_i)$ and $C = (c_i)$ be two integral vectors, $S = (S_k)$, where $1 \leq k \leq l$, be a set of binary images, and G be the network associated with R, C and S. Let us assume that there exists a binary matrix $B = (b_{i,j})$ such that row and column projections of B are R and C respectively, and

$$B = arg[\min_{T \in \Gamma_{(R,C)}} \sum_{k=1}^{l}(|T - S_k|)]$$

let us first prove that R and C are consistent and max flow value for the network G corresponds to R, C and S is $|f| = \sum_{i=1}^{m} r_i$.
Consider the following flow for the network G
For each $1 \leq i \leq m$ and $1 \leq j \leq n$,
$f(s, u_i) = r_i$, $f(w_j, t) = c_j$,
$f(u_i, w_j) = 1$ if $b_{i,j} = 1$,
$f(u_i, w_j) = 0$ if $b_{i,j} = 0$,
$f(u_i, s) = -f(s, u_i)$, $f(t, w_j) = -f(w_j, t)$,
$f(w_j, u_i) = -f(u_i, w_j)$
and $f(e) = 0$ for all $e \in V \times V$ such that $f(e)$ is not defined above. The flow f has the following properties:

Capacity constraint: $f(e) \leq C''(e)$ for all $e \in V \times V$. From our definition of f, capacity constraint is evident.

Skew symmetry: $f(u, v) = -f(v, u)$ for all $(u, v) \in V \times V$. Skew symmetry is also evident from our definition of flow f.

Flow conservation: For all $u \in V - \{s, t\}$ $\sum_{v \in V} f(u, v) = 0$. Since for each $1 \leq i \leq m$, row i has r_i $1's$, the number of outgoing edges with capacity 1 from u_i is r_i. Hence total amount outgoing flow from u_i is r_i. The only incoming flow to vertex u_i is from the source, which is also r_i. Since f is skew symmetric and the incoming flow is same as the outgoing flow at node u_i, $\sum_{v \in V} f(u_i, v) = 0$ where $1 \leq i \leq m$. Since for each $1 \leq j \leq n$, column j has c_j $1's$, the number of incoming edges with capacity 1 to w_j is c_j. Hence total amount of the incoming flow from w_j is c_j. The only outgoing flow from vertex w_j is to the sink, which is also c_j. Since f is skew symmetric and the incoming flow is same as the outgoing flow at node w_j, $\sum_{v \in V} f(w_j, v) = 0$ where $1 \leq j \leq n$. Hence the flow conservation follows.

The value of flow f is $|f| = \sum_{v \in V} f(s, v) = \sum_{i=1}^{m} f(s, u_i)$. Since $|f| \leq \sum_{v \in V} C''(s, v)$ for any f and for our f, $|f| = \sum_{v \in V} C(s, v)$, $|f| = \sum_{i=1}^{m} r_i$ is the maximum flow. Since $|f| = \sum_{i=1}^{m} f(s, u_i)$ and $|f| = \sum_{i=1}^{n} f(w_j, t)$, $\sum_{i=1}^{m} r_i = \sum_{j=1}^{n} c_j$. Hence R and C are consistent.

Let us now prove that the cost of the flow f is minimum.

Suppose f is not minimum, then there exists an another flow f' such that cost of f' is less than f. ie $\sum_{i=1}^{m} \sum_{j=1}^{n} f'(u_i, w_j) C'(u_i, w_j) < \sum_{i=1}^{m} \sum_{j=1}^{n} f(u_i, w_j) C'(u_i, w_j)$.

Let us construct a binary matrix B' as follows:

$B'(i, j) = 1$ if $f'(u_i, w_j) = 1$.
$B'(i, j) = 0$ otherwise

Since f' is a flow of network G associated with R and C, the row and column projections of B' are R and C respectively.

Claim: $\sum_{k=1}^{l} (|B' - S_k|) < \sum_{k=1}^{l} (|B - S_k|)$

As cost of f' min ,

$$\sum_{(i,j) \in B_1'} \sum_{k=1}^{l} S_k(i,j) > \sum_{(i,j) \in B_1} \sum_{k=1}^{l} S_k(i,j)$$

$$\implies \sum_{(i,j) \in B_1} \sum_{k=1}^{l} S_k(i,j) - \sum_{(i,j) \in B_1'} \sum_{k=1}^{l} S_k(i,j) < 0$$

$$\implies (\sum_{(i,j) \in B_0'} \sum_{k=1}^{l} S_k(i,j) - \sum_{(i,j) \in B_0} \sum_{k=1}^{l} S_k(i,j))$$

$$+ (\sum_{(i,j) \in B_1} \sum_{k=1}^{l} S_k(i,j) - \sum_{(i,j) \subset B_1'} \sum_{k=1}^{l} S_k(i,j)) < 0 \text{ (By Lemma 1)}$$

$$\implies \sum_{(i,j) \in B_0'} \sum_{k=1}^{l} S_k(i,j) + (l \sum_{i=1}^{m} r_i) - \sum_{(i,j) \in B_1'} \sum_{k=1}^{l} S_k(i,j)$$

$$< \sum_{(i,j) \in B_0} \sum_{k=1}^{l} S_k(i,j) + (l \sum_{i=1}^{m} r_i) - \sum_{(i,j) \in B_1} \sum_{k=1}^{l} S_k(i,j)$$

$$\implies \sum_{(i,j) \in B_0'} \sum_{k=1}^{l} |B'(i,j) - S_k(i,j)| + \sum_{(i,j) \in B_1'} \sum_{k=1}^{l} |B'(i,j) - S_k(i,j)|$$

$$< \sum_{(i,j) \in B_0} \sum_{k=1}^{l} |B(i,j) - S_k(i,j)| + \sum_{(i,j) \in B_1} \sum_{k=1}^{l} |B(i,j) - S_k(i,j)|$$

Hence the claim.

Since the claim contradicts

$$B = arg[\min_{T \in \Gamma_{(R,C)}} \sum_{k=1}^{l} (|T - S_k|)],$$

the cost of the flow f is minimum.

$\Longleftarrow$:

Let us assume that R and C are consistent and max flow value for the network G corresponds to R, C and S is $|f| = \sum_{i=1}^{m} r_i$, and cost of the flow is minimum.

Let us prove that there exists a binary matrix $B = (b_{i,j})$ such that row and column projections of B are R and C respectively, and

$$B = arg[\min_{T \in \Gamma_{(R,C)}} \sum_{k=1}^{l} (|T - S_k|)]$$

Let us construct binary matrix B from flow as follows:
$B(i,j) = 1$ if $f(u_i, w_j) = 1$, $B(i,j) = 0$ otherwise
As the cost of flow is minimum,

$$\sum_{i=1}^{m} \sum_{j=1}^{n} f(u_i, w_j) C'(u_i, w_j) \tag{3}$$

is min (By Lemma 2)

Since f is max-flow for G which is associated with R and C, and R and C are consistent, the row and column projections of B are R and C.

Claim : $B = arg[\min_{T \in \Gamma_{(R,C)}} \sum_{k=1}^{l} (|T - S_k|)]$

Suppose not, there exists B' such that

$$\sum_{k=1}^{l} |B' - S_k| < \sum_{k=1}^{l} |B - S_k|$$

$$\Longrightarrow \sum_{(i,j) \in B'_0} \sum_{k=1}^{l} |B'(i,j) - S_k(i,j)| + \sum_{(i,j) \in B'_1} \sum_{k=1}^{l} |B'(i,j) - S_k(i,j)|$$

$$< \sum_{(i,j) \in B_0} \sum_{k=1}^{l} |B(i,j) - S_k(i,j)| + \sum_{(i,j) \in B_1} \sum_{k=1}^{l} |B(i,j) - S_k(i,j)|$$

$$\Longrightarrow \sum_{(i,j) \in B'_0} \sum_{k=1}^{l} S_k(i,j) + l \sum_{i=1}^{m} r_i - \sum_{(i,j) \in B'_1} \sum_{k=1}^{l} S_k(i,j)$$

$$< \sum_{(i,j) \in B_0} \sum_{k=1}^{l} S_k(i,j) + l \sum_{i=1}^{m} r_i - \sum_{(i,j) \in B_1} \sum_{k=1}^{l} S_k(i,j)$$

$$\Longrightarrow (\sum_{(i,j)\in B_0'} \sum_{k=1}^{l} S_k(i,j) - \sum_{(i,j)\in B_0} \sum_{k=1}^{l} S_k(i,j))$$

$$+ (\sum_{(i,j)\in B_1} \sum_{k=1}^{l} S_k(i,j) - \sum_{(i,j)\in B_1'} \sum_{k=1}^{l} S_k(i,j)) < 0$$

$$\Longrightarrow (\sum_{(i,j)\in B_1} \sum_{k=1}^{l} S_k(i,j) - \sum_{(i,j)\in B_1'} \sum_{k=1}^{l} S_k(i,j)) < 0 \text{ (By Lemma 1)}$$

Let us construct flow f' as follows
$f'(u_i, w_j) = 1$ if $B'(i,j) = 1$
$f'(u_i, w_j) = 0$ otherwise
It is easy to verify that f' is a max-flow for the network.

$$\sum_{i=1}^{m} \sum_{j=1}^{n} f'(u_i, w_j) C'(u_i, w_j) = \sum_{(i,j)\in B_1'} \sum_{k=1}^{l} -S_k(i,j)$$

(4)

$$\sum_{i=1}^{m} \sum_{j=1}^{n} f(u_i, w_j) C(u_i, w_j) = \sum_{(i,j)\in B_1} \sum_{k=1}^{l} -S_k(i,j)$$

(5)

From (3), (4) and (5), $\sum_{i=1}^{m} \sum_{j=1}^{n} f'(u_i, w_j) C'(u_i, w_j) < \sum_{i=1}^{m} \sum_{j=1}^{n} f(u_i, w_j) C'(u_i, w_j)$

which contradicts that the cost of f is minimum. Hence the claim.

Algorithm: *Binary image reconstruction*
Input: A set of images $S = (S_k)$ and row and column projections $R = (r_1, r_2, \ldots, r_m)$ and $C = (c_1, c_2, \ldots, c_n)$ of A
Output: Bi level image $B = (b_{i,j})$ such that R and C are the row and column projections of $B = (b_{i,j})$ respectively, and

$$B = arg[\min_{T \in \Gamma_{(R,C)}} \sum_{k=1}^{l} (|T - S_k|)]$$

Initialization: $m :=$ the number of components in R, $n :=$ the number of components in C. For each $1 \le i \le m$ and $1 \le j \le n$, $b_{i,j} := 0$

Step 1: Compute cost matrix $C' = (c_{i,j}')$ where $c_{i,j}' = -\sum_{k=1}^{l} S_k(i,j)$ and $1 \le i \le m$ and $1 \le j \le n$,

Step 2: Construct Net work using projections and cost matrix as given below

$G = (V, E, C', C'')$ be a weighted directed graph where
$V = U \cup W \cup \{s, t\}$

$U = \{ u_i \mid 1 \le i \le m \}$
$W = \{ w_j \mid 1 \le j \le n \}$
$E = \{ (u_i, w_j) \mid 1 \le i \le m, \ 1 \le j \le n \}$
$\quad \cup \{ (s, u_i) \mid 1 \le i \le m \} \cup \{ (w_j, t) \mid 1 \le j \le n \}$

We define the cost associated with each edge as follows: For each $1 \le i \le m$ and $1 \le j \le n$,
$C'(u_i, w_j) = -c_{i,j}$, $C'(s, u_i) = -1$ and $C'(w_j, t) = -1$

We define the capacity associated with each edge as follows: For each $1 \le i \le m$ and $1 \le j \le n$,
$C''(u_i, w_j) = 1$, $C''(s, u_i) = r_i$ and $C''(w_j, t) = c_j$ where s is the source and t is the sink.

Step 3. Compute Min-cost Max-flow for the network constructed in step 2.

Step 4. Construct image from flow obtained in step 3 as follows For each $1 \le i \le m$ and $1 \le j \le n$, $b_{i,j} = 1$ if flow from u_i and w_j is 1

4 Simulation Studies

We have taken a real image of a vascular system of size 64×64 (Fig 1.) and created artificial blocks in the blood vessels at random locations. A set of ten images is synthesized from the real image by creating artificial blocks, and those images are considered as similar images (Fig 2(a) through Fig 2(j)). Another set of five images is synthesized from the real image by creating artificial blocks at random locations, and they are considered as test images (Fig 3(a). through Fig 3(e)). We considered each test image as unknown image and computed row and column projections of each test image. We reconstructed each test image without any error from its row and column projections using those similar images we have considered. But, the reconstructed images are not same as the unknown images(test images) when we considered some subsets of Fig 2(a) through Fig 2(j) as shown in the following table. We define the reconstruction **error** as $|A - B|$ where A is the unknown image (test image) B is the reconstructed image. It may be noted that A and B have same row and column projections. We have shown some experimental results for the reconstruction of a test image namely Fig 3(a) using various subsets of Fig 2(a through j). From the experimental results shown in the table, we can infer that more the a priori images or closer the a priori image to the unknown image lesser will be the reconstruction error. Note that when a priori images Fig 2. b,f,h,j are considered separately, the reconstructed errors are 68, 76,50,208 pixels

Fig. 1. Real vascular image

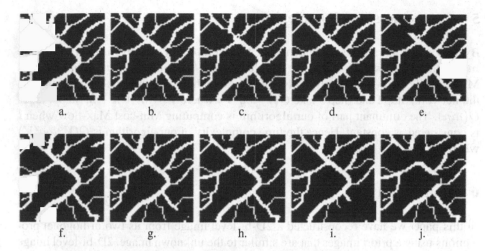

Fig. 2. The a priori images that are similar to the unknown image

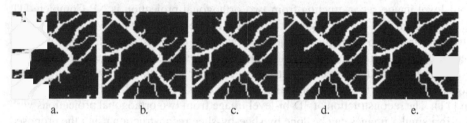

Fig. 3. The unknown images

respectively, but no error while all of them considered together. The error is 32 pixels When Fig 2. f and j are considered together, and 8 pixels are erroneous while Fig 2. b,f,j are considered together. It may also be noted that there is no reconstruction error when Fig 2(d) alone is considered as Fig 2(d) is very close to the unknown image.

Experimental Results

Table 1. Reconstruction error for various set of a priori images

S	error	S	error	S	error
{ a }	4	{ i }	4	{a, b, c }	0
{ b }	68	{ j }	208	{d, e, f }	0
{ c }	14	{ a, b }	4	{g, h, i }	0
{ d }	0	{ c, d }	0	{b, f, j }	8
{ e }	22	{ e, f }	0	{b, f, h, j }	0
{ f }	76	{ g, h }	0	{a, b, c, d, e }	0
{ g }	12	{ i, j }	0	{f, g, h, i, j }	0
{ h }	50	{ f, j }	32		

5 Correctness and Complexity

By Theorem 1, correctness is evident. The time complexity of Step 1 is $O(mnl)$ where m is number of rows , n is number of columns and l is number of similar images. As Min-cost Max-flow can be solved in $O(V^3 log^2 V)$ where V is the number of vertices in the network, Step2 and Step 3 take $O(V^3 log^2 V)$ where $V = m + n + 2$. Step 4 takes $O(mn)$. The dominant part of our algorithm is computing Min-cost Max-flow when l is considered as constant. Hence the time complexity of our algorithm is $O(V^3 log^2 V)$ where $V = m + n + 2$.

6 Conclusion

In this paper we have reconstructed a 2D-bi-level image from its two orthogonal projections using a priori images that are similar to the unknown image. 2D-bi-level image reconstruction from two orthogonal projections has polynomial time algorithm [11], 2D-tri-level image reconstruction from two orthogonal projections is still open, 2D-four-level image reconstruction from two orthogonal projection is NP-Complete [1]. The problem that we have solved is more complex than 2D-bi-level image reconstruction from two orthogonal projections and less complex than 2D-tri-level image reconstruction from two orthogonal projections. We implemented our algorithm and compared the quality of reconstructed image by our algorithm with the quality of reconstructed image by algorithm, which does not consider a priori similar images, given in [11]. The reconstruction of 3D-bi-level image from two orthogonal projections with a priori similar images can be done by slice-by-slice reconstruction using the proposed algorithm. One of the possible areas in which our algorithm can be used is medical imaging. Another area of application is crystallography. Our algorithm can be used to reconstruct crystalline structure from two projections without damaging the crystal. Though our algorithm always constructs an image whose orthogonal projections are the same as the orthogonal projections of the unknown image, the reconstructed image may be distorted from the unknown image if the similar images are not close to the unknown image or the number of a priori images is very small.

References

1. Chrobak, M., Durr, C.: Reconstructing polyatomic structures from discrete X-rays: NP-completeness proof for three atoms. Theoretical Computer Science 259, 81–98 (2001)
2. Gale, D.: A theorem on flows in networks. Pacific. J. Math. 7, 1073–1082 (1957)
3. Gardner, R.J., Gritzmann, P.: Discrete tomography: Determination of finite sets by X-rays. Trans. Amer. Math. Soc. 349(9), 2271–2295 (1997)
4. Hermann, G.T., Kuba, A.: Discrete tomography in medical imaging. Proceedings of the IEEE 91(10), 1612–1626 (2003)
5. Hermann, G.T., Kuba, A.: Discrete tomography: Foundations, algorithms and applications. Birkhäuser, Basel (1999)
6. Irving, R.W., Jerrum, M.R.: Three-dimensional statistical data security problems. SIAM Journal of Computing 23(1), 170–184 (1994)

7. Kiesielolowski, C., Schwander, P., Baumann, F.H., Seibt, M., Kim, Y., Ourmazd, A.: An approach to quantitative high-resolution transmission electron microscopy of crystalline materials. Ultramicroscopy 58(9), 131–135 (1995)
8. Kuba, A., Rusko, L., Rodek, L., Kiss, Z.: Preliminary studies of discrete tomography in neutron imaging. IEEE Transactions on Nuclear Science 52(1), 375–379 (2005)
9. Matej, S., Vardi, A., Hermann, G.T., Vardi, E.: Binary tomography using Gibbs priors. In: Discrete tomography: Foundations, algorithms, and applications, Birkhauser, Basel (1999)
10. Prause, G.P.M., Onnasch, D.G.W.: Binary reconstruction of the heart chambers from biplane angiographic image sequence. IEEE Transactions on Medical Imaging 15(4), 532–559 (1996)
11. Ryser, H.J.: Combinatorial properties of matrices of zeroes and ones. Canad. J. Math. 9, 371–377 (1957)
12. Shliferstien, A.R., Chien, Y.T.: Switching components and the ambiguity problem in the reconstruction of pictures from their projections. Pattern Recognition 10(5), 327–340 (1978)

Comparison of Local and Global
Region Merging in the Topological Map[*]

Alexandre Dupas[1] and Guillaume Damiand[2]

[1] Université de Poitiers, SIC, Bâtiment SP2MI, F-86962 Futuroscope, France
`dupas@sic.univ-poitiers.fr`
[2] LaBRI, Université de Bordeaux 1, UMR 5800, F-33405 Talence, France
`damiand@labri.fr`

Abstract. The topological map is a model that represents 2D and 3D images subdivision. It aims to allow the use of topological and geometrical features of the subdivision in image processing operations. When handling regions in an image, one of the main operation is the region merging, for example in segmentation process. This paper presents two algorithms of region merging in 3D topological maps: one local which modifies locally the map around merged regions, and another one global which runs through all the elements of the map. We study their complexities and present experimental results to compare both approaches.

Keywords: Combinatorial maps, Intervoxel boundaries, Region merging, Image segmentation.

1 Introduction

Region based segmentation consists in partitioning an image into connected sets of pixels or voxels called regions. One of the approaches, derived from the *split-and-merge* [9] methods and called *bottom-up*, begins with an over-partition of the image, and decreases the number of regions using successive region merging operations. These algorithms require a model that describes images, and operations onto these models like the region merging.

Many works have studied models representing partitions of an image. Topological data structures describe images as a set of elements and their adjacency relations. The most famous example is the Region Adjacency Graph (RAG) [13] which represents each region by a vertex, and where neighboring regions are connected by an edge. But the RAG suffers from several drawbacks as it does not represent multiple adjacency or makes no differences between inclusion and adjacency relations. To solve these issues, the RAG model has been extended, for instance in dual-graph structure to represent 2D images [11] or in topological maps [1,2,5,8] used to represent 2D and 3D images.

The aim of our work is to provide image processing operations using 3D topological maps. We need algorithms that maintain and update the information

---

[*] Partially supported by the ANR program ANR-06-MDCA-008-05/FOGRIMMI.

V.E. Brimkov, R.P. Barneva, H.A. Hauptman (Eds.): IWCIA 2008, LNCS 4958, pp. 420–431, 2008.
© Springer-Verlag Berlin Heidelberg 2008

stored, like relations between regions (adjacency, inclusion, etc.). In a previous work [7], a region merging method has been defined on 3D topological maps, but it is limited to the merging of two adjacent regions.

In this paper, we present and compare two algorithms, one *local* and one *global*, for region merging. The first one is dedicated to interactive processing of 3D topological maps as it aims to minimize the number of modifications of the map while merging together a set of connected regions. The global algorithm is designed for automated processing for example in segmentation operations. It allows the merging of any number of regions by connected components. Some experiments show that the two proposed algorithms behave as intended. The *local region merging* is slower when used intensively contrary to the *global region merging*.

In Sect. 2, we first present topological maps, which are combinatorial maps verifying specific properties used to represent 3D images. Then, Sect. 3 studies the local approach of region merging in topological maps. We explain the algorithm and give its complexity. In Sect. 4, we present the global approach. Section 5 shows the experimentation results, and compares both approaches in different cases. Lastly, we conclude and give some perspectives in Sect. 6.

2 Recalls on 3D Topological Maps

A 3D topological map is an extension of a combinatorial map used to represent a 3D image partition. Let us recall the notions on combinatorial maps, 3D images, intervoxel elements and topological maps that are used in this work.

A combinatorial map is a mathematical model describing the subdivision of a space, based on planar maps. A combinatorial map encodes all the cells of the subdivision and all the incidence and adjacency relations between the different cells, and so describe the topology of this space.

The single basic elements used in the definition of combinatorial maps are called *darts*, and adjacency relations are defined onto darts. We call β_i the relation between two darts that describes an adjacency between two i-dimensional cells (see Fig. 1 B for one example of combinatorial map and [12] for more details on maps and comparison with other combinatorial models). Intuitively, with this model, the notion of cells is represented by a set of darts linked by specific β_i relations. For example, a face incident to a dart d is represented by the set of darts accessible using any combination of β_1 and β_3 relations. Moreover, given a dart d, which belongs to an i-cell c, we can find the i-cell adjacent to c along the $(i-1)$-cell which contains d by using $\beta_i(d)$. For example, given a dart d that belongs to a face f and a volume v, the volume adjacent to v along f is the 3-cell containing $\beta_3(d)$.

Let us now present some usual notions about image and intervoxels elements. A voxel is a point of discrete space $\mathbb{Z}^3$ associated with a value which could be a color or a gray level. A three dimensional image is a finite set of voxels. In this work, combinatorial maps are used to represent voxel sets having the same labeled value and that are 6-connected. The label of a voxel is given by a labeled

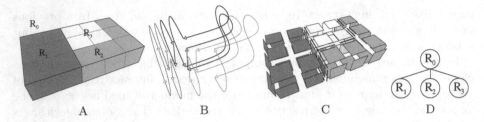

Fig. 1. The different parts of the topological map used to represent an image. (A) 3D image. (B) Minimal combinatorial map. (C) Intervoxel matrix (embedding). (D) Inclusion tree of regions.

function $l : \mathbb{Z}^3 \to L$ that gives for each voxel its label (a value in the finite set L). We speak about region for a maximal set of 6-connected voxel having the same label.

To avoid particular process for the image border voxels, we consider an infinite region R_0 that surrounds the image. If a region R_j is completely surrounded by a region R_i we say that R_j is *included* in R_i.

In the intervoxel framework [10], an image is considered as a subdivision of a 3-dimensional space in a set of cells: voxels are the 3-cells, surfels the 2-cells between two 3-cells, linels the 1-cells between two 2-cells and pointels the 0-cells between two 1-cells (see example in Fig. 1 C where 2-cells, 1-cells and 0-cells are drawn).

The *topological map* is a data structure used to represent the subdivision of an image into regions. It is composed of three parts:

- a minimal combinatorial map representing the topology of the image;
- an intervoxel matrix used to retrieve geometrical information associated to the combinatorial map. The intervoxel matrix is called the *embedding* of the combinatorial map;
- an inclusion tree of regions.

Figure 1 presents an example of topological map. The 3D image, composed of three regions plus the infinite region R_0 (Fig. 1 A), is represented by the topological map which is divided in three parts labeled B, C and D. The minimal combinatorial map extracted from this image is shown in Fig. 1 B. The embedding of the map is represented in Fig. 1 C, and the inclusion tree of regions in Fig. 1 D.

The combinatorial map allows the representation of all the incidence and adjacency relations between cells of an object. In the topological map framework, we use the combinatorial map as a topological representation of the partition of an image in regions. Each face of the topological map is separating two adjacent regions and two adjacent faces do not separate the same two regions. With these rules, we ensure the minimality (in number of cells) of the topological map (see [7,4] for more details on topological maps).

The intervoxel matrix is the embedding of the combinatorial map. Each cell of the map is associated with intervoxel elements representing geometrical information of the cell. A face, in the combinatorial map, is embedded by a set of surfels separating voxels of the two incident regions. The edges, which are the border of faces, are represented by a set of linels. The vertices, which are the border of edges, are embedded by pointels. Thus the intervoxel matrix allows to retrieve the geometry of the labeled image represented by the combinatorial map.

The inclusion tree of regions represents the inclusion relations. Each region in the topological map is associated to a node in the inclusion tree. The nodes are linked together by the inclusion relation previously defined. To link the tree with the combinatorial map, each dart d of the map knows its belonging region (called $region(d)$). Each region R knows one of its dart called *representative dart* (called $rep(R)$). $rep(R)$ has to belong to the external surface of R and its other incident region R_2, given by $region(\beta_3(rep(R)))$, has to be a smaller region than R considering the sweeping order of the image voxels (i.e. R_2 is found before R when we run through the image with a scan line algorithm).

3 Local Region Merging

The objective of the local merging approach is to modify the topological map in a local way to reflect the merging of the selected regions. As its a local operation, we do not want to run through all the darts of the topological map. We also aim to locally transform the inclusion tree of regions. Lastly, it is necessary to respect the minimality property of the topological map and update the geometrical embedding of the map.

3.1 Algorithm

The region merging algorithm takes in input a topological map M and a set S of connected regions to merge. The algorithm modifies the topological map M such as all the regions of S are merged together in the *resulting region*.

Algorithm 1 presents the local approach of the region merging operation. An overview of the local process divides it into three main steps:

- initialize the main region and mark the internal faces (i.e. faces incident to two regions that will be merged);
- update the inclusion tree to take into account the possible modifications of the inclusion relations;
- update the combinatorial map and the embedding by removing internal faces and simplifying their incident edges and vertices.

Before starting the merging process itself, we choose in set S a region, called *main region* (line 1 of Algo. 1). Instead of creating a new region, we use the main region as the resulting region of the merging algorithm. Thus, we choose the main region so its representative dart respects the definition of the representative dart (see Sect. 2) for the resulting region.

Algorithm 1. Local approach of the region merging operation

   Data: Topological map M ; Connected set of regions S
   Result: Merge together all the regions of S in M.

1 Choose the main region among the region of S;
2 **foreach** *dart d belonging to regions of* S *in the map* **do**
      if region($\beta_3(d)$) $\in$ S **then**
         └ Mark all the darts belonging to the face incident to d;

3 Update the inclusion tree;
4 Remove all internal faces (previously marked);
5 Simplify the cells incident to the removed faces;

The next step is the marking of the internal faces between the regions of set S. We run through the darts belonging to regions of S in the topological map (line 2). Each dart d such as region(d) $\in$ S and region($\beta_3(d)$) $\in$ S is incident to an internal face. We mark all the darts incident to the face incident to such a dart. In order to check if a dart belongs to a region of set S in a constant time, we use a second mark applied on regions that belong to S.

To update the inclusion tree (line 3) we remove the selected regions of the tree and then we run through all the darts of the selected regions to find newly internal surfaces. For each one, we build a new connected component of regions in the inclusion tree.

The next step concerns the two other structures of the topological map: the combinatorial map and its embedding, the intervoxel matrix. We want to remove internal faces from the combinatorial map (line 4). This is achieved by using the face-removal (defined in [6] like any other cell-removal operations used in this work) on each internal face previously marked.

The last step of this operation is the map simplification (line 5). Indeed, the removal of the internal faces modifies the degree of the incident edges. Degree two edges are not minimal according to the constraints of the topological map: they are removed with the edge-removal operation. If a degree-two edge is removed, the degree of the two incident vertices changes, so they are simplified if needed.

When a cell is removed in the combinatorial map by using cell-removal operation, the intervoxel matrix is also modified in order to remove the corresponding embedding. For instance, when a face (2-cell) is removed in the combinatorial map, all the corresponding surfels are removed in the embedding.

3.2 Complexity

Now we are going to study the time complexity of the local region merging algorithm presented in Algo. 1. The selection of the main region is achieved in $O(|S|)$ where $|S|$ is the number of regions to merge. Checking the validity of the representative dart is a constant time operation.

The loop marking internal faces process each dart belonging to the regions of S exactly once. Testing if region($\beta_3(d)$) $\in$ S is a constant time operation, so this step has time complexity $O(D_{selected})$ where $D_{selected}$ is the number of darts belonging to the regions of S.

Updating the inclusion tree is the most expensive operation regarding the number of covered darts. To find new internal surfaces, we run through all the darts belonging to selected regions. This step has complexity $O(D_{selected})$. If no inclusion appears, $D_{included}$, the number of darts belonging to newly included regions, is equal to zero. Otherwise all the included darts are covered. Thus this operation has time complexity $O(D_{selected} + D_{included})$ and in the worst case, when all regions are newly included in the resulting region, we cover the darts of the whole topological map.

The cell-removal operation in the combinatorial map has linear time complexity in number of darts incident to the corresponding cell. We only remove cells belonging to the surface of the merged regions. Thus, the complexity is linear in number of darts of the merged region: $D_{selected}$. Updating the embedding is linear in the number of intervoxel elements representing the cells. The whole process may be upper-bounded by $S_{removed}$, the number of surfels removed during the face-removal operation (because this number is always greater than the number of linels and the number of pointels). This stage has time complexity $O(D_{selected} + S_{removed})$.

Algorithm 1 has time complexity $O(D_{selected} + D_{included} + S_{removed})$. It can be upper bounded by the number of darts and the number of surfels of the whole topological map.

4   Global Region Merging

As a first approach of the region merging in the topological map, the local merge does not provide an effective way to deal with multiple sets of connected regions to merge. To overcome this drawback, we look for another approach allowing several merges at the same time.

The principle of the global merging algorithm is to separate the modifications of the topological map from the merging of regions. The aim is at first to handle the regions at an high level, and then to translate the high level merging into an effective one by removing cells from the topological map.

The first part of the algorithm concerns the symbolic merging of regions. To handle the high level merging and the representation of region sets, we use a disjoint-set forest [3] of regions. Disjoint-sets are used to represent multiple sets. The main operations are the retrieval of the belonging set of an element (find), and the merging of two sets (union). We use union-find trees to represent disjoint-sets. In [14], R. Tarjan shows that the union and find operations can be considered as constant time operations in practical cases.

During the symbolic merging, all regions in a same connected component are merged together in the same union-find tree. Then the merge operation, defined on disjoint-set forest, is used and some internal features of regions like number of voxels, or mean color are propagated. The root of each tree in the disjoint-set forest will be the resulting region for the merge of the underlying connected set of regions. When all the sets have been processed, each root of the disjoint-set forest is a remaining region of the merging process.

Algorithm 2. Global approach of the region merging operation

Data: Topological map M ; `Oracle` function
Result: Merge all the regions by connected components according to `Oracle` in M.

1 **foreach** *dart d of* M **do**
2   if `Oracle(region(`d`),region(`$\beta_3(d)$`))` **then**
        ⌊ Merge the union-find trees of `region(`d`)` and `region(`$\beta_3(d)$`)`;

3 Remove all internal faces;
4 Simplify the cells incident to the removed faces;
5 Rebuild the inclusion tree of regions;

The second part of the algorithm removes all the internal faces. Then, we simplify the topological map and rebuild the inclusion tree.

4.1   Algorithm

The global approach of the region merging algorithm takes in input a topological map M and an oracle function that tells how regions have to be merged. The algorithm modifies the topological map M such that all the connected regions that have to be merged according to the oracle are actually merged.

Algorithm 2 presents the global approach of the region merging operation, which is divided into three main steps:

- compute the disjoint-set forest of regions: this is the symbolic merging;
- remove the internal faces and simplify the incident cells;
- build the new inclusion tree.

The first step of the algorithm (line 1 of Algo. 2) concerns the symbolic merging. If the oracle merges r and its neighboring region r_2 in a same connected component, we merge the union-find trees containing the two regions. At the end of the symbolic merging, since only neighboring regions have been merged, each union-find tree represents a connected component of regions. The oracle function could be for example a labelling function which merges regions having the same label.

The next step (line 3) concerns the removal of the internal faces in the topological map. This is the same process as the one used in the internal face removal for the local approach, but without using a mark on internal faces. Indeed, a dart d belongs to an internal face if $\mathtt{find}(\mathtt{region}(d)) = \mathtt{find}(\mathtt{region}(\beta_3(d)))$ (i.e. both adjacent regions are in the same disjoint-set). We run through all the darts of the topological map, and for each dart d validating the previous assertion, we use the face-removal operation.

Then, we use the map simplification, presented in the local approach, to obtain the minimal combinatorial map and its corresponding embedding (line 4).

The last step of this approach concerns the building of the new inclusion tree of regions (line 5). This operation is processed with the same algorithm as the one used during the extraction of the topological map (see [4] for more details).

Contrary to the local approach, the inclusion tree is completely destroyed and then rebuilt ; we do not modify the previous tree. This is justified because the global approach is generally used to merge many regions, and in such a case it is more expensive to update the tree than to rebuild it (see experiments in Sect. 5).

4.2 Complexity

The first step of the global region merging (presented in Algo. 2) is the symbolic merging. The test line 2 of Algo. 2, which has constant time complexity using a mark on regions, is performed on all the darts of the topological map. The merging of the disjoint-sets containing the two implied regions is considered as a constant time operation. This step has time complexity $O(|D|)$ where $|D|$ is the number of darts of the topological map.

The complexity of the internal face removal is $O(|D|)$ where $|D|$ is the total number of darts in the map. The map simplification algorithm and the updating of the embedding are the same as the ones used in the local approach. So this whole step, covering the face removals and the simplification of the map, has complexity $O(|D| + S_{removed})$ where $S_{removed}$ is the number of removed surfels in the embedding. The complexity of the inclusion tree building is in $O(|D|)$ which give us the time complexity of Algo. 2 in $O(|D| + S_{removed})$. We can notice that the complexity of the two approaches is the same in the worst case, but they have different processing times depending on the number of merged regions.

5   Experiments and Analysis

In this section, we are interested in comparing the two approaches of the region merging in topological maps. We study the processing time of the two methods. All the following experiments use as input a same topological map representing 32^3 regions in a 32^3 voxels image (which means each voxel belongs to a different region). In this topological map, we select several regions to merge, in order to study some specific configurations. The goal of the experiments is to compare the behaviour of both algorithms and thus showing the more interesting approach in different cases. For this reason, we use small artificial images to facilitate the comparison without depending on the content of the image.

5.1   Merging of a Connected Component of Regions

We have compared the two approaches when the selected regions form only one connected component. We have introduced two protocols of experimentation in this case. For the first one, we have merged an increasing number of regions using both approaches. Figure 2 A shows the total processing time of both methods in function of the number of darts belonging to merged regions. This shows that the processing time of the local approach increases linearly with the number of darts. The processing time of the global algorithm tends to decrease as the number of merged regions increases. Indeed, during the merging operation, most of the

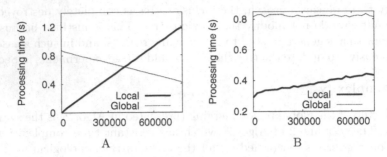

Fig. 2. Processing time comparison of the two region merging approaches (processing time given in seconds). (A) By the number of darts belonging to merged regions. (B) By the number of darts belonging to newly included region.

darts are removed, and the simplification and the building of the inclusion tree are cheaper when the number of darts is smaller. To conclude this experiment, we can observe that the local approach is faster than the global one when the number of merged regions remains small. On the contrary, the global approach is faster when the number of regions is bigger.

Table 1 presents the processing times of the three different steps of both algorithms in this experiment. The chosen steps are the same as the ones used in the overview of both algorithms. We can observe, in the local approach results, that each step takes an increasing time as the number of merged regions increases. The global approach behave differently. The symbolic merging growth slowly, but the face-removal, the simplification, and the building of the inclusion tree take less time as the number of merged regions increases. These values show the differences between the two approaches and explain the behavior of their processing times.

The second experiment, presented in Fig. 2 B, compares the processing time of the two methods when the number of included regions increases. We have

Table 1. Processing time of different steps of both approaches in function of the number of darts belonging to merged regions

Merged darts	24576	196608	393216	589824	786432
Local approach					
Initialization	0.008	0.052	0.100	0.148	0.196
Face-removal & simplification	0.028	0.156	0.208	0.316	0.388
Inclusion tree updating	0.012	0.120	0.304	0.480	0.616
Total	0.048	0.328	0.612	0.944	1.200
Global approach					
Initialization	0.060	0.064	0.064	0.072	0.072
Face-removal & simplification	0.520	0.476	0.424	0.380	0.336
Inclusion tree building	0.288	0.228	0.152	0.084	0.008
Total	0.868	0.768	0.640	0.536	0.416

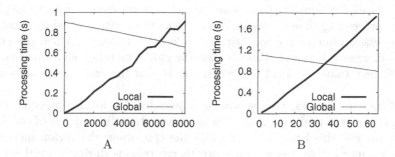

Fig. 3. Processing time comparison of the two region merging approaches (time given in seconds). (A) By the number of sets (size 2) merged. (B) By the size of the 256 sets merged.

merged a constant number of regions (5768), but these regions are merged in order to include a specified number of regions. This example shows the merging of a small number of regions regarding to the topological map size. This explain why the global approach is slower than the local one in this experiment. The main point we can observe is that the processing time of the global approach remains constant whereas in the local approach where it slowly increases. This is due to the updating of the inclusion tree of regions which depends on the number of included darts. The global approach does not depend on the configuration of these included regions so its processing time is more reliable in general cases.

5.2 Merging of Several Connected Components of Regions

We have also compared the two approaches when the selected regions form several connected components. We have studied the impact of the number of sets on the merging process, as well as the impact of their size. We have used the same kind of topological map representing 32^3 regions. In this image, we have selected several regions that belongs to different connected components. To merge such sets of connected regions with the local approach, we have used the local algorithm on each set, and we have computed the total processing time. The same result is obtained by giving to the global method the union of all the sets.

Firstly, we have merged an increasing number of sets of connected regions. Each set contains two regions. Figure 3 A presents the processing time of the two approaches in function of the number of merged sets. It shows that the local approach processing time increases with the number of merged set whereas the global method one decreases. The local approach behaves as expected since it is the addition of the processing time of a single merge in the topological map. The global approach behavior is explained, as previously, by the fact that the number of remaining darts decreases and thus the map simplification cost decreases. Figure 3 A shows that the global approach becomes more efficient than the local one as the number of sets growth.

Secondly, we are looking into the influence of the set size on both approaches. In the experiment, we have always merged 256 sets of regions, but we increase the size of each set from 2 to 64 regions. As in the previous experiment, the global merging processing time decreases as the number of merged regions increases. The processing time of the local approach is more expensive when the size of the sets increases. Figure 3 B shows that the global merging operation becomes more efficient than the local one when the size of the connected components increases.

In all the experiments, we have a same conclusion. The local approach is better when the number of selected regions is small and when the number of connected components remains low. This is mainly the case when the region merging is used in an interactive way or when we merge regions during a local process. In the case of a more global process, we prefer to use the global approach of the region merging as it will be generally more efficient and also have a more predictable processing time in function of the image size.

6   Conclusion

In this paper, we have presented two different approaches of the region merging operation in the 3D topological map. This work shows that the topological map could be efficiently modified in order to represent the evolution of the representation of 3D images during an image processing operation.

We have detailed the local approach of the region merging in 3D topological maps. It processes by applying local modifications to the three components of the topological map: the combinatorial map, the intervoxel matrix and the inclusion tree. The local method allows an efficient processing when the number of merged regions remains small compared to the map size. The configuration of the selected regions also influences the processing time of the region merging when there is included regions. Another drawback of this approach is that the selected regions to merge have to be in only one connected component. If it is not the case, the local region merging have to be applied to each connected component.

To overcome issues of the local approach, we have proposed a second method of region merging. It aims to merge by connected component any number of regions by processing the whole map at once. As shown in the experiments, the processing time does not change even if there are included regions or if there are several connected components. On the contrary, the processing time of the global approach tends to decrease when the number of merged regions increases. This method gives a better way to merge regions in an automated way.

Thus, we have two region merging algorithms on 3D topological maps that allow users to process regions interactively or automatically. These are the first needed tools for image processing with the 3D topological maps.

The next step of our work is to use these operations in a real bottom-up segmentation process. Our idea is to change the symbolic merging step of the global approach to merge regions according to a criterion, and then applying the last steps of the algorithm to produce the segmented image. We also want to study the opposite operation of the merge called region splitting. The aim of this operation is to divide a region into several smaller regions given a criterion, which may be a cut surface or an homogeneity measure. A splitting operation will gives the ability to implement a split-and-merge segmentation in the 3D topological map.

References

1. Bertrand, Y., Damiand, G., Fiorio, C.: Topological map: Minimal encoding of 3d segmented images, Ischia, Italy, IAPR-TC15, pp. 64–73 (2001)
2. Braquelaire, J.P., Domenger, J.P.: Representation of segmented images with discrete geometric maps 17(10), 715–735 (1999)
3. Cormen, T.H., Leiserson, C.E., Rivest, R.: Introduction to Algorithms. MIT Press, Cambridge (1990)
4. Damiand, G.: Définition et étude d'un modèle topologique minimal de représentation d'images 2d et 3d. Thèse de doctorat, Université Montpellier II (2001)
5. Damiand, G., Bertrand, Y., Fiorio, C.: Topological model for two-dimensional image representation: Definition and optimal extraction algorithm. Computer Vision and Image Understanding 93(2), 111–154 (2004)
6. Damiand, G., Lienhardt, P.: Removal and contraction for n-dimensional generalized maps. In: Nyström, I., Sanniti di Baja, G., Svensson, S. (eds.) DGCI 2003. LNCS, vol. 2886, pp. 408–419. Springer, Heidelberg (2003)
7. Damiand, G., Resch, P.: Split and merge algorithms defined on topological maps for 3d image segmentation. Graphical Models 65(1-3), 149–167 (2003)
8. Fiorio, C.: A topologically consistent representation for image analysis: The frontiers topological graph, Lyon, France, vol. 1176, pp. 151–162 (1996)
9. Horowitz, S., Pavlidis, T.: Picture segmentation by a directed split and merge procedure, 424–433 (1974)
10. Khalimsky, E., Kopperman, R., Meyer, P.: Boundaries in digital planes. Journal of Applied Mathematics and Stochastic Analysis 3(1), 27–55 (1990)
11. Kropatsch, W., Macho, H.: Finding the structure of connected components using dual irregular pyramids. In: Discrete Geometry for Computer Imagery, pp. 147–158 (1995) (invited lecture)
12. Lienhardt, P.: Topological models for boundary representation: a comparison with n-dimensional generalized maps. Computer-Aided Design 23(1), 59–82 (1991)
13. Rosenfeld, A.: Adjacency in digital pictures. Information and Control 26(1), 24–33 (1974)
14. Tarjan, R.: Efficiency of a good but not linear set union algorithm. Journal of the ACM 22(2), 215–225 (1975)

Novel Edge Detector

Imran Touqir and Muhammad Saleem

Department of Electrical Engineering, University of Engineering
and Technology, Lahore, Pakistan
imran@uet.edu.pk, drmsaleem@uet.edu.pk

Abstract. A major dilemma in edge detections is the choice of optimum threshold which lacks generality. The problem is further amplified by the presence of false edges in the image due to noise. Addressing this dilemma the paper presents a novel technique by exploiting scale correlation with in wavelet subband for two dimensional signals with a view to retain structural information. The image is decomposed by dyadic wavelet transform up to 4th level through multilevel wavelet decomposition. The detail coefficients in concordant bands are multiplied after interpolation and then synthesized. Quartic root of resultant product yields edge map of the image coupled with noise suppression. Experimental results reveal that the proposed algorithm outperforms the classical edge detectors for real, synthetic and noisy images while it is simpler to implement.

Keywords: Edge detection, wavelet scale correlation, image denoising, multiresolution analysis, entropy reduction.

1 Introduction

Edge detection is a process of detecting areas of abrupt changes or discontinuities in some visual property [9]. It is a critical preprocessing step towards high-level image understanding. Edges are essentially surface discontinuities, thus they hold important feature information about objects in an image (e.g. size, shape and location) on which subsequent processing highly depends. In typical images, edges characterize object boundaries and are therefore useful for segmentation, registration, and identification of objects in a scene. Edge detection is an essential process in image analysis and many techniques have been proposed. Edge detection as a preprocessing stage is application dependent. On ground of different needs, distinct edges should be extracted. These factors make edge detection difficult in general. Edges can be determined from the image by processing directly in the spatial domain, or by transforming to a different domain.

Edge is a local feature therefore, highpass filtering through Fourier Analysis is inadequate for edge detection due to its global nature. Roberts operator [4] gains in execution speed and loses fidelity because of its compact support of 2x2 neighborhood. Consequently it is very sensitive to noise. Extending the mask of Sobel beyond 3x3 reduces the fidelity of the final edge image and increases computation time. Laplace operator [18],[16] is extremely sensitive to noise. Zero Crossing

V.E. Brimkov, R.P. Barneva, H.A. Hauptman (Eds.): IWCIA 2008, LNCS 4958, pp. 432–443, 2008.

detector looks for values in the Laplacian of Gaussian [14] of an image crossing the level, i.e. points where the Laplacian changes its sign. Such points often occur at edges in images but they also occur at places which are difficult to associate with edges. Zero crossings [11] always lie on closed contours so the output of the zero crossing detectors is usually a binary image with single pixel. Canny [3] uses different values of the parameter σ for edge detection with different precision. A thresholding operation is required to bring the resulting image to binary from the derivative operation. The main problem in edge thresholding is how to choose a proper threshold value so that a better edge image can be obtained. The selection of good threshold value is usually a disturbing problem because there is insignificant knowledge about the nature of edges in the image. Optimum threshold for Canny is not unique and varies for image to image and the noise model, therefore, lacks generality. Hybrid edge detector by cascading smoothing capabilities of wavelets [17] prior to classical operators improves the edge map of an image but choice of optimal wavelet filter depends upon noise model and lacks generality. Multiscale Canny edge detector is equivalent to detecting modulus maxima in a two dimensional dyadic wavelet transform [8]. The Lipschitz regularity of edge points is derived from the decay of wavelet modulus maxima across scales. Tony Lindeberg [6] defined edges as connected points in scale-space that allows the scale to vary along the edge. Fine scales are selected for sharp edges and coarse scales are selected for diffuse edges, such that an edge model constitutes a valid abstraction of the intensity profile across the edge. In [5] a statistical approach at multiple scales using data driven probability distributions is envisaged and results are evaluated using Chernoff information and conditional entropy. In [15] a multiscale edge detection algorithm that can fuse multiscale data to generate an edge map at the image pixel level has been presented. Many techniques have been proposed for multiscale edge detection, however there is less agreement on precisely how to combine the results at different scales. In this paper the detail coefficients of the concordant bands up to 4th level are thresholded to suppress noise, interpolated to 1st level of wavelet decomposition which recaptures the missing edge pixels, concordant bands are multiplied which strengthens the edge pixels as the structural details exist at all resolutions and then resultant detail bands are synthesized to give the edge map of the image.

Edge images in real world applications are diverse. The edge map of the image can be visualized by humans, a preprocessing stage for segmentation/ registration or can be used in machine vision. Therefore, the figure of merit for edge detection [1] is also very challenging job and is applications dependent. The convolution kernels of the gradient or second derivative operators delocalize image edge map. The distinction between true edges and noise is an ill posed problem. Therefore conventional error measure based on measuring distance between original and the synthesized image is not advocated. In this paper Distance Transform (DT) [2],[13] has been used as similarity measure. Difference in DT of edge maps with respect to original image has been used as measure of error for estimation of peak signal to noise ratio (PSNR). The entropy of the edge map of the classical operators and the proposed algorithm are estimated which supports psycho-visual comparison.

The rest of the paper is formulated such that section 2 develops the theory of wavelet transform. In section 3 the proposed algorithm has been discussed. Section 4 highlights experimental results and section 5 concludes the paper.

2 Wavelet Transform

Wavelets are functions generated from one single function called mother wavelet by dilatations and translations in time domain. If mother wavelet is denoted by $\psi(t)$ the other wavelets $\psi_{a,b}(t)$ can be represented as

$$\psi_{a,b} = \frac{1}{\sqrt{|a|}}\psi(\frac{t-b}{a})\tag{1}$$

where a and b are two arbitrary real numbers and represent dilations and translations respectively. Based on this definition of wavelets, the wavelet transform of a function f(t) is mathematically represented as

$$W(a,b) = \int_{-\infty}^{\infty}\psi_{a,b}(t)f(t)dt\tag{2}$$

The inverse transform to reconstruct $f(t)$ from $W(a,b)$ is represented as

$$\tag{3}$$

$$f(t) = \frac{1}{c}\int_{-\infty}^{\infty}\int_{-\infty}^{\infty}W(a,b)\psi_{a,b}(t)dadb$$

where

$$C = \int_{-\infty}^{\infty}\frac{|\psi(w)|^2}{|w|}dw\tag{4}$$

and $\psi(w)$ is the Fourier transform of mother wavelet $\psi(t)$.

It is prudent to discretize a and b and then represent discrete wavelets [12], [10] accordingly. The most popular approach of discretizing a and b is

$$a = a_0^m\tag{5}$$

$$b = nb_0a_0^m\tag{6}$$

where m and n are integers. Hence DWT can be represented as

$$\psi_{m,n}(t) = a_0^{-\frac{m}{2}}\psi(a_0^{-m}t - nb_0)\tag{7}$$

The scaling function and the wavelets in one-dimensional space can be given by the following general formula:

$$\varphi_{a,b}(x) = a^{-1/2}\varphi(\frac{x-b}{a})K......a \succ 0, b \in R\tag{8}$$

$$\psi_{a,b}(x) = a^{-1/2}\psi(\frac{x-b}{a})K......a \succ 0, b \in R\tag{9}$$

Where, $\varphi_{a,b}(x)$ is the family of scaling function at scale a and translated by b, $\psi_{a,b}(x)$ is the family of wavelets at scale a and translated by b, φ and ψ are $\varphi_{0,0}$ and $\psi_{0,0}$ respectively.

Analysis of images with Quadrature Mirror Filtering (QMF) [12],[4],[7] has been exploited. With this methodology better approximations and details could be obtained using orthogonal and bi-orthogonal wavelets. The scaling function and the three wavelet functions for two dimensional signals are defined as

$$\varphi(x, y) = \varphi(x)\varphi(y) \tag{10}$$

$$\psi^1(x, y) = \varphi(x)\psi(y) \tag{11}$$

$$\psi^2(x, y) = \psi(x)\varphi(y) \tag{12}$$

$$\psi^3(x, y) = \psi(x)\psi(y) \tag{13}$$

The horizontal ψ^1, diagonal ψ^2 and vertical ψ^3 are nothing but the gradient of image along x, y and diagonal directions . It is however necessary condition for wavelets that the higher dimension functions should be separable into lower dimensions.

3 Proposed Algorithm

The above wavelet decomposition equations are classical to find the approximations and details in two dimensional signals which can be further exploited by wavelet scale correlation as following

1. A pair of QMF is operated on the gray level image in vertical followed by horizontal direction.
2. To maintain the size of the data at each level with respect to original image, decimation by two is applied after each filtering stage.
3. High frequency details at level-1 are extracted and used to get the magnitude image of vertical and diagonal image.
4. On the magnitude image so obtained thresholding is performed to obtain the edge map at level-1.
5. Lowpass residue is taken for analysis to get second level decomposition.
6. Steps 1, 2, 3 and 4 are performed to obtain edge detected image at level 2.
7. Lowpass residue is carried over from previous level to iterate up to level-4.
8. The edge details of different precision are obtained at each level of decomposition. Thresholding is taken as four times the mean value of the respective band coefficients.
9. Due to downsampling by a factor of two at each stage, the lower resolution images are interpolated by nearest neighborhood up to wavelet level-1 decomposition to facilitate matrix multiplications.
10. Due to noise suppression and downsampling, a few edge pixels are also diminished which are automatically recaptured by interpolation.

11. All matrix multiplications are carried out at resolution of level-1. The synthesis of product bands is then re-interpolated to match the original size of the image.
12. Quartic root of product filter is taken which yields the edge map of the image.

The algorithm is an iterative process. Image being passed on to the following stage every time gets smoothed as high frequency details are extracted at every level. In this paper results have been compiled up to level-4 wavelet decomposition. The approximations are further decomposed and the concordant band coefficients are multiplied after interpolation as follows to extract the better approximation and detail coefficients along with noise rejection.

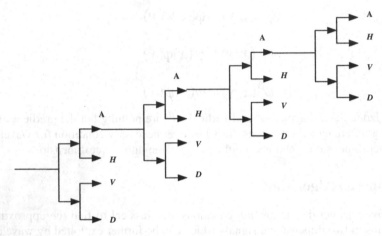

Fig. 1. Wavelet decomposition up to level-4. In each stage **A** denotes approximations, **H** horizontal details, **V** vertical details and **D** diagonal details.

$$D^h = \psi^h_{v+1} * \psi^h_{v+2} * \psi^h_{v+3} * \psi^h_{v+4} \tag{14}$$

$$D^v = \psi^v_{v+1} * \psi^v_{v+2} * \psi^v_{v+3} * \psi^v_{v+4} \tag{15}$$

$$D^d = \psi^d_{v+1} * \psi^d_{v+2} * \psi^d_{v+3} * \psi^d_{v+4} \tag{16}$$

$$D = D^h + D^v + D^d \tag{17}$$

$$E = D^{1/4} \tag{18}$$

ψ represents detail coefficients band after interpolation to the resolution of level-1, φ represents the approximations band coefficients after interpolation to the resolution level-1, super script h, v and d denotes horizontal, vertical and diagonal band coefficients and the subscripts denote the level of decomposition. D^h, D^v and D^d are the horizontal, vertical and diagonal detail coefficients respectively after multiplication of the interpolated concordant band coefficients. D is the augment of detail coefficients which corresponds to the gradient of image. E is the resultant edge map of the image.

4 Experimental Results

Edges detected by the proposed algorithm and conventional filters that include Robert, Prewitt, Sobel, Laplacian, Laplacian of Gaussian, Canny, Wavelet edge detectors and through multiresolution analysis using different orthogonal and bi-orthogonal wavelets at different precision are shown in figure-2. The proposed edge detector using db1 has outperformed the conventional operators. The missing edge pixels as compared to other detectors are recaptured due to interpolation prior to multiplications of concordant bands. However the process of interpolation blurs the contours. Figure 3 depicts the same for low SNR image and reveals the high supremacy of proposed algorithm using db1 due to its compact support which favors local singularity detection. The threshold for edge detection in proposed algorithm is taken at default value which is four times the image mean value. Adequate edge detection results were achieved even for noise variance as high as 0.4 (normalized). Db1 scale correlation up to 4th level gave the optimum detection. The edge blurring occurs with increase of the length of wavelet filter coefficients as depicted in figure-3(aa) and figure-4(aa). Results for uniform noise induced in the image are trivial due to wavelets in built approximating and detailing characteristics. Further comparison of natural and synthetic images for edge detection for different noise models exports similar results and are found to be more significant for high resolution images.

As the edges are delocalized, therefore legitimate mean square error does not correlate with psycho-visual comparison.

$$MSE = \frac{1}{N} \| I - I_R \|_2^2 \tag{19}$$

Where I is the original image and I_R is the reconstructed image and N is the total number of pixels in the image. To cater for delocalization of the boundary pixels DT [13] of both edge detected images i.e. edges detected from original image and edges detected from its noisy version are taken with original image as similarity measure.

The absolute difference in their DTs is taken as measure of error. PSNR based on the DT is evaluated as

$$PSNR = 10 \log_{10} \frac{\chi}{|DT_1 - DT_2|} \tag{20}$$

Where χ is the peak signal value which is 255 in our experiments. DT_1 and DT_2 are DTs of the edge detected image from original image and its noisy version respectively.

Entropy of the edge maps are calculated as

$$H(X) = \sum_i P(x_i) \ln P(x_i) \tag{21}$$

Where $P(x_i)$ is the probability of ith pixel value. The entropy of edge gives the amount of information in the image contrary to entropy reduction method for edge detection [19]. However its criteria supports within the family of wavelets. The entropy of the image, reconstructed image and edge maps of building image with varying SNR are shown in figure-4 and figure-5.

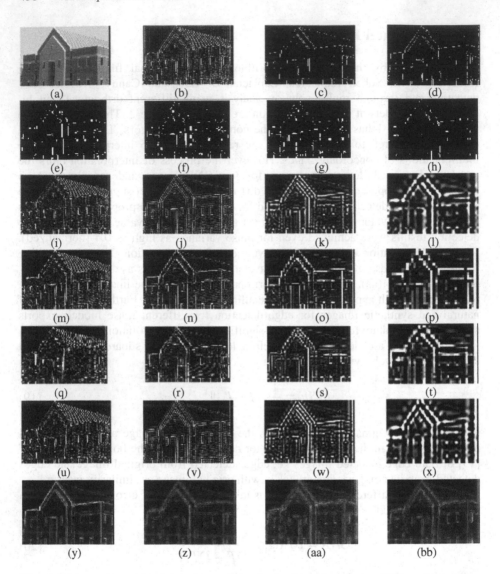

Fig. 2. Edges extracted by classical, wavelet and wavelet scale correlated filters. (a) building image used for extracting edges through (b) Fourier (c) Robert (d) Prewitt (e) Sobel (f)Laplacian (g) Log (h) Canny (i) DWT level-1 using Bior 3.7 (j) DWT level-2 using Bior 3.7 (k) DWT level-3 using Bior 3.7 (l) DWT level-4 using Bior 3.7 (m) DWT level-1 using Haar (n) DWT level-2 using Haar (o) DWT level-3 using Haar (p) DWT level-4 using Haar (q) DWT level-1 using db2 (r) DWT level-2 using db2 (s) DWT level-3 using db2 (t) DWT level-4 using db2 (u)DWT level-1 using db8 (v) DWT level-2 using db8 (w) DWT level-3 using db8 (x) DWT level-4 using db8 (y) Wavelet Scale correlation up to level-4 using Haar (z) Wavelet Scale correlation up to level-4 using db2 (aa) Wavelet Scale correlation up to level-4 using db8 (bb)Wavelet Scale correlation up to level-4 using Bior3.7 filters.

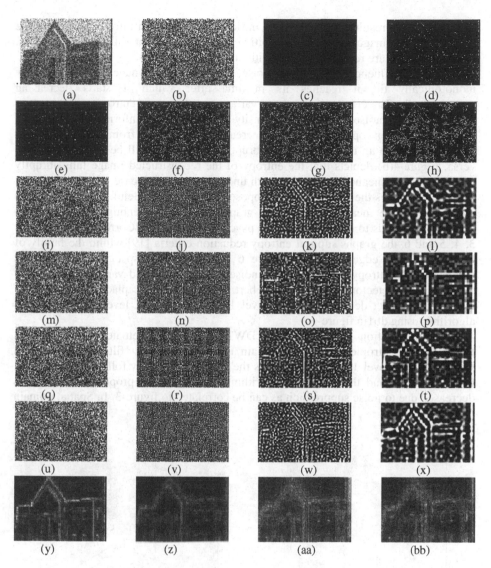

Fig. 3. Edges extracted by classical, wavelet and wavelet scale correlated filters. (a) Gaussian noise of 0 mean & .05 variance induced in building image used for extracting edges through (b)Fourier (c) Robert (d) Prewitt (e) Sobel (f) Laplacian (g) Log (h) Canny (i) DWT level-1 using Bior 3.7 (j) DWT level-2 using Bior 3.7 (k) DWT level-3 using Bior 3.7 (l) DWT level-4 using Bior 3.7 (m) DWT level-1 using Haar (n) DWT level-2 using Haar (o) DWT level-3 using Haar (p) DWT level-4 using Haar (q) DWT level-1 using db2 (r) DWT level-2 using db2 (s) DWT level-3 using db2 (t) DWT level-4 using db2 (u) DWT level-1 using db8 (v) DWT level-2 using db8 (w) DWT level-3 using db8 (x) DWT level-4 using db8. (y) Wavelet Scale correlation up to level-4 using Haar (z) Wavelet Scale correlation up to level-4 using db2 (aa) Wavelet Scale correlation up to level-4 using db8 (bb) Wavelet Scale correlation up to level-4 using Bior3.7 filters.

Experimental results reveal that it is intricate to distinguish information and noise contents in an image. However exploiting the correlation at different resolutions, structural details are retained coupled with noise suppression.

Under the influence of Gaussian noise, the entropy of image increases strictly monotonically till it reaches its maxima after which it starts decreasing monotonically. The effects of mean and variance on the image entropy are eminent in figure 4(a). High noise saturates the intensity values and hides information contents of the image. The entropy maxima with increase of noise vary from image to image. More information an image has more abruptly the maxima will be reached and vice versa. Figure-4(b) depicts that the entropy of the reconstructed image falls abruptly with increase of mean but decreases with undulation with increase of noise variance. Figure 5 illustrates the entropy of the proposed algorithm. Sudden fall of entropy with increase of noise mean is trivial. The graphs do not give enough information in isolation and needs to be correlated with psycho-visual results. Correlating the figures 3, 4, 5 and 6, the graphs support entropy reduction criteria [19] within the family of DWT for better edge detection. Figure 6 highlights this aspect more clearly by comparison of entropy under Gaussian noise of zero mean and varying variance for different edge detector that includes Robert, Prewitt, Sobel, Laplacian, Laplacian of Gaussian, Wavelet decomposition at level 1, level-2, level-3, level-4 and proposed algorithm using db1in figure-3.

More information is preserved by DWT filters and fluctuates around 6 bits. Difference of entropies of spatial domain filters and wavelet filters is eminent in figure-6. DWT level-1 edge detector has the maximum entropy followed by level -2, level-3, level-4 and the proposed algorithm. Entropy of the proposed algorithm is decreased due to noise suppression as can be correlated to figure-3. In Spatial domain

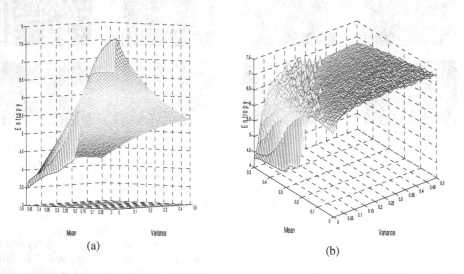

(a)

(b)

Fig. 4. Entropy of the images under Gaussian noise of varying mean and variance (figure-2) when induced in the building image. (a) Entropy of the noisy image. (b) Entropy of reconstructed image after wavelet scale correlation.

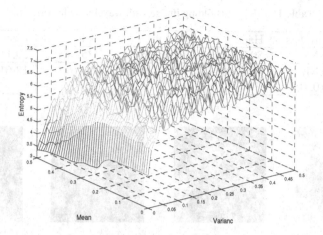

Fig. 5. Entropy of proposed edge detector under Gaussian noise

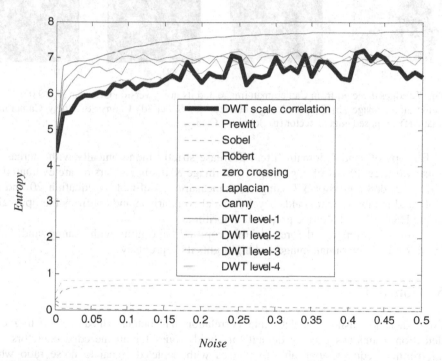

Fig. 6. Entropy of edge detected images of figure 5 under Gaussian noise of zero mean and varying variance

filters maximum entropy is preserved by Canny. The optimal results of Canny depend upon the selection of optimal threshold for edge detection, however, in this work all the experimental results are based on default threshold values.

Table 1. PSNR of Building image with wavelet scale correlation

Wavelets	Haar	Db2	Db8	Bior3.7
PSNR	50.991	39.1919	25.044	29.1158
PSNR N(0,.05)	18.6349	18.6185	18.618	18.618

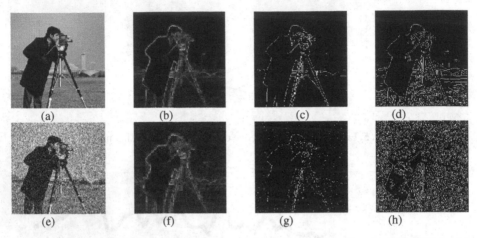

Fig. 7. Edges detected from Cameraman image and its noisy version Gaussian N(0,0.05).   (a) Cameraman image (b) Proposed edge detector (c) Sobel (d) Canny (e) Noisy Cameraman image (f) Proposed edge detector (g) Sobel (h) Canny.

Entropy of spatial domain filters decrease strictly monotonically with increase of noise variance. PSNR of edge maps from image & it's noisy version are evaluated for db1, db2, db8 and Bior-3.7 using error measure as defined by equation 20 and are tabulated in table-1 which aids psycho-visual comparison and confirms the optimality of the Haar wavelets for the proposed algorithm.

Figure 7 compares the results of proposed algorithm with Canny and Sobel operator for Cameraman image and highlights its supremacy.

5   Conclusion

Edge detection using wavelet scale correlation has outperformed the existing edge detection techniques keeping default threshold values for all the edge operators. The algorithm is equally applicable to images with depleted signal to noise ratio where conventional filters fall short to give adequate edge map on default values. Numerous techniques exist in spatial domain and for multiscale edge detection which vary in terms of synthesis of final edge map. The interpolation in the proposed algorithm reinstates the weak edge pixels depleted in the edge map due to thresholding. The algorithm is advocated for edge detection where noise model or noise intensity in the image either varies or not predictable prior to image processing. It works without the user's interaction and can be elegantly cascaded in pre processing stage in segmentation/feature extraction or matching. The reconstructed image through scale correlation gracefully suppresses noise, reduces image entropy and favors further processing in diverse applications such as image compression or multiple description coding.

References

1. Abdou, I.E., Pratt, W.K.: Quantitative design end evaluation of enhancement/ thresholding of edge detectors. Proceedings of IEEE 67(5), 753–763 (1979)
2. Breu, H., Gil, J., Kirkpatrik, D., Werman, M.: Linear time Euclidean distance transform algorithms. IEEE Trans on Pattern Analysis and Machine Intelligence 17(5), 209–226 (1995)
3. Canny, J.: A computational approach to Edge detection. IEEE Trans. PAMI 8, 250–468 (1986)
4. Gonzalez, R.C., Woods, R.E., Eddins, S.L.: Digital Image Processing using Matlab. Pearson Prentice Hall, Saddle River, NJ (2004)
5. Konish, S., Yuille, A.L., Coughlan, J.M.: A statistical approach to multi-scale edge detection. In: Proc. Workshop Generative Model Based Vision GMBV 2002 (2002)
6. Lindeberg, T.: Edge detection and ridge detection with automatic scale selection. International Journal of computer vision 32(2), 77–116 (1996)
7. Mallat, S.G.: A Theory for multiresolution signal decomposition the wavelet representation. IEEE Trans. on Pattern and Machine intelligence 11 (1989)
8. Mallat, S.G., Zhong, S.: Characterization of signals from multiscale edges. IEEE Trans. Patt Anal and Mach. Intl. 14(7), 710–732 (1992)
9. Marr, D., Hildreth, E.: Theory of Edge detection. Proc. Royal Soc. London 207, 187–217 (1980)
10. Park, D.J., Nam, K.N., Park, R.H.: Multiresolution edge detection techniques. Pattern Recognition Letters 28 (1995)
11. Rosenfeld, A., Thurston, M.: Edge and curve detection for visual scene analysis. IEEE Trans on computers, 562–569 (1971)
12. Sadler, B.M., Swami, A.: Analysis of multiscale products for step detection and estimation. IEEE Trans. on Information Theory 45, 1043–1051 (1999)
13. Saleem, M., Touqir, I., Siddiqui, A.M.: Novel edge detection. In: 4th International Conference on Information Technology (ITNG 2007), pp. 175–180. IEEE computer society, Los Alamitos (2007)
14. Sharifi, M., Fathy, M., Mahmoudi, M.T.: A classified and comparative study of edge-detection algorithms. In: International Conference on Information Technology, Coding and Computing (ITCC 2002), pp. 117–120 (2002)
15. Sun, J., Gu, D., Chen, Y., Zhang, S.: A multiscale edge detection algorithm based on wavelet domain vector hidden Markov tree Model. Pattern Recognition Society 37(7), 1315–1324 (2004)
16. Torre, V., Poggio, T.: Edge detection. IEEE Trans. Pattern Anal. Mach. Intel. PAMI 2, 147–163 (1986)
17. Touqir, I., Saleem, M., Siddiqui, A.M.: Hybrid edge detector. In: International Conference on Electrical Engineering (ICEE) (2007)
18. William, D.J., Shah, M.: Edge characterization using normalized edge detector. CVGIP 5(4), 311–318 (1993)
19. Wu, Y., He, Y., Cai, H.M.: Optimal threshold selection algorithm in edge detection based on wavelet transform. Intl. conf. on Image and Vision Computing 23(13), 1159–1169 (2005)

Author Index

Alba, Alfonso   306
Annadurai, S.   352
Arce-Santana, Edgar Román   306
Asano, Tetsuo   250

Balázs, Péter   112
Batenburg, K. Joost   372
Bhattacharya, Bhargab B.   124
Bhowmick, Partha   124
Bishnu, Arijit   1, 250
Biswas, Arindam   124
Brimkov, Valentin E.   87
Buzer, Lilian   205

Charrier, Emilie   205
Chu, Chee-Hung Henry   197
Corso, Jason J.   172, 295

Damiand, Guillaume   63, 420
Darbon, Jérôme   229
Dare, V.R.   352
de Vieilleville, François   26
Debled-Rennesson, Isabelle   160
Di Gesù, Vito   384
Dupas, Alexandre   420

Fahiem, Muhammad Abuzar   342
Falcão, Alexandre X.   136
Faure, Alexandre   148
Feschet, Fabien   148
Fuchs, Laurent   63

Geethalakshmi, M.   330
Gerard, Yan   363

He, Qiang   197

Imiya, Atsushi   262

Kalyani, T.   352
Kameda, Yusuke   262
Kenmochi, Yukiko   99
Klamroth, Kathrin   217

Komuravelli, Anvesh   1
Koroutchev, Kostadin   286
Korutcheva, Elka   286
Kosters, Walter A.   372
Krithivasan, Kamala   408

Lachaud, Jacques-Olivier   26
Lo Bosco, Giosuè   384
Lu, Yinghua   396

Malcolm, James   185
Martín del Rey, Angel   318
Mascarenhas, Nelson D.A.   136
Masilamani, Vedhanayagam   408
Millonzi, Filippo   384
Morales, Sandino   13

Nagar, Atulya K.   330
Nagy, Benedek   51
Nwogu, Ifeoma   295

Ohnishi, Naoya   262

Papa, João P.   136
Peltier, Samuel   63
Pfeuffer, Frank   217
Prasad, Bishal   250
Provot, Laurent   160
Pu, Dongbing   396

Qi, Shuang   396

Rathi, Yogesh   185

Saleem, Muhammad   432
Sarkar, Moumita   124
Schulz, Henrik   38
Sinha, Arnab   1
Stelldinger, Peer   274
Stiglmayr, Michael   217
Strand, Robin   51
Subramanian, K.G.   330
Sugimoto, Akihiro   99
Suzuki, Celso.T.N.   136

Tajine, Mohamed   75
Tannenbaum, Allen   185
Teelen, Kristof   238
Thibault, Yohan   99
Thomas, D.G.   352
Touqir, Imran   432
Tu, Zhuowen   172

Valenti, Cesare   384
Veelaert, Peter   238

Wiederhold, Petra   13

Yuille, Alan   172

Zhou, Chunguang   396

Lecture Notes in Computer Science

Sublibrary 6: Image Processing, Computer Vision, Pattern Recognition, and Graphics

Vol. 4958: V.E. Brimkov, R.P. Barneva, H.A. Hauptman (Eds.), Combinatorial Image Analysis. XVI, 446 pages. 2008.

Vol. 4931: G. Sommer, R. Klette (Eds.), Robot Vision. XI, 468 pages. 2008.

Vol. 4901: D. Zhang (Ed.), Medical Biometrics. XII, 324 pages. 2007.

Vol. 4844: Y. Yagi, S.B. Kang, I.S. Kweon, H. Zha (Eds.), Computer Vision – ACCV 2007, Part II. XXVIII, 915 pages. 2007.

Vol. 4843: Y. Yagi, S.B. Kang, I.S. Kweon, H. Zha (Eds.), Computer Vision – ACCV 2007, Part I. XXVIII, 969 pages. 2007.

Vol. 4842: G. Bebis, R. Boyle, B. Parvin, D. Koracin, N. Paragios, S.-M. Tanveer, T. Ju, Z. Liu, S. Coquillart, C. Cruz-Neira, T. Müller, T. Malzbender (Eds.), Advances in Visual Computing, Part II. XXXIII, 827 pages. 2007.

Vol. 4841: G. Bebis, R. Boyle, B. Parvin, D. Koracin, N. Paragios, S.-M. Tanveer, T. Ju, Z. Liu, S. Coquillart, C. Cruz Neira, T. Müller, T. Malzbender (Eds.), Advances in Visual Computing, Part I. XXXIII, 831 pages. 2007.

Vol. 4815: A. Ghosh, R.K. De, S.K. Pal (Eds.), Pattern Recognition and Machine Intelligence. XIX, 677 pages. 2007.

Vol. 4814: A. Elgammal, B. Rosenhahn, R. Klette (Eds.), Human Motion – Understanding, Modeling, Capture and Animation. X, 329 pages. 2007.

Vol. 4792: N. Ayache, S. Ourselin, A. Maeder (Eds.), Medical Image Computing and Computer-Assisted Intervention – MICCAI 2007, Part II. XLVI, 988 pages. 2007.

Vol. 4791: N. Ayache, S. Ourselin, A. Maeder (Eds.), Medical Image Computing and Computer-Assisted Intervention – MICCAI 2007, Part I. XLVI, 1012 pages. 2007.

Vol. 4781: G. Qiu, C. Leung, X.-Y. Xue, R. Laurini (Eds.), Advances in Visual Information Systems. XIII, 582 pages. 2007.

Vol. 4778: S.K. Zhou, W. Zhao, X. Tang, S. Gong (Eds.), Analysis and Modeling of Faces and Gestures. X, 305 pages. 2007.

Vol. 4756: L. Rueda, D. Mery, J. Kittler (Eds.), Progress in Pattern Recognition, Image Analysis and Applications. XXI, 989 pages. 2007.

Vol. 4738: A. Paiva, R. Prada, R.W. Picard (Eds.), Affective Computing and Intelligent Interaction. XVIII, 781 pages. 2007.

Vol. 4729: F. Mele, G. Ramella, S. Santillo, F. Ventriglia (Eds.), Advances in Brain, Vision, and Artificial Intelligence. XVI, 618 pages. 2007.

Vol. 4713: F.A. Hamprecht, C. Schnörr, B. Jähne (Eds.), Pattern Recognition. XIII, 560 pages. 2007.

Vol. 4679: A.L. Yuille, S.-C. Zhu, D. Cremers, Y. Wang (Eds.), Energy Minimization Methods in Computer Vision and Pattern Recognition. XII, 494 pages. 2007.

Vol. 4678: J. Blanc-Talon, W. Philips, D. Popescu, P. Scheunders (Eds.), Advanced Concepts for Intelligent Vision Systems. XXIII, 1100 pages. 2007.

Vol. 4673: W.G. Kropatsch, M. Kampel, A. Hanbury (Eds.), Computer Analysis of Images and Patterns. XX, 1006 pages. 2007.

Vol. 4642: S.-W. Lee, S.Z. Li (Eds.), Advances in Biometrics. XX, 1216 pages. 2007.

Vol. 4633: M. Kamel, A. Campilho (Eds.), Image Analysis and Recognition. XII, 1312 pages. 2007.

Vol. 4584: N. Karssemeijer, B. Lelieveldt (Eds.), Information Processing in Medical Imaging. XX, 777 pages. 2007.

Vol. 4569: A. Butz, B. Fisher, A. Krüger, P. Olivier, S. Owada (Eds.), Smart Graphics. IX, 237 pages. 2007.

Vol. 4538: F. Escolano, M. Vento (Eds.), Graph-Based Representations in Pattern Recognition. XII, 416 pages. 2007.

Vol. 4522: B.K. Ersbøll, K.S. Pedersen (Eds.), Image Analysis. XVIII, 989 pages. 2007.

Vol. 4485: F. Sgallari, A. Murli, N. Paragios (Eds.), Scale Space and Variational Methods in Computer Vision. XV, 931 pages. 2007.

Vol. 4478: J. Martí, J.M. Benedí, A.M. Mendonça, J. Serrat (Eds.), Pattern Recognition and Image Analysis, Part II. XXVII, 657 pages. 2007.

Vol. 4477: J. Martí, J.M. Benedí, A.M. Mendonça, J. Serrat (Eds.), Pattern Recognition and Image Analysis, Part I. XXVII, 625 pages. 2007.

Vol. 4472: M. Haindl, J. Kittler, F. Roli (Eds.), Multiple Classifier Systems. XI, 524 pages. 2007.

Vol. 4466: F.B. Sachse, G. Seemann (Eds.), Functional Imaging and Modeling of the Heart. XV, 486 pages. 2007.

Vol. 4418: A. Gagalowicz, W. Philips (Eds.), Computer Vision/Computer Graphics Collaboration Techniques. XV, 620 pages. 2007.

Vol. 4417: A. Kerren, A. Ebert, J. Meyer (Eds.), Human-Centered Visualization Environments. XIX, 403 pages. 2007.

Vol. 4391: Y. Stylianou, M. Faundez-Zanuy, A. Esposito (Eds.), Progress in Nonlinear Speech Processing. XII, 269 pages. 2007.

Vol. 4370: P.P. Lévy, B. Le Grand, F. Poulet, M. Soto, L. Darago, L. Toubiana, J.-F. Vibert (Eds.), Pixelization Paradigm. XV, 279 pages. 2007.

Vol. 4358: R. Vidal, A. Heyden, Y. Ma (Eds.), Dynamical Vision. IX, 329 pages. 2007.

Vol. 4338: P.K. Kalra, S. Peleg (Eds.), Computer Vision, Graphics and Image Processing. XV, 965 pages. 2006.

Vol. 4319: L.-W. Chang, W.-N. Lie (Eds.), Advances in Image and Video Technology. XXVI, 1347 pages. 2006.

Vol. 4292: G. Bebis, R. Boyle, B. Parvin, D. Koracin, P. Remagnino, A. Nefian, G. Meenakshisundaram, V. Pascucci, J. Zara, J. Molineros, H. Theisel, T. Malzbender (Eds.), Advances in Visual Computing, Part II. XXXII, 906 pages. 2006.

Vol. 4291: G. Bebis, R. Boyle, B. Parvin, D. Koracin, P. Remagnino, A. Nefian, G. Meenakshisundaram, V. Pascucci, J. Zara, J. Molineros, H. Theisel, T. Malzbender (Eds.), Advances in Visual Computing, Part I. XXXI, 916 pages. 2006.

Vol. 4245: A. Kuba, L.G. Nyúl, K. Palágyi (Eds.), Discrete Geometry for Computer Imagery. XIII, 688 pages. 2006.

Vol. 4241: R.R. Beichel, M. Sonka (Eds.), Computer Vision Approaches to Medical Image Analysis. XI, 262 pages. 2006.

Vol. 4225: J.F. Martínez-Trinidad, J.A. Carrasco Ochoa, J. Kittler (Eds.), Progress in Pattern Recognition, Image Analysis and Applications. XIX, 995 pages. 2006.

Vol. 4191: R. Larsen, M. Nielsen, J. Sporring (Eds.), Medical Image Computing and Computer-Assisted Intervention – MICCAI 2006, Part II. XXXVIII, 981 pages. 2006.

Vol. 4190: R. Larsen, M. Nielsen, J. Sporring (Eds.), Medical Image Computing and Computer-Assisted Intervention – MICCAI 2006, Part I. XXXVVIII, 949 pages. 2006.

Vol. 4179: J. Blanc-Talon, W. Philips, D. Popescu, P. Scheunders (Eds.), Advanced Concepts for Intelligent Vision Systems. XXIV, 1224 pages. 2006.

Vol. 4174: K. Franke, K.-R. Müller, B. Nickolay, R. Schäfer (Eds.), Pattern Recognition. XX, 773 pages. 2006.

Vol. 4170: J. Ponce, M. Hebert, C. Schmid, A. Zisserman (Eds.), Toward Category-Level Object Recognition. XI, 618 pages. 2006.

Vol. 4153: N. Zheng, X. Jiang, X. Lan (Eds.), Advances in Machine Vision, Image Processing, and Pattern Analysis. XIII, 506 pages. 2006.

Vol. 4142: A. Campilho, M. Kamel (Eds.), Image Analysis and Recognition, Part II. XXVII, 923 pages. 2006.

Vol. 4141: A. Campilho, M. Kamel (Eds.), Image Analysis and Recognition, Part I. XXVIII, 939 pages. 2006.

Vol. 4122: R. Stiefelhagen, J.S. Garofolo (Eds.), Multimodal Technologies for Perception of Humans. XII, 360 pages. 2007.

Vol. 4109: D.-Y. Yeung, J.T. Kwok, A. Fred, F. Roli, D. de Ridder (Eds.), Structural, Syntactic, and Statistical Pattern Recognition. XXI, 939 pages. 2006.

Vol. 4091: G.-Z. Yang, T. Jiang, D. Shen, L. Gu, J. Yang (Eds.), Medical Imaging and Augmented Reality. XIII, 399 pages. 2006.

Vol. 4073: A. Butz, B. Fisher, A. Krüger, P. Olivier (Eds.), Smart Graphics. XI, 263 pages. 2006.

Vol. 4069: F.J. Perales, R.B. Fisher (Eds.), Articulated Motion and Deformable Objects. XV, 526 pages. 2006.

Vol. 4057: J.P.W. Pluim, B. Likar, F.A. Gerritsen (Eds.), Biomedical Image Registration. XII, 324 pages. 2006.

Vol. 4046: S.M. Astley, M. Brady, C. Rose, R. Zwiggelaar (Eds.), Digital Mammography. XVI, 654 pages. 2006.

Vol. 4040: R. Reulke, U. Eckardt, B. Flach, U. Knauer, K. Polthier (Eds.), Combinatorial Image Analysis. XII, 482 pages. 2006.

Vol. 4035: T. Nishita, Q. Peng, H.-P. Seidel (Eds.), Advances in Computer Graphics. XX, 771 pages. 2006.

Vol. 3979: T.S. Huang, N. Sebe, M. Lew, V. Pavlović, M. Kölsch, A. Galata, B. Kisačanin (Eds.), Computer Vision in Human-Computer Interaction. XII, 121 pages. 2006.

Vol. 3954: A. Leonardis, H. Bischof, A. Pinz (Eds.), Computer Vision – ECCV 2006, Part IV. XVII, 613 pages. 2006.

Vol. 3953: A. Leonardis, H. Bischof, A. Pinz (Eds.), Computer Vision – ECCV 2006, Part III. XVII, 649 pages. 2006.

Vol. 3952: A. Leonardis, H. Bischof, A. Pinz (Eds.), Computer Vision – ECCV 2006, Part II. XVII, 661 pages. 2006.

Vol. 3951: A. Leonardis, H. Bischof, A. Pinz (Eds.), Computer Vision – ECCV 2006, Part I. XXXV, 639 pages. 2006.

Vol. 3948: H.I. Christensen, H.-H. Nagel (Eds.), Cognitive Vision Systems. VIII, 367 pages. 2006.

Vol. 3926: W. Liu, J. Lladós (Eds.), Graphics Recognition. XII, 428 pages. 2006.

Vol. 3872: H. Bunke, A.L. Spitz (Eds.), Document Analysis Systems VII. XIII, 630 pages. 2006.

Vol. 3852: P.J. Narayanan, S.K. Nayar, H.-Y. Shum (Eds.), Computer Vision – ACCV 2006, Part II. XXXI, 977 pages. 2006.

Vol. 3851: P.J. Narayanan, S.K. Nayar, H.-Y. Shum (Eds.), Computer Vision – ACCV 2006, Part I. XXXI, 973 pages. 2006.

Vol. 3832: D. Zhang, A.K. Jain (Eds.), Advances in Biometrics. XX, 796 pages. 2005.

Vol. 3736: S. Bres, R. Laurini (Eds.), Visual Information and Information Systems. XI, 291 pages. 2006.

Vol. 3667: W.J. MacLean (Ed.), Spatial Coherence for Visual Motion Analysis. IX, 141 pages. 2006.

Vol. 3417: B. Jähne, R. Mester, E. Barth, H. Scharr (Eds.), Complex Motion. X, 235 pages. 2007.

Vol. 2396: T.M. Caelli, A. Amin, R.P.W. Duin, M.S. Kamel, D. de Ridder (Eds.), Structural, Syntactic, and Statistical Pattern Recognition. XVI, 863 pages. 2002.

Vol. 1679: C. Taylor, A. Colchester (Eds.), Medical Image Computing and Computer-Assisted Intervention – MICCAI'99. XXI, 1240 pages. 1999.